Lecture Notes
in Business Information Processing   342

Series Editors

Wil van der Aalst
  RWTH Aachen University, Aachen, Germany
John Mylopoulos
  University of Trento, Trento, Italy
Michael Rosemann
  Queensland University of Technology, Brisbane, QLD, Australia
Michael J. Shaw
  University of Illinois, Urbana-Champaign, IL, USA
Clemens Szyperski
  Microsoft Research, Redmond, WA, USA

More information about this series at http://www.springer.com/series/7911

Florian Daniel · Quan Z. Sheng
Hamid Motahari (Eds.)

Business Process Management Workshops

BPM 2018 International Workshops
Sydney, NSW, Australia, September 9–14, 2018
Revised Papers

 Springer

Editors
Florian Daniel (iD)
Politecnico di Milano
Milan, Italy

Quan Z. Sheng (iD)
Macquarie University
Sydney, NSW, Australia

Hamid Motahari
Global Technology Innovation at EY
EY AI Lab
San Jose, CA, USA

ISSN 1865-1348   ISSN 1865-1356   (electronic)
Lecture Notes in Business Information Processing
ISBN 978-3-030-11640-8   ISBN 978-3-030-11641-5   (eBook)
https://doi.org/10.1007/978-3-030-11641-5

Library of Congress Control Number: 2018967941

This Springer imprint is published by the registered company Springer Nature Switzerland AG
The registered company address is: Gewerbestrasse 11, 6330 Cham, Switzerland

Foreword

This volume contains the proceedings of the workshops held on September 10, 2018, in conjunction with the 16th International Conference on Business Process Management (BPM 2018), which took place in Sydney, Australia. The proceedings are so-called post-workshop proceedings, as the authors were allowed to revise and improve their papers after the actual workshops to take into account the feedback obtained from the audience during their presentations.

Due to its interdisciplinary nature that naturally involves researchers and practitioners alike, the BPM conference has traditionally been perceived as a premium event for co-locating workshops with. The 2018 edition of the conference was no exception: Its call for workshop proposals attracted a good number of workshop proposals with topics ranging from traditional BPM concerns like requirements engineering and business process mining to emerging topics like data science and artificial intelligence. The following eight workshops were selected for co-location with BPM 2018:

- 14th International Workshop on Business Process Intelligence (BPI) – organized by Boudewijn van Dongen, Jan Claes, Jochen De Weerdt, Andrea Burattin.
  This year's BPI Workshop focused particularly on process mining in the context of big data. The workshop has a long tradition at the BPM conference and, as before, featured: the presentation of interesting research papers in the BPI domain; the BPI Challenge 2018, with data provided by the German company Data Experts; the IEEE Task Force meeting; and the Process Mining Reception.
- 11th Workshop on Social and Human Aspects of Business Process Management (BPMS2) – organized by Rainer Schmidt, Selmin Nurcan.
  BPMS 2018 explored how social software interacts with business process management, how business process management has to change to comply with weak ties, social production, egalitarianism and mutual service, and how business processes may profit from these principles. Furthermore, the workshop investigated human aspects of business process management such as new user interfaces, e.g., augmented reality and voice bots.
- First International Workshop on Process-Oriented Data Science for Health Care (PODS4H) – organized by Jorge Munoz-Gama, Carlos Fernandez-Llatas, Niels Martin, Owen Johnson.
  PODS4H 2018 aimed at providing a high-quality forum for interdisciplinary researchers and practitioners (both data/process analysts and medical audience) to exchange research findings and ideas on health-care process analysis techniques and practices. PODS4H research includes a wide range of topics from process mining techniques adapted for health-care processes, to practical issues on implementing PODS methodologies in health-care centers' analysis units.
- First International Workshop on Artificial Intelligence for Business Process Management (AI4BPM) – organized by Richard Hull, Riccardo De Masellis, Krzysztof Kluza, Fabrizio Maria Maggi, Chiara Di Francescomarino.

The goal of AI4BPM was to establish a forum for researchers and professionals interested in understanding, envisioning, and discussing the challenges and opportunities of moving from current, largely programmatic approaches for BPM, to emerging forms of AI-enabled BPM. The workshop represents the union of two workshops held at BPM 2017, namely, Business Process Innovation with Artificial Intelligence (BPAI) and Cognitive Business Process Management (CBPM).

- First International Workshop on Emerging Computing Paradigms and Context in Business Process Management (CCBPM) – organized by Jianmin Wang, Michael Sheng, Shiping Chen, Xiao Liu, James Xi Zheng.

The goal of CCBPM 2018 was to promote the role of emerging computing paradigms such as mobile-cloud computing, edge/fog computing, and context in business process management (BPM) by discussing what opportunities and challenges the emerging computing paradigms and context-aware technologies can bring to BPM, and what are the novel use cases and state-of-the-art solutions.

- Joint Business Processes Meet the Internet-of-Things/Process Querying Workshop (BP-Meet-IoT/PQ) – organized by Agnes Koschmider, Massimo Mecella, Estefanía Serral, Victoria Torres, Artem Polyvyanyy, Arthur ter Hofstede, Claudio Di Ciccio. This joint BP-Meet-IoT/PQ Workshop brought together practitioners and researchers interested in IoT-based business processes (state of ongoing research, industry needs, future trends, and practical experiences) and process querying (automated methods for the inquiry, manipulation, and update of models and data of observed and envisioned processes).

- First Declarative/Decision/Hybrid Mining and Modeling for Business Processes (DeHMiMoP) – organized by Claudio Di Ciccio, Jan Vanthienen, Tijs Slaats, Dennis Schunselaar, Sóren Debois.

DeHMiMoP aimed at providing a platform for the discussion, introduction, and integration of ideas related to the decision and rule perspectives on process modeling and mining. The objectives were to extend the reach of the BPM audience toward the decisions and rules community, and increase the integration between imperative, declarative, and hybrid modeling perspectives.

- Joint Requirements Engineering and Business Process Management Workshop/Education Forum (REBPM/EdForum) – organized by Banu Aysolmaz, Rüdiger Weißbach, Onur Demirörs, Fethi Rabhi, Wasana Bandara, Helen Paik, Cesare Pautasso.

This joint workshop brought together practitioners and researchers interested in requirements engineering and education in BPM. The focus of the workshop was on the interrelations between RE and BPM domains with a focus on agile and flexible BPM, and on effective education and training methods for developing BPM professionals.

The selected workshops formed an extraordinary and balanced program of high-quality events. We are confident the reader will enjoy this volume as much as we enjoyed organizing this outstanding program and assembling its proceedings.

Of course, we did not organize everything on our own. Many people from the BPM 2018 Organizing Committee contributed to the success of the workshop program. We would particularly like to thank the general chairs of BPM 2018, Boualem Benatallah and Jian Yang, for involving us in this unique event, the local organizers for the smooth management of all on-site issues, the workshop organizers for managing their workshops and diligently answering the numerous of e-mails we sent around, and, finally, the authors for presenting their work and actually making all this possible.

November 2018

Florian Daniel
Hamid Motahari
Michael Sheng

Contents

**First International Workshop on Process-Oriented Data Science
for Healthcare (PODS4H)**

**First International Workshop on Artificial Intelligence for Business
Process Management (AI4BPM)**

**First International Workshop on Emerging Computing Paradigms
and Context in Business Process Management (CCBPM)**

**Joint Requirements Engineering and Business Process
Management Workshop/Education Forum (REBPM/EdForum)**

14th International Workshop
on Business Process Intelligence (BPI)

14th International Workshop on Business Process Intelligence (BPI)

Business process intelligence (BPI) is a growing area both in industry and academia. BPI refers to the application of data- and process-mining techniques to the field of business process management. In practice, BPI is embodied in tools for managing process execution by offering several features such as analysis, prediction, monitoring, control, and optimization.

The main goal of this workshop is to promote the use and development of new techniques to support the analysis of business processes based on run-time data about the past executions of such processes. We aim at bringing together practitioners and researchers from different communities, e.g., business process management, information systems, database systems, business administration, software engineering, artificial intelligence, and data mining, who share an interest in the analysis and optimization of business processes and process-aware information systems. The workshop aims at discussing the current state of research and sharing practical experiences, exchanging ideas, and setting up future research directions that better respond to real needs. In a nutshell, it serves as a forum for shaping the BPI area.

The 14th edition of this workshop attracted eight international submissions. Each paper was reviewed by at least three members of the Program Committee. From these submissions, the top five were accepted as full papers for presentation at the workshop. The papers presented at the workshop provide a mix of novel research ideas, evaluations of existing process mining techniques, as well as new tool support.

Rehse and Fettke propose a four-step approach for vertically clustering event logs in order to discover reference model components from complex event logs. Their approach is based on proximity scoring of activities as an input for the hierarchical subprocess construction. Van Eck, Sidorova, and van der Aalst focus on process discovery in complex systems with multiple artifacts and corresponding lifecycles. The paper presents a mutli-instance mining technique to discover lifecycle models and their interactions with many-to-many relations between artifact types. The technique is implemented as the Multi-Instance Miner plugin in ProM. Lee, Munoz-Gama, Verbeek, van der Aalst, and Sepúlveda address the recomposition step when applying decomposition for alignment-based conformance checking by proposing several strategies to improve the performance of such an iterative alignment approach. Their technique is shown to improve on existing techniques based on both synthetic as well as real-life data. Van Dongen also focusses on the efficient computation of alignments by presenting an algorithm and memory structures of the extended marking equation approach. Both the time complexity of the algorithm as well as the properties of the different data structures are scrutinized. Deeva and De Weerdt look at a more practical application of process mining techniques by investigating the use of local process mining techniques to the context of learning processes and the potential of understanding feedback in these processes based on the analysis of event data.

As with previous editions of the workshop, we hope that the reader will find this selection of papers useful to keep track of the latest advances in the BPI area. We look forward to keep bringing new advances in future editions of the BPI workshop.

November 2018

Boudewijn van Dongen
Jochen De Weerdt
Andrea Burattin
Jan Claes

Organization

Program Committee

Ahmed Awad	Cairo University, Egypt
Josep Carmona	Universitat Politècnica Catalunya, Spain
Raffaele Conforti	Queensland University of Technology, Australia
Johannes De Smedt	The University of Edinburgh, UK
Benoit Depaire	Universiteit Hasselt, Belgium
Claudio Di Ciccio	Vienna University of Economics and Business, Austria
Chiara Di Francescomarino	Fondazione Bruno Kessler - IRST, Italy
Luciano García-Bañuelos	University of Tartu, Estonia
Gianluigi Greco	University of Calabria, Italy
Gert Janssenswillen	Universiteit Hasselt, Belgium
Toon Jouck	Universiteit Hasselt, Belgium
Anna Kalenkova	Higher School of Economics, Russia
Michael Leyer	University of Rostock, Germany
Jorge Munoz-Gama	Pontificia Universidad Católica de Chile, Chile
Pnina Soffer	University of Haifa, Israel
Suriadi Suriadi	Queensland University of Technology, Australia
Seppe vanden Broucke	KU Leuven, Belgium
Eric Verbeek	Eindhoven University of Technology, The Netherlands
Matthias Weidlich	Humboldt-Universität zu Berlin, Germany
Wil van der Aalst	RWTH Aachen University, Germany

Clustering Business Process Activities for Identifying Reference Model Components

Jana-Rebecca Rehse[✉] and Peter Fettke

Institute for Information Systems (IWi), German Research Center for Artificial Intelligence (DFKI GmbH) and Saarland University,
Campus D3.2, Saarbrücken, Germany
{Jana-Rebecca.Rehse,Peter.Fettke}@iwi.dfki.de

Abstract. Reference models are special conceptual models that are reused for the design of other conceptual models. They confront stakeholders with the dilemma of balancing the size of a model against its reuse frequency. The larger a reference model is, the better it applies to a specific situation, but the less often these situations occur. This is particularly important when mining a reference model from large process logs, as this often produces complex and unstructured models. To address this dilemma, we present a new approach for mining reference model components by vertically dividing complex process traces and hierarchically clustering activities based on their proximity in the log. We construct a hierarchy of subprocesses, where the lower a component is placed the smaller and the more structured it is. The approach is implemented as a proof-of-concept and evaluated using the data from the 2017 BPI challenge.

Keywords: Reference model mining · Activity clustering ·
Reference components · Reference modeling · Process mining

1 Introduction

Reference models can be considered as special conceptual models that serve to be reused for the design of other conceptual models. By providing a generic template for the design of new process models in a certain industry, reference process models allow organizations to adapt and implement the respective processes in a resource-efficient way [3]. The introduction of a common terminology and the subsequent simplification of communications along with the industry-specific experience contained in a reference model yield higher-quality processes and process models, while also reducing the required time, cost, and personnel resources required for business process management [8].

When developing a model for the purpose of reuse, model designers are faced with the dilemma of balancing the scope of a model, i.e. its size, specificity, and degree of coverage against its reuse potential, i.e. the number of situations where it can be applied. The larger and the more specific a model is, the less

© Springer Nature Switzerland AG 2019
F. Daniel et al. (Eds.): BPM 2018 Workshops, LNBIP 342, pp. 5–17, 2019.
https://doi.org/10.1007/978-3-030-11641-5_1

adaptations it needs to be applied in a certain model context, but the less often these situations occur. On the contrary, smaller reference models for subprocesses, named reference components, can be directly applied to many different situations, but do not suffice to cover the entire modeling domain.

We consider reference components as frequently appearing process model building blocks, i.e. temporally and logically isolated activity sets within a process [16]. For process designers, such building blocks strike a balance between the necessity to find a reference model for their exact use case and the disadvantages that come with modeling a process from scratch. Due to only a limited number of predefined interaction points with other process parts, they are frequently reusable and highly flexible to be combined into new process models [16]. By using pre-defined domain-specific collections of process building blocks, such as the PICTURE method for public administration modeling, process designers are able to leverage the multiple benefits of reference modeling and simultaneously address the specific challenges of their own process domain [4].

However, constructively using reference model components for all stakeholders' advantage first requires finding the right degree of specificity versus reusability. As this decision depends on the intended domain and purpose, it cannot be universally determined, but needs to be decided individually for each use case. In order to provide process designers with useful and reliable data to support the decision-making process, this contribution presents a novel approach for mining reference model components from instance-level data. A given input log is vertically divided and the activities are hierarchically clustered based on their spatial proximity in the log, determining the groups of activities that form a reference component. The components are mined for each cluster individually, resulting in a subprocess hierarchy, where the lower a component is placed, the smaller but the more structured and frequent it is.

This article is based on ideas for subprocess identification sketched in a report submitted to the 2017 BPI challenge [6]. We describe the conceptual design in Sect. 2. The realization in the RefMod-Miner research prototype is treated in Sect. 3, along with an experimental evaluation. We report on related work in Sect. 4, before concluding the article with a discussion in Sect. 5.

2   Conceptual Design of the Approach

2.1   Illustrating Example and Outline

The objective of this paper is to provide a data-based solution to determining the appropriate degree of specificity versus reusability when designing reference model components, i.e. to overcome the dilemma that the larger and more specific a reference model is, the less situations it applies to. Figure 1 illustrates our solution by means of an exemplary company-specific invoice handling process, executed e.g. by an accounting clerk. Once an invoice is received and processed, it is checked. If no further action is required (e.g. for a pro-forma-invoice), the invoice is archived directly. If the payment amount exceeds a certain limit, the invoice gets forwarded to the superior, as the clerk is not authorized for payment. If the limit is not exceeded, the clerk pays and archives the invoice.

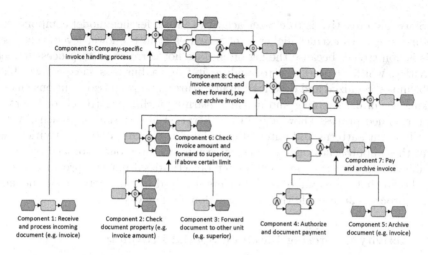

Fig. 1. Illustrating example for mining reference components

While the complete process model can be used for the design of other models, its application scope is limited to invoice handling. If, however, we divide it into its subprocesses, we see that they can be generalized to apply in other contexts. This is illustrated by the cluster structure in Fig. 1, which gradually divides the specific process into smaller parts. The smaller the parts get, the more generic they are. For example, Component 1, where the invoice is received and processed, could be part of any other invoice handling process, but could also be slightly abstracted to serve as a generic document handling subprocess.

Our approach consists of four major steps, described in the following subsections. Individual activities are identified from the provided event log and clustered hierarchically based on spatial proximity, resulting in a tree structure, where the higher a cluster is located, the more and less interrelated activities it contains. For each cluster, a reference component is mined, constructing a model hierarchy in analogy to the cluster structure. The higher a reference component is located, the more activities it contains and thus, the more specific it is.

2.2   Identifying Activities from Event Logs

Definition 1 (Activities, Traces, Event Logs [1]). *For a set S, let $\mathcal{B}(S)$ be the set of all multisets over S. Let $\mathcal{A}$ be the activity universe and $A \subseteq \mathcal{A}$ a set of activities. A trace $t = \langle t_1, \ldots, t_n \rangle \in A^*$ is a finite sequence of activities, with t_i denoting the activity at the ith position. A log $L \in \mathcal{B}(A^*)$ is a multiset of traces. $\mathcal{A}(L) = \bigcup_{t \in L} \bigcup_{i=1}^{|t|} t_i$ is the set of activities in L. For $A \subseteq \mathcal{A}(L)$, $T_A = \{t \in L \,|\, \forall a \in A : a \in t\}$ is the set of traces that contain all activities in A.*

Since the objective is to cluster activities into reference model components, the first step is to extract higher-level activities from the lower-level event log. This is non-trivial, because the log often does not correspond to process model activities, which is necessary to analyze it from a business perspective. The problem is recognized and described in literature, with proposed solutions either making use of data mining and machine learning techniques [10,11] or leveraging predefined process knowledge in a supervised abstraction approach [14] to identify event patterns and automatically label the corresponding activities. As all of these approaches offer promising results, we base our approach on their capabilities to extract a meaningful set of activities from the provided event log, should that be required. Stakeholders involved in the reference component design process can provide the necessary domain knowledge.

2.3   Activity Clustering Based on Spatial Proximity

For clustering the activities into a hierarchical structure, we measure their spatial proximity, assuming that activities which often appear close to each other form a logical and structured unit. To measure activity proximity, we select the set of traces containing both activities at least once. We count the number of steps between the two activities, divide it by the length of the trace to get a normalized value, and deduct the result from 1. If the activities appear multiply within one trace, the minimum distance is calculated.

Definition 2 (Trace-based Spatial Proximity). *Let L be a log, $\mathcal{A}(L)$ its activity set, $a, b \in \mathcal{A}(L)$ two activities, and $T_{ab} \subseteq L$ the set of traces that contain both a and b. For a trace $t \in T_{ab}$, let $I_t = \{i \in \mathbb{N} | t_i = a\}$ and $J_t = \{j \in \mathbb{N} | t_j = b\}$ be the event index sets for activities a and b. The trace-based spatial proximity $p : \mathcal{A} \times \mathcal{A} \to [0,1]$ between two activities is defined as*

$$p(a,b) = \begin{cases} \dfrac{\sum_{t \in T_{ab}} \left(1 - \dfrac{min(\{|i-j| \mid i \in I_t, j \in J_t\})}{|t|}\right)}{|T_{ab}|} & \text{if } |T_{ab}| \geq 1, \\ 0 & \text{otherwise.} \end{cases}$$

The pairwise spatial proximity in form of a similarity matrix is used as input for clustering. A hierarchical-agglomerative clustering approach allows us to inspect activity clusters on different size and specificity levels [7]. The result is a strict cluster hierarchy, with the singular activities as leaves and the complete activity set as root cluster. Each internal node cluster contains the union of activities that are contained in its two child clusters. To get a precise cluster result, we do not parametrize the expected number of clusters, which increases the runtime complexity. Compared to e.g. trace clustering, where a few thousand clustering objects are still computationally feasible [15], activity clustering typically contains not more than a hundred objects, so computation times should not become a problem. The result of a clustering is an activity hierarchy.

Definition 3 (Activity Hierarchy). *Let L be a log, $\mathcal{A}(L)$ its set of activities. An* activity hierarchy $H_{\mathcal{A}(L)}$ *is a connected, directed, acyclic graph $H = (C, D)$ (i.e. a tree), where $2^{\mathcal{A}(L)} \supset C = \{A_1, \ldots, A_n\}$ is a set of subsets of $\mathcal{A}(L)$ and $D \subset C \times C$ is a set of edges connecting them, such that:*

- $\mathcal{A}(L) \in C$ *is the cluster root, $\forall a \in \mathcal{A}(L) : \{a\} \in C$ are the cluster leaves.*
- $\forall A_i : |\{(A_i, x) \in D\}| = 0 \wedge |\{(A_i, x) \in D\}| = 2$, *i.e. an internal cluster has exactly two children.*
- $(A_i, A_j) \in D \Rightarrow A_j \subset A_i$, *i.e. a set is fully contained in its parent set,*
- $\forall i : \bigcup_{(A_i, A_k) \in D} A_k = A_i$, *i.e. a set is equal to the unification of its children.*
- $(A_i, A_j) \in D \wedge (A_i, A_k) \in D \Rightarrow A_i \cap A_k = \emptyset$, *i.e. sibling sets are disjoint.*

2.4 Mining Reference Model Components

After obtaining the cluster hierarchy, we mine a reference model component for each identified activity set. Therefore, we use our RMM-2 approach for reference model mining based on execution semantics, adapted to work with process traces instead of process models [15]. It analyzes the represented process semantics in terms of behavioral profiles and computes a reference model subsuming the specified behavior. To apply the adapted RMM-2 for successfully mining reference components, the input data has to be modified, such that the reference components contain the same activities as the associated cluster.

Definition 4 (Log Projection [1]). *Let L be a log, $\mathcal{A}(L)$ its set of activities, $A \subseteq \mathcal{A}(L)$ a set of activities, and $T_A \subseteq L$ the set of traces that contain all activities in A. The* log projection $L_A = \{t\lceil_A | t \in L\}$ *contains only the activities in A. For a trace $t \in T_A$, the trace projection function $t\lceil_A$ is defined recursively: (1)$\langle\rangle\lceil_A = \langle\rangle$ and (2) for $t_1 \in \mathcal{A}(L), t' \in \mathcal{A}(L)^*$:*

$$(\langle t_1 \rangle \circ t')\lceil_A = \begin{cases} \langle t_1 \rangle \circ (t'\lceil_A) & \text{if } t_1 \in A, \\ t'\lceil_A & \text{if } t_1 \notin A. \end{cases}$$

For an activity set A and its projected log L_A, the reference component is mined as follows:

1. Compute Behavioral Profiles: For each unique trace in L_A, a separate behavioral profile is computed using the trace-based activity relations.
2. Integrate Behavioral Profiles: The set of individual profiles is merged into an integrated behavioral profile, choosing the most frequent relation above a certain confidence level, such that it representing typical behavior. The noise level parameter specifies the minimum threshold for a relation to be included in the integrated profile. Conflicts are manually resolved.
3. Derive Reference Component: Finally, the semantics represented in the behavioral profile are conveyed into a process model in form of an event-driven process chain (EPC).

Formally, mining the reference components assigns a reference component in form of a process model (in this case, an EPC) to each activity set contained in the activity hierarchy. In the following, we define an *process model* as a tuple $P = (N, E, type)$, according to the definition by [19, p. 92].

Definition 5 (Reference Model Hierarchy). *Let L be a log, $\mathcal{A}(L)$ its set of activities, and $H_{\mathcal{A}(L)} = (C, D)$ an activity hierarchy. The corresponding reference model hierarchy $RM_{H_{\mathcal{A}(L)}}$ is a function that assigns each activity set $A_i \in C$ to a reference model RM_i in form of an process model $P_i = (A_i, E_i, type_i)$.*

2.5   Evaluating Reference Model Components

Finally, we need to provide stakeholders with concise data to find an optimal solution for the dilemma of reusability for their use case. Reference components are supposed to contain frequently appearing model parts, so that they can be reused and applied in different contexts. They are also supposed to be internally coherent, i.e. the contained activities should exhibit some kind of correlation with one another. Applying this to our reference component hierarchy, we need to find those cluster integration steps, where combining two clusters into one does not produce a viable reference component, because the activity set is either not sufficiently frequent or not sufficiently interconnected anymore. While the former can be assessed by measuring the difference in relative frequency of the activity sets in the log, the latter can be assessed by comparing the size of the integrated reference component with the sum of the individual component sizes. Therefore, we define the following measures.

Definition 6 (Diffusion Rate and Inflation Rate). *Let L be a log, $\mathcal{A}(L)$ its set of activities, $H_{\mathcal{A}(L)} = (C, D)$ an activity hierarchy, and $RM_{H_{\mathcal{A}(L)}}$ a reference model hierarchy. Let $A_i \in C$ be a an activity set, $A_j, A_k \in C$ its children, and RM_i, RM_j, RM_k the corresponding reference models.*

The diffusion rate $\mathcal{D}_{A_i}$ defines the ratio between the number of traces that contain A_i, $T_{A_i} = T_{A_j} \cap T_{A_k}$ and the number of traces that contain its respective children, $T_{A_j} \cup T_{A_k}$.

$$\mathcal{D}_{A_i} = \frac{|T_{A_i}|}{|T_{A_j} \cup T_{A_k}|}$$

The inflation rate $\mathcal{I}_{A_i}$ defines the ratio between the size of RM_i, $|RM_i| = |A_i|$ and the added sizes of RM_j, RM_k.

$$\mathcal{I}_{A_i} = \frac{|RM_i|}{|RM_k| + |RM_j|}$$

By measuring these two relations for each reference component, we determine those points where an integration is not useful anymore, because the two components to be merged are so different from one another. They either conjointly appear in very few traces, such that the intersection of the two trace sets is much smaller than the union, resulting in low a diffusion rate, or the size of the merged component is much larger than the sum of the child components, meaning that many connector nodes are required to merge them, resulting in an inflation rate considerably higher than 1. One could argue that if many connector nodes are required for merging two components, there is a high degree of interconnection between them, requiring many interfaces. We don't want to completely rule out such a scenario, but the question arises why the two components were not clustered together in the first place, if they are so closely connected. The components' eligibility as reference components would be limited in such a case, due to their complicated structure. In general, these are only structural metrics for assessing the degree of interrelation between process model activities and may not replace the final decision by the reference model designer.

3   Proof-of-Concept and Experimental Evaluation

In order to demonstrate the capabilities of the suggested approach, it was prototypically implemented as a proof-of-concept in the RefMod-Miner research prototype, a Java-based software tool for process model analysis developed in our research group (https://refmod-miner.dfki.de). This implementation was used for an evaluation using the "application event log" from the 2017 BPI Challenge [17], describing a loan application process from a Dutch financial institute. We distinguish A-type events (subprocess of application handling), O-type events (offer creation), and W-type events (workflow activities). The log contains 561,671 events from 31,509 individual loan application cases.

The first step is to identify the activities. As we see from inspecting the log, the events are recorded on a lower level, but are clearly associated with a higher level activity ("concept:name"). The event itself is specified by the attribute "lifecycle:transition" and the separate ID. Table 1 lists the extracted activities.

Table 1. Identified activities from the BPI 2017 log

A_Accepted	A_Submitted	O_Returned	W_Handle leads
A_Cancelled	A_Validating	O_Sent (mail and online)	W_Personal Loan collection
A_Complete	O_Accepted	O_Sent (online only)	W_Shortened completion
A_Concept	O_Cancelled	W_Assess potential fraud	W_Validate application
A_Denied	O_Create Offer	W_Call after offers	A_Create Application
A_Incomplete	O_Created	W_Call incomplete files	
A_Pending	O_Refused	W_Complete application	

The spatial proximity measure is used to compute a matrix between all pairs of activities, which serves as input for the clustering, using the readily available implementation *hclust* in the statistical computing language R. We did not specify a maximum size or height of the clusters and determined the distance between two clusters by means of complete linkage. The details of the cluster assignments and merging steps are shown by the dendrogram in Fig. 2.

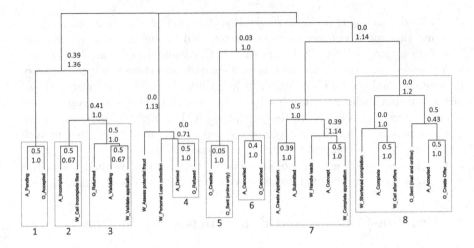

Fig. 2. Dendrogram for activity clustering (Color figure online)

Next, we mine a reference component for each cluster by applying the trace-based RMM-2 approach. We apply the standard parametrization (noise level 0.2, minimum frequency 0.1) to ensure a sufficient frequency. Finally, diffusion and inflation rates are computed for each reference component. In Fig. 2, each cluster is annotated with its diffusion rate (top) and inflation rate (bottom). Ideally, both numbers should be close to 1. Lower diffusion and higher inflation rates indicate that merging two child components might not make sense. We see inflation rates smaller than 1, indicating that the merged component is smaller than the combined sizes of its children. This can be explained by control flow structures. For example, a self-loop with three nodes (one activity, two connectors) might be frequent in a larger log, but the connectors might not be contained in a merged component, e.g. a two-activity sequence.

The diffusion rates seem comparably low across all clusters. This can be attributed to the large log, where trace structures differ in frequency and variability. For example, the component where a loan is denied ("A_Denied, O_Refused") has low diffusion rates (rounded down to 0.0) from the lowest level on upwards, which suggests that it is not contained in a lot of traces. On the other hand, the component which describes the completion and submission of the application has fairly high diffusion rates. Based on the evaluation data, we selected eight reference components, marked by red borders in Fig. 2 and shown in Fig. 3.

Each selected reference component is a subprocess of the loan application process, with six small and two larger components. The latter are coherent subprocesses, namely the application creation (top) and application acceptance (bottom). The application creation appears straightforward. Offer creation and acceptance is kept separate, increasing its reusability, as it removes the strict association between applications and offers. The same applies to calling incomplete files; incomplete applications may be automatically denied. On the other hand, the subprocess is potentially incomplete and not directly usable. The second subprocess, application acceptance, is more specific to the individual process and therefore directly applicable, but less reusable. It describes the relation between offer and application, with the offer creation directly following the application acceptance. Counterintuitively, the offer creation is not part of this process, but this is not be supported by the clustering data.

Most small components describe the activities in a very specific situation, i.e. when an application is denied or canceled, if it is incomplete, or if the offer is accepted. These four components are applicable, but also fairly reusable due to their small size. They all associate offers with applications, which may impact their reusability. The remaining two components are the offer creation, which should be part of the application acceptance, and the application validation, which can be regarded as a subprocess in itself. The components' usability is impeded by back-loops, which often appear unnecessary (e.g. "O_Cancelled") and should be removed for a more generic component. Also, the lack of operators except XOR is apparent, as the components lack clear semantics.

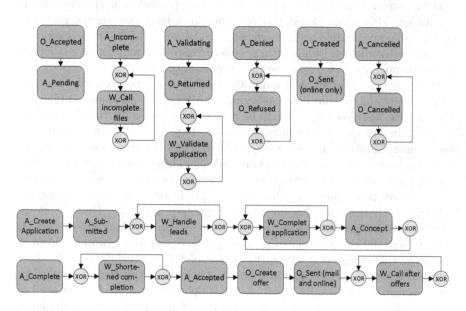

Fig. 3. Components mined for clusters 1–6 (top) and clusters 7 and 8 (bottom)

4   Related Work

In this contribution, we mine reusable reference model components from instance-level event logs. Comparable approaches mine complete reference models, such as the approach by Gottschalk et al., which is set out to mine configurable reference models and according configurations from log files of well-running IT systems [9]. Instead of ensuring better reusability by mining smaller model components, the authors construct configurable reference models, which contain all potential process variants. This might work well, but a configurable model could become quite large, when covering rather diverse process behavior. Our own work on mining reference process models from large instance data works in a similar way, applying a clustering approach onto the traces of an execution log. The objective of this approach is to determine reference models depicting the whole process execution, but with differing degrees of generality or domain specificity [15], as opposed to reference components depicting process fragments. For this purpose, the event log is divided and clustered horizontally along the trace similarity instead of vertically according to activity proximity, as we do here.

All other automated approaches towards inductive reference modeling rely on type-level process models instead of instance-level events logs as input data. A contribution by Li et al. presents two approaches [13]. The first one uses a heuristic search algorithm and an approximation of the graph-edit distance for evolving an existing reference model such that it better fits a set of its derived variants. The second one uses an iterative clustering technique with a process-semantic proximity measure to construct a reference model from a predetermined activity set. This is also closely related to our work here, however the goal is to combine all activities into a single reference model, solely based on their process-semantic order relations, while we are explicitly set out to mine smaller reference model components, based on an arbitrary proximity measure.

In this article, we rely on several others BPM areas. For example, our spatial proximity can be considered as a similarity measure. The more similar two activities are, the closer they should be associated within a cluster. Whereas we limit ourselves to spatial proximity, approaches to process model matching combine as many similarity measures as possible to determine the most appropriate correspondences between two process models [2]. These correspondences (called matches) are then used to determine the similarity between process models [5].

Clustering techniques are often used in BPM, for example for identifying high-level activities from low-level event logs (see Sect. 2.2). Günther et al. describe a technique that clusters events into recognizable patterns based on temporal and data-object-related proximity [11] as well as a new and improved technique, clustering event classes by means of trace segmentation [10]. This approach is very similar to ours in terms that it builds an event hierarchy solely based on a spatial proximity measure, but it operates on a much lower level of detail and is not directed towards the reusability of model components. Similarly, the POD-Discovery tool by Weber et al. also uses a hierarchical clustering approach to enable the application of process mining tools on low-level operational process logs [18] and Verbeek et al. present an approach that uses clustering for decomposed and therefore more efficient process mining.

5 Discussion and Conclusion

The contribution at hand has the objective to provide an automated and data-centered approach for supporting stakeholder in designing reference components. It is set out to address the dilemma of reusability, where the better a reference model applies to a situation, the fewer those situations are. We define reference components on spatial proximity, following the idea that semantically related activities are typically executed in close proximity to one another. This assumption can be questioned, as some empirical work suggests otherwise [12]. Measuring proximity also complicates the detection of choice constructs, as those activities never appear jointly in a trace. Moreover, results may be adulterated by rare or noisy activities, or similar activities placed differently by different organizations. So, for more dependable results, other activity similarity measures, such as temporal proximity, and filtering approaches could be taken into account.

Generally, the reference component's domain and characterization are determined by the input data, which, in turn, depends on the intended usage of the reference components. If considering data from one company and one process only, the components will be process-specific, i.e. focus on patterns within this process. Considering the same process across multiple companies discovers cross-organizational similarities or patterns, i.e. industry-specific components. Organization-specific components describe similarities across multiple processes, i.e. common process patterns in one company. Finally, when analyzing data from multiple processes and multiple companies, we obtain generic, domain-independent components. All of these scenarios have meaningful applications, but this choice should be considered prior to determining the components, as it will influence the appropriate values for diffusion and inflation rate.

One could also argue that the input data needs to be carefully chosen with regard to the represented process. The main motivation of this article is the dilemma of reusability, i.e. the difficulties that appear when reusing models or model parts in the design of a new conceptual model. There is no point in designing reference components for a domain in which no subprocess is sufficiently frequent or relevant such that there is a general interest in reusing it. This means, that repetitive processes, like the loan application process in our case study, are particularly suitable for such an analysis. This is also evident, as those processes are most likely supported by information systems, such that process logs exist.

While it is also possible to mine reference components from type-level model data, we have decided to base our approach on instance-level execution data, which offers a more realistic perspective on the process than type-level data. By computing the activity proximity based on factually executed process traces associated with timestamps, resources, or related data objects, we are able to provide a realistic view on the process that is actually executed, increasing the probability for component reuse. However, this means that our result is a set of *descriptive* reference components instead of *normative* ones. They depict as-is process behavior instead of to-be recommendations. Both have realistic application scenarios. Companies typically strive to standardize support processes, such that more resources are available to focus on their core processes as a competitive advantage. Hence, the former, which are often automated and IT-supported, can be designed by reusing a descriptive reference model.

As reference model design always depends on the intended model (re-)use and purpose, our objective is not to design a completely automated approach, but rather a helpful tool for decision-making support. Full automation is mainly difficult, because constructing completely generic components requires some abstraction of either activities or the entire process model (e.g. by changing the label to describe a more generic task like "check document" instead of "check invoice"). Those abstractions are context-dependent and hard to find by algorithmic means, but fairly easy for a human process designer to make.

References

1. van der Aalst, W.M.P.: Decomposing petri nets for process mining: a generic approach. Distrib. Parallel Databases **31**(4), 471–507 (2013)
2. Antunes, G., Bakhshandelh, M., Borbinha, J., Cardoso, J., Dadashnia, S., et al.: The process matching contest 2015. In: Kolb, J., Leopold, H., Mendling, J. (eds.) Proceedings of the 6th International Workshop on Enterprise Modelling and Information Systems Architectures (EMISA-15), 3–4 September, Innsbruck, Austria. Köllen Druck+Verlag GmbH, Bonn, September 2015
3. Becker, J., Meise, V.: Strategy and organizational frame. In: Becker, J., Kugeler, M., Rosemann, M. (eds.) Process Management. A Guide for the Design of Business Processes, pp. 91–132. Springer, Heidelberg (2011). https://doi.org/10.1007/978-3-642-15190-3_4
4. Becker, J., Pfeiffer, D., Räckers, M.: Domain specific process modelling in public administrations – the PICTURE-approach. In: Wimmer, M.A., Scholl, J., Grönlund, Å. (eds.) EGOV 2007. LNCS, vol. 4656, pp. 68–79. Springer, Heidelberg (2007). https://doi.org/10.1007/978-3-540-74444-3_7
5. Becker, M., Laue, R.: A comparative survey of business process similarity measures. Comput. Ind. **63**(2), 148–167 (2012)
6. Dadashnia, S., Fettke, P., Hake, P., Klein, S., et al.: Exploring the potentials of artificial intelligence techniques for business process analysis (2017). http://www.win.tue.nl/bpi/lib/exe/fetch.php?media=2017:bpi2017_paper_41.pdf
7. Everitt, B.S., Landau, S., Leese, M., Stahl, D.: Hierarchical clustering. Cluster Analysis, pp. 71–110. Wiley, London (2011)
8. Fettke, P., Loos, P.: Perspectives on reference modeling. In: Fettke, P., Loos, P. (eds.) Reference Modeling for Business Systems Analysis, pp. 1–20. Idea Group Publishing, London (2007)
9. Gottschalk, F., van der Aalst, W.M.P., Jansen-Vullers, M.H.: Mining reference process models and their configurations. In: Meersman, R., Tari, Z., Herrero, P. (eds.) OTM 2008. LNCS, vol. 5333, pp. 263–272. Springer, Heidelberg (2008). https://doi.org/10.1007/978-3-540-88875-8_47
10. Günther, C.W., Rozinat, A., van der Aalst, W.M.P.: Activity mining by global trace segmentation. In: Rinderle-Ma, S., Sadiq, S., Leymann, F. (eds.) BPM 2009. LNBIP, vol. 43, pp. 128–139. Springer, Heidelberg (2010). https://doi.org/10.1007/978-3-642-12186-9_13
11. Günther, C., van der Aalst, W.: Mining activity clusters from low-level event logs. In: BETA Working Paper Series, WP 165, Eindhoven University of Technology, Eindhoven (2006)

12. Klinkmüller, C., Weber, I.: Analyzing control flow information to improve the effectiveness of process model matching techniques. Decis. Support Syst. **100**, 6–14 (2017)

13. Li, C., Reichert, M., Wombacher, A.: Mining business process variants: challenges, scenarios, algorithms. Data Knowl. Eng. **70**(5), 409–434 (2011)

14. Mannhardt, F., de Leoni, M., Reijers, H.A., van der Aalst, W.M.P., Toussaint, P.J.: From low-level events to activities - a pattern-based approach. In: La Rosa, M., Loos, P., Pastor, O. (eds.) BPM 2016. LNCS, vol. 9850, pp. 125–141. Springer, Cham (2016). https://doi.org/10.1007/978-3-319-45348-4_8

15. Rehse, J.-R., Fettke, P.: Mining reference process models from large instance data. In: Dumas, M., Fantinato, M. (eds.) BPM 2016. LNBIP, vol. 281, pp. 11–22. Springer, Cham (2017). https://doi.org/10.1007/978-3-319-58457-7_1

16. Rupprecht, C., Fünffinger, M., Knublauch, H., Rose, T.: Capture and dissemination of experience about the construction of engineering processes. In: Wangler, B., Bergman, L. (eds.) CAiSE 2000. LNCS, vol. 1789, pp. 294–308. Springer, Heidelberg (2000). https://doi.org/10.1007/3-540-45140-4_20

17. Van Dongen, B.: BPI challenge (2017). https://data.4tu.nl/repository/uuid:5f3067df-f10b-45da-b98b-86ae4c7a310b

18. Weber, I., Li, C., Bass, L., Xu, X., Zhu, L.: Discovering and visualizing operations processes with POD-Discovery and POD-Viz. In: 45th Annual IEEE/IFIP International Conference on Dependable Systems and Networks (DSN), pp. 537–544. IEEE (2015)

19. Weske, M.: Business Process Management. Concepts. Languages, Architectures. Springer, Heidelberg (2007). https://doi.org/10.1007/978-3-642-28616-2

Multi-instance Mining: Discovering Synchronisation in Artifact-Centric Processes

Maikel L. van Eck[1]([⊠]), Natalia Sidorova[1], and Wil M. P. van der Aalst[1,2]

[1] Eindhoven University of Technology, Eindhoven, The Netherlands
{m.l.v.eck,n.sidorova}@tue.nl
[2] RWTH Aachen University, Aachen, Germany
wvdaalst@pads.rwth-aachen.de

Abstract. In complex systems one can often identify various entities or artifacts. The lifecycles of these artifacts and the loosely coupled interactions between them define the system behavior. The analysis of such artifact system behavior with traditional process discovery techniques is often problematic due to the existence of many-to-many relationships between artifacts, resulting in models that are difficult to understand and statistics that are inaccurate. The aim of this work is to address these issues and enable the calculation of statistics regarding the synchronisation of behaviour between artifact instances. By using a Petri net formalisation with step sequence execution semantics to support true concurrency, we create state-based artifact lifecycle models that support many-to-many relations between artifacts. The approach has been implemented as an interactive visualisation in ProM and evaluated using real-life public data.

1 Introduction

Process discovery is the automated creation of process models that explain the behaviour captured in event data [1]. Over the years, various algorithms and tools have been developed that support process discovery. However, traditional process discovery techniques are not always suitable for every type of process.

In processes where we can identify *artifacts*, key entities whose lifecycles and interactions define the overall process [6], traditional process discovery techniques often fail [4,7,8,11,12]. Such *artifact-centric processes* form naturally in business environments supported by information systems based on Entity-Relationship models and databases [3], e.g. a procurement process with sales orders, invoices and deliveries. They can also occur in complex environments, e.g. a medical procedure with surgeons, nurses and medical systems. Additionally, software is often developed according to object-oriented programming paradigms, so processes describing the behaviour of such software systems can also be decomposed into artifacts representing the software objects and components.

This research was performed in the context of the IMPULS collaboration project of Eindhoven University of Technology and Philips: "Mine your own body".

F. Daniel et al. (Eds.): BPM 2018 Workshops, LNBIP 342, pp. 18–30, 2019.
https://doi.org/10.1007/978-3-030-11641-5_2

The existence of *many-to-many* relationships [3,13] between process artifacts is an important reason why traditional process discovery approaches have difficulties with artifact-centric processes [7,8]. These relationships make it difficult to identify a unique process instance notion to group related events. Enforcing a flat grouping leads to data convergence and divergence problems when calculating statistics related to event occurrences and causal relations between events [8,13].

Recently, artifact-centric process discovery approaches have been developed that aim to discover models that are not affected by data convergence and divergence issues [7,8]. However, these approaches have difficulties identifying synchronisation points and flexible interactions between loosely coupled artifacts. Examples of synchronisation points in artifact lifecycles are milestone patterns, e.g. the payment of all invoices before a delivery, and collaborative efforts, e.g. several people meeting to create a project plan.

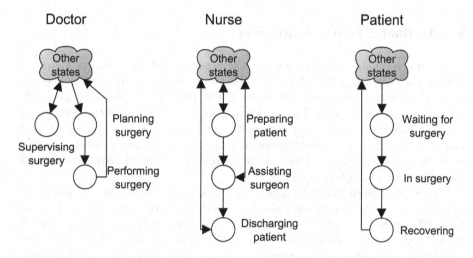

Fig. 1. Partial lifecycle models showing the possible states and transitions between states of three types of artifacts involved in a hospital process: doctors, nurses and patients.

Consider the example of a simplified artifact-centric hospital process involving doctors, nurses and patients, all modelled as artifacts with their lifecycles shown in Fig. 1. This example process has several possible synchronisation points where people interact: while *preparing patients* two nurses are needed to lift the patient onto the operating table, and the patient can only start *recovering* if the surgeon has finished *performing surgery*. However, the interaction is flexible: the nurses *preparing* a patient are not necessarily the same as those *discharging* the patient, a doctor can be *supervising* multiple surgeries in a single day while doing other things in between, and not every surgery has an additional doctor

supervising. Such loosely coupled interactions are difficult to analyse using existing process discovery techniques, e.g. because statistics and relations from the
viewpoint of individual doctors are not the same as from the viewpoint of individual patients. Therefore, we have developed an approach to provide accurate
statistics and insights in complex artifact-centric processes.

In this paper we describe a process discovery approach suitable for the analysis of synchronous artifact behaviour. This approach builds on ideas presented
in [4] for the discovery of state-based models for artifact-centric processes. The
state machine formalisation in that work supports the analysis of synchronous
artifact behaviour, but only for relations between pairs of artifact instances. We
show that by using a Petri net formalisation with step sequence execution semantics to support true concurrency, we can create state-based models that support
many-to-many relations between artifacts. This approach has been implemented
as a plug-in for the ProM process mining framework and we have evaluated it
using the public datasets of the BPI Challenge of 2017.

2 Artifact System Modelling

Our goal is to analyse situations where the process of interest involves a number
of artifacts interacting. In such an *artifact system* context we distinguish between
artifact *types*, e.g. doctors or sales orders, and artifact *instances*, e.g. a specific
person working as a doctor. For a given artifact type there is a set of *type states*,
i.e. the possible states that artifacts of this type can have. An artifact *lifecycle
model* is a graphical representation of an artifact type, its states and the possible
transitions between states. The conceptual representation of artifact systems is
shown as a class diagram in Fig. 2a.

There exist various modelling languages to describe operational processes [1].
In this work we aim to explicitly analyse the states of artifacts and their interactions. Therefore, we model artifact lifecycles as *state machines*, building on ideas
from [4]. State machines are a subclass of Petri nets with the property that each
transition has exactly one incoming and one outgoing edge, enabling choice but
not concurrency [9]. In a state machine artifact lifecycle model the places represent the type states and a token represents the current state of an artifact
instance. To model the beginning and finishing of a lifecycle, we use interface
transitions [10] that represent connections to the environment. For simplicity,
we represent transitions in lifecycle models using only edges, as in Fig. 1.

Definition 1. *A* state machine *artifact lifecycle model* $\mathcal{A}$ *is a Petri net tuple*
(T^c, T^b, T^f, P, E), *where* T^c *is a set of transitions to change states,* T^b *is a set
of input interface transitions,* T^f *is a set of output interface transitions,* P *is a
finite set of places,* $E \subseteq ((T^c \cup T^b) \times P) \cup (P \times (T^c \cup T^f))$ *is a set of directed edges
between transitions and places, with* $\forall t \in (T^c \cup T^b) : \left|\{p \in P | (t,p) \in E\}\right| = 1$,
and $\forall t \in (T^c \cup T^f) : \left|\{p \in P | (p,t) \in E\}\right| = 1$.

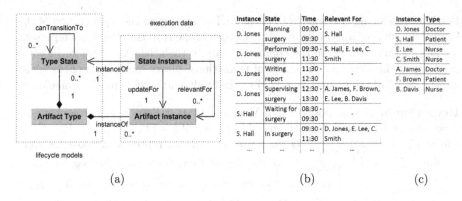

(a)   (b)   (c)

Fig. 2. (a) Class diagram providing a conceptual representation of an artifact system. (b) A list of state instances from execution data of the artifact-centric example process. (c) A mapping of artifact instances to types

We discover artifact lifecycle models based on process execution data containing *state instances*, i.e. moments in time where a specific artifact instance obtains a certain state. In Fig. 2b each row is a state instance updating the state of a specific instance of an artifact in the example hospital process. A mapping of instances to artifact types for this execution data is given in Fig. 2c.

Definition 2. *A log of process execution data $\mathcal{L}$ is a tuple $(\mathbb{S}, \mathcal{I}, \mathbb{A}, S, \mathbb{T}, itype)$, where $\mathbb{S}$ is a set of state instances, $\mathcal{I}$ is a set of artifact instances, $\mathbb{A}$ is a set of artifact types, S is a set of type states, $\mathbb{T}$ is a time domain, and $itype : \mathcal{I} \to \mathbb{A}$ is a mapping from instances to types.*

Each state instance $\varsigma \in \mathbb{S}$ is a tuple (ι, s, τ, I), where $\iota \in \mathcal{I}$ is the primary artifact instance that obtains state $s \in S$ at time $\tau \in \mathbb{T}$, and $I \subset \mathcal{I}$ is the set of secondary artifact instances for which it is relevant. The end time of ς is an attribute derived by ordering all state instances for ι by their timestamp. We assume that $\forall \iota' \in \mathcal{I}, \tau' \in \mathbb{T} : \left| \{ (\iota, s, \tau, I) \in \mathbb{S} | \iota = \iota' \wedge \tau = \tau' \} \right| \leq 1.$

Each state instance can be relevant for a number of secondary artifact instances, which means it can be used to determine synchronisation points or calculate interaction statistics from the perspective of the secondary artifact lifecycles. As shown in Fig. 2, performing surgery involves a patient and nurses, so this state instance can be used to determine per nurse what doctors they work with and for how long. On the other hand, writing reports only concerns one doctor, so it is not relevant from the perspective of other artifact instances. Note that this relation is not necessarily symmetric, e.g. the patient S. Hall is waiting while doctor D. Jones is planning the surgery, but the waiting time is not considered relevant for the doctor's artifact lifecycle.

By modelling artifact lifecycles as state machines there can be no concurrency in the state of a single artifact instance. However, in the context of an artifact system there are multiple interacting artifact instances that each concurrently have a certain state. Therefore, we define a *Composite State Machine* (CSM)

Petri net modelling the behaviour of an artifact system as the composition of the lifecycle models of a number of artifact types. Tokens in this Petri net represent artifact instance states, so the combined state of the artifact system is the total marking of the Petri net. The sets of transitions, places, and edges of the CSM are the union of the respective sets of the individual artifact type state machines.

Definition 3. *A* Composite State Machine $\mathcal{M}$ *is a Petri net representing the product of a set of n state machines $\mathcal{A}_1, \ldots, \mathcal{A}_n$. $\mathcal{M} = (T_{\mathcal{M}}^c, T_{\mathcal{M}}^b, T_{\mathcal{M}}^f, P_{\mathcal{M}}, E_{\mathcal{M}})$, where $T_{\mathcal{M}}^c = \bigcup_{i \in \{1,\ldots,n\}} T_i^c$ is the set of transitions representing possible state changes, $T_{\mathcal{M}}^b = \bigcup_{i \in \{1,\ldots,n\}} T_i^b$ is the set of transitions where lifecycles begin, $T_{\mathcal{M}}^f = \bigcup_{i \in \{1,\ldots,n\}} T_i^f$ is the set of transitions where lifecycles finish, $P_{\mathcal{M}} = \bigcup_{i \in \{1,\ldots,n\}} P_i$ is the set of places, $E_{\mathcal{M}} = \bigcup_{i \in \{1,\ldots,n\}} E_i$ is the set of edges,*

There is no explicit modelling of synchronisation points in the CSM definition above. Therefore, we model synchronous behaviour by adopting step execution semantics [2,10] instead of the common atomic firing of transitions. Intuitively, a finite multiset of transitions, a *step*, may fire if the artifact instances represented by Petri net tokens can change their state at the same moment in time.

Definition 4. *Given a Petri net with transitions T, places P, edges E and a current state s (or marking) as a multiset of tokens $s \in \mathbb{N}^P$, then a step $\mathcal{F} \in \mathbb{N}^T$ is fireable iff $\forall p \in P : \sum_{\{t \in T | (p,t) \in E\}} \mathcal{F}(t) \leq s(p)$, i.e. all input places contain enough tokens. Firing the multiset of transitions in step $\mathcal{F}$ results in the state s' where $\forall p \in P : s'(p) = s(p) - \sum_{\{t \in T | (p,t) \in E\}} \mathcal{F}(t) + \sum_{\{t \in T | (t,p) \in E\}} \mathcal{F}(t)$.*

The use of step execution semantics results in Petri net behaviour with true concurrency, instead of interleaving of transitions [2]. This means that artifact instances can change states independently from each other, but simultaneous state changes can also be analysed to find synchronisation points.

3   Multi-instance Mining Approach

An overview of the process discovery approach to enable the multi-instance mining of Composite State Machines (CSMs) is shown in Fig. 3.

3.1   State Instance Creation

The executions of many processes are recorded in the form of event logs [1]. To use these logs in our approach they need to be transformed into the format described in Definition 2. This involves choosing a state representation and determining the set of artifact instances for which a state instance is relevant.

For a given log of process event traces, every position in a trace corresponds to a state of the process [1]. A state representation is a function that, given a sequence of events and a number, produces the process state after the specified number of events in the sequence have occurred. Examples of state representation

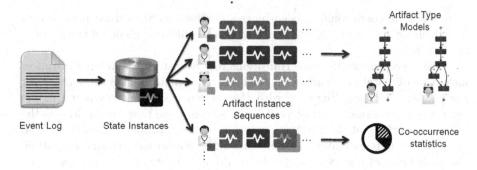

Fig. 3. The overall approach for multi-instance mining of CSMs.

functions are e.g. the last event executed, the set of events executed, or the full ordered list of events executed. We assume that similar artifact type state representations have been chosen as a preprocessing step, to transform sequences of artifact instance lifecycle events into sequences of state instances.

By linking related artifact instances it is possible to determine the set of relevant secondary artifact instances for a state instance. Such information, relating e.g. doctors to patients or sales orders to deliveries, can be found in relational databases in enterprise settings [3], in ERP systems [8] or in the event log itself [11]. Using domain knowledge, some state instances may get a different set of relevant instances, e.g. they only affect the primary artifact instance.

3.2   Composite State Machine Creation

Given process execution data, we create a time-ordered sequence per artifact instance of all state instances where this artifact is the primary artifact instance. This sequence represents the lifecycle of the given artifact instance.

Individual state machine lifecycle models can be discovered per artifact type by grouping all sequences from instances of the same type and applying the discovery approach described in [4]. The creation of the CSM describing the system is the union of these models, as described in Definition 3.

3.3   Multi-instance Mining Statistics

We use the notion of a primary instance to group related state instances and calculate accurate sojourn time and co-occurrence statistics in the presence of many-to-many relations. For each artifact instance, we create a time-ordered sequence of state instances where the artifact is either the primary or a secondary instance. That is, given execution data $\mathcal{L} = (\mathbb{S}, \mathcal{I}, \mathbb{A}, S, \mathbb{T}, \mathsf{itype})$ and instance $\iota' \in \mathcal{I}$ we create a sequence from $\{(\iota, s, \tau, I) \in \mathbb{S} | \iota' = \iota \vee \iota' \in I\}$. State instances with the same timestamp τ represent co-occurring transitions, i.e. they are part of a step in the execution of the mined CSM. For example, given the data in Fig. 2b, for artifact instance *S. Hall* there is a transition to *In surgery* co-occurring with a transition from *D. Jones* to Performing surgery. This enables us to calculate conditional co-occurrence statistics similar to those presented in [5], which we map to states and transitions in the lifecycle model of the primary artifact instance.

When the primary instance transitions to a new state, the related instances have a certain state depending on their last known state instance relevant for the primary instance. Together with the related transitions occurring in the same step, this forms a partial view of the system state that we consider as the changing marking of the CSM co-occurring with the primary transition. From this, we calculate probability estimates conditional on the primary transition: the probability of observing a specific marking or multiset of transitions given the execution of the primary transition, and the probability of having a specific number of instances in a given state when the primary transition is executed.

Similarly, during a system step of related instances not including the primary artifact instance there is a changing partial system marking. From this, we calculate probability estimates for the time spent in certain states co-occurring with the current state of the primary artifact. For example, if the surgery in Fig. 2b had one nurse present for the entire procedure and one nurse present only for the first half hour then based on this the estimated conditional probability for other patients *In surgery* would be that 25% of the time is spent with two nurses in the state *Assisting surgeon* and 75% with one nurse. If the second nurse was e.g. preparing another patient during the later part of the surgery then this is not relevant from the point of view of the current primary patient instance. In this way, the presence of many-to-many relations does not affect the calculation of the conditional co-occurrence statistics.

4   Implementation and Evaluation

In this section we discuss the implementation and evaluation of the Multi-Instance Miner, a plug-in[1] in the open-source process mining framework ProM.

4.1   Multi-instance Miner

A screenshot of the interactive visualisation of the Multi-Instance Miner is shown in Fig. 4. For each artifact type, a lifecycle model is shown with its type states and transitions. The user can click on states and transitions, which cause the tool to highlight the other states and transitions that can co-occur with the selected element. Moving the mouse to one of the highlighted elements creates a pop-up window that shows how often or for what duration the co-occurrence was observed. Below the main visualisation is a table that provides a detailed list of the statistics mentioned in Sect. 3.3. The user can select what type of co-occurrence relation they want to investigate and filter the results.

---

[1] Contained in the *MIMiner* package of the ProM 6 nightly build, available at http://www.promtools.org/.

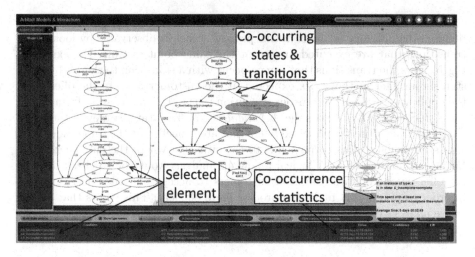

Fig. 4. The interactive visualisation of a discovered CSM. The selected state is circled in red and its co-occurring states and transitions are marked in orange. (Color figure online)

4.2 Case Study

The Multi-Instance Miner has been used to analyse the BPI Challenge 2017 data set [14]. This real-life event log concerns a loan application process at a Dutch financial institute.

The event log contains information on the status and the activities performed for all loan applications filed in a single year at the institute. There are 31509 applications, with between 1 and 10 offers related to each application for a total of 42995 offers in the dataset. We filtered out 98 ongoing applications that did not have a final application status and their related offers and workflow activities. There are 26 distinct activity types, divided into three categories: application state changes (A), offer state changes (O), and workflow activities (W). We consider each category as an artifact type, with application and offer IDs referring to specific instances.

The three lifecycle models are shown in Fig. 4. The application lifecycle includes its creation, complete delivery of the required information, a validation and one of three final results: accepted and pending payment, denied by the institute or cancelled by the customer. The offer lifecycle involves their creation, the sending to and return from the customer, and equivalent final results as for the application. The workflow lifecycle concerns the manual activities performed, so it represents the state of the employee currently working on the application. These activities are changes made to the applications and offers and calls made to contact the customer e.g. regarding missing files. The application and offer models are simple, with the application model containing only a single cycle and the offer model being acyclic. The workflow model is more complex, as workflow activities can be suspended and resumed multiple times before completion.

Explorative analysis revealed that there are many synchronisation points between the different artifacts in this process, which is observable by looking at transitions that are executed in step at the same point in time. For example, Fig. 5 shows the transitions that can co-occur with a transition to the O_Accepted state in an offer instance. As expected, if the customer accepts the offer then the application has a simultaneous transition to the A_Pending state from either the A_Incomplete (28.7%) or the A_Validating (71.3%) state. However, there are also 5797 transitions from related offers that are cancelled at the same time. Further study showed this corresponds to the situation where the customer asked for more than one offer. A logical follow-up analysis would be to check if there are offer characteristics that lead customers to accept one offer over another, in order to potentially make the process more efficient. This shows that explorative analysis can lead to new research questions and possibly ideas for process improvement.

Consequence	Value	Confidence	Lift
a:(A_Incomplete+complete,A_Pending+complete)	4936	0,287	11,217
a:(A_Validating+complete,A_Pending+complete)	12284	0,713	11,217
o:(O_Created+complete,O_Cancelled+complete)	177	0,01	4,855
o:(O_Returned+complete,O_Cancelled+complete)	1409	0,082	8,215
o:(O_Sent (mail and online)+complete,O_Cancelled+complete)	3752	0,218	4,938
o:(O_Sent (online only)+complete,O_Cancelled+complete)	459	0,027	6,802
w:(W_Call incomplete files+resume,W_Call incomplete files+complete)	40	0,002	0,546
w:(W_Call incomplete files+start,W_Call incomplete files+complete)	19	0,001	5,92
w:(W_Validate application+resume,W_Validate application+complete)	3093	0,18	6,907
w:(W_Validate application+start,W_Validate application+complete)	255	0,015	8,829

Fig. 5. The transitions that co-occur with a transition to O_Accepted.

During the challenge, the company was also asking analysts to answer several different business-relevant questions. Two questions we can answer with our tool that they asked were: (1) What are the throughput times per part of the process? and (2) How many customers ask more than one offer, either in a single conversation or subsequently, and how does this affect conversion?

Splitting the log into sequences of state instances for different artifact types separates the events related to employee activities from events related to input by the applicant. As a result, we can determine throughput times that show the difference between the time spent in the company's systems waiting for processing by an employee and the time spent waiting on the applicant. For example, in Fig. 6a there are on average 15 days between the point where an offer is first sent to the customer and the return of a response. By contrast, in Fig. 6b there is on average 1 day and 3 h between a suspension of the validation due to incomplete files and the first subsequent action e.g. to contact the customer.

Customers can ask for more than one offer, sometimes in the same conversation and sometimes as a result of follow-up calls by the institute. A count of the number of O_Created state instances per application is shown in Table 1.

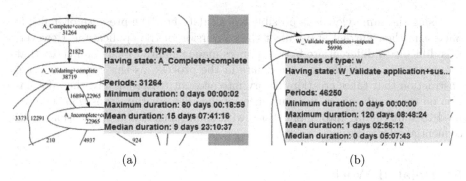

(a)   (b)

Fig. 6. (a) Sojourn time of applications in state *A_Complete*. (b) Sojourn time of the workflow in state *W_Validate application+suspend*

To determine whether customers ask multiple offers in one conversation or sequentially, we can look at the state marking for all related artifact instances at the time of an offer creation. This view is given in Fig. 7, showing e.g. that there are 11464 offer creations while the related application is in state *A_Accepted* and the related workflow is in state *W_Complete application+start*. From this, we can calculate that given a total of 31411 applications there are at most 3602 applications (31411 − 11464 − 9244 − 5441 − 1660) where the customer initially asked for more than one offer. This information can also be used to determine the conditions under which a subsequent offer is created, e.g. 690 offers were created after the initial offer was returned with insufficient information and 641 offers were created after two initial offers and a follow-up call.

Table 1. A count of applications by the number of related offers.

Offers	1	2	3	4	5	6	7	8	9	10
Count	22900	6549	1337	438	125	29	16	12	3	2

Consequence	Value	Confidence ...
[o:{[Initial State],O_Created+complete], a:{A_Accepted+complete}, w:{W_Complete application+start}]	11464	0,268 ...
[o:{[Initial State],O_Created+complete], a:{A_Accepted+complete}, w:{W_Complete application+suspend}]	9244	0,216 ...
[o:{[Initial State],O_Created+complete}, w:{W_Complete application+resume}, a:{A_Accepted+complete}]	5441	0,127 ...
[o:{[Initial State],O_Created+complete}, w:{W_Call after offers+suspend}, o:{O_Sent (mail and online)+complete}, a:{A_Complete+complete}]	2835	0,066 ...
[o:{[Initial State],O_Created+complete}, a:{A_Accepted+complete}]	1660	0,039 ...
[o:{[Initial State],O_Created+complete} x 2, a:{A_Accepted+complete}, w:{W_Complete application+start}]	1264	0,03 ...
[o:{[Initial State],O_Created+complete}, o:{O_Created+complete}, w:{W_Complete application+start}]	1262	0,028 ...
[o:{[Initial State],O_Created+complete} x 2, a:{A_Accepted+complete}, w:{W_Complete application+suspend}]	876	0,02 ...
[o:{[Initial State],O_Created+complete}, a:{A_Accepted+complete}, o:{O_Created+complete}, w:{W_Complete application+suspend}]	847	0,02 ...
[o:{[Initial State],O_Created+complete}, w:{W_Call incomplete files+suspend}, o:{O_Returned+complete}, a:{A_Incomplete+complete}]	690	0,016 ...
[o:{[Initial State],O_Created+complete}, w:{W_Call after offers+suspend}, o:{O_Sent (mail and online)+complete} x 2, a:{A_Complete+complete}]	641	0,015 ...

Fig. 7. A partial view of the state markings that co-occur with a transition to *O_Created*, showing that applications had multiple offers created simultaneously.

Using the same view it is possible to calculate how the presence of multiple offers and their current state affect the conversion of the application. However, it is difficult to differentiate between offers made simultaneously and those that are created sequentially at later points in the process, as this requires a state abstraction that takes creation history into account. Alternatively, the information on the simultaneous co-occurrences of the offer creation transitions may be used to split the log into applications with simultaneous and applications with sequential offer creations.

5   Related Work

As mentioned in the introduction, several discovery approaches have been developed that can be applied in the context of artifact-centric processes.

Initially, work focussed on enabling the discovery of lifecycles of individual artifact types [12]. These techniques provided lifecycle models and information on which specific artifact instances are related. They did not discover interaction or synchronisation between artifact instances, which is needed to determine how the lifecycle progression of one instance influences the lifecycle of another instance.

In [11] work was presented to discover synchronisation conditions between artifacts. These synchronisation conditions specify the number of instances of a given type that need to be in a specific state to enable a transition to a specific state for a related instance of another type. As such, the synchronisation points are only discovered for pairs of artifact types, the technique produces no lifecycle models, and the synchronisation conditions do not cover simultaneous state changes in different artifact types.

In [8] an approach was presented to discover causal relations between events for artifact instances related through many-to-many relations from e.g. ERP tables. The resulting lifecylce models show clear causal relations between events from different artifact types, but in settings with loosely coupled artifacts this can result in graphical models with a large number of arcs between the models. This approach does not identify the synchronous execution of activities in different artifacts and unfortunately there is no publicly available implementation.

In [4] we presented a state-based approach for the discovery of artifact interactions. However, the artifact interactions were limited to pairs of artifact instances and only suitable for processes with one-to-one relations between artifact types instead of one-to-many or many-to-many relationships.

In [7] a technique is presented to discover Declare-like constraints with cardinalities on the number of artifact instances involved. This approach supports the calculation of statistics unaffected by convergence or divergence problems and shows many-to-many relations between instances. However, for real-life processes the number of constraints discovered is large and not easy to explore.

6   Conclusion

This paper presented a multi-instance mining approach to discover lifecycle models and their interactions in the context of artifact-centric processes with many-to-many relations between artifact types. In this approach state machine Petri net lifecycle models are discovered from process execution data and combined into a Composite State Machine with true concurrency support through step sequence execution semantics. For the calculation of co-occurrence statistics that identify synchronisation points we used the notion of a primary artifact instance to map state instances onto the lifecycle models.

The developed Multi-Instance Miner supports interactive exploration that allows the user to point at a lifecycle state or transition and see what can happen with a certain estimated probability while an artifact instance is in the selected state or executing the specified transition. It also provides a list of all co-occurrence statistics for more detailed analysis. During the evaluation on the public BPI Challenge 2017 data we were able to answer business-relevant analysis questions and provide starting points for more detailed investigations.

The correlation based statistics are more suitable for loosely coupled artifacts than strict interaction rules, but the list of correlations can grow large. Therefore, a major challenge remains with visualising the co-occurrence results, e.g. by highlighting the most important results in the models. We also aim to use the approach to analyse the artifact-centric BPI Challenge 2018 data.

References

1. van der Aalst, W.M.P.: Process Mining. Data Science in Action, 2nd edn. Springer, Heidelberg (2016). https://doi.org/10.1007/978-3-662-49851-4
2. Best, E., Devillers, R.R.: Sequential and concurrent behaviour in petri net theory. Theor. Comput. Sci. **55**(1), 87–136 (1987)
3. Chen, P.P.: The entity-relationship model - toward a unified view of data. ACM Trans. Database Syst. **1**(1), 9–36 (1976)
4. van Eck, M.L., Sidorova, N., van der Aalst, W.M.P.: Discovering and exploring state-based models for multi-perspective processes. In: La Rosa, M., Loos, P., Pastor, O. (eds.) BPM 2016. LNCS, vol. 9850, pp. 142–157. Springer, Cham (2016). https://doi.org/10.1007/978-3-319-45348-4_9
5. van Eck, M.L., Sidorova, N., van der Aalst, W.M.P.: Guided interaction exploration in artifact-centric process models. In: 19th IEEE Conference on Business Informatics, CBI 2017, Thessaloniki, Greece, 24–27 July 2017, Volume 1: Conference Papers, pp. 109–118 (2017)
6. Hull, R., et al.: Business artifacts with guard-stage-milestone lifecycles: managing artifact interactions with conditions and events. In: Proceedings of the Fifth ACM International Conference on Distributed Event-Based Systems, DEBS 2011, New York, NY, USA, 11–15 July 2011, pp. 51–62 (2011)
7. Li, G., de Carvalho, R.M., van der Aalst, W.M.P.: Automatic discovery of object-centric behavioral constraint models. In: Abramowicz, W. (ed.) BIS 2017. LNBIP, vol. 288, pp. 43–58. Springer, Cham (2017). https://doi.org/10.1007/978-3-319-59336-4_4

8. Lu, X., Nagelkerke, M., van de Wiel, D., Fahland, D.: Discovering interacting artifacts from ERP systems. IEEE Trans. Serv. Comput. **8**(6), 861–873 (2015)
9. Murata, T.: Petri nets: properties, analysis and applications. Proc. IEEE **77**(4), 541–580 (1989)
10. Nielsen, M., Priese, L., Sassone, V.: Characterizing behavioural congruences for petri nets. In: Lee, I., Smolka, S.A. (eds.) CONCUR 1995. LNCS, vol. 962, pp. 175–189. Springer, Heidelberg (1995). https://doi.org/10.1007/3-540-60218-6_13
11. Popova, V., Dumas, M.: Discovering unbounded synchronization conditions in artifact-centric process models. In: Lohmann, N., Song, M., Wohed, P. (eds.) BPM 2013. LNBIP, vol. 171, pp. 28–40. Springer, Cham (2014). https://doi.org/10.1007/978-3-319-06257-0_3
12. Popova, V., Fahland, D., Dumas, M.: Artifact lifecycle discovery. Int. J. Coop. Inf. Syst. **24**(1) (2015)
13. Raichelson, L., Soffer, P., Verbeek, E.: Merging event logs: combining granularity levels for process flow analysis. Inf. Syst. **71**, 211–227 (2017)
14. Van Dongen, B.: BPI challenge 2017 (2017). https://data.4tu.nl/repository/uuid:5f3067df-f10b-45da-b98b-86ae4c7a310b

Improving Merging Conditions
for Recomposing Conformance Checking

Wai Lam Jonathan Lee[1]([✉]), Jorge Munoz-Gama[1], H. M. W. Verbeek[2],
Wil M. P. van der Aalst[3], and Marcos Sepúlveda[1]

[1] Pontificia Universidad Católica de Chile, Santiago, Chile
{walee,jmun}@uc.cl, marcos@ing.puc.cl
[2] Eindhoven University of Technology, Eindhoven, The Netherlands
h.m.w.verbeek@tue.nl
[3] RWTH Aachen University, Aachen, Germany
wvdaalst@pads.rwth-aachen.de

Abstract. Efficient conformance checking is a hot topic in the field of
process mining. Much of the recent work focused on improving the scalability of alignment-based approaches to support the larger and more
complex processes. This is needed because process mining is increasingly
applied in areas where models and logs are "big". Decomposition techniques are able to achieve significant performance gains by breaking down
a conformance problem into smaller ones. Moreover, recent work showed
that the alignment problem can be resolved in an iterative manner by
alternating between aligning a set of decomposed sub-components before
merging the computed sub-alignments and recomposing sub-components
to fix merging issues. Despite experimental results showing the gain of
applying recomposition in large scenarios, there is still a need for improving the merging step, where log traces can take numerous recomposition
steps before reaching the required merging condition. This paper contributes by defining and structuring the recomposition step, and proposes
strategies with significant performance improvement on synthetic and
real-life datasets over both the state-of-the-art decomposed and monolithic approaches.

Keywords: Recomposition · Conformance checking · Process mining

1 Introduction

In today's organizations, it is important to ensure that process executions follow
the protocols prescribed by process stakeholders so that compliance is maintained. Conformance checking in process mining compares event data with the
corresponding process model to identify commonalities and discrepancies [2].
Detailed diagnostics provide novel insights into the magnitude and effect of
deviations. The state-of-the-art in conformance checking are alignment-based
techniques that provide detailed explanations of the observed behavior in terms
of modeled behavior [4].

© Springer Nature Switzerland AG 2019
F. Daniel et al. (Eds.): BPM 2018 Workshops, LNBIP 342, pp. 31–43, 2019.
https://doi.org/10.1007/978-3-030-11641-5_3

However, one of the limitations of alignment-based approaches is the explosion of state-space during the alignment computation. For example, the classic cost-based alignment approach [4] in the worst case is exponential with respect to the model size [5].

One research line focuses on decomposition techniques which break down a conformance problem into smaller sub-problems [1]. Experimental results have shown that decomposed approaches can be several times faster than their monolithic counterparts and can compute alignments for datasets that were previously infeasible. But until recently, decomposition techniques have been limited to resolving the decision problem of deciding if a log trace is perfectly fitting with the model. As a result, reliable diagnostics are missing. However, recent work has shown that overall alignment results can be computed under decomposed conformance checking by using the so-called *recomposition* approach. A framework that computes overall alignment results in a decomposed manner was presented in [10, 11].

A key result of the work is in defining and proving the border agreement condition which permits the merging of sub-alignment results as an overall result. If the condition is not met, the decomposed sub-components are "recomposed" to encourage the merging condition in the next alignment iteration. Experimental results have shown significant performance gains using recomposition, but they have also shown that the merging aspect of the framework can become a performance bottleneck where log traces may require numerous recompositions to reach the merging condition. Under this context, this paper is a step towards that direction by defining and structuring the recomposition step, proposing different recomposition strategies, and evaluating their impact to the overall computation time. The experimental results show that by applying the presented recomposition strategies, exact alignment results can be computed on synthetic and real-life datasets much faster.

The remainder of the paper is structured as follows: Sect. 2 introduces the required notations and concepts. In particular, Sect. 2.2 presents the recomposition approach as the focus of the paper. Section 3 defines and structures the recomposition step and sheds light on the limitations of the existing recomposition strategies. Section 4 presents four recomposition strategies that can be used in the recomposition step. Section 5 details the experimental setup for the evaluation of the proposed strategies, and Sect. 6 analyzes the experimental results. Section 7 presents the related work. Finally, Sect. 8 presents some conclusions and future work.

2   Preliminaries

This section introduces basic concepts related to process models, event logs, and alignment-based conformance checking techniques.

Let X be a set. $\mathcal{B}(X)$ denotes the set of all possible multisets over set X, and X^* denotes the set of all possible sequences over set X. $\langle \rangle$ denotes the empty sequence. Concatenation of sequences $\sigma_1 \in X^*$ and $\sigma_2 \in X^*$ is denoted

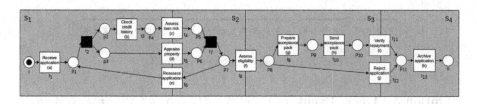

Fig. 1. System net S that models a loan application process (adapted from [6])

as $\sigma_1 \cdot \sigma_2$. Given a tuple $\boldsymbol{x} = (x_1, x_2, \ldots, x_n) \in X_1 \times X_2 \times \ldots \times X_n$, $\pi_i(\boldsymbol{x}) = x_i$ denotes the projection operator for all $i \in \{1, \ldots, n\}$. This operator is extended to sequences so that given a sequence $\boldsymbol{\sigma} \in (X_1 \times X_2 \times \ldots \times X_n)^*$ of length m with $\boldsymbol{\sigma} = \langle (x_{1_1}, x_{2_1}, \ldots, x_{n_1}), (x_{1_2}, x_{2_2}, \ldots, x_{n_2}), \ldots, (x_{1_m}, x_{2_m}, \ldots, x_{n_m}) \rangle$, $\pi_i(\boldsymbol{\sigma}) = \langle x_{i_1}, x_{i_2}, \ldots, x_{i_m} \rangle$ for all $i \in 1, \ldots, n$. Projection is also defined over sets and functions recursively. Given $Y \subseteq X$ and a sequence $\sigma \in X^*$, $\langle \rangle {\upharpoonright}_Y = \langle \rangle$, and $(\langle x \rangle \cdot \sigma) {\upharpoonright}_Y = \langle x \rangle \cdot \sigma {\upharpoonright}_Y$ if $x \in Y$, and $(\langle x \rangle \cdot \sigma) {\upharpoonright}_Y = \sigma {\upharpoonright}_Y$ if $x \notin Y$. Similarly, given a function $f : X \to Y$ and a sequence $\sigma = \langle x_1, x_2, \ldots, x_n \rangle \in X^*$, $f(\sigma) = \langle f(x_1), f(x_2), \ldots, f(x_n) \rangle$.

2.1 Preliminaries on Petri Net, Event Log, and Net Decomposition

In this paper, Petri nets are used to represent process models.

Definition 1 (Labeled Petri net). *Let P denote a set of places, T denote a set of transitions, and $F \subseteq (P \times T) \cup (T \times P)$ denote the flow relation. A labeled Petri net $N = (P, T, F, l)$ is a Petri net (P, T, F) with labeling function $l \in T \nrightarrow \mathcal{U}_A$ where $\mathcal{U}_A$ is some universe of activity labels.*

In a process setting, there is typically a well-defined start and end to an instance of the process. This can be denoted with the initial and final marking of a system net.

Definition 2 (System net). *A system net is a triplet $S = (N, I, O)$ where $N = (P, T, F, l)$ is a labeled Petri net, $I \in \mathcal{B}(P)$ is the initial state and $O \in \mathcal{B}(P)$ is the final state. $\phi_f(S)$ is the set of transition sequences that reach the final state when started in the initial state. If σ is a transition sequence, then $l(\sigma{\upharpoonright}_{dom(l)})$ is an activity sequence.*

$T_v(S) = dom(l)$ *is the set of* visible transitions *in S. $T_v^u(S) = \{t \in T_v(S) \mid \forall_{t' \in T_v(S)} l(t) = l(t') \Rightarrow t = t'\}$ is the set of unique visible transitions in S.*

Figure 1 presents a system net S that models a loan application process (ignore the grey boxes in the background for now). [i] is the initial marking and [o] is the final marking. An example activity sequence is $\langle \mathsf{a}, \mathsf{b}, \mathsf{c}, \mathsf{d}, \mathsf{f}, \mathsf{g}, \mathsf{h}, \mathsf{i}, \mathsf{k} \rangle$ which corresponds to the occurred events of a successful loan application. The process executions in real-life are recorded as event data and can be expressed as an event log.

$$L = [\overbrace{\langle a, b, c, d, f, g, h, i, k \rangle}^{\sigma_1}{}^5, \overbrace{\langle a, c, b, d, f, g, i, h, k \rangle}^{\sigma_2}{}^{10}, \overbrace{\langle a, c, b, d, f, g, j, k \rangle}^{\sigma_3}{}^5]$$

Fig. 2. Running example: event log L

Definition 3 (Trace, Event log). *Let* $A \subseteq \mathcal{U}_A$ *be a set of activities. A trace* $\sigma \in A^*$ *is a sequence of activities. An event log* $L \in \mathcal{B}(A^*)$ *is a multiset of traces.*

Figure 2 presents an event log L corresponding to the system net in Fig. 1. Log L has 20 cases in total with 5 cases following trace σ_1, 10 cases following trace σ_2, and 5 cases following trace σ_3. In cost-based alignment conformance checking, a trace is aligned with the corresponding system net to produce an alignment.

Definition 4 (Alignment [4]). *Let* $L \in \mathcal{B}(A^*)$ *be an event log with* $A \subseteq \mathcal{U}_A$, *let* $\sigma_L \in L$ *be a log trace and* $\sigma_M \in \phi_f(S)$ *a complete transition sequence of system net* S. *An alignment of* σ_L *and* σ_M *is a sequence of pairs* $\gamma \in ((A \cup \{\gg\}) \times (T \cup \{\gg\}))^*$ *where* $\pi_1(\gamma)\restriction_A = \sigma_L$, $\pi_2(\gamma)\restriction_T = \sigma_M$, $\forall_{(a,t)\in\gamma} \ a \neq \gg \vee t \neq \gg$, *and* $\forall_{(a,t)\in\gamma} \ a \neq \gg \wedge (t = \gg \vee a = l(t))$.

Each pair in an alignment is a *legal move*. There are four types of legal moves: a *synchronous move* (a, t) means that the activity matches the activity of the transition, i.e., $a = l(t)$, a *log move* $(a, \gg)$ means that there is a step in the log that is not matched by a corresponding step in the model, a *model move* $(\gg, t)$ where $t \in dom(l)$ means that there is a step in the model that is not matched by a corresponding step in the log, and an *invisible move* $(\gg, t)$ where $t \in T \setminus dom(l)$ means that the step in the model corresponds to an invisible transition that is not observable in the log.

Definition 5 (Valid decomposition [1] and Border activities [11]). *Let* $S = (N, I, O)$ *with* $N = (P, T, F, l)$ *be a system net.* $D = \{S_1, S_2, \ldots, S_n\}$ *is a valid decomposition if and only if the following properties are fulfilled:*

- $S_i = (N_i, I_i, O_i)$ *is a system net with* $N_i = (P_i, T_i, F_i, l_i)$ *for all* $1 \le i \le n$.
- $l_i = l\restriction_{T_i}$ *for all* $1 \le i \le n$.
- $P_i \cap P_j = \emptyset$ *and* $T_i \cap T_j \subseteq T_v^u(S)$ *for all* $1 \le i < j \le n$.
- $P = \bigcup_{1 \le i \le n} P_i$, $T = \bigcup_{1 \le i \le n} T_i$, *and* $F = \bigcup_{1 \le i \le n} F_i$.

$\mathcal{D}(S)$ *is the set of all valid decompositions of* S.

$A_b(D) = \{l(t) \mid \exists_{1 \le i < j \le n} \ t \in T_i \cap T_j\}$ *is the set of border activities of the valid decomposition* D. *To retrieve the sub-nets that share the same border activity, for an activity* $a \in rng(l)$, $S_b(a, D) = \{S_i \in D \mid a \in rng(l_i)\}$ *is the set of sub-nets that contain* a *as an observable activity.*

Figure 1 presents a valid decomposition D of net S where sub-nets are marked by the grey boxes. For example, sub-net S_1 consists of the transitions t_1, t_2, t_3, t_4, t_5, and t_6. Border activities can be identified as the activities of the transitions that are shared between two sub-nets. They are t_4, t_5, t_6, t_8, t_{11}, and t_{12}. Under the recomposition approach framework, overall alignments can be computed in a decomposed manner.

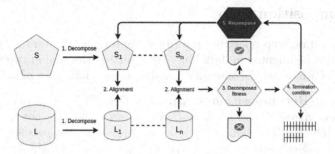

Fig. 3. Recomposing conformance checking framework with the recomposition step highlighted in dark blue (Color figure online)

2.2   Recomposing Conformance Checking

Figure 3 presents an overview of the recomposing conformance checking framework [10,11] which consists of the following five steps: (1) The net and log are decomposed using a decomposition strategy, e.g., maximal decomposition [1]. (2) Alignment-based conformance checking is performed per sub-net and sub-log to produce a set of sub-alignments for each log trace. (3) Since sub-components overlap on border activities, the set of sub-alignments for each log trace also overlap on moves involving border activities. In [11], it was shown that if the sub-alignments synchronize on these moves, then they can be merged as an overall optimal alignment using the merging algorithm presented in [18]. This condition was formalized as the *total border agreement* condition. Log traces that do not meet the requirement are either rejected or left for the next iteration. As such, only border activities can cause merge conflicts. (4) User-configured termination conditions are checked at the end of each iteration. If the framework is terminated before computing the overall optimal alignments for all log traces, then an approximate overall result is given. The results of the framework consist of a fitness value and a set of alignments corresponding to the log traces. In the case of an approximate result, the fitness value would be an interval bounding the exact fitness value and the set of alignments would have pseudo alignments. (5) If there are remaining log traces to be aligned and the termination conditions are not reached, then a recomposition step is taken to produce a new net decomposition and a corresponding set of sub-logs. The next iteration of the framework then starts from Step (2).

While experimental results have shown significant performance gains from the recomposition approach over its monolithic counterpart, large scale experimentation has shown that recomposition is a potential bottleneck. In particular, the strategies used at the recomposition step can have a significant impact. The following section takes a more detailed look at the recomposition step and discusses the limitations of the current recomposition strategies.

3   Recomposition Step

The recomposition step refers to Step (5) of the framework overview presented in Fig. 3 and is highlighted in dark blue. We formalize the step in two parts: the production of a new net decomposition and a corresponding set of sub-logs.

Definition 6 (Recomposition step). *Let $D \in \mathcal{D}(S)$ be a valid decomposition of system net S and let $L = \mathcal{B}(A^*)$ be an event log. For $1 \leq i \leq n$, where $n = |D|$, let $M_i = (A_i \cup \{\gg\}) \times (T_i \cup \{\gg\})$ be the possible alignment moves for a sub-component so that $\Gamma_D = [(\gamma_{i_1}, \ldots, \gamma_{i_n}) \in M_1^* \times \ldots \times M_n^* \mid \exists_{\sigma_i \in L} \forall_{j \in \{1, \ldots, n\}} \pi_1(\gamma_{i_j}) \lceil_{A_j} = \sigma_i \lceil_{A_j}]$ contains the latest sub-alignments for all log traces. Given the valid decomposition, and the latest sub-alignments, $R_S : \mathcal{D}(S) \times \mathcal{B}(M_1^* \times \ldots \times M_n^*) \to \mathcal{D}(S)$ creates a new valid decomposition $D' \in \mathcal{D}(S)$ where $m = |D'| < |D|$. Then, given the new and current net decompositions, the event log, and the latest sub-alignments, $R_L : \mathcal{D}(S) \times \mathcal{D}(S) \times \mathcal{B}(A^*) \times \mathcal{B}(M_1^* \times \ldots \times M_n^*) \nrightarrow \mathcal{B}(A_1'^*) \times \ldots \times \mathcal{B}(A_m'^*)$ creates a set of sub-logs to align in the following iteration of the recomposition approach. Overall, the recomposition step R creates a new net decomposition and a corresponding set of sub-logs, $R : \mathcal{D}(S) \times \mathcal{B}(A^*) \times \mathcal{B}(M_1^* \times \ldots \times M_n^*) \nrightarrow \mathcal{D}(S) \times \mathcal{B}(A_1'^*) \times \ldots \times \mathcal{B}(A_m'^*)$.*

The current recomposition strategy involves recomposing on the most frequent conflicting activities (MFC) and constructing sub-logs that contains to-be-aligned traces which carry conflicting activities that have been recomposed upon (IC).

Most frequent conflict (MFC) recomposes the current net decomposition on the activity set $A_r = \{a \in A_b(D) \mid a \in \arg \max_{a' \in A_b(D)} \sum_{\gamma_i \in \mathrm{Supp}(\Gamma_D)} C(\gamma_i)(a')\}$ where $\Gamma_D \in \mathcal{B}(M_1^* \times \ldots \times M_n^*)$ are the latest sub-alignments and $C : M_1^* \times \ldots \times M_n^* \to \mathcal{B}(A)$ is a function that gives the multiset of conflicting activities of sub-alignments. Hence, A_r contains the border activities with the most conflicts.

Inclusion by conflict (IC) then creates a log $L_r = [\sigma_i \in L \mid \exists_{a \in A_b(D)} C(\gamma_i)(a) > 0 \wedge a \in A_r]$ where $\gamma_i \in \Gamma_D$ are the sub-alignments of trace $\sigma_i \in L$ and net decomposition $D \in \mathcal{D}(S)$. As such, log L_r includes to-be-aligned log traces which have conflicts on at least one of the border activities that have been recomposed upon. Later, log L_r is then projected onto the new net decomposition to create the corresponding sub-logs.

3.1   Limitations to the Current Recomposition Strategies

To explain the limitations, we refer to the set of optimal sub-alignments in Fig. 4 from aligning net decomposition D in Fig. 1 and log L in Fig. 2. We first note that for the conflicting activities which are highlighted in grey: $\sum_{\gamma \in \Gamma_D} C(\gamma)(\mathrm{c}) = 2$, $\sum_{\gamma \in \Gamma_D} C(\gamma)(\mathrm{i}) = 1$, and $\sum_{\gamma \in \Gamma_D} C(\gamma)(\mathrm{j}) = 1$, where $\Gamma_D = \{\gamma_1, \gamma_2, \gamma_3\}$. With activity c being the most frequent conflicting activity, MFC recomposes the current net decomposition on $A_r = \{\mathrm{c}\}$ and IC creates the corresponding sub-logs containing $L_r = \{\sigma_2, \sigma_3\}$ since both traces have activity c as a conflicting

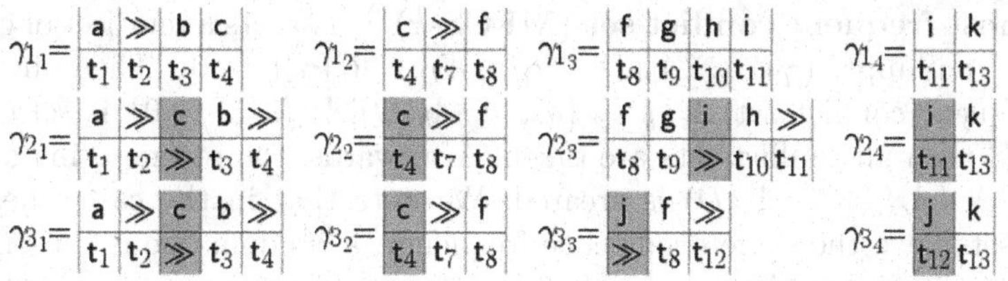

$$\gamma_{1_1}=\begin{array}{|c|c|c|c|} a & \gg & b & c \\ \hline t_1 & t_2 & t_3 & t_4 \end{array} \quad \gamma_{1_2}=\begin{array}{|c|c|c|} c & \gg & f \\ \hline t_4 & t_7 & t_8 \end{array} \quad \gamma_{1_3}=\begin{array}{|c|c|c|c|} f & g & h & i \\ \hline t_8 & t_9 & t_{10} & t_{11} \end{array} \quad \gamma_{1_4}=\begin{array}{|c|c|} i & k \\ \hline t_{11} & t_{13} \end{array}$$

$$\gamma_{2_1}=\begin{array}{|c|c|c|c|c|} a & \gg & c & b & \gg \\ \hline t_1 & t_2 & \gg & t_3 & t_4 \end{array} \quad \gamma_{2_2}=\begin{array}{|c|c|c|} c & \gg & f \\ \hline t_4 & t_7 & t_8 \end{array} \quad \gamma_{2_3}=\begin{array}{|c|c|c|c|c|} f & g & i & h & \gg \\ \hline t_8 & t_9 & \gg & t_{10} & t_{11} \end{array} \quad \gamma_{2_4}=\begin{array}{|c|c|} i & k \\ \hline t_{11} & t_{13} \end{array}$$

$$\gamma_{3_1}=\begin{array}{|c|c|c|c|c|} a & \gg & c & b & \gg \\ \hline t_1 & t_2 & \gg & t_3 & t_4 \end{array} \quad \gamma_{3_2}=\begin{array}{|c|c|c|} c & \gg & f \\ \hline t_4 & t_7 & t_8 \end{array} \quad \gamma_{3_3}=\begin{array}{|c|c|c|} j & f & \gg \\ \hline \gg & t_8 & t_{12} \end{array} \quad \gamma_{3_4}=\begin{array}{|c|c|} j & k \\ \hline t_{12} & t_{13} \end{array}$$

Fig. 4. Sub-alignments $\gamma_1 = (\gamma_{1_1}, \gamma_{1_2}, \gamma_{1_3}, \gamma_{1_4})$, $\gamma_2 = (\gamma_{2_1}, \gamma_{2_2}, \gamma_{2_3}, \gamma_{2_4})$, and $\gamma_3 = (\gamma_{3_1}, \gamma_{3_2}, \gamma_{3_3}, \gamma_{3_4})$ of log L_1 and net decomposition D_1 with merge conflicts highlighted in grey

activity. The new net decomposition will contain three sub-nets rather than four where sub-net S_1 and sub-net S_2 are recomposed upon activity c as a single sub-net. The corresponding sub-log set is created by projecting log L_r onto the new net decomposition.

While one merge conflict is resolved by recomposing on activity c, the merge conflicts at activity i and j will remain in the following iteration. In fact, under the current recomposition strategy, trace σ_2 and σ_3 have to be aligned three times each to reach the required merging condition to yield overall alignments. This shows the limitation of MFC in only partially resolving merge conflicts on the trace level and IC in leniently including to-be-aligned log traces whose subsequent sub-alignments are unlikely to reach the necessary merging condition.

As such, the key to improving the existing recomposition strategies is in lifting conflict resolution from the individual activity level to the trace level so that the net recomposition strategy resolves merge conflicts of traces rather than activities and the log recomposition strategy selects log traces whose merge conflicts are likely to be fully resolved with the latest net recomposition. In the following section, three net recomposition strategies and one log recomposition strategy are presented. These strategies improve on the existing ones by looking at merge conflict sets, identifying co-occurring conflicting activities, and minimizing the average size of the resulting recomposed sub-nets. The later experimental results show that the strategies lead to significant performance improvements in both synthetic and real-life datasets.

4   Recomposition Strategies

In this section, three net recomposition strategies and one log recomposition strategy are presented.

4.1   Net Recomposition Strategies

As previously shown, resolving individual conflicting activities may only partially resolve the merge conflicts of traces. This key observation motivates the following net recomposition strategies which target conflicts at the trace level.

Top k most frequent conflict set (MFCS-k) constructs a multiset of conflict sets $A_{cs} = [\text{Supp}(C(\gamma)) \subseteq A_b(D) \mid \gamma \in \Gamma_D \wedge |C(\gamma)| > 0]$. Then the top k most frequent conflict set $A_{cs,k} \subseteq \{a_{cs} \subseteq A_b(D)|A_{cs}(a_{cs}) > 0\}$ is selected. If $|A_{cs}| < k$, then all conflict sets are taken. Afterwards, the recomposing activity set $A_r = \cup(A_{cs,k}) \subseteq A_b(D)$ is created. We note that in the case where two conflict sets have the same occurrence frequency, a random one is chosen. This secondary criterion avoids bias, and gives better performances empirically than any other straightforward criteria.

Merge conflict graph (MCG) recomposes on conflicting activities that co-occur on the trace level by constructing a weighted undirected graph $G = (V, E)$ where $E = \{\{a_1, a_2\} \mid \exists_{\gamma \in \Gamma_D} \; a_1 \in C(\gamma) \wedge a_2 \in C(\gamma) \wedge a_1 \neq a_2\}$ with a weight function $w : E \rightarrow \mathbb{N}^+$ such that $w((a_1, a_2)) = |\{\gamma \in \Gamma_D \mid C(\gamma)(a_1) > 0 \wedge C(\gamma)(a_2) > 0\}|$ and $V = \{a \in A_b(D) \mid \exists_{(a_1,a_2) \in E} \; a = a_1 \vee a = a_2\}$. Then, with a threshold $t \in [0, 1]$, edges are filtered so that $E_f = \{e \in E \mid w(e) \geq t \times w_{\max}\}$ where $w_{\max}$ is the maximum edge weight in E. The corresponding vertex set and filtered graph can be created as $V_f = \{a \in A_b(D) \mid \exists_{(a_1,a_2) \in E_f} a = a_1 \vee a = a_2\}$ and $G_f = (V_f, E_f)$. Finally, the current net decomposition is recomposed on activity set $A_r = V_f$.

Balanced. This recomposition strategy extends the MFCS-k strategy but also tries to minimize the average size of the sub-nets resulting from the recomposition. For a border activity $a \in A_b(D)$, $|(a, D)| = | \cup_{S_i \in S_b(a, D)} A_v(S_i)|$ approximates the size of the recomposed sub-net on activity a. The average size of the recomposed sub-nets for a particular conflict set can then be approximated by $|(A_c, D)| = \frac{\sum_{a \in A_c} |(a,D)|}{|A_c|}$. The score of the conflict set can be computed as a weighted combination $\beta(A_c, D) = w_0 \times \frac{m(A_c)}{\max_{A_c' \in A_{cs}} m(A_c')} + w_1 \times (1 - \frac{|(A_c, D)|}{\max_{A_c' \in A_{cs}} |(A_c', D)|})$ where higher scores are assigned to frequent conflict sets that do not recompose to create large sub-nets. The activities of the conflict sets with the highest score, $A_r = \{a \in A_c \mid A_c \in \arg \max_{A_c' \in A_{cs}} \beta(A_c', D)\}$, are then recomposed upon to create a net decomposition.

4.2   Log Recomposition Strategy

Similar to the net recomposition strategies, the existing IC strategy can be too lenient in including log traces which have conflicting activities that are unlikely to be resolved in the following decomposed replay iteration.

Strict include by conflict (SIC) increases the requirement for a to-be-aligned log trace to be selected for the next iteration. This addresses the limitation of IC which can include log traces whose merge conflicts are only partially covered by the net recomposition. Given the recomposed activity set A_r, SIC includes log traces as $L_r = [\sigma_i \in L \mid \forall_{a \in C(\gamma_i)} \; a \in A_r]$ with merge conflict if the corresponding conflict set is a subset of set A_r. However, this log strategy only works in conjunction with the net strategies that are based on conflict sets, i.e., MFCS-k and Balanced, so that at least one to-be-aligned log trace is included.

5   Experiment Setup

Both synthetic and real-life datasets are used to evaluate the proposed recomposition strategies. Dataset generation is performed using the PTandLogGenerator [8] and information from the empirical study [9]; it is reproducible as a RapidProM workflow [3]. The BPIC 2018 dataset is used [16] as the real-life dataset. Moreover, two baseline net recomposition strategies are used: **All** recomposes on all conflicting activities, and **Random** recomposes on a random number of conflicting activities. Similarly, a baseline log recomposition **All** which includes all to-be-aligned log traces is used. For the sake of space, the full experimental setup and datasets are available at the GitHub repository[1] so that the experimental results can be reproduced.

6   Results

The results shed light on two key insights: First, the selection of the recomposition approach may lead to very different performances. Second, good performance requires both selecting appropriate conflicting activities and well-grouped to-be-aligned log traces.

Figure 5 presents the experimental results for both synthetic and real-life datasets. For each of the synthetic models, there are three event logs of different noise profiles described as net$X$-*noise probability-dispersion over trace* where $X \in \{1, 2, 3\}$. For the sake of readability, we only show the results of three out of five synthetic datasets, but the results are consistent across all five synthetic datasets). Readers interested in more details are referred to the GitHub link for a detailed explanation on noise generation and the rest of the experimental results. For the MFCS-k and Balanced strategies, only configurations using the SIC log strategy are shown; results showed that the SIC log strategy provides a better performance. For the others where SIC is not applicable, only configurations using the IC log strategy are shown as results indicated better performances. Overall, the results show that for both the monolithic and recomposition approach, it is more difficult to compute alignment results for less fitting datasets.

Different Approaches Give Different Performances. Comparing the monolithic and recomposition approach, it is clear that the recomposition approach provides a better performance than the monolithic counterpart under at least one recomposition strategy configuration. Furthermore, performance can vary significantly across different recomposition approaches. For example, the existing MFC strategy is the worst performing strategy where it is not able to give exact results for the real-life dataset and both the net$X$-10-60 and net$X$-60-10 noise scenarios of the synthetic datasets. The MFCS-k and Balanced strategies are shown to be the best performing strategies. While for high fitness scenarios, i.e., net$X$-10-10, MFCS-k give better performances with a high $k = 10$. This is because when there is little noise, it becomes simply a "race" to aligning traces

---

[1] See https://github.com/wailamjonathanlee/Characterizing-recomposing-replay.

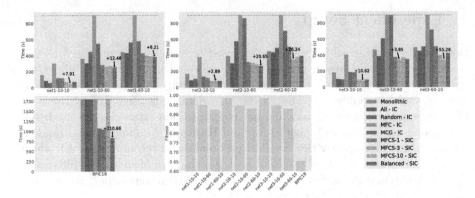

Fig. 5. Bar chart showing fitness and overall time per net recomposition strategy (including the monolithic approach). The time limit is shown as a dashed red line and indicates infeasible replays. Best performing approaches and their time gains from the second fastest times are specified by black arrows. (Color figure online)

with similar merge conflicts. Conversely, for low fitness scenarios, because merge conflicts are potentially much more difficult to resolve, the Balanced strategy avoids quickly creating large sub-components that take longer to replay. In these cases, the time differences between the different feasible strategies can go up to three minutes. For all the experiments, the proposed recomposition strategies outperform the baseline strategies. Lastly, for the real-life dataset BPIC18, only the MFCS-1, Balanced, and MCG recomposition strategies are able to compute exact alignment results and the Balanced strategy outperforms MFCS-1 by more than three minutes.

Both Net and Log Recomposition Strategies Matter. Figure 6 presents the number of aligned traces and percentage of valid alignments per iterations under All, IC, and SIC log strategies with net strategy fixed as Balanced on BPIC18. We first note that only the SIC log strategy resulted with exact alignment results. While all strategies start with aligning all traces in the first iteration, there are significant differences in the number of aligned traces across iterations. Similar to the All strategy, the existing IC strategy includes a high number of traces to align throughout all iterations; the number of aligned traces only tapered off in the later iterations as half of the traces have resulted as valid alignments. This confirms the hypothesis that the existing IC strategy can be too lenient with the inclusion of traces to align. Furthermore, up until iteration 13, none of the aligned traces reaches the necessary merging condition to result as a valid alignment; this means that both the All and IC strategies are "wasting" resources aligning many traces. Conversely, the SIC strategy keeps the number of traces to align per iteration comparatively lower. Moreover, at the peak of the number of traces to align at iteration 21, almost 80% of the ~300 aligned traces resulted as valid alignments. These are likely to explain why only the SIC log strategy is able to compute an exact result.

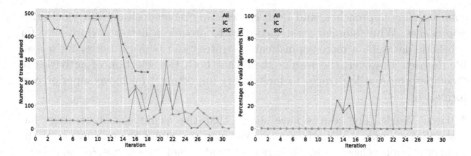

Fig. 6. Comparing log strategies by showcasing the number of aligned traces (left) and percentage of valid alignments (right) per iteration on the real-life dataset BPIC18.

7 Related Work

Performance problems related to alignment-based conformance checking form a well-known problem. A large number of conformance checking techniques have been proposed to tackle this issue. Approximate alignments have been proposed to reduce the problem complexity by abstracting sequential information from segments of log traces [14]. The notion of indication relations has been used to reduce the model and log prior to conformance checking [15]. Several approaches have been proposed along the research line of decomposition techniques. This include different decomposition strategies, e.g., maximal [1], and SESE-based [12]. Moreover, different decomposed replay approaches such as the hide-and-reduce replay [17] and the recomposition approach [11] have also been investigated. Compared to the existing work, this paper investigates different strategies for the recomposition approach in order to improve the overall performance in computation time.

Other than the alignment-based approach, there are also other conformance checking approaches. This includes the classical token replay [13], behavioral profile approaches [19] and more recently approaches based on event structures [7].

8 Conclusions and Future Work

This paper investigated the recomposition aspect of the recomposing conformance checking approach which can become a bottleneck to the overall performance. By defining the recomposition problem, the paper identifies limitations of the current recomposition strategy in not fully resolving merge conflicts on the trace level and also being too lenient in the inclusion of log traces for the subsequent decomposed replay iteration. Based on the observations, three net recomposition strategies and one log recomposition strategy have been presented. The strategies were then evaluated on both synthetic and real-life datasets with two baseline approaches. The results show that different recomposition strategies can significantly impact the overall performance in computing alignments. Moreover, they show that the presented approaches provide a better performance than baseline approaches, and both the existing recomposition and monolithic approaches. While simpler strategies tend to provide a better performance for synthetic datasets, a more sophisticated strategy can perform better for a real-life dataset. However, the results show that both the selection of activities to recompose on and log traces to include are important to achieve superior performances.

Future Work. The results have shown that the recomposition strategy has a significant impact on performance. We plan to extend the evaluation of the presented approaches to a larger variety of models, noise scenarios, initial decomposition strategies, and other real-life datasets. For the current and presented approaches, new net decompositions are created by recomposing the initial decomposition on selected activities. Entirely different net decompositions can be created using the merge conflict information from the previous iteration; however, our preliminary results showed that this may be difficult. Lastly, in the current framework, the same strategies (both decomposition and recomposition) are used in all iterations; higher level meta-strategies might be useful. For example, it might be good to switch to the monolithic approach for a small number of log traces that cannot be aligned following many iterations.

Acknowledgments. This work is partially supported by *CONICYT-PCHA/ Doctorado Nacional/2017-21170612*, FONDECYT Iniciación 11170092, *CONICYT Apoyo a la Formación de Redes Internacionales Para Investigadores en Etapa Inicial REDI170136*, the *Vicerrectoría de Investigación de la Pontificia Universidad Católica de Chile/Concurso Estadías y Pasantías Breves 2016*, and the *Departamento de Ciencias de la Computación UC/Fond-DCC-2017-0001*. The authors would like to thank Alfredo Bolt for his comments on the data generation details.

References

1. van der Aalst, W.M.P.: Decomposing Petri nets for process mining: a generic approach. Distrib. Parallel Databases **31**(4), 471–507 (2013)
2. van der Aalst, W.M.P.: Process Mining - Data Science in Action. Springer, Heidelberg (2016). https://doi.org/10.1007/978-3-662-49851-4
3. van der Aalst, W.M.P., Bolt, A., van Zelst, S.J.: RapidProM: mine your processes and not just your data. CoRR abs/1703.03740 (2017)
4. Adriansyah, A.: Aligning Observed and Modeled Behavior. Ph.D. thesis, Technische Universiteit Eindhoven (2014)
5. van Dongen, B., Carmona, J., Chatain, T., Taymouri, F.: Aligning modeled and observed behavior: a compromise between computation complexity and quality. In: Dubois, E., Pohl, K. (eds.) CAiSE 2017. LNCS, vol. 10253, pp. 94–109. Springer, Cham (2017). https://doi.org/10.1007/978-3-319-59536-8_7
6. Dumas, M., Rosa, M.L., Mendling, J., Reijers, H.A.: Fundamentals of Business Process Management. Springer, Heidelberg (2013). https://doi.org/10.1007/978-3-642-33143-5
7. García-Bañuelos, L., van Beest, N., Dumas, M., Rosa, M.L., Mertens, W.: Complete and interpretable conformance checking of business processes. IEEE Trans. Softw. Eng. **44**(3), 262–290 (2018). https://doi.org/10.1109/TSE.2017.2668418

8. Jouck, T., Depaire, B.: PTandLogGenerator: a generator for artificial event data. In: BPM (Demos). CEUR Workshop Proceedings, vol. 1789, pp. 23–27. CEUR-WS.org (2016)

9. Kunze, M., Luebbe, A., Weidlich, M., Weske, M.: Towards understanding process modeling – the case of the BPM academic initiative. In: Dijkman, R., Hofstetter, J., Koehler, J. (eds.) BPMN 2011. LNBIP, vol. 95, pp. 44–58. Springer, Heidelberg (2011). https://doi.org/10.1007/978-3-642-25160-3_4

10. Lee, W.L.J., Verbeek, H.M.W., Munoz-Gama, J., van der Aalst, W.M.P., Sepúlveda, M.: Replay using recomposition: alignment-based conformance checking in the large. In: Proceedings of the BPM Demo Track and BPM Dissertation Award, Barcelona, Spain, 13 September 2017. CEUR Workshop Proceedings, vol. 1920. CEUR-WS.org (2017)

11. Lee, W.L.J., Verbeek, H., Munoz-Gama, J., van der Aalst, W.M.P., Sepúlveda, M.: Recomposing Conformance: Closing the Circle on Decomposed Alignment-Based Conformance Checking in Process Mining (2017, under review). processmininguc.com/publications

12. Munoz-Gama, J., Carmona, J., van der Aalst, W.M.P.: Single-entry single-exit decomposed conformance checking. Inf. Syst. **46**, 102–122 (2014)

13. Rozinat, A., van der Aalst, W.M.P.: Conformance checking of processes based on monitoring real behavior. Inf. Syst. **33**(1), 64–95 (2008)

14. Taymouri, F., Carmona, J.: A recursive paradigm for aligning observed behavior of large structured process models. In: La Rosa, M., Loos, P., Pastor, O. (eds.) BPM 2016. LNCS, vol. 9850, pp. 197–214. Springer, Cham (2016). https://doi.org/10.1007/978-3-319-45348-4_12

15. Taymouri, F., Carmona, J.: Model and event log reductions to boost the computation of alignments. In: Ceravolo, P., Guetl, C., Rinderle-Ma, S. (eds.) SIMPDA 2016. LNBIP, vol. 307, pp. 1–21. Springer, Cham (2018). https://doi.org/10.1007/978-3-319-74161-1_1

16. van Dongen, B.F., Borchert, F.: BPI Challenge 2018 (2018)

17. Verbeek, H.M.W.: Decomposed replay using hiding and reduction as abstraction. In: Koutny, M., Kleijn, J., Penczek, W. (eds.) Transactions on Petri Nets and Other Models of Concurrency (ToPNoC) XII. LNCS, vol. 10470, pp. 166–186. Springer, Heidelberg (2017). https://doi.org/10.1007/978-3-662-55862-1_8

18. Verbeek, H.M.W., van der Aalst, W.M.P.: Merging alignments for decomposed replay. In: Kordon, F., Moldt, D. (eds.) PETRI NETS 2016. LNCS, vol. 9698, pp. 219–239. Springer, Cham (2016). https://doi.org/10.1007/978-3-319-39086-4_14

19. Weidlich, M., Polyvyanyy, A., Desai, N., Mendling, J., Weske, M.: Process compliance analysis based on behavioural profiles. Inf. Syst. **36**(7), 1009–1025 (2011)

Efficiently Computing Alignments

Algorithm and Datastructures

Boudewijn F. van Dongen[(✉)]

Eindhoven University of Technology, Eindhoven, Netherlands
`B.F.v.Dongen@tue.nl`

Abstract. Conformance checking is considered to be anything where observed behaviour needs to be related to already modelled behaviour. Fundamental to conformance checking are alignments which provide a precise relation between a sequence of activities observed in an event log and a execution sequence of a model. However, computing alignments is a complex task, both in time and memory, especially when models contain large amounts of parallelism.

In this tool paper we present the actual algorithm and memory structures used for the experiments of [15]. We discuss the time complexity of the algorithm, as well as the space and time complexity of the main data structures. We further present the integration in ProM and a basic code snippet in Java for computing alignments from within any tool.

Keywords: Alignments · Conformance checking · Process mining

1 Introduction

Conformance checking is considered to be anything where observed behaviour needs to be related to already modelled behaviour. Conformance checking is embedded in the larger contexts of Business Process Management and Process Mining, where conformance checking is typically used to compute metrics such as fitness, precision and generalization to quantify the relation between a log and a model. Fundamental to conformance checking are alignments [2]. Alignments provide a precise relation between a sequence of activities observed in an event log and a execution sequence of a model. For each trace, this precise relation is expressed as a sequence of "moves". Such a move is either a "synchronous move" referring to the fact that the observed event in the trace corresponds directly to the execution of a transition in the model, a "log move" referring to the fact that the observed event has no corresponding transition in the model, or a "model move" referring to the fact that a transition occurred which was not observed in the trace. Computing alignments is a complex task, both in time and memory.

In [15] a technique is presented to efficiently compute alignments for Petri nets, using the extended marking equation. In this paper, we present the algorithm in pseudocode and the datastructures used. We discuss the time and

© Springer Nature Switzerland AG 2019
F. Daniel et al. (Eds.): BPM 2018 Workshops, LNBIP 342, pp. 44–55, 2019.
https://doi.org/10.1007/978-3-030-11641-5_4

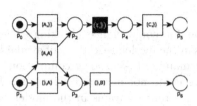

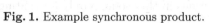

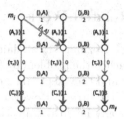

Fig. 1. Example synchronous product.   **Fig. 2.** Example search space.

memory complexity for the algorithm and we provide a code snippet for actually using the code which is publicly available. We assume the reader to be familiar with some basic notions of Petri nets.

The problem of finding alignments can be expressed the following way: *Given a synchronous product Petri net and a cost function, provide the cheapest firing sequence from the initial marking to the final marking.* Consider the example in Fig. 1. Here, a synchronous product for a sequential Petri net with three transitions, labelled A, τ and C and a trace containing two events A and B is depicted. The reachability graph of this model is shown in Fig. 2 and for an alignment, we look for the shortest path from the top-left marking $m_i = [p_0, p_1]$ to the bottom-right marking $m_f = [p_5, p_6]$. The distances at each edge are determined by a cost function which associates costs to firing transitions in the synchronous product. For a complete formal specification of this question, we refer to [15], however the following key elements are important for alignments:

- The synchronous product is a Petri net with both an initial and final marking. The final marking is assumed to be reachable from the initial marking through at least one firing sequence,
- Each transition in the synchronous product corresponds to a so-called move in the alignment. These moves are "model-moves", "log-moves", "synchronous-moves" or "τ-moves",
- Synchronous and τ moves typically have cost 0, while model and log moves have cost >0,
- The cost to reach the final marking can be underestimated using the (extended) marking equation. This requires solving a (integer) linear program.
- The size of the reachability graph of the synchronous product equals the size of the reachability graph of the model times the length of the trace that is being aligned.

The remainder of this paper is structured as follows. In Sect. 2, we discuss some related work. Then in Sect. 3, we discuss the algorithm in detail and in Sect. 4 the data structures needed. In Sect. 5 we discuss the inclusion in ProM and we conclude the paper in Sect. 6.

2   Related Work

In [12], Rozinat et. al. laid the foundation for conformance checking. They approached the problem by firing transitions in the model, regardless of available tokens and they kept track of missing and remaining tokens after completion of an observed trace. However, these techniques could not handle duplicate labels (identically labelled transition occurring at different locations in the model) or "invisible" transitions, i.e. τ-labelled transitions in the model purely for routing purposes.

As an improvement to token replay, alignments were introduced in [3]. The work proposes to transform a given Petri net and a trace from an event log into a synchronous product net, and solve the shortest path problem using A^* [8] on the its reachability graph. This graph may be considerable in size as it is worst-case exponential in the size of the synchronous product, in some cases, the proposed algorithm has to investigate the entire graph despite of tweaks identified in [16].

In [15], an incremental technique is presented to compute alignments more efficiently using so-called splitpoints and the extended marking equation. However, no pseudo code is presented, nor a discussion on the main algorithm and data structures.

Several techniques exist to compute alignments. Planner-based techniques [6] are available for safe Petri nets. For safe, acyclic models, constraint satisfaction [9] can be used. When not using a model, but its reachability graph as input, automata matching [11] can be applied. Other conformance checking work includes decomposition-based approaches [1,10] and approximation schemes [13].

3   Algorithm

The core algorithm for alignments is an A^* search technique, which is shown in Algorithm 1. In lines 2 to 10, the algorithm is initialized. Then the mean search loop between 11 and 47 iteratively expands the search space. In every iteration, a marking is selected for expansion (line 12). This marking m is chosen such that it minimizes the cost to reach that marking from the initial marking $g(m)$ plus the estimated remaining cost $h(m)$. The algorithm stops if the final marking m_f is reached.

As explained in [15], estimating the remaining distance is done in two stages. First, an underestimate is "guessed" (line 40) and second, it is computed exactly (line 21). Lines 16 through 27 handle the case in which the selected marking m has a "guessed" estimate in which cast it is part of the set Y. The incremental nature of the work presented in [15] is shown in lines 17 through 20. In this part, a splitpoint is added to the set K after which the algorithm is restarted from scratch. The value added to the set K is s which represents the index of the last event explained by the closed markings in set A. This can be implemented without any recursion, so there is no memory overhead here.

Lines 28 through 46 make up the regular A^* algorithm. The selected marking m is expanded by iterating over all enabled transitions and firing them to obtain new markings m'. If a marking m' was already closed, we continue with the next marking. If it was not closed, then it is added the open set (if it wasn't already in there) in line 33 and a temporary value for g is computed in line 34. Each marking m' is then assigned a value for $g(m')$ and $h(m')$, where the latter depends on whether it can be derived from $h(m)$ (line 38) or whether it has to be "guessed" (lines 40–41). Furthermore, the predecessor relation p is updated in line 43.

In this algorithm, the set of splitpoints K is automatically updated whenever the algorithm reaches a marking which does not have an exact heuristic, but only a guessed value and s is currently not in the set of splitpoints. This incremental way of computing alignments has proven to be the most efficient way to date [15]. However, removing lines 17 through 20 from the algorithm and ignoring the set K would yield the alignment algorithm as presented in [2,3].

3.1   Time Complexity

A^* relies on a heuristic function to underestimate the remaining cost. In traditional alignments, this underestimation function is an (integer) linear program with the number of rows equal to the number of places in the synchronous product and the number of columns equal to the number of transitions. For many markings reached in the search for an optimal alignment such a linear program is solved. This leads to the fact that for the experiments in [15], 99% of the time of the classic A^* is spent on solving such linear programs using LPSolve [5].

In the incremental A^* technique of [15], an incremental heuristic function is proposed that uses the extended marking equation with a number of splitpoints. The heuristic function is a linear program of which the number of rows and columns scale linearly in the size of K. Solving a linear equation system is polynomial in the rank of its matrix (hence in the number of rows and columns). The algorithm proposed in [15] and detailed in Algorithm 1 maximizes reuse of previously computed solutions to these linear programs. while the linear programs are more complex than for classical A^*, far fewer have to be solved. The total fraction of the time spent in the LP Solver for the experiments in [15] was 77% for the incremental A^*. The remaining 23% of the time is spent on the internal computations within Algorithm 1, therefore it is important to choose the right data structures for use with this algorithm.

As the algorithm computes an alignment for a single trace, it is rather straightforward to parallelize it when aligning an entire log. By simply distributing traces over multiple cores of a CPU, walltime can be gained. However, as usual with multi-threading on a single CPU, this comes at a cost of synchronization. For the experiments presented in [15], going from 1 thread to 4 on a CPU with 4 physical cores and 4 hypercores, causes the wallclock time to reduce to 27%. The sum over the CPU times per trace increases by 35% in that case.

In the algorithm, there are several core data structures each with a specific set of requirements. In the next section, we discuss each data structure independently.

Procedure 1. $A^\star$ Algorithm for Alignments with estimated heuristics

Let $SN = ((P, T, F, \lambda), m_i, m_f)$ be a synchronous product and let $c : T \to \mathbb{N}$ be a cost function and $h : \text{RS} \to \mathbb{N}$ be a heuristic underestimating the cost of getting from any marking to the final marking.

1: **function** ASTAR(SN, O, c, K)
2:      $A \leftarrow \emptyset$ ▷ Initialise closed set
3:      $X \leftarrow \{m_i\}$ ▷ Initialise open set as a priority queue favouring markings for which the exact heuristic is known.
4:      $Y \leftarrow \emptyset$ ▷ Initialise estimated heuristics
5:      $s \leftarrow 0$ ▷ Initialise the number of events explained
6:      $p(m_i) = (\tau, \tau)$ ▷ Initialise predecessor function
7:      $\forall_{m \in \text{RS}(SN)} \; d(m) = \infty$ ▷ Initialise cost so far function d
8:      $g(m_i) = 0$
9:      $\forall_{m \in \text{RS}(SN)} \; f(m) = \infty$ ▷ Initialise estimated total cost function f
10:      $f(m_i) = h(m_i, K)$ ▷ Compute estimate for initial marking with the splitpoints in K
11:      **while** $X \neq \emptyset$ **do**   ▷ While not all states visited
12:          $m \leftarrow m \in X$ minimizing $f(m)$ ▷ Get the most promising marking m
13:          **if** $m = m_f$ **then**   ▷ final marking reached
14:              **break while**
15:          **end if**
16:          **if** $m \in Y$ **then**   ▷ Heuristic of m is not exact
17:              **if** $s \notin K$ **then**   ▷ Check if s is not already a splitpoint in K
18:                  $K \leftarrow K \cup \{s\}$ ▷ Add the maximum events explained to k
19:                  **return** Astar(SN, O, c, K)   ▷ Restart with a longer list of splitpoints.
20:              **end if**
21:              $x \leftarrow h(m)$ ▷ Compute the true estimate
22:              $Y \leftarrow Y \setminus \{m\}$ ▷ Remove estimated heuristic
23:              **if** $x > \hat{h}(m)$ **then**   ▷ Heuristic increased
24:                  $f(m) \leftarrow g(m) + h(m)$ ▷ Update estimated total cost function
25:                  **continue while**   ▷ Note: m may not be minimizing f any more
26:              **end if**   ▷ If heuristic did not chance, continue with m
27:          **end if**
28:          $A \leftarrow A \cup \{m\}$ ▷ Add m to the closed set
29:          $X \leftarrow X \setminus \{m\}$ ▷ Remove m from the open set
30:          $s \leftarrow max(s, \text{events explained by } m)$ ▷ keep track of the maximum number of events explained
31:          **for all** $t \in T'$ with $m[t\rangle m'$ **do**   ▷ For each relevant transition enabled in m
32:              **if** $m' \notin A$ **then**   ▷ Reaching a marking not yet visited, or found shorter path
33:                  $X \leftarrow X \cup \{m'\}$ ▷ Add m' to the open set
34:                  $a \leftarrow g(m) + c(t)$ ▷ Compute the cost so far of reaching m' via m
35:                  **if** $a < g(m')$ **then**   ▷ If this current cost is better than known cost so far
36:                      $g(m') \leftarrow a$ ▷ Update cost so far function
37:                      **if** $h(m', K)$ can be derived from $h(m, K)$ **then**
38:                          $f(m') \leftarrow g(m') + h(m', K)$ ▷ Update estimated total cost function
39:                      **else**
40:                          $Y \leftarrow Y \cup \{m'\}$ ▷ Add m' to the estimated heuristics set
41:                          $f(m') \leftarrow g(m') + h(m, K) - c(t)$ ▷ Update estimated total cost function
42:                      **end if**
43:                      $p(m') \leftarrow (t, m)$ ▷ Update predecessor function
44:                  **end if**
45:              **end if**
46:          **end for**
47:      **end while**
48:      $\gamma \leftarrow \langle t_0, \ldots, t_n \rangle$ such that $t_n = \#_1(p(m_f))$, $t_{n-1} = \#_1(p(\#_2(p(m_f))))$ etc. until the initial marking is reached recursively.
49:      **return** $f(m_f), \gamma$ ▷ Return distance and alignment
50: **end function**

4   Data Structures

The alignment algorithm contains a number of core data structures on which it operates. In this section, we discuss each data structure and we provide insights into the operations performed on these structures as well as the memory requirements for them. We start with the most prominent concept in the algorithm, namely the marking.

4.1   Marking m

A marking is represented by a 31 bit number (the Java type of signed integer). Apart from the initial and final marking, which are represented as arrays, no marking is ever represented explicitly in memory. Instead, whenever marking m is used in the code, it is constructed by following the p relation back to the origin of the search graph. Markings are simply numbered from 0 upwards. This implies that there is a bit of computational overhead in line 12 of the algorithm when marking m is retrieved from the open set. Here, the integer id of marking m is returned after which the marking itself is constructed into an array with a byte value for each place (we assume no more than 128 tokens in a place at any point in time).

For each marking, we store the value of the g function, the value of the h function and the predecessor relation p in a total of 128 bits in an array indexed by the marking id (of which 2 bits are still unused). These values only have to be retrieved or written for a given marking hence using an array with the marking id as index makes these operations $O(1)$.

4.2   Predecessor Relation p

The predecessor relation p stores the transition t in the synchronous product that was fired to reach a marking as well as the marking in which t was fired. We use this predecessor relation in two parts. First, we use it to compute the actual alignment (line 48). Furthermore, we use it throughout the code to construct a marking explicitly as we only store marking ids, not explicit markings. We store the predecessor transition in 31 bits unsigned (where one value is reserved for indicating a non-existing predecessor). The predecessor marking is also stored using its 31 bit id.

4.3   Cost-so-far Function g

The cost through which a marking is reached is stored as a 31 bit unsigned value. This implies that the cost to reach a marking can be at most 2,147,483,647. This is important when selecting a cost function for non-synchronous moves. Selecting too high cost values for certain moves will lead to internal overflow (which is detected and reported as such by the Java code).

4.4 Heuristic Function h

Like the g function, we also store the h function in 31 unsigned bits. In addition, we store a flag of 1 bit to indicate if the heuristic is exact (line 24 and line 42) or if the heuristic is estimated (line 45).

Together, p, g and h make up 125 bits. We use one more bit to store a flag indicating if the marking is in the closed set or not. The closed set is not represented in another way and in line 33 of the algorithm, this flag is checked to check if m' is in the closed set.

To store these functions we allocate 16 kByte arrays initially. Whenever 1,024 markings have been visited, we allocate an additional block of 16 kByte.

4.5 Open Set X

Perhaps the most interesting data structure is the open set X. This set should allow for efficient removal of the element minimizing f and favouring exact heuristics over estimated ones (line 12), efficient checking of inclusion (line 33) and efficient insertion (line 34). Furthermore, because of the estimated heuristics, it should allow for updating the locations of markings in the set whenever their estimated heuristic function is overwritten by an exact heuristic (line 24).

This set is implemented as a hashtable-backed, balanced binary heap using a sorting procedure which uses the f score as a first order sorting criterion. The second order sorting criterion is the availability of an exact estimate of a marking and the third order sorting is based on the g score (favouring markings with a higher g score). The heap itself is an array of markings (array of integers). The hash-table is needed to efficiently check at what location a marking is in the queue.

The memory use of this heap is limited to the array of markings (4 bytes per marking in the queue) and the hashtable. The latter uses an implementation from an external package which consumes 13 bytes per element for which the table has capacity. With the default load factor of 50%, this implies approximately 26 bytes per marking in the queue.

Checking if a marking exists in the queue as well as inspecting the head element are $O(1)$ operations. All others (insertion, removal, updating) are $O(\log(n))$ operations. However, the performance of this set depends on the hash function used. We experimented with a large collection of hash functions for arrays of bytes (recall markings are arrays of bytes with one byte per place). The default Java function performed equally well as others, both in terms of minimal hash collisions and time.

As markings are not represented explicitly, computing hash codes for a specific marking is a more complex task than usual. Each marking for which the hashcode needs to be computed needs to be instantiated fully first by following the predecessor function p. This implies that the open set should aim to minimize the hash collisions so that, for any insertion, only one hash code has to be computed, i.e. the hashcode of the marking to be inserted.

Table 1. Time and collision test for hashcode on 1,048,576 semi-random arrays, expecting 128 hash collisions.

Reference	hashCode name	CPU time (s)	Collisions
[4]	AdHASH	0.506	115
[7]	FNV1	0.519	137
[7]	FNV1a	0.532	119
https://tanjent.livejournal.com/755617.html	MurMur3	0.270	135
http://www.burtleburtle.net/bob/c/lookup3.c	Jenkings	0.309	125
http://www.cse.yorku.ca/~oz/hash.html	Bernstein	0.315	113
Default Java	JAVA31	0.311	126

4.6 Hashing Markings

To maximize the speed of the algorithm, we tested a large collection of hashcodes. The results are shown in Table 1. For a total of 1,048,576 semi-random arrays, we expect 128 hash collisions, i.e. for 128 arrays, we expect that they have the same hashcode as another array, whilst not being equal. As shown, the tested hashcodes perform equally well in terms of collisions. The CPU time shows that the default Java implementation for hashing an array is in the lower range, so we decided to use this. Murmur3 is slightly faster, but with more collisions. Note that the speed of the Java default may have to do with the fact that there is a bytecode translation available in Java which is optimized for speed.

4.7 Closed Set A

The closed set A contains all markings which have been visited before. In the code, each marking carries a flag to indicate if it was closed or not. The set A is therefore not directly represented. Instead, the code uses a hashset to store the union of A and X, i.e. the set of all visited markings. The two operations on this set are insertion (line 28) and checking the presence of an element (line 32). The memory use is limited to 4 bytes per element for which the set has capacity, which, with a default load factor of 50%, implies 8 bytes per element. Insertion is a $O(\log(n))$ operation, whereas checking presence of an element is amortized $O(1)$. As for the open set, hash collisions should be avoided as much as possible as computing hash codes for stored markings is relatively expensive.

4.8 Solution Vector Cache for Function h

As indicated in [15], the solution vectors for the linear programs can be reused to obtain new solutions. This requires us to store the solution vectors for the linear programs in memory. We do so in a hashmap, mapping markings to these vectors. Each vector is represented by a minimal number of bits. For each solution vector, we need as many bits per transition as the maximum number in the vector requires (i.e. if the vector contains a value 4, then we need 3 bits per transition to store this vector). We limit the number of firings to 255, so the maximum number of bits we need per element of the vector is 8. A header is used to store a flag of 1 bit indicating if the solution was computed (line 10, line 21) or derived (line 38) and we use 3 bits to store the number of bits per element of the vector, i.e. we have a 4 bit header per vector. The vectors are stored in a hashmap mapping markings to the arrays.

4.9   Set of Estimated Heuristics Y

The set of estimated heuristics Y does not need explicit representation. Instead, a 1-bit flag is stored in the function h to indicate if an estimate is exact or estimated.

4.10   Memory Use of the LP Solver

LPSolve [5] is used as an LP solver to obtain values for the heuristic function. LPSolve stores the linear programs in memory. This linear program is essentially a large, but very sparse, matrix of double values. We estimate that, for each non-zero element 16 bytes are used, 4 to store the row, 4 to store the column and 8 for the actual value. Then some internal arrays are needed to perform the calculations, but is limited.

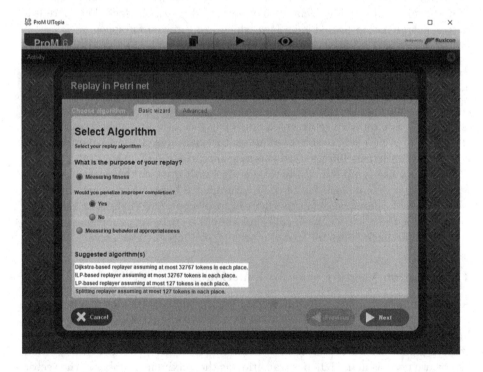

Fig. 3. User interface of ProM showing the algorithm of this paper selected.

Table 2. Java code for standalone execution of the Replayer.

```
public static void doReplay(XLog log, Petrinet net, Marking initialMarking,
                  Marking finalMarking, XEventClasses classes, TransEvClassMapping mapping) {
   // Setup default parameters with 2 threads
   ReplayerParameters parameters = new ReplayerParameters.Default(2, Debug.NONE);
   // Setup the replayer
   Replayer replayer = new Replayer(parameters, net, initialMarking, finalMarking, classes,
                                                          mapping, false);
   // Set a timeout per trace in milliseconds
   int toms = 10 * 1000;
   // preprocessing time to be added to the statistics if necessary
   long preProcessTimeNanoseconds = 0;
   // Setup a threadpool for the multithreaded execution
   ExecutorService service = Executors.newFixedThreadPool(parameters.nThreads);
   // Setup an array to store the results
   Future<TraceReplayTask>[] futures = new Future[log.size()];
    for (int i = 0; i < log.size(); i++) {
      // Setup the trace replay task
      TraceReplayTask task = new TraceReplayTask(replayer, parameters, log.get(i), i, toms,
                             parameters.maximumNumberOfStates, preProcessTimeNanoseconds);
      // submit for execution
      futures[i] = service.submit(task);
   }
   // initiate shutdown and wait for termination of all submitted tasks and obtain results.
   service.shutdown();
   for (int i = 0; i < log.size(); i++) {
      TraceReplayTask result;
      result = futures[i].get();
   }
   switch (result.getResult()) {
      case DUPLICATE :
         assert false; // cannot happen in this setting
         throw new Exception("Result cannot be a duplicate in per-trace computations.");
      case FAILED :
         // internal error in the construction of synchronous product or other error.
         throw new RuntimeException("Error in alignment computations");
      case SUCCESS :
         // process succcesful execution of the replayer
         SyncReplayResult replayResult = result.getSuccesfulResult();
         // obtain the exit code of the replay algorithm
         int ec = replayResult.getInfo().get(Replayer.TRACEEXITCODE).intValue();
         if ((ec & Utils.OPTIMALALIGNMENT) == Utils.OPTIMALALIGNMENT) {
            // Optimal alignment found.
            // Handle further processing here.
         } else if ((ec & Utils.FAILEDALIGNMENT) == Utils.FAILEDALIGNMENT) {
            // failure in the alignment. Error code shows more details.
            // Handle further processing here.
         }
         // Additional exitcode information for failed alignments:
         if ((ec & Utils.ENABLINGBLOCKEDBYOUTPUT) == Utils.ENABLINGBLOCKEDBYOUTPUT) {}
            // in some marking, there were too many tokens in a place.
         if ((ec & Utils.COSTFUNCTIONOVERFLOW) == Utils.COSTFUNCTIONOVERFLOW) {}
            // in some marking, the cost function went through the upper limit of 2^24
         if ((ec & Utils.HEURISTICFUNCTIONOVERFLOW) == Utils.HEURISTICFUNCTIONOVERFLOW) {}
            // in some marking, the heuristic function went through the upper limit of 2^24
         if ((ec & Utils.TIMEOUTREACHED) == Utils.TIMEOUTREACHED) {}
            // alignment failed with a timeout
         if ((ec & Utils.STATELIMITREACHED) == Utils.STATELIMITREACHED) {}
            // alignment failed due to reacing too many states.}
         if ((ec & Utils.COSTLIMITREACHED) == Utils.COSTLIMITREACHED) {}
            // no optimal alignment found with cost less or equal to the given limit.
         if ((exitCode & Utils.CANCELED) == Utils.CANCELED) {}
            // user-cancelled.
         break;
} }
```

Overall, the memory overhead of the algorithm is very low. For all experiments conducted in [15], the actual memory used internally in the algorithm presented as Algorithm 1 was 8 Megabytes, not counting the overhead of storing the Petri net or the event log in Java memory.

5   Implementation

The algorithm presented as Algorithm 1 has been implemented in ProM [14] as part of the "Alignment" package. It is one of the algorithms available for the plugin entitled "Replay a log on Petri net for Conformance Analysis" and it is selected by default if it is installed. Figure 3 shows a screenshot of ProM with the plugin highlighted.

The use of this plugin is not different from the traditional alignment plugin presented in [2,3]. Both the input and the output are the same and the implementation seamlessly integrates with ProM.

The algorithm can also be used in a stand-alone fashion from the command line. The code snippet depicted in Table 2 shows how to do this in Java code.

6   Conclusions and Future Work

This paper presents, in detailed pseudo-code, the efficient alignment algorithm presented in [15]. Computing an alignment for a given process model and trace is, in general, a complex and time consuming task. In part, this complexity stems from the fact that the search space of the A^* algorithm is worst case the size of the reachability graph of the model times the length of the trace. A second contributing factor to the complexity is the heuristic function used, but also the choice of data structures in the core algorithm.

We discuss the time complexity of the algorithm as well as the properties the various data structures have. As far as we know, this is the first time the actual algorithm and the properties of these data structures have been presented.

References

1. van der Aalst, W.M.P.: Decomposing Petri nets for process mining: a generic approach. Distrib. Parallel Databases **31**(4), 471–507 (2013)
2. van der Aalst, W.M.P., Adriansyah, A., van Dongen, B.F.: Replaying history on process models for conformance checking and performance analysis. Wiley Interdisc. Rev. Data Min. Knowl. Discov. **2**(2), 182–192 (2012)
3. Adriansyah, A.: Aligning Observed and Modeled Behavior. Ph.D. thesis, Eindhoven University of Technology, Department of Computer Science (2014)
4. Bellare, M., Micciancio, D.: A new paradigm for collision-free hashing: incrementality at reduced cost. In: Fumy, W. (ed.) EUROCRYPT 1997. LNCS, vol. 1233, pp. 163–192. Springer, Heidelberg (1997). https://doi.org/10.1007/3-540-69053-0_13
5. Berkelaar, M., Eikland, K., Notebaert, P.: lpsolve: open source (mixed-integer) linear programming system

6. de Leoni, M., Marrella, A.: Aligning real process executions and prescriptive process models through automated planning. Expert Syst. Appl. **82**, 162–183 (2017)
7. Eastlake, D., Fowler, G., Vo, K.-P., Noll, L.: The FNV non-cryptographic hash algorithm (2015)
8. Hart, P.E., Nilsson, N.J., Raphael, B.: A formal basis for the heuristic determination of minimum cost paths. IEEE Trans. Syst. Sci. Cybern. **4**(2), 100–107 (1968)
9. Gómez López, M.T., Borrego, D., Carmona, J., Gasca, R.M.: Computing alignments with constraint programming: the acyclic case. In: Proceedings of ATAED 2016, Torun, Poland, 20–21 June 2016. CEUR Workshop Proceedings, vol. 1592, pp. 96–110. CEUR-WS.org (2016)
10. Munoz-Gama, J., Carmona, J., van der Aalst, W.M.P.: Single-entry single-exit decomposed conformance checking. Inf. Syst. **46**, 102–122 (2014)
11. Reißner, D., Conforti, R., Dumas, M., La Rosa, M., Armas-Cervantes, A.: Scalable conformance checking of business processes. In: Panetto, H., et al. (eds.) OTM 2017 Conferences, Part I. LNCS, vol. 10573, pp. 607–627. Springer, Cham (2017). https://doi.org/10.1007/978-3-319-69462-7_38
12. Rozinat, A., van der Aalst, W.M.P.: Conformance checking of processes based on monitoring real behavior. Inf. Syst. **33**(1), 64–95 (2008)
13. Taymouri, F., Carmona, J.: A recursive paradigm for aligning observed behavior of large structured process models. In: La Rosa, M., Loos, P., Pastor, O. (eds.) BPM 2016. LNCS, vol. 9850, pp. 197–214. Springer, Cham (2016). https://doi.org/10.1007/978-3-319-45348-4_12
14. van Dongen, B.F., de Medeiros, A.K.A., Verbeek, H.M.W., Weijters, A.J.M.M., van der Aalst, W.M.P.: The ProM framework: a new era in process mining tool support. In: Ciardo, G., Darondeau, P. (eds.) ICATPN 2005. LNCS, vol. 3536, pp. 444–454. Springer, Heidelberg (2005). https://doi.org/10.1007/11494744_25
15. van Dongen, B.F.: Efficiently computing alignments using the extended marking equation. In: Accepted for BPM 2018, Sydney Australia (2018)
16. van Zelst, S.J., Bolt, A., van Dongen, B.F.: Tuning alingment computation: an experimental evaluation. In: Proceedings of ATAED 2017, Zaragoza, Spain, 25–30 June 2017, pp. 1–15 (2017)

Understanding Automated Feedback in Learning Processes by Mining Local Patterns

Galina Deeva[✉] and Jochen De Weerdt

Faculty of Economics and Business, Department of Decision Sciences
and Information Management, KU Leuven, Leuven, Belgium
galina.deeva@kuleuven.be

Abstract. Process mining, and in particular process discovery, provides useful tools for extracting process models from event-based data. Nevertheless, certain types of processes are too complex and unstructured to be able to be represented with a start-to-end process model. For such cases, instead of extracting a model from a complete event log, it is interesting to zoom in on some parts of the data and explore behavioral patterns on a local level. Recently, local process model mining has been introduced, which is a technique in-between sequential pattern mining and process discovery. Other process mining methods can also used for mining local patterns, if combined with certain data preprocessing. In this paper, we explore discovery of local patterns in the data representing learning processes. We exploit real-life event logs from JMermaid, a Smart Learning Environment for teaching Information System modeling with built-in feedback functionality. We focus on a specific instance of feedback provided in JMermaid, which is a reminder to simulate the model, and locally explore how students react to this feedback. Additionally, we discuss how to tailor local process model mining to a certain case, in order to avoid the computationally expensive task of discovering all available patterns, by combining it with other techniques for dealing with unstructured data, such as trace clustering and window-based data preprocessing.

Keywords: Process discovery · Local process models ·
Automated feedback · Trace clustering

1 Introduction

Nowadays, most educational institutions use a variety of information systems to support educational processes. In most cases, these information systems have a logging functionality that allows for monitoring and analyzing the process it supports. These data can be analyzed from a variety of different perspectives, showing different aspects of learning. Traditional data mining techniques have been used to build predictive models, acquire better understanding of learning

© Springer Nature Switzerland AG 2019
F. Daniel et al. (Eds.): BPM 2018 Workshops, LNBIP 342, pp. 56–68, 2019.
https://doi.org/10.1007/978-3-030-11641-5_5

processes, or give recommendations to students and educators. However, the majority of traditional data mining techniques do not have an objective to analyze, discover and visually represent a complete educational process. Process mining does have this objective, as it aims to extract process-related knowledge from event logs stored by information systems [1].

Process mining provides useful tools for extracting knowledge from event-based data [2]. One of the most insightful tasks of process mining is process discovery, i.e. extracting a process model that represents the event log from start to end. However, while providing useful results in many cases, process discovery is not always able to represent complex and unstructured processes. In learning analytics, the data oftentimes contains a large number of activity types, making it hard to be represented by a single process model and resulting in so-called spaghetti models or flower models. An example of this can be found in our previous study [3], in which we discussed that discovering a process model from behavioral data in a Massive Open Online Course (MOOC) in a global way is a rather challenging task, while methods that work on a more local level, such as sequence mining, yield more insightful results.

Recently introduced in [4], Local Process Model (LPM) mining can be positioned in-between of process and sequence mining. As such, LPM mining can cope with unstructured processes on a local level, avoiding the difficulty of representing the process as a whole. In addition, LPM mining is also capable to grasp concepts that are hard to represent with most sequential pattern mining approaches, such as concurrency, choice, loop and sequential composition [4]. The goal of local process model discovery is to find patterns that occur in the event log on a local level. Such local models can be very insightful, especially if the task of describing the behavior in the complete event log is too complex, but also in cases when it is interesting to focus on the local behavior. Another advantage of LPM discovery is that it is capable to grasp relations between more than 3 items, which is often not possible with most of sequence mining techniques.

In this work, we explore discovery of local patterns from event-based data from JMermaid[1], a Smart Learning Environment for teaching Information System (conceptual) modeling, enriched with a feedback mechanism that provides students with real-time automated feedback. Our goal is to employ LPM mining and other techniques in order to study the behavior of students after they receive feedback. We focus on a certain instance of feedback, which is a reminder to simulate the created conceptual model, and then locally explore the patterns that follow this feedback on real-life datasets from JMermaid.

As discussed in [5], it is a computationally difficult task to discover local patterns in the event log with too many activity types. Due to that, in this paper we discuss and compare possible ways of discovering LPM that are tailored to a specific problem, as to reduce the computational complexity of discovering all available local models. We explore combinations of different techniques for dealing with unstructured data, such as trace clustering and window-based data preprocessing.

---

[1] http://merode.econ.kuleuven.ac.be/mermaid.aspx.

The paper is structured as follows. In Sect. 2, recent studies on process mining in an educational context and LPM mining are reviewed. Next, the JMermaid learning environment, research questions and scenarios for mining local patterns in the context of a smart learning environment (SLE) with automated feedback are discussed in Sect. 3. Subsequently, the data preparation and results are presented in Sect. 4. Finally, Sect. 5 outlines our findings and gives directions for future work.

2   Related Work

2.1   Process Mining in an Educational Context

There has been a variety of studies that applied process mining within the field of education, in which cases it has been frequently addressed as Educational Process Mining (EPM). EPM aims to build complete educational process models that are able to reproduce the observed behavior, check if the modeled behavior matches the behavior observed, manage extracted information to make the tacit knowledge explicit and to facilitate a better understanding of the processes [1].

Recently, there has been an increasing number of studies that applied EPM to real-life cases. The objectives of such applications vary widely. For example, Weerapong, Porouhan and Premchaiswadi [6] analyzed the control flow perspective of student registration at the university, with the goal to solve issues that might occur during this process. Vahdat et al. [7] used Fuzzy Miner and a complexity metric to estimate the understandability of process models of engineering laboratory sessions. Trcka and Pechenizkiy [1] explored online-assessment data to investigate how students navigate between multiple choice questions, and whether this process can be improved with automated feedback. More recently, Juhaňák, Zounek and Rohlíková [8] analyzed students' quiz-taking behavior patterns in a learning management system Moodle.

A few studies aimed to explore process mining in a more global setting of Massive Open Online Courses (MOOCs). For example, Mukala et al. [9] applied the dotted chart, process discovery and conformance checking techniques to a Coursera MOOC dataset. In another study, Deeva et al. [3] investigated the applicability of process discovery techniques for dropout prediction with a case study on a MOOC from the EdX platform. Additionally, Maldonado-Mahauad et al. [10] used process mining for exploring frequent interaction sequences in three Coursera MOOCs.

A common goal in EPM research is to find behavioral patterns typical for certain groups of learners, or to compare the behavior of student clusters. For example, Schoor and Bannert [11] aimed to find discriminative process patterns for high and low performing groups in a collaborative learning task, showing that successful students perform regulatory activities with a higher frequency and in a different order than less successful students. Another comparison of different student groups was performed by van der Aalst, Guo and Gorissen [12], where records of watching video lectures are analyzed with comparative process mining using process cubes, which allowed to discriminate between the learning behavior of student subgroups, such as successful vs. unsuccessful, male and female, local and foreign, as well as the behavior within different chapters of the course. Moreover, Papamitsiou and Economides [13] exploited comprehensive process models with concurrency patterns in order to detect and model guessing behavior in computer-based testing, revealing common patterns for students with different goal-orientation levels.

Previous research involving the JMermaid learning environment can be found in [14] and [15], where process mining was used for revealing modeling behavior patterns that can be related to certain learning outcomes. We expect that more insightful patterns can be observed in event logs from JMermaid by mining local models instead of complete process models due to the complexity of underlying processes. Thus, it is anticipated that more sophisticated methods for mining local patterns can facilitate deeper understanding of these data.

More information regarding EPM can be found in the most recent survey of this topic by Bogarín, Cerezo and Romero [16].

2.2   Local Process Model Mining

Local process model mining can be positioned in-between process discovery and episode/sequential pattern mining. The concept and the procedure of LPM discovery was introduced by Tax et al. [4], where it was also compared with other techniques for mining local patterns in unstructured event logs, such as process mining algorithms Declare miner, Fuzzy miner and Episode Miner, and the sequential pattern mining algorithm PrefixSpan. The authors showed that LPM discovery is capable of deriving insightful patterns that in some cases cannot be discovered with aforementioned techniques. Additionally, they proposed metrics for assessing the quality of local process models.

The same authors in [5] expanded their findings by introducing heuristic approaches for coping with computational difficulties of discovering LPMs. These approaches are Markov clustering, log entropy and the relative information gain heuristics, which are used to create projections of event logs. The example event log from this study contained 1734 activity types, which is too difficult to deal with a straightforward approach described in [4], since the computational complexity grows significantly for such substantial amount of variety in the logs. To solve this problem, Tax et al. exploited the idea of discovering local models from the projections of event logs, containing only activities of interest for a particular LPM.

Dalmas, Tax and Norre [21] introduced the heuristics for high-utility LPM discovery. The authors aimed to reduce computational complexity of the LPM mining task by specifying a utility function based on business insights. It was concluded, however, that the search space of LPMs cannot be reduced without loss. Similarly, Tax et al. [22] presented goal-driven discovery of LPMs based on utility functions and constraints for addressing particular business questions.

For fine-granular event logs it is useful to combine events to a higher level of abstraction, which is typically done with clustering techniques. In the study by Mannhardt and Tax [23], local process models are used for automated event abstractions, resulting in overall process models with more balanced precision and fitness scores.

3   Mining Local Patterns in a Smart Learning Environment

3.1   Automated Feedback in JMermaid

In this work, we analyze event-based data from the JMermaid learning environment, developed in our Management Informatics Research Group at the Faculty of Business and Economics, KU Leuven for teaching Information Systems modeling. It is based on MERODE, a method for Enterprise Systems development [24], and used in the Architecture and Modeling of Management Information Systems (AMMIS) course[2].

JMermaid is enriched with a feedback mechanism that provides personalized immediate feedback in an automated way. Based on the findings of a previous study of JMermaid [15], which indicated that frequent simulation of a conceptual model is strictly correlated with the successful learning outcome of a student, we implemented a learning dialog that reminds students to simulate their model after a certain number of actions is conducted in the tool (Fig. 1).

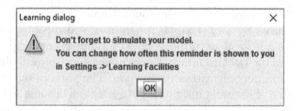

Fig. 1. Reminder to simulate the model provided as feedback to students in the JMermaid tool

We analyze local patterns that involve this instance of feedback, which is addressed below as the *simulation reminder* (SR). The study aims to tackle the following research questions:

(1) How to discover typical patterns of student reactions to automated feedback in a smart learning environment?

(2) What are the optimal ways to discover those patterns, which are both insightful and computationally efficient?

(3) How to grasp differences in reaction to feedback between low and high-performing students? Is there any correlation between students' reactions to feedback and their final scores?

3.2   Window-Based Preprocessing for Detecting Automated Feedback

To see the immediate student reactions to automated feedback, we focus on events that directly follow SR. We choose 10 following events; however, this

---

[2] http://onderwijsaanbod.kuleuven.be/syllabi/e/D0I71AE.htm.

number is not restrictive and can be adjusted in the future analyses. Next, we disregard a few outlier traces that contain less than 10 events after SR, since it most likely means that a student stopped working in the tool instead of reacting on feedback. Thus, SR is always acting as a start event in a set of traces containing 11 events (SR and 10 following events). Subsequently, an artificial end event is added to each trace. As a result of such window-based preprocessing, the obtained data consists of traces with 12 events, which are $<SR, a, b, c, d, e, f, g, h, i, j, End>$. The data in this format is further addressed as the *filtered data*.

3.3 Scenarios for Detecting Automated Feedback

Five scenarios for extracting local patters containing SR are discussed.

(1) LPM Mining Applied to the Complete Event Logs. The first approach is to apply the LPM mining algorithm to the complete event logs without window-based preprocessing, as to extract all possible LPMs, and subsequently filter them focusing on the models with the simulation reminder. As discussed above, this brute-force approach is expected to be computationally expensive, and might benefit from further optimization. Nevertheless, it is also possible that reducing the search space will cause an information loss [21].

(2) LPM Mining Applied to the Filtered Data. The second scenario is to apply LPM discovery to the filtered event logs to investigate if data preprocessing can reduce computational complexity and facilitate LPM discovery.

(3) LPM Mining Combined with Trace Clustering and the Filtered Data. Similarly to LPM mining, trace clustering techniques aim to resolve the issue of overly unstructured process models that are discovered from event logs with, e.g., a large number of activity types [26,27]. In trace clustering, similar traces are grouped together so as to focus on similar behavioral scenarios within an event log. Trace clustering techniques could potentially work well on the data representing learning processes, in which different groups of students might follow several distinct learning paths. Nevertheless, not all event data can be potentially clustered.

The third approach is to apply a trace clustering technique k-gram [28], available in Guide Tree Miner plugin in ProM, to the filtered data, and subsequently apply LPM mining, to investigate whether trace clustering can facilitate data exploration in the context of unstructured learning processes.

(4) Process Discovery Applied to the Filtered Data. The fourth approach is to apply process discovery techniques Inductive Miner and Alpha miner on the filtered data. We want to investigate whether plain process discovery is more or at least equally capable of extracting meaningful local patterns than LPM mining, if being combined with intelligent subsetting of the data.

(5) Process Discovery Combined with Trace Clustering and the Filtered Data. Similarly to (3), the last approach is to cluster the traces and subsequently apply process discovery.

4 Experimental Evaluation

4.1 Data Description and Preparation

We analyze event logs of the students performing 23 distinct modeling tasks during in-class exercise sessions. An overview of the data is given in Table 1. To compare the behavioral patterns of low and high-performing students, the students are divided into two groups according to their performance. For this we apply k-means clustering with their final scores for the course and the grades for two intermediate assignments (which are not part of the final score) as features, and obtain two clusters of 43 and 21 students. Dataset 1 (D1) and Dataset 2 (D2) are the complete event logs for low-performing (Group 1) and high-performing (Group 2) students, respectively. Filtered Dataset 1 (FD1) and Filtered Dataset2 (FD2) are the event logs preprocessed as described in Sect. 3.2. Note that in case of D1 and D2 we use *User_id* as a case id, thus analyzing data from the student perspective, and in case of FD1 and FD2 we use each trace with the simulation reminder as a separate case, thus analyzing each particular reaction to this feedback.

Table 1. An overview of the datasets used in the experimental evaluation

Dataset	Performance	# of students	# of activity types (L)	# of activity types (H)	# of events	# of SR
D1	Low	43	63	16	24296	276
FD1	Low	40	52	16	3076	276
D2	High	21	64	16	21789	232
FD2	High	21	49	16	2684	232

An example of an event log from JMermaid is shown in Fig. 2. *ActivityL* represents activities on a more fine-granular level, i.e. at the low level of abstraction, which has around 60 (for D1 and D2) or 50 (for FD1 and FD2) activity types. The JMermaid tool also logs some aspects of the modeling process, such as "view" and "category" (structural (S) or behavioral (B)), as well as a type of performed action (Feedback, Create, Delete, Edit, Customize, Error, Check, Save). A combination of the "type" of action with structural or behavioral aspect gives 16 variations, including the simulation reminder. This less fine-granular view, i.e. high level of activity abstraction, is referred in the logs as *ActivityH*.

Timestamp	User_id	View	Category	Type	ActivityL	ActivityH
2017-03-03 14:28:23.645	User1	EDG	S	FEEDBACK	Simulation reminder	Simulation reminder
2017-03-03 14:28:23.661	User1	EDG	S	FEEDBACK	Dependency feedback	FEEDBACK+S
2017-03-03 14:28:23.661	User1	EDG	S	CREATE	Create dependency	CREATE+S
2017-03-03 14:28:38.654	User1	EDG	S	CUSTOMIZE	Move object	CUSTOMIZE+S
2017-03-03 14:28:42.773	User1	EDG	S	CUSTOMIZE	Move object	CUSTOMIZE+S

Fig. 2. An example of an event log from JMermaid

4.2   Results

The scenarios described in Sect. 3.3 are applied to the data with low (L) and high (H) levels of activity abstraction for Group 1 and Group 2. The results are summarized in Table 2.

Table 2. A summary of the results of the five scenarios

#	Group 1 (L)	Group 2 (L)	Group 1 (H)	Group 2 (H)
1	The algorithm returned no results	The algorithm returned no results	The algorithm returned no results	The algorithm returned no results
2	The algorithm returned no results	The algorithm returned no results	LPMs are obtained in a very long time (more than 1 h)	The algorithm returned no results
3	The algorithm returned no results	The algorithm returned no results	LPMs are obtained very fast (less than 1 min)	LPMs are obtained very fast (less than 1 min)
4	The obtained models are too unstructured and flower-like	The obtained models are too unstructured and flower-like	The models provide insightful patterns	The models provide insightful patterns
5	The models provide insightful patterns	The models provide insightful patterns	The models provide insightful patterns	The models provide insightful patterns

(1) LPM Mining. For the first scenario, LPM mining applied to a complete event log was not able to discover local patterns for any level of event abstraction. This result is expected and can be explained by very high levels of computational complexity. For low level of abstraction, LPM mining was still not able to discover local models even if combined with data filtering or trace clustering. Since the amount of activity types in this case is close to 50, this result is also explained by a high variety of activity combinations. As discussed in [5], more than 17 activity types might already be too many for the plain LPM mining to handle, requiring further data optimization. Therefore high level of activity abstraction with 16 activity types was expected to be easier to analyze. Nevertheless, in the second scenario, LPM mining was able to discover LPMs for Group 1, but not for Group 2. As also seen from the results of the other techniques, the event logs for Group 2 might contain more distinct activity combinations, for which it is more challenging to discover local patterns. Finally, LPM mining combined with trace clustering discovered LPMs very fast, which indicated that trace clustering is capable of combining traces of learning behavior to meaningful clusters, making it easier to a discovery technique to deal with such data.

Examples of discovered LPMs are provided in Fig. 3. First of all, the local patterns discovered for both Group 1 and 2 are similar, which, given the large amount of the patterns containing SR, can be difficult to interpret in terms of addressing our goal of distinguishing between two groups of students. Second, most of the LPMs contain a choice between SR and other activities, which is not useful for analyzing pattern that follow SR. To conclude, LPM mining is capable to provide interesting patterns in learning processes if combined with other data optimization techniques, but it is not optimal for the purpose of analyzing feedback.

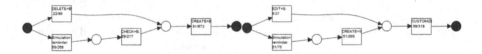

Fig. 3. LPMs discovered in the second (left) and third (right) scenarios

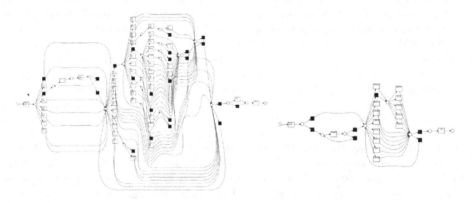

Fig. 4. Process models discovered by Inductive Miner for low (left) and high (right) levels of abstraction in the event logs

(2) Process Discovery. For experimental evaluation, we apply both Inductive Miner and Alpha++ miner available in ProM process mining toolkit. Since the results are similar, we provide the examples of models discovered by Inductive Miner (Fig. 4). The process models discovered from the data with the larger number of activity types are too unstructured and flower-like, even with data preprocessing. On the other hand, the models derived from the logs with higher level of activity abstraction are more structured and can generally provide insights into student reactions to feedback.

(3) Trace Clustering Techniques. The models discovered by Inductive Miner after applying k-gram trace clustering are capable of giving useful insights in case of both high and low levels of activity abstractions (Fig. 5).

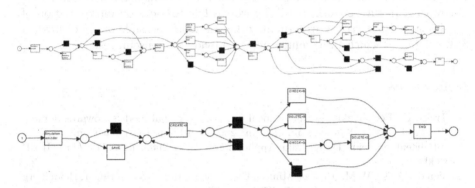

Fig. 5. Process models discovered in the fifth scenario for the low (top) and high (bottom) levels of abstraction in the event logs

5   Conclusion and Future Work

In this paper, the possible ways to discover students reactions to automated feedback in a Smart Learning Environment (SLE) are investigated. We explored Local Process Mining (LPM) discovery and its combinations with other techniques for working with unstructured data, as well as window-based preprocessing of the data. The discussion contained five scenarios for discovering local patterns tailored to a specific case, which included (1) LPM mining on complete event logs, (2) LPM mining on filtered data, (3) LPM mining combined with trace clustering and filtered data, (4) process discovery on filtered data, and (5) trace clustering on filtered data. These scenarios are evaluated on two datasets with log data of high and low-performing students, with the purpose of finding behavioral patterns typical for certain student groups. Two setups with different levels of activity granularity are investigated; one containing 50 activity types and the other with 16 aggregated activity types.

The results reveal that plain LPM discovery is hardly capable to deal with processes with low levels of activity abstraction (50 activity types in our case). In case of less variety in the logs (16 activity types), LPM discovery still requires some adequate data preparation to be able to discover local models. Similarly, process discovery on filtered data is able to achieve meaningful results only in case of less variety in the logs. However, the models discovered with process discovery are more suitable for addressing our research questions, since they give more insights into patterns that follow the feedback. Finally, trace clustering combined with filtered data is capable to achieve meaningful results in case of high as well as low levels of activity granularity. The models discovered on clusters of traces are the most insightful for our task.

This study provides initial steps for exploring reactions to automated feedback in SLEs. Given the limited scope of the paper, we do not focus on a detailed interpretation of the discovered patterns, but rather show possible ways of their discovering. In future work, it will be worthwhile to focus on interpretation of the discovered patterns. Furthermore, other tasks are possible in the context of SLE's data, for which LPM mining might provide more useful results.

References

1. Trcka, N., Pechenizkiy, M.: From local patterns to global models: towards domain driven educational process mining. In: 2009 Ninth International Conference on Intelligent Systems Design and Applications, ISDA 2009, pp. 1114–1119. IEEE (2009)
2. Van der Aalst, W.M.: Process Mining: Data Science in Action. Springer, Heidelberg (2016). https://doi.org/10.1007/978-3-662-49851-4
3. Deeva, G., De Smedt, J., De Koninck, P., De Weerdt, J.: Dropout prediction in MOOCs: a comparison between process and sequence mining. In: Teniente, E., Weidlich, M. (eds.) BPM 2017. LNBIP, vol. 308, pp. 243–255. Springer, Cham (2018). https://doi.org/10.1007/978-3-319-74030-0_18
4. Tax, N., Sidorova, N., Haakma, R., van der Aalst, W.M.: Mining local process models. J. Innov. Dig. Ecosyst. 3(2), 183–196 (2016)
5. Tax, N., Sidorova, N., van der Aalst, W.M., Haakma, R.: Heuristic approaches for generating local process models through log projections. In: 2016 IEEE Symposium Series on Computational Intelligence (SSCI), pp. 1–8. IEEE (2016)
6. Weerapong, S., Porouhan, P., Premchaiswadi, W.: Process mining using α-algorithm as a tool (a case study of student registration). In: 2012 10th International Conference on ICT and Knowledge Engineering (ICT and Knowledge Engineering), pp. 213–220. IEEE (2012)
7. Vahdat, M., Oneto, L., Anguita, D., Funk, M., Rauterberg, M.: A learning analytics approach to correlate the academic achievements of students with interaction data from an educational simulator. In: Conole, G., Klobučar, T., Rensing, C., Konert, J., Lavoué, É. (eds.) EC-TEL 2015. LNCS, vol. 9307, pp. 352–366. Springer, Cham (2015). https://doi.org/10.1007/978-3-319-24258-3_26
8. Juhaňák, L., Zounek, J., Rohlíková, L.: Using process mining to analyze students' quiz-taking behavior patterns in a learning management system. Comput. Hum. Behav. (2017)

9. Mukala, P., Buijs, J., Van Der Aalst, W.: Exploring students learning behaviour in MOOCs using process mining techniques. Department of Mathematics and Computer Science, University of Technology, Eindhoven, The Netherlands (2015)
10. Maldonado-Mahauad, J., Pérez-Sanagustín, M., Kizilcec, R.F., Morales, N., Munoz-Gama, J.: Mining theory-based patterns from big data: identifying self-regulated learning strategies in massive open online courses. Comput. Hum. Behav. **80**, 179–196 (2018)
11. Schoor, C., Bannert, M.: Exploring regulatory processes during a computer-supported collaborative learning task using process mining. Comput. Hum. Behav. **28**(4), 1321–1331 (2012)
12. van der Aalst, W.M.P., Guo, S., Gorissen, P.: Comparative process mining in education: an approach based on process cubes. In: Ceravolo, P., Accorsi, R., Cudre-Mauroux, P. (eds.) SIMPDA 2013. LNBIP, vol. 203, pp. 110–134. Springer, Heidelberg (2015). https://doi.org/10.1007/978-3-662-46436-6_6
13. Papamitsiou, Z., Economides, A.A.: Process mining of interactions during computer-based testing for detecting and modelling guessing behavior. In: Zaphiris, P., Ioannou, A. (eds.) LCT 2016. LNCS, vol. 9753, pp. 437–449. Springer, Cham (2016). https://doi.org/10.1007/978-3-319-39483-1_40
14. Sedrakyan, G., Snoeck, M., De Weerdt, J.: Process mining analysis of conceptual modeling behavior of novices-empirical study using jmermaid modeling and experimental logging environment. Comput. Hum. Behav. **41**, 486–503 (2014)
15. Sedrakyan, G., De Weerdt, J., Snoeck, M.: Process-mining enabled feedback: tell me how to do it right. Comput. Hum. Behav. **57**, 352–376 (2016)
16. Bogarín, A., Cerezo, R., Romero, C.: A survey on educational process mining. Wiley Interdisc. Rev. Data Min. Knowl. Discov. **8**(1), e1230 (2018)
17. Maggi, F.M., Mooij, A.J., van der Aalst, W.M.: User-guided discovery of declarative process models. In: 2011 IEEE Symposium on Computational Intelligence and Data Mining (CIDM), pp. 192–199. IEEE (2011)
18. Günther, C.W., van der Aalst, W.M.P.: Fuzzy mining – adaptive process simplification based on multi-perspective metrics. In: Alonso, G., Dadam, P., Rosemann, M. (eds.) BPM 2007. LNCS, vol. 4714, pp. 328–343. Springer, Heidelberg (2007). https://doi.org/10.1007/978-3-540-75183-0_24
19. Leemans, M., van der Aalst, W.M.P.: Discovery of frequent episodes in event logs. In: Ceravolo, P., Russo, B., Accorsi, R. (eds.) SIMPDA 2014. LNBIP, vol. 237, pp. 1–31. Springer, Cham (2015). https://doi.org/10.1007/978-3-319-27243-6_1
20. Han, J., et al.: PrefixSpan: mining sequential patterns efficiently by prefix-projected pattern growth. In: Proceedings of the 17th International Conference on Data Engineering, pp. 215–224 (2001)
21. Dalmas, B., Tax, N., Norre, S.: Heuristics for high-utility local process model mining. In: Proceedings of the International Workshop on Algorithms and Theories for the Analysis of Event Data, pp. 106–121 (2017)
22. Tax, N., Dalmas, B., Sidorova, N., van der Aalst, W.M., Norre, S.: Interest-driven discovery of local process models. arXiv preprint arXiv:1703.07116 (2017)
23. Mannhardt, F., Tax, N.: Unsupervised event abstraction using pattern abstraction and local process models. arXiv preprint arXiv:1704.03520 (2017)
24. Snoeck, M.: Enterprise Information Systems Engineering: The MERODE Approach. Springer, Cham (2014)
25. Serral, E., De Weerdt, J., Sedrakyan, G., Snoeck, M.: Automating immediate and personalized feedback taking conceptual modelling education to a next level. In: 2016 IEEE Tenth International Conference on Research Challenges in Information Science (RCIS), pp. 1–6. IEEE (2016)

26. Song, M., Günther, C.W., van der Aalst, W.M.P.: Trace clustering in process min-
    ing. In: Ardagna, D., Mecella, M., Yang, J. (eds.) BPM 2008. LNBIP, vol. 17, pp.
    109–120. Springer, Heidelberg (2009). https://doi.org/10.1007/978-3-642-00328-
    8_11
27. De Weerdt, J., vanden Broucke, S., Vanthienen, J., Baesens, B.: Active trace clus-
    tering for improved process discovery. IEEE Trans. Knowl. Data Eng. **25**(12),
    2708–2720 (2013)
28. Bose, R.J.C., Van der Aalst, W.M.: Context aware trace clustering: towards
    improving process mining results. In: Proceedings of the 2009 SIAM International
    Conference on Data Mining, SIAM, pp. 401–412 (2009)

11th Workshop on Social and Human Aspects of Business Process Management (BPMS2)

11th Workshop on Social and Human Aspects of Business Process Management (BPMS2)

Social software[1,2] is a new paradigm that is spreading quickly in society, organizations, and economics. It enables social business that has created a multitude of success stories. More and more enterprises use social software to improve their business processes and create new business models. Social software is used both in internal and external business processes. Using social software, the communication with the customer is increasingly bi-directional. For example, companies integrate customers into product development to capture ideas for new products and features. Social software also creates new possibilities to enhance internal business processes by improving the exchange of knowledge and information, to speed up decisions, etc. Social software is based on four principles: weak ties, social production, egalitarianism, and mutual service provisioning.

To date, the interaction of social and human aspects with business processes has not been investigated in depth. Therefore, the objective of the workshop is to explore how social software interacts with business process management, how business process management has to change to comply with weak ties, social production, egalitarianism, and mutual service, and how business processes may profit from these principles.

The workshop discussed the following three topics:

1. Social business process management (SBPM), i.e., the use of social software to support one or multiple phases of the business process lifecycle
2. Social business: Social software supporting business processes
3. Human aspects of business process management

Based on the successful BPMS2 series of workshops since 2008, the goal of the 11th BPMS2 workshop was to promote the integration of business process management with social software and to enlarge the community pursuing the theme.

The workshop started with a keynote from Selmin Nurcan and Rainer Schmidt. They gave a retrospective view of 10 years of BPMS2 at the BPM conference. During the workshop, six teams presented the results of their research.

Paul Mathiesen, Jason Watson, and Wasana Bandara use a technology affordance perspective to identify and conceptualize affordances of enterprise social technology within the context of process improvement activities.

---

[1] Schmidt, R., Nurcan, S.: BPM and Social Software. In: Ardagna, D., Me-cella, M., Yang, J., Aalst, W., Mylopoulos, J., Rosemann, M., Shaw, M.J., and Szyperski, C. (eds.) Business Process Management Workshops. pp. 649–658. Springer Berlin Heidelberg (2009).

[2] Bruno, G., Dengler, F., Jennings, B., Khalaf, R., Nurcan, S., Prilla, M., Sarini, M., Schmidt, R., Silva, R.: Key challenges for enabling agile BPM with social software. Journal of Software Maintenance and Evolution: Research and Practice. 23, 297–326 (2011).

In their paper, Shamsul Duha and Mohammad E. Rangiha propose critical success factors (CSF) for a social BPM implementation that allows businesses to adapt and be flexible to ever-changing demands.

Michael Möhring, Rainer Schmidt, Barbara Keller, Jennifer Hamm, Sophie Scherzinger, and Ann-Kristin Vorndran describe the possibilities of business processes integrating in location-based services in the tourism sector.

"The Evaluation of WfMC Awards for Case Management" is presented by Johannes Tenschert and Richard Lenz with a literature review and an analysis of case studies in regard to targeted knowledge workers, advertised features, and type of system.

Jeroen Bolle and Jan Claes investigate the trade-off between the effectiveness and efficiency of process modeling. By analyzing modeling sessions a provisional explanation for the difference in syntactic quality of process models is given.

"The Repercussions of Business Process Modeling Notations on Mental Load and Mental Effort" is investigated in the paper from Michael Zimoch, Rüdiger Pryss, Thomas Probst, Winfried Schlee, and Manfred Reichert.

We wish to thank all the people who submitted papers to BPMS2 2018 for having shared their work with us, the many participants for creating fruitful discussion, as well as the members of the BPMS2 2018 Program Committee, who made a remarkable effort in reviewing the submissions. We also thank the organizers of BPM 2018 for their help with the organization of the event.

November 2018   Rainer Schmidt
                                                                             Selmin Nurcan

Organization

Program Committee

Renata Araujo	UNIRIO, Brazil
Jan Bosch	Chalmers University of Technology, Sweden
Marco Brambilla	Politecnico di Milano, Italy
Lars Brehm	Munich University of Applied Science, Germany
Claudia Cappelli	UNIRIO, Brazil
Norbert Gronau	University of Potsdam, Germany
Monique Janneck	Fachhochschule Lübeck, Germany
Barbara Keller	Munich University of Applied Sciences, Germany
Ralf Klamma	RWTH Aachen University, Germany
Sai Peck Lee	University of Malaya, Malaysia
Michael Möhring	Munich University of Applied Sciences, Germany
Selmin Nurcan	Université Paris 1 Panthéon - Sorbonne, France
Mohammad Ehson Rangiha	City University, UK

Gustavo Rossi   LIFIA-F. Informatica. UNLP, Argentina
Flavia Santoro   NP2Tec/UNIRIO, Brazil
Rainer Schmidt   Munich University of Applied Sciences, Germany
Miguel-Angel Sicilia   University of Alcala, Spain
Pnina Soffer   University of Haifa, Israel
Irene Vanderfeesten   Eindhoven University of Technology, The Netherlands
Moe Wynn   Queensland University of Technology, Australia

Social Technology Affordances for Business Process Improvement

Paul Mathiesen[⊠], Jason Watson[⊠], and Wasana Bandara[⊠]

Information Systems School, Queensland University of Technology,
Brisbane, QLD 4000, Australia
p.mathiesen@connect.qut.edu.au,
{ja.watson,w.bandara}@qut.edu.au

Abstract. Organisations across diverse industries have started to embed Enterprise Social Technology (EST) to create collaborative, human-centric environments in their day-to-day operations. With this growing trend, the use of EST within process improvement initiatives is gaining popularity. While the potential that EST brings (in particular with better connecting and influencing people's participation) to process improvements is widely acknowledged, research providing insights into how this actually takes place and specifically contributes towards process improvement efforts is very limited. This study adopts a 'technology affordance' perspective to identify and conceptualise affordances of EST within the context of process improvement activities. Based on forming theory on this topic, a process improvement effort that applied EST was investigated through a series of interviews. The interviews were rigorously designed, and carefully executed and analyzed via a tool-supported data coding and analysis approach. The study outcomes resulted with a refined and partially validated 'EST affordances for process improvements' model with 9 EST affordances and 3 'contingency variables'.

Keywords: Business process improvement ·
Enterprise Social Technology · Qualitative research ·
Affordance · NVivo

1 Introduction and Background

Enterprise Social Technology (EST) can be positioned as *"software that supports the interaction of human beings and production of artefacts by combining the input from independent contributors"* (Schmidt and Nurcan 2009, p. 633). EST can be applied throughout a process improvement lifecycle (Becker et al. 2001; Mathiesen et al. 2011), mainly to support process stakeholder collaboration (Magdaleno et al. 2008). Increasingly, organizational approaches to process improvement is being enhanced by a range of social technology (Mathiesen et al. 2011). These approaches to crowd-sourcing and solving process improvement opportunities have been discussed by numerous researchers (Dollmann et al. 2009; Rossi and Vitali 2009; Silva et al. 2010)

Regular Paper. Topic 2 - Social Business: Social software supporting business processes.

© Springer Nature Switzerland AG 2019
F. Daniel et al. (Eds.): BPM 2018 Workshops, LNBIP 342, pp. 73–84, 2019.
https://doi.org/10.1007/978-3-030-11641-5_6

in support of collaborative process design (Erol et al. 2010) to support a more human – centric approach (Mathiesen et al. 2011).

It is recognised that social technology has the potential to "*extend the reach and impact of process improvement efforts*" (Gottanka and Meyer 2012, p. 94). This opportunity is best realized by conceptualizing how social technology and people can be "*woven together*" (Zammuto et al. 2007, p. 753). On this basis, a 'technology affordance' perspective has been applied in this study to uncover the affordances of Enterprise Social Technology within the context of process improvement activities. This research adopts the perspective of Volkoff and Strong (2013, p. 823) who describe affordances as "*the potential for behaviours associated with achieving an immediate concrete outcome and arising from the relation between an object (e.g., an IT artefact) and a goal-oriented actor or actors.*" As stated by Riemer (2010), there is some appreciation for the benefits of digital collaboration tools, but little understanding of this potential within an organization. Furthermore, academia lacks the theory, models and frameworks which describe this relationship (Niehaves and Plattfaut 2011; Walsh and Deery 2006). This study contributes towards addressing this knowledge gap and is driven by the research question: "*what is the role of Enterprise Social Technology affordances in a business process improvement context?*"

The paper first introduces relevant literature (which informs the motivation for this research, and provides the theoretical foundations), next presents the research method followed by the findings, and concludes with a presentation of an empirically supported model of EST affordances in the context of business process improvement.

2 Literature Review and Theoretical Foundations

2.1 Social Technology for Process Improvements

Process management literature recognizes the importance of interactions between the various stakeholders involved in a process both within and outside of the organisation, and this remains a key challenge (Abbate and Coppolino 2010; Balzert et al. 2012; Martinho and Rito-Silva 2011; Niehaves and Henser 2011). There are numerous studies discussing the benefits of social technology to Business Process Management e.g. (Brambilla et al. 2011; Dengler et al. 2011; Erol et al. 2010). Early research was primarily focused on several distinct topics such as user collaboration (Schmidt and Nurcan 2009); model-reality divide (Schmidt and Nurcan 2009); trust (Koschmider et al. 2009); and bottom-up modelling (Neumann and Erol 2009; Schmidt and Nurcan 2009; Silva et al. 2010). Literature from the broader domain of Information Systems (e.g. Akkermans and van Helden 2002; Fiedler et al. 1995; Gasson 2006; Newman and Zhao 2008; Niehaves and Plattfaut 2011; Tarafdar and Gordon 2007) present how social technology can be used as an enabler for organizational transformations, especially addressing the opportunities that social technologies could bring to process improvement efforts (Mathiesen et al. 2013).

Social technology can support process improvement initiatives with its collaboration and communication benefits (Gottanka and Meyer 2012), and offer improved and adaptable business process design (Erol et al. 2010). A typical process improvement lifecycle consists of process identification, discovery, analysis, re-design, implementation and continuous evaluation and control (Dumas et al. 2013). Social technology can be of use for process identification - to collect and collate insights on areas of issues and opportunities; in the discovery and analysis phases - to obtain input about the current as-is processes from multiple stakeholders; in the redesign phase - to obtain innovative ideas from multiple stakeholders; in the implementation phase - to communicate process changes, and for continuous evaluation and control - to receive input on the process's ongoing performance and ideas for continuous improvement.

2.2   The Affordance Concept

Gibson (1977) was the first scholar to present the concept of an 'affordance', and positioned affordances as relating to *"perceptual cues of an environment or object that indicate possibilities for action"* (Lübbe 2011, p. 2). This initial definition was adapted by McLoughlin and Lee (2007) in a social technology context who posit affordances as a 'can do' statement that does not pertain to specific functionality or platform. In recent years there has been robust discussion on using the affordance theory to develop theories pertaining to technology related organizational change (Volkoff and Strong 2013). Given that most technology implementations results in process changes and many process improvements deploy technology/automation for process enhancements and efficiency, the body of literature on technology affordances is also arguably relevant for the context of process improvement studies like this. Researchers purport that taking an affordance perspective enables one to build better theories on the effects of introducing new systems (and processes) into organizations (following Volkoff and Strong 2013).

2.3   The Selected Theoretical Base

A thorough literature review was conducted in search of theories or frameworks that described the role of EST for process improvement. There have been very limited attempts to conceptualize social technology for BPI. Many attempts investigate specific social technologies in the wider context of business process management and are not reusable or independent from those technologies. Recently a reusable meta-model for executing processes in a collaborative way was proposed by Ariouat et al. (2017) but this is not specific to BPI and focused on assisting rule-based-reasoning (computation). This study selects an EST affordance perspective that has potential to assist strategic alignment of BPI activities with organisational goals.

This review resulted in the adoption of the (literature derived) a-priori model of Mathiesen et al. (2013) as our a priori theoretical base. This is a thorough synthesis of reported EST affordances across the Information Systems domain and already positioned within a business process improvement context, and the most relevant work on this topic to date. They present the seven affordances of; (i) Participation, (ii) Collective Effort (iii) Transparency, (iv) Independence, (v) Persistence, (vi) Emergence, and (vii) Connectivity, and describe these with evidence from prior literature, but do not provide precise definitions nor progress any further in the conceptualization of these. Revisiting the cited literature by Mathiesen et al. (2013) and complimenting this with new literature found; this study formed our own initial definitions (see in Table 1) for each affordance in order to derive a stronger a priori base for the planned empirical work (see Sects. 3 and 4).

Table 1. A priori EST affordances for process improvements (adopted from Mathiesen et al. 2013)

Affordances	Definition derived and used in this study
Participation	Participation increases the understanding and adoption of a process by the wider stakeholder community (Brambilla et al. 2012b)
Collective effort	"*Collaboration activities in a shared context*" (Abbate and Coppolino 2010, p. 5). This concept of "*collective creativity*" as put forward by Helms et al. (2012, p. 2), refers to the crowd-sourcing of solutions to specific problems or issues and capturing the collective intelligence (Lee and McLoughlin 2008) of the organisation (Erol et al. 2010)
Transparency	Brambilla et al. (2012a, p. 223) state the goal of transparency is to make the "*decision procedures internal to the process more visible to the affected stakeholders*"
Independence	The notion of egalitarian contributions so that participants can contribute without the coordination of other participants and regardless of physical location (Bradley 2009) or organisational boundaries (Lee and McLoughlin 2008)
Persistence	The capacity for social technology to retain, share and augment contextualised information is an affordance that BPI will benefit from as all historical process model changes are retained (Erol et al. 2010; Gottanka and Meyer 2012)
Emergence	Previously unidentified expertise, informal organisational structures or work processes (Bradley 2009)
Connectivity	This notion of connectivity may also "*supplement existing relationships, and help build a greater sense of community*" (Treem and Leonardi 2012, p. 31)

3   Method

A single process improvement case context, the 'Debtor-Finance customer on-boarding' initiative at the Bank of Queensland (BOQ), an Australian Financial Institution, was investigated to achieve the goals of the study. An Enterprise Social Technology; Microsoft Yammer[1], was used by geographically distributed business process professionals and key stakeholders to communicate and collaborate on this process improvement initiative. Overall 35 BOQ staff were involved, out of which 5 were selected for semi-structured interviews (namely; the Senior Manager - Business Excellence (P1), Business Excellence Analyst (P2), Client Manager (P3), Senior Client Manager (P4), and Senior Risk Manager (P5)). These 5 interview participants were chosen due to accessibility, availability and interest in the study. The team was

---

[1] See www.yammer.com for further details.

dispersed nationally and most unable to meet face to face. The semi-structured interviews were conducted in person, audio recorded (for transcription), supplemented by Researcher notes (to capture insights) and were on average 45 min in length. Only staff who used Yammer on a regular basis were included in this study.

The interview questions were partly based on the a-priori model adopted from Mathiesen et al. (2013) but also prompted participants to openly discuss their experiences and perspectives of using EST during the process improvement initiative. Process documentation and actual participant conversations maintained in Yammer (the Enterprise Social Technology) were used as other *"sources of evidence"* for triangulation purposes. Additional observations were recorded in field notes.

This study applied a hybrid approach to thematic analysis (mixing both deductive and inductive coding), similar to Fereday and Muir-Cochrane (2008). This approach allowed the validation of the a-priori model and refinement and extension through inductive reasoning. A guiding protocol [including a coding rule book following Saldaña (2012)] was derived, tested and used; and NVivo was applied throughout the analysis as a support tool to maintain rigor and transparency.

4 Findings

All 7 of the priori model constructs (See Table 1) were instantiated by the coding process through the identification of supportive themes. Four new constructs emerged inductively from the data. Initial themes were captured, first as 'free nodes' using the in-vivo[2] coding technique. These were then grouped to form coding families and then into higher level nodes forming the new constructs.

Coder notes in the form of annotations and memos were used at all times to assist with maintaining the trail-of-evidence. Inter-coder-comparison-queries were run and corroboration sessions [where approaches such as *"think out loud coding demonstrations"* (Saldaña 2012)] were undertaken to understand potential differences in interpretation and to sharpen and refine the constructs. Overlaps between the data constructs were analyzed and removed both through manual observations and through the support of a series of NVivo matrix intersection[3] 'AND' searches and several detailed corroboration sessions between the two coders. Removal or merger of constructs was achieved by following agreed protocols between the two coders. This resulted in two previously identified constructs, Participation and Independence (from the original a-priori model), being removed from the final list of constructs. This action was taken as the 'in-vivo' driven data codes were reallocated across other constructs, which demonstrated a better definitional alignment. A final inter-coder check of the coded data resulted in strong outcomes, with kappa scores between 0.75 and 0.99[4]. By the

---

[2] In-vivo coding: the coding technique of *"assigning a label to a section of data, such as an interview transcript, using a word or short phrase taken from that section of the data"* (King 2008, p. 3).

[3] A two dimensional Boolean search.

[4] It is generally considered (Fleiss et al. 1981; Seigel et al. 1992) that a Kappa score between 0.4 and 0.6 is accepted as 'fair', a score in the range of 0.6 and 0.8 is deemed 'good' and above 0.8 as excellent.

completion of the analysis, the 5 a-priori constructs were confirmed, 4 new affordances discovered and 3 contingency variables (variables that might have an influence on how the EST affordances behaved) discovered. These are presented with summary descriptions and their sub constructs in Table 2.

Table 2. The final model constructs and sub-constructs

Model constructs	Sub-constructs (total # of interviews, total # of citations)
Confirmed EST affordances	
Collective effort *Enables* **collaboration** *and* **knowledge exchange** *(group think)*	Knowledge sharing (2, 2)
	Request for input (2, 2)
	Breaking down (communication) silos (1, 1)
Transparency *Enables the ability to* **see more about** *the process*	Understand the current process (1, 1)
	Gives deeper insight of the stakeholder role in the process (1, 1)
	Understand the potential future state (1, 3)
Persistence *Enables the potential to* **retain and reuse** *the digital artefact*	Forms an evidence base (1, 1)
	Supports recollection (2, 3)
	Traceability of discussions (1, 1)
Emergence *Enables* **new** *ideas to surface*	Unique new ideas (3, 3)
	Feel more open with sharing ideas (1, 1)
	Volume of ideas (2, 3)
	Mature an idea (3, 3)
Connectivity *Enables better use of* **current** *relationships*	Build new relationships (3, 3)
	Better use of current relationships (2, 2)
Discovered EST affordances	
Agility *Enables the ability to* **contribute beyond traditional means** *(regardless of time zones, cycle time, physical locations etc.)*	Less dependency on face-2-face workshops (3, 4)
	Reduced cycle time by not having to wait (3, 4)
	Reduces impact on business due to virtual environment (3, 3)
	Removes geographic boundaries (2, 2)
Empowerment *Provides a* **voice** *to the people who would not normally contribute*	People having the chance to have a say (4, 7)
	Gives a new channel-mode to have a say (3, 4)
	Sense of belonging (4, 4)
	Enabling people who would normally not contribute (4, 9)

(*continued*)

Table 2. (*continued*)

Model constructs	Sub-constructs (total # of interviews, total # of citations)
Ownership	Ownership of ideas (2, 2)
*Provides the ability to **own contributions**, Ideas and Change*	Recognition (2, 2)
Visibility	See WHAT the varying
*Enables staff to see the **contribution of others***	contributions-perspectives (3, 9)
Discovered Contingency Variables (and # of citations)	
Stakeholder Authority (3)	Their own status
The perceived authority of process stakeholders	Reluctant to challenge
	Who they were and their status
Trust (3)	Difficult relations between groups
The status of relationships between process stakeholders	Safe environment
	Not a lot trust between teams
Voluntary Contribution (2)	Choice to participate
The ability to contribute without coercion	Contribute if they want to

5   The Revised Model and Discussion

This study builds on prior research (Mathiesen et al. 2013) and through a carefully designed and implemented case study approach; (1) re-specifies already discovered EST affordances to improve their conceptualization; (2) identifies new EST affordances and clearly defines these; and (3) identifies and conceptualizes contingency variables that influence the way EST affordances manifest in practice. The resulting conceptual model with nine (9) identified EST affordances, four (4) of which are new, and three (3) contingency variables, (observed to have a moderating or mediating impact) is presented in Fig. 1.

Future exploration of the potential relationships between the contingency variables and the affordances (and also between the different affordances) is planned as future research. The preliminary observations points to interesting interaction effects. For example, the contingency variable 'Stakeholder Authority', appears to have an impact upon the affordances of 'Agility', 'Collective Effort', and 'Visibility'. Also the data indicated that when stakeholders in a position of authority made a visible contribution, it appeared to hinder the collective effort of other participants; 'blocking' others from freely commenting and editing content. We acknowledge that the different EST affordances can have diverse implications within different organizational, process and process improvement contexts. For example, large, geographically dispersed organisations are likely to benefit most from the incorporation of ESTs. And ESTs are likely to be more useful where the process participants have some experience with ESTs, and there is more of a 'technology driven' and 'engagement friendly' culture (as observed within the case context of this study). An investigation of external environmental and contingency factors that can further impact the application of EST within process improvements initiatives is planned as future research.

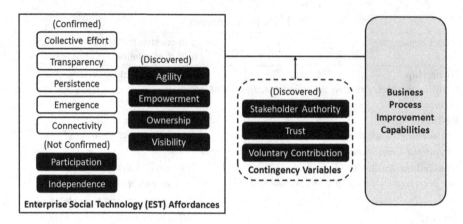

Fig. 1. Perceived EST affordances in the context of business process improvement

As discussed by Barki (2008, p. 9), researchers can make significant contributions to research and practice, by "*introducing new constructs*" and "*by better conceptualising existing constructs*". They position construct conceptualization as a very important contribution in theory development. Correctly conceptualized constructs are a prerequisite for 'good' theory building (Wacker 2004). Wacker (2004) explains how; conceptual definitions are needed for all theory-building empirical research and are necessary for content, criterion, convergent and discriminant validity and vehemently argue for construct definitions to take place 'before' any statistical tests are performed, as any statistical tests are meaningless until the concepts are formally defined.

This study did not only look at the construct definitions of EST affordances, but also looked at construct definitions of related contingency variables that could have a moderating/mediating effect on the EST affordances. (Frazier et al. 2004) strongly encourages the identification of such variables very early on, as the theories built in their absence can lead to weak results diminishing the impact of the specific research and impeding the progression of the research field as a whole.

Given stakeholders' engagement is key to the ultimate success of any BPM initiative (Hailemariam and vom Brocke 2011), a deeper understanding on how ESTs can assist to overcome this, is valuable to practitioners leading process changes. Study outcomes depicted how ESTs could involve diverse stakeholders (especially if geographically dispersed) through 'conversations', which supports the perceived degree of inclusion and participation; which are known key challenges with process improvement efforts. Conversations within ESTs can complement traditional workshops, as extended discussions post-workshops or as preparatory (or 'warm-up') work for upcoming workshop activities. They create a more social and casual environment which can enhance stakeholder responsiveness and openness to emergent ideas and contributions that surfaces through the EST communications. These conversations are also useful to discover the 'hidden networks' of individuals. Further, knowledge of the three contingency variables (Stakeholder Authority, Trust and Voluntary Contribution), can inform overarching management aspects. For instance, removing the impact (real or perceived) of any 'Stakeholder Authority' amongst workshop participants will allow improved outcomes and remove possible restrictions on participation; establishing 'Trust' amongst process improvement participants will also foster better outcomes; and it is important to establish that the stakeholders had a choice ('Voluntary Contribution').

6 Conclusion

Organizations are increasingly adopting EST as an approach to crowd-source innovative solutions to organisational improvement opportunities. Additionally, process improvement practice is also leveraging these technologies to *"extend the reach and impact of process improvement efforts"* (Gottanka and Meyer 2012, p. 94). Recognizing that Academia (and industry) lack an understanding of the theory, models and frameworks (Niehaves and Plattfaut 2011; Walsh and Deery 2006), that explains the applicability of Enterprise Social Technology, especially in the context of process improvements, this study embarked to contribute towards addressing this gap. Applying the literature-based model of Mathiesen et al. (2013), this study investigates how EST affordances can contribute to process improvement initiatives. Through empirical data collected from well designed and executed interviews within a single case setting this study presents a further revised, empirically supported EST affordances model for process improvement contexts. Amongst the academic contributions of this study are the establishment of literature-based and empirically derived EST affordance constructs (both identification and operationalisation) that contribute towards building BPM capabilities. From an applied (practical) perspective, this study provides substantial contributions for both BPM practitioners (and other process stakeholders) and the software vendors who design and create ESTs.

The findings presented here is a preliminary step towards further empirical work in this direction. It was based on five interviews across one organization, and though the interviews were in-depth and well planned, and other supporting documentation was reviewed, the analysis was primarily based on interviews of five selected stakeholders. Other potential limitations of the study such as researcher bias in data collection and analysis have been mitigated with the coding procedures applied (i.e. coding guidelines and two coders working towards strong inter-coder reliability). Though we acknowledge these may impact the completeness and generalizability of the findings presented in this paper, this is a first empirical step towards identifying the affordances of EST for process improvements. Future research will be conducted to investigate the potential relationships between these identified EST affordances and contingency variables and expand the operationalization that this work provides a basis for.

References

Abbate, T., Coppolino, R.: Open innovation and creativity: conceptual framework and research propositions. In: Paper Presented at the VII Conference of the Italian Chapter of AIS (itAIS 2010) (2010)

Akkermans, H., van Helden, K.: Vicious and virtuous cycles in ERP implementation: a case study of interrelations between critical success factors. Eur. J. Inf. Syst. **11**(1), 35–46 (2002)

Ariouat, H., Hanachi, C., Andonoff, E., Benaben, F.: A conceptual framework for social business process management. Procedia Comput. Sci. **112**, 703–712 (2017). https://doi.org/10.1016/j.procs.2017.08.151

Balzert, S., Fettke, P., Loos, P.: A framework for reflective business process management. In: 2012 45th Hawaii International Conference on Paper presented at the System Science (HICSS), 4–7 Jan 2012 (2012)

Barki, H.: Thar's gold in them thar constructs. ACM SIGMIS Database **39**(4), 90 (2008)

Becker, J., Kugeler, M., Rosemann, M.: Business Process Lifecycle Management. White paper (2001)

Bradley, A.: The Six Core Principles of Social-Media-Based Collaboration (2009). Gartner.com

Brambilla, M., Fraternali, P., Vaca, C.: A notation for supporting social business process modeling business process model and notation. In: Dijkman, R., Hofstetter, J., Koehler, J. (eds.) BPMN 2011. LNBIP, vol. 95, pp. 88–102. Springer, Heidelberg (2011). https://doi.org/10.1007/978-3-642-25160-3_7

Brambilla, M., Fraternali, P., Ruiz, C.K.V.: Combining social web and BPM for improving enterprise performances: the BPM4People approach to social BPM. In: Proceedings of the 21st international conference companion on World Wide Web, Lyon, France (2012a)

Brambilla, M., Fraternali, P., Vaca, C.: BPMN and design patterns for engineering social BPM solutions. In: Daniel, F., Barkaoui, K., Dustdar, S. (eds.) Business Process Management Workshops, vol. 99, pp. 219–230. Springer, Heidelberg (2012b). https://doi.org/10.1007/978-3-642-28108-2_22

Dengler, F., Koschmider, A., Oberweis, A., Zhang, H.: Social software for coordination of collaborative process activities business process management workshops. In: zur Muehlen, M., Su, J. (eds.) BPM 2010. LNBIP, vol. 66, pp. 396–407. Springer, Berlin (2011). https://doi.org/10.1007/978-3-642-20511-8_37

Dollmann, T., Fettke, P., Loos, P.: Web 2.0 enhanced automation of collaborative business process model management in cooperation environments. In: ACIS 2009 Proceedings, vol. 41 (2009)

Dumas, M., La Rosa, M., Mendling, J., Reijers, H.A.: Fundamentals of Business Process Management. Springer, Heidelberg (2013). https://doi.org/10.1007/978-3-642-33143-5

Erol, S., et al.: Combining BPM and social software: contradiction or chance? J. Softw. Maint. Evol.: Res. Pract. **22**(6–7), 449–476 (2010). https://doi.org/10.1002/smr.460

Fereday, J., Muir-Cochrane, E.: Demonstrating rigor using thematic analysis: a hybrid approach of inductive and deductive coding and theme development. Int. J. Qual. Methods **5**(1), 80–92 (2008)

Fiedler, K.D., Grover, V., Teng, J.T.C.: An empirical study of information technology enabled business process redesign and corporate competitive strategy. Eur. J. Inf. Syst. **4**, 17–30 (1995)

Fleiss, J.L., Levin, B., Paik, M.C.: The measurement of interrater agreement. Stat. Methods Rates Proportions **2**, 212–236 (1981)

Frazier, P.A., Tix, A.P., Barron, K.E.: Testing moderator and mediator effects in counseling psychology research. J. Couns. Psychol. **51**(1), 115–134 (2004). https://doi.org/10.1037/0022-0167.51.1.115

Gasson, S.: A genealogical study of boundary-spanning IS design. Eur. J. Inf. Syst. **15**(1), 26–41 (2006)

Gibson, J.J.: The theory of affordances. In: Perceiving, Acting and Knowing: Toward an Ecological Psychology, pp. 67–82 (1977)

Gottanka, R., Meyer, N.: ModelAsYouGo: (Re-) design of S-BPM process models during execution time. In: Stary, C. (ed.) S-BPM ONE 2012. LNBIP, vol. 104, pp. 91–105. Springer, Berlin (2012). https://doi.org/10.1007/978-3-642-29133-3_7

Hailemariam, G., vom Brocke, J.: What is sustainability in business process management? A theoretical framework and its application in the public sector of Ethiopia. In: zur Muehlen, M., Su, J. (eds.) BPM 2010. LNBIP, vol. 66, pp. 489–500. Springer, Berlin (2011). https:// doi.org/10.1007/978-3-642-20511-8_45

Helms, R.W., Booij, E., Spruit, M.R.: Reaching out: involving users in innovation tasks through social media. In: ECIS 2012 Proceedings, Paper 193 (2012)

King, A.: In vivo coding. In: The SAGE Encyclopedia of Qualitative Research Methods. SAGE, Thousand Oaks (2008)

Koschmider, A., Song, M., Reijers, H.A.: Social software for modeling business processes. In: Ardagna, D., Mecella, M., Yang, J. (eds.) BPM 2008. LNBIP, vol. 17, pp. 666–677. Springer, Berlin (2009). https://doi.org/10.1007/978-3-642-00328-8_67

Lee, M.J., McLoughlin, C.: Harnessing the affordances of web 2.0 and social software tools: can we finally make "student-centered" learning a reality? In: World Conference on Educational Multimedia, Hypermedia and Telecommunications (2008)

Lübbe, A.: Principles for business modelling with novice users. In: Proceedings of the Participatory Innovation Conference, Sønderborg, Denmark, pp. 318–322 (2011)

Magdaleno, A.M., Cappelli, C., Baiao, F.A., Santoro, F.M., Araujo, R.: Towards collaboration maturity in business processes: an exploratory study in oil production process. Inf. Syst. Manage. **25**(4), 302–318 (2008)

Martinho, D., Rito-Silva, A.: ECHO an evolutive vocabulary for collaborative BPM discussions. In: zur Muehlen, M., Su, J. (eds.) BPM 2010. LNBIP, vol. 66, pp. 408–419. Springer, Berlin (2011). https://doi.org/10.1007/978-3-642-20511-8_38

Mathiesen, P., Bandara, W., Watson, J.: The affordances of social technology: a BPM perspective. In: Proceedings of the 34th International Conference on Information Systems, Milan (2013)

Mathiesen, P., Watson, J., Bandara, W., Rosemann, M.: Applying social technology to business process lifecycle management. In: Daniel, F., Barkaoui, K., Dustdar, S. (eds.) BPM 2011. LNBIP, vol. 99, pp. 231–241. Springer, Berlin (2011). https://doi.org/10.1007/978-3-642-28108-2_23

McLoughlin, C., Lee, M.J.W.: Social software and participatory learning: pedagogical choices with technology affordances in the web 2.0 era. In: Proceedings of ascilite Singapore 2007 Paper presented at the ICT: Providing Choices for Learners and Learning (2007)

Neumann, G., Erol, S.: From a social Wiki to a social workflow system. In: Ardagna, D., Mecella, M., Yang, J. (eds.) BPM 2008. LNBIP, vol. 17, pp. 698–708. Springer, Berlin (2009). https://doi.org/10.1007/978-3-642-00328-8_70

Newman, M., Zhao, Y.: The process of enterprise resource planning implementation and business process re-engineering: tales from two Chinese small and medium-sized enterprises. Inf. Syst. J. **18**(4), 405–426 (2008)

Niehaves, B., Henser, J.: Boundary spanning practices in BPM: a dynamic capability perspective. In: AMCIS 2011 Proceedings - All Submissions, Paper 230 (2011)

Niehaves, B., Plattfaut, R.: Collaborative business process management: status quo and quo vadis. Bus. Process Manage. J. **17**(3), 384–402 (2011)

Riemer, K., Alexander, R.: Tweet inside: microblogging in a corporate context. In: BLED 2010 Proceedings Paper 41 (2010)

Rossi, D., Vitali, F.: Workflow enactment in a social software environment. In: Ardagna, D., Mecella, M., Yang, J. (eds.) BPM 2008. LNBIP, vol. 17, pp. 716–722. Springer, Berlin (2009). https://doi.org/10.1007/978-3-642-00328-8_72

Saldaña, J.: The Coding Manual for Qualitative Researchers. Sage, Beverly Hills (2012)

Seigel, D.G., Podgo, M.J., Remaley, N.A.: Acceptable values of kappa for comparison of two groups. Am. J. Epidemiol. **135**(5), 571–578 (1992)

Schmidt, R., Nurcan, S.: BPM and social software. In: Ardagna, D., Mecella, M., Yang, J. (eds.) BPM 2008. LNBIP, vol. 17, pp. 649–658. Springer, Berlin (2009). https://doi.org/10.1007/978-3-642-00328-8_65

Silva, A.R., Meziani, R., Magalhães, R., Martinho, D., Aguiar, A., Flores, N.: AGILIPO: embedding social software features into business process tools. In: Rinderle-Ma, S., Sadiq, S., Leymann, F. (eds.) BPM 2009. LNBIP, vol. 43, pp. 219–230. Springer, Berlin (2010). https://doi.org/10.1007/978-3-642-12186-9_21

Tarafdar, M., Gordon, S.: Understanding the influence of information systems competencies on process innovation: a resource-based view. J. Strateg. Inf. Syst. 16(4), 353–392 (2007)

Treem, J., Leonardi, P.: Social media use in organizations: exploring the affordances of visibility, editability, persistence, and association. Commun. Yearb. 36, 143–189 (2012)

Volkoff, O., Strong, D.M.: Critical realism and affordances: theorizing IT-associated organizational change processes. MIS Q. 37(3), 819–834 (2013)

Wacker, J.G.: A theory of formal conceptual definitions: developing theory-building measurement instruments. J. Oper. Manage. 22(6), 629–650 (2004)

Walsh, J., Deery, S.: Refashioning organizational boundaries: outsourcing customer service work. J. Manage. Stud. 43(3), 557–582 (2006)

Zammuto, R.F., Griffith, T.L., Majchrzak, A., Dougherty, D.J., Faraj, S.: Information technology and the changing fabric of organization. Organ. Sci. 18(5), 749–762 (2007)

Social Business Process Management (SBPM)

Critical Success Factors (CSF)

Shamsul Duha and Mohammad E. Rangiha(⊠)

Department of Computer Science, School of Mathematics, Computer Science
and Engineering, City, University of London, London, UK
{Shamsul.Duha,Mohammad.Rangiha.2}@city.ac.uk

Abstract. Social BPM allows for businesses to adapt and be flexible to ever changing demands. This flexibility is created by the participation and collaboration between users. These interactions are achieved through the successful implementation of Social BPM. This paper will propose critical success factors (CSF) which lead to a successful Social BPM implementation such that these benefits are realised. This is a progress paper which is part of a broader method which will validate against the literature, expert opinions and case studies in order to produce a definitive set of CSFs for Social BPM.

Keywords: Social BPM · Critical Success Factors · BPM · Social software

1 Introduction

Businesses produce goods and services using business process management (BPM). BPM is a management discipline which improves organisational performance through the structuring of business processes [22]. These business processes are becoming ever more complex and needing to adapt to highly dynamic environments. However, BPM is rigid as it involves "processes right from the outset of their initiation until the end" [14]. This creates a set of pre-defined steps which are aligned to structured business processes. This rigidity is at odds with the frequent customisation of goods and services and does not support the case for when "exceptions become the rule" [7].

With this in mind, Social Business Process Management was formed. Social BPM is defined by Brambilla as the fusing of "business process management practices with social networking applications, with the aim of enhancing the enterprise performance by means of a controlled participation of external stakeholders to process design and enactment" [1]. This is in contrast to traditional BPM which provides a "platform for the management, measurement and improvement of business processes" [12]. The latter faces limitations such as "lack of information fusion, model reality divide, information pass-on threshold and lost innovation, strict access-controls, lack of context" [12].

Social BPM on the other hand, has been designed to address these limitations such as the 'reality-model divide' to ensure that those designing the process and those executing the process are in synchronization. This is particularly important for scenarios where flexibility is required as "substantial contribution to these processes comes

© Springer Nature Switzerland AG 2019
F. Daniel et al. (Eds.): BPM 2018 Workshops, LNBIP 342, pp. 85–95, 2019.
https://doi.org/10.1007/978-3-030-11641-5_7

from human knowledge, while knowledge related to the processes is perishable and quickly outdated" [22]. Another benefit of using Social BPM is that businesses do not lose innovation as executors of the processes can engage in a feedback loop to achieve continuous improvement. This contrasts with BPM whereby a process executor may never have communication with the process designer therefore any improvements are limited to tacit knowledge by the process executor and not shared for the collective benefit.

By adopting Social BPM, organisations can significantly improve their processes to be more collaborative and increase participation of stakeholders, however there is a lack of clarity and consensus as to what exactly is required for the successful implementation of Social BPM.

To fill this gap in the literature, we will be proposing CSFs for Social BPM to maximise the chances of successful implementation. This will allow for a greater breath of implementation experience within different sectors and highlight challenges in the real world which may not feature in the literature at present.

The proposed CSFs will be developed using three methods, which will be as follows;

1. *Literature method*
2. *Academic method*
3. *Practitioner method*

Firstly, the literature method will be the focus of this working paper, secondly the academic method will consist of surveying experts and finally the practitioner method, will examine case studies of Social BPM implementation. To ensure that the CSFs are comprehensive, three methods have been selected and will allow for the exploration of the intersection, in order to produce a multi-faceted set of CSFs for Social BPM.

This paper will achieve its aim by conducting a literature review in Sect. 2, produce a preliminary set of CSFs for Social BPM in Sect. 3, ensure the validity of them in Sect. 4 and finally conclude with the CSFs that have been discovered for Social BPM and further work to be conducted.

2   Literature Review

To begin the literature review, we will look at BPM CSFs. BPM CSFs have been selected above other types of CSFs as they tie closely with Social BPM. Social BPM allows for "software that supports the interaction of human beings and production of artifacts by combining the input from independent contributors without predetermining the way to do this" [17]. Both BPM and Social BPM focus heavily on business improvement however BPM focuses on experts designing these improvement processes [18] whereas Social BPM embeds a collaborative and egalitarian approach to business improvement.

This review of BPM CSFs will also provide us with a solid understanding of how BPM has been successfully implemented [2]. This is important as it will provide us with a list of CSFs which have been demonstrated to work within businesses, [10] these can then be used as a benchmark for CSFs for Social BPM.

Research into Business Process Management began in the late 1980s and was triggered by the seminal work published by Davenport & Short and Hammer & Champy [11]. Much of the work that has been carried out within the literature can be synthesized into six categories. These are *people, culture, information technology, methods, governance and strategic alignment.* These six categories are based on the principles of BPM. Each CSF listed below is needed to ensure the success of BPM [16]:

Governance: BPM governance ensures that roles and responsibilities are clearly defined based on the BPM level being implemented whether that is from portfolio all the way to operational level. In addition, the process of decision making, and reward process is focused upon.

Methods: BPM methods are the tools and techniques that support the BPM implementation across the lifecycle such as process modelling or process improvement techniques.

Information Technology: The system which allows for BPM to work i.e. process aware information systems (PAIS) These are integral to BPM as the software is needs to be process aware to understand the processes that require execution.

People: People are individuals and groups of users who improve and apply their process and process management expertise, so they can better business performance. This is the knowledge base of the business and as such is the human capital.

Culture: BPM culture means that there is a shared belief in a process driven organisation and for continual improvement. This is by far, the hardest CSF to change however to not have the right culture prior to implementation could lead to failure. Therefore, preparing the organisation for BPM and making sure that the environment is conducive has a clear impact on the successful BPM implementation.

Strategic Alignment: The need for BPM to be linked to strategy within the organisation. The synchronization of strategic priorities to the action of improving of business processes to improve business performance.

To understand the CSFs for BPM further, Fig. 1 shows the high-level categories and the link with the capability areas underneath each category. Take for example the category of People, this BPM CSF includes five sub categories which include sub categories such as the expertise of the stakeholders against the specific requirements of a process. This is incredibility important as the lack of expertise or a subject matter expert could mean the failure of implementation. This category also discusses process collaboration and communication, for example how groups work together and how process knowledge is "discovered, explored and disseminated" [16]. This shares commonality with Social BPM and Social BPM is designed to very much facilitate for this.

Within this section, we reviewed the history of BPM and identified six CSFs; strategic alignment, governance, methods, information technology, people and culture from the literature. These six show the complex nature of a successful implementation of BPM. We will use the CSFs found in this section as the baseline for our proposed Social CSFs in the next section.

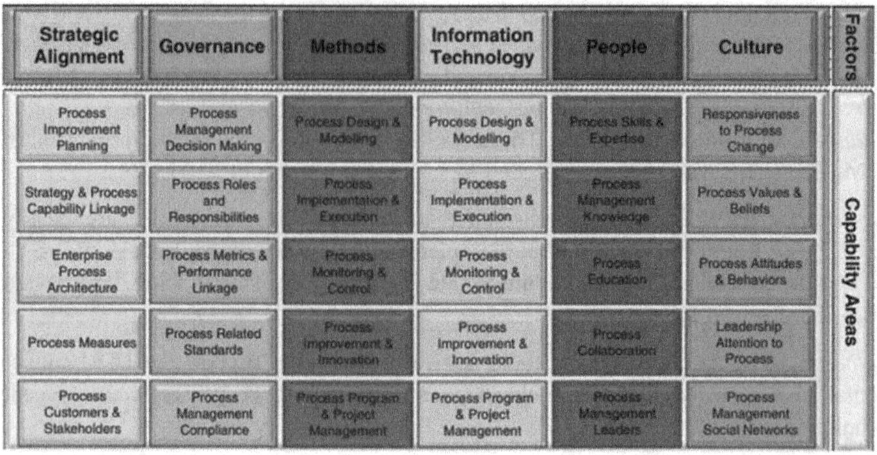

Strategic Alignment	Governance	Methods	Information Technology	People	Culture	Factors
Process Improvement Planning	Process Management Decision Making	Process Design & Modelling	Process Design & Modelling	Process Skills & Expertise	Responsiveness to Process Change	
Strategy & Process Capability Linkage	Process Roles and Responsibilities	Process Implementation & Execution	Process Implementation & Execution	Process Management Knowledge	Process Values & Beliefs	
Enterprise Process Architecture	Process Metrics & Performance Linkage	Process Monitoring & Control	Process Monitoring & Control	Process Education	Process Attitudes & Behaviors	
Process Measures	Process Related Standards	Process Improvement & Innovation	Process Improvement & Innovation	Process Collaboration	Leadership Attention to Process	
Process Customers & Stakeholders	Process Management Compliance	Process Program & Project Management	Process Program & Project Management	Process Management Leaders	Process Management Social Networks	

Capability Areas

Fig. 1. The six core elements of BPM [16]

3 Proposed CSFs for Social BPM

Having established six CSFs within BPM in Sect. 2, we will now determine if there is homogeneity between CSFs of BPM and CSFs of Social BPM within this section.

To do this we will first conduct a professional search, the following databases will be used: Emerald, JSTOR, ProQuest, Wiley Online Library, ScienceDirect and Web of Science. The methodology of the search will use an advanced search, utilizing operators to help improve the accuracy of the search results. The construction of the search query will be as follows:

'Critical Success Factors' AND 'Social BPM'
Upon conducting the search, 1 relevant paper was found from the search query 'Social BPM' AND 'Critical Success Factors'. This meant that a broadening of search queries was conducted. Figure 2 displays the process flow and the two ways in which the search was broadened.

The first way in which the search was broadened was by searching for 'Social BPM' and searching across the six databases whether there was literature which identified factors which are required for Social BPM without explicitly identifying them as CSFs. This proved useful as the literature identifies specific areas of concern when implementing Social BPM however these were often in isolation and very few of the journals looked at Social BPM in as broad prospective as the framework set out by Brocke and Rosemann [16]. The second search query was replacing 'Social BPM' with 'BPM' to identify the body of work that has already been researched. This proved vast and helped to compare against the research conducted in the first two queries. It became quite apparent that there is much overlap between the CSFs for Social BPM and BPM.

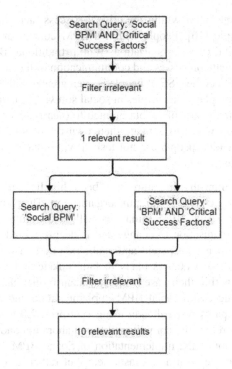

Fig. 2. Search strategy

The table below shows ten papers which feature Social CSFs such as the importance of a collaborative environment, the need for a reward mechanism or the requirement of the technology which underpins Social BPM (Fig. 3).

Proposed Social BPM CSFs						
Literature	People	Information technology	Methods	Governance	Culture	Strategic Alignment
P1 [22]	x	x	x		x	x
P2 [6]	x	x			x	
P3 [15]	x				x	
P4 [7]	x	x			x	x
P5 [8]	x	x			x	
P6 [3]		x	x			
P7 [13]	x			x	x	x
P8 [9]	x		x	x		
P9 [20]	x	x	x	x	x	x
P10 [21]	x	x	x	x		x

Fig. 3. Social BPM CSFs mentioned by paper i.e. People as a CSF mentioned in ten papers

People – The dominant factor within the critical success factors outlined by Brocke and Rosemann is people [16]. People within a BPM context are important as they reflect the human capital of an organisation. Within traditional BPM, subject matter experts are trusted to create processes and communication of the design is controlled by a limited few. To this effect the CSF of people is even greater within Social BPM given that "trust and reputation play crucial roles in social software. Changes are not initiated or authorized by hierarchic structures, but granted to (nearly) everybody, based on the assumption that nobody wants to damage their own reputation" [7]. This requires a different approach in which people are not tasked with creating processes but rather motivated to contribute to them.

Culture – Culture within the literature has been highlighted frequently and the importance of embracing changes within the organisation. This is because without it, it is likely that BPM or Social BPM initiatives will fail. Culture is one of the most difficult facets of an organisation to change and is deeply rooted. Therefore, the idea that implementing an open, transparent and egalitarian system within an organisation that is not aligned to these values is likely to fail. Vukšić and Vugec appreciate the importance of culture within their case study and identify that clan organizational is a "very good base for successful social BPM implementation and usage" [22]. Within their case study participants were already using enterprise 2.0 tools to create process content and context therefore the commitment to collaboration and knowledge sharing was already present prior to the implementation of Social BPM. This is important as Social BPM should not be seen as a drastic leap but rather an extension of what is already in place. Culture as a CSF is relevant to Social BPM just as it has been to BPM however the criticality of it in context to the others is difficult to evaluate given that there are very few successful Social BPM case studies within the literature.

Information Technology – The transformation of BPMS to support Social BPM is critical in the implementation of Social BPM. The need for systems to be able to facilitate for customisation "when unknown solutions to problems must be found or when the precise ordering of activities cannot be established beforehand" [7] is needed. One particular implementation that is suggested within the literature is that of wiki-enabled workflows [7]. Rather than having a highly modelled workflow which is unfit for rapid changes, it would be advantageous to expose to a community a wiki-based framework which would be adaptive and "workflow changes will be reached and exceptions can be detected and repaired in a collaborative manner" [7]. This need for BPMS to go further demonstrates that the system still underpins the ability for Social BPM to succeed.

Methods – One of the more comprehensive works of methods which could be used would be Gokaldas and Rangiha whereby they employ a three-level framework to improve the engagement of users of Social BPM. [8] The first level is organisational whereby the onus is on the managers to drive the engagement, the second level is that of social software whereby attention to usability is particularly important and finally the tasks should provide a value add. This framework supports the nature of web 2.0 whereby users are empowered to make contributions and methods need to change to facilitate for this. Another suggestion is to use honour points for rewards. In most processes, "users carry out their activities because they are instructed to do so by their superiors. In most social software, on the other hand, participation is voluntary" [8] therefore a form of gamification could be used as a method to support the usage of Social BPM. This CSF is valid for Social BPM however the nature of the method has differed from its BPM roots.

Governance – The way in which governance is conducted has changed with the advent of Social BPM. With BPM the structure of decision-making was much more controlled as it was understood who would make the decision however this might have come at the cost of the speed of the decision-making process and the ability to respond. With Social BPM, decisions can be made quickly however this is put down to the wisdom of the crowd. This poses a problem as the wisdom of the crowd might not be sufficient for the outcome required or the contextual information provided within a social software may only provide a one-dimensional outlook. Erol suggests "building difficult checking processes cannot be the answer as effects of speed, feedback, authenticity and direct-ness are ignored and hence one motivation of active usage is destroyed. New kinds of risk management and governance rules are needed with different levels of inference and strictness" [7]. The issue of governance as a CSF is debatable as it could be argued that Social BPM is self-governing and the participants of the platform ultimately decide.

Strategic Alignment – To bring about competitive advantage it is important to have BPM/Social BPM aligned to the strategic goals of an organisation. This CSF is applicable to Social BPM and ensures that process improvement initiatives are going to meet strategically prioritised goals. Strategic alignment for Social BPM is difficult for two reasons. Firstly, Social BPM for users who are encountering it for the first time, will have a steep learning curve. This learning curve needs to be accepted by the organisation as a time when productive will drop however if strategies are thought of in the context of business quarters and take a short-term horizon then Social BPM will fail before it has had a chance to make an impact. Secondly the strategic alignment of Social BPM is difficult to evaluate as the benefits of collaboration, transparency and distributed decision making are difficult to put into ROI terms. Erol identifies that it is difficult even to demonstrate it "adds value and is attractive to the members" [7] of it. Despite the drawbacks, the need for alignment to strategy for Social BPM is needed to ensure that social software is used to support the business and not as an end to itself.

In summary, this section has used the CSFs identified for BPM in Sect. 2 to evaluate whether they have a place within the preliminary framework for CSFs for Social BPM. It has been argued that the BPM framework is still broad enough that it covers the scope of Social BPM. This is not to say that with further methods such as feedback from experts or case studies, that new CSFs will not be found or debated. The very human elements of people and culture have come up frequently within the papers analysed as a primary concern when adopting Social BPM. Furthermore, the area which is of most discussion is around the CSF of governance and whether it is a valid CSF for Social BPM. It has been argued in this paper that it is still relevant however when employing other methods, this may become an area of further discussion. In the next section we will look at the validity of the method used and the three methods that will be used.

4 Validation

Social BPM is a recent development in the field of Business Process Management and is fragmented within the literature. We aim to address the issue of validity by triangulating the results from the three methods highlighted earlier. This is useful as it shows us the convergence of results as well as the contradictions when searching for CSFs for Social BPM [4, 19].

To ensure that all CSFs for Social BPM are captured, we will tackle the question through the methods below:

1. *Literature method*

   • Propose CSFs for Social BPM from the literature review

2. *Academic method*

   • Survey academic experts through qualitative and quantitative questions to find out their views on the CSFs for Social BPM

3. *Practitioner method*

   • Use case studies and industry reports to identify CSFs for Social BPM from practitioners

Once we have the results from each method, we will collate them and identify commonalities as well as see whether there are new CSFs in one method which are missing in another. We will then start a discussion as to why they may be missing and finally rank them by frequency across the three methods.

Within this paper, we have conducted a literature review in Sect. 2 to inform us of CSFs within BPM, we then used these CSFs in Sect. 3 whereby we introduced our proposed CSFs for Social BPM and conducted a professional search. The search for relevant literature was then evaluated against the proposed CSFs for Social BPM. We also identified from the professional search the most common CSFs across ten papers in order to understand the homogeneity of the CSFs. This allows us to see the most mentioned to the least mentioned CSF, in order to understand what the literature reflects as the most important CSF.

This section has firstly explained the purpose of using three methods and how they will be triangulated to maintain the validity of the research. Secondly it has provided the steps taken to produce valid results.

5 Conclusion

Within this paper, we have identified what Social BPM is, why it is desirable, conducted a literature review and proposed CSFs from BPM. In Sect. 3, we proposed CSFs for Social BPM and ten papers were analysed to see which CSFs were made mention of within the literature. We then used Sect. 4 to explain our validation process. This would see the results of this working paper triangulated with other methods of expert opinions from academics and case studies to establish CSFs for Social BPM.

Social BPM has many features which make it desirable for a changing world. The ability to implement it successfully is still an area of research which is in its infancy. In the absence of CSFs for Social BPM, we have taken a wider view of and used BPM as a benchmark to start to understand whether these CSFs hold true for Social BPM as they do for BPM. We have identified that some CSFs such as process collaboration, leadership's attention to process and process management social networks do hold value and can be defined as CSFs for Social BPM.

The broad category of People is the most frequently cited within the literature. The change in behaviour of people to becoming more open, transparent and collaborative is difficult to achieve overnight however the research suggests that this a determining factor for implementation of BPM and Social BPM alike. Therefore, with a change that will see not only a group of experts build business processes but rather anyone in the organisation, this would be even more of a CSF for Social BPM. The ability for users to participate doesn't mean they will and the nature of social means that a network effect is desirable. A network effect "occur when the probability that an actor will adopt a practice is an increasing function of the number or proportion of persons in the actor's social network who already have adopted that" [5]. It is therefore important that each user contributes in order to add value to the entire platform.

Although similarities have been found with the CSF of people, on the other hand, not all CSFs are as aligned to Social BPM as they are to BPM. Social BPM elicits opinions from all participants which contribute to decision making therefore it is meant to be self-governing. However, having process roles clearly defined goes against the egalitarian principle of Social BPM. In addition, it is clear that process management decision making is a critical challenge for BPM, which Social BPM aims to resolve as participants have ultimate control over what decisions are made. This could be at odds with the organisation. Therefore, some capability areas are ill fitting and some are at odds with Social BPM completely.

However, this is one method and we shall be getting the opinions of experts as well as case studies to find out which CSFs are applicable to Social BPM and possibly new ones that are not featured in the literature.

In conclusion, some CSFs that have been identified within BPM are highly suitable as Social BPM CSFs however there are many that are not. Therefore, further methods need to be employed to produce a definitive set of CSFs for Social BPM.

The CSFs identified within this paper are but one method that is drawn from the existing literature. This paper is designed to be the starting point of a three method approach to establishing what are the CSFs for Social BPM implementation. To that end, there needs to be further primary research conducted in the form of asking experts from academia their opinions through surveys and evaluating industry reports in order to learn about additional CSFs which have not been identified by the literature as well as validate those which are found within the literature. This will help practitioners of Social BPM build far more collaborative business processes that take into account the collection intelligence of the organisation.

References

1. Brambilla, M., Fraternali, P., Vaca Ruiz, C.K.: Combining social web and BPM for improving enterprise performances. In: Proceedings of the 21st International Conference Companion on World Wide Web - WWW 2012, Companion New York, USA, p. 223. ACM Press (2012). https://doi.org/10.1145/2187980.2188014
2. Buh, B., Kovačič, A., Štemberger, I.M.: Critical success factors for different stages of business process management adoption – a case study. Econ. Res. Ekon. Istraž. **28**(1), 243–258 (2015)
3. Cerenkovs, R., Kirikova, M.: Supporting introduction of social interaction in business processes. In: Johansson, B., Andersson, B., Holmberg, N. (eds.) BIR 2014. LNBIP, vol. 194, pp. 187–201. Springer, Cham (2014). https://doi.org/10.1007/978-3-319-11370-8_14
4. Creswell, J.W.: Research Design: Qualitative, Quantitative, and Mixed Methods Approaches, Fourth, International Student edn., p. 15. SAGE, Los Angeles (2014)
5. Di Maggio, P., Garip, F.: Network effects and social inequality. Annu. Rev. Sociol. **38**, 93–118 (2012). http://0-www.jstor.org.wam.city.ac.uk/stable/23254588
6. Eikebrokk, T.R., Iden, J., Olsen, D.H., Opdahl, A.L.: Understanding the determinants of business process modelling in organisations. Bus. Process Manag. J. **17**(4), 639–662 (2011)
7. Erol, S., et al.: Combining BPM and social software: contra-diction or chance? J. softw. Maint. Evol. Res. Pract. **22**, 492–507 (2010)
8. Gokaldas, V., Rangiha, M.E.: A framework for improving user engagement in social BPM. In: Teniente, E., Weidlich, M. (eds.) BPM 2017. LNBIP, vol. 308, pp. 391–402. Springer, Cham (2018). https://doi.org/10.1007/978-3-319-74030-0_30
9. Kovačič, A., Hauc, G., Buh, B., Štemberger, M.I.: BPM adoption and business transformation at Snaga, a public company: critical success factors for five stages of BPM. In: vom Brocke, J., Mendling, J. (eds.) Business Process Management Cases. MP, pp. 77–89. Springer, Cham (2018). https://doi.org/10.1007/978-3-319-58307-5_5
10. Lückmann, P., Feldmann, C.: Success factors for business process improvement projects in small and medium sized enterprises – empirical evidence. Procedia Comput. Sci. **121**, 439–445 (2017)
11. Rangiha, M.E.: A framework for Social BPM based on social tagging. School of Mathematics, Computer Science and Engineering & City University. Department of Computer Science, p. 23 (2016)
12. Rangiha, M.E., Karakostas, B.: A socially driven, goal-oriented approach to business process management. Int. J. Adv. Comput. Sci. Appl. (IJACSA), 8–13 (2013). Special Issue on Extended Papers from Science and Information Conference 2018-03-09
13. Ravesteyn, P., Batenburg, R.: Surveying the critical success factors of BPM-systems implementation. Bus. Process Manag. J. **16**(3), 492–507 (2010)
14. Lee, R.G., Dale, B.G.: Business process management: a review and evaluation. Bus. Process Manag. J. **4**(3), 214–225 (1998)
15. Rode, H.: To share or not to share: the effects of extrinsic and intrinsic motivations on knowledge-sharing in enterprise social media platforms. J. Inf. Technol. **31**, 152–165 (2016)
16. Rosemann, M., vom Brocke, J.: The six core elements of business process management. In: Brocke, J., Rosemann, M. (eds.) Handbook on Business Process Management 1. International Handbooks on Information Systems, pp. 107–122. Springer, Heidelberg (2010). https://doi.org/10.1007/978-3-642-00416-2_5
17. Schmidt, R., Nurcan, S.: BPM and social software. In: Ardagna, D., Mecella, M., Yang, J. (eds.) BPM 2008. LNBIP, vol. 17, pp. 649–658. Springer, Heidelberg (2009). https://doi.org/10.1007/978-3-642-00328-8_65

18. Sikdar, A., Payyazhi, J.: A process model of managing organizational change during business process redesign. Bus. Process Manag. J. **20**(6), 975 (2014)

19. Tashakkori, A.: Mixed Methodology: Combining Qualitative and Quantitative Approaches, p. 43. Sage, London (1998)

20. Trkman, P.: The critical success factors of business process management. Int. J. Inf. Manag. **30**(2), 125–134 (2010)

21. Ariyachandra, T.R., Frolick, M.N.: Critical success factors in business performance management—striving for success. Inf. Syst. Manag. **25**(2), 113–120 (2008). https://doi.org/10.1080/10580530801941504

22. Vukšić, V., Vugec, S., Lovrić, A.: Social business process management: Croatian IT company case study. Bus. Syst. Res. **8**, 60–70 (2017). https://0-search-proquest-com.wam.city.ac.uk/docview/1923642069?pq-origsite=summon

Enabling Co-creation in Product Design Processes Using 3D-Printing Processes

Michael Möhring[1]($\boxtimes$), Rainer Schmidt[1], Barbara Keller[1],
Jennifer Hamm[2], Sophie Scherzinger[2], and Ann-Kristin Vorndran[2]

[1] Munich University of Applied Sciences, Munich, Germany
{michael.moehring, rainer.schmidt,
barbara.keller}@hm.edu
[2] University of Applied Sciences Würzburg-Schweinfurt, Würzburg, Germany

Abstract. For a long time, geographical distances restricted competition, nowadays competition is global. Thus, companies must build up strategies to cope with this situation. Individualized products could help enterprises to retain customers and their market position due to a differentiated supply. In this research paper we discuss 3D-printing processes as enabler for Co-Creation in product design processes. It enables enterprises to react quickly to customer preferences and changing trends e.g., in design. Furthermore, 3D-Printing enables the integration of customers into product innovation processes. This ends up in Co-Creation and the emergence of related advantages (e.g., customer-centric products or production processes). However, operational processes when using 3D-printing processes for co-creation were not investigated in depth so far. But nevertheless, the improvement of manufacturing processes is important in BPM practice and research as well. Therefore, we address this gap in our paper.

Keywords: BPM · Production processes · 3D printing · Co-creation · Customer integration

1 Introduction

For a long time, geographical distances restricted competition, nowadays competition is global [1]. Thus many enterprises face increasing challenges to compete with competitors having substantial cost advantages e.g. by producing in Asia [1]. For many products nearly equivalent alternatives from different suppliers around the globe are available [2]. A possibility to face cope with these challenges is the upcoming trend of individual and customized products [3]. Customer requirements are becoming more specialized especially in terms of individualized products [4]. Companies react to these changes by individualizing products and related business processes according to individual preferences and thus improving their own supply and retaining a competitive position [4]. A second strategy to cope with competitors from low-wage companies is quick response to the customer [5]. Companies such as Trigema [6] show that being able to quickly produce parts and adjust related processes according to customer specifications near to the target market is a viable strategy.

© Springer Nature Switzerland AG 2019
F. Daniel et al. (Eds.): BPM 2018 Workshops, LNBIP 342, pp. 96–106, 2019.
https://doi.org/10.1007/978-3-030-11641-5_8

In manufacturing, 3D-printing [7] is a key technology for implementing strategies aiming at responsiveness and individualization. They enable manufacturers producing economically even small quantities down to lot-size 1 [8]. Therefore, it does not surprise that according to Gartner Research 3D-printing [9] will change many production business models and related business processes [10].

One of the main advantages of the 3D-printing is, that a multitude of parts can be produced from the same base material, thus warehousing and logistics are simplified [11]. This is a huge difference to traditional manufacturing techniques like milling or carving, where a supply of different raw material must be maintained in order to react quickly to customer requirements [11]. Another advantage of additive manufacturing processes is for instance, that 3D-printing have the possibility to quickly produce parts according to customer specifications and to adapt changes in the design quickly.

The combination of quick response and individualization is the key to Co-Creation, the integration of the customer into product innovation processes [12], especially open innovation [13]. By quickly providing prototypes to the customers, collecting feedback and using it for redesign an improvement cycle can be initiated that is not possible with traditional manufacturing technologies due to their high latency [12].

Existing research on the 3D-printing focused either on technical aspects of development or high-level. strategic (management) questions (e.g., [14, 15]). However, there is a gap between these two research areas. The operational processes when using 3D-printing processes for co-creation were not investigated in depth so far. But nevertheless, the improvement of manufacturing processes are important in BPM practice and research as well [16–18]. We address this gap in our paper investigating *"The benefits and influencing factors gained by enabling customer integration into product design by using 3D-printing processes."* as part of an ongoing research project.

Our paper is structured as follows: Sect. 2 after introducing the subject a background of 3D-printing processes and co-creation aspects is given, Sect. 3 the research model as well as the pre-study design is defined, Sect. 4 Research methods and data collection are described, followed by Sect. 5 were results are shown and Sect. 6 a conclusion is given.

2 Background

Product design processes and product lifecycle management are an important area of research on business process management [16–19]. Now, changes of the consumer's role in product design and design relevant technologies such as 3D-printing increasingly impact the product design processes. To demonstrate these impacts 3D-printing will be investigated. Afterwards the influence of co-creation on product design process will be analyzed.

2.1 3D-Printing

3D-printing is an additive manufacturing approach [7]. Contrary to subtractive processes such as milling or drilling 3D-printing is depositing material to create parts. Its potential to revolutionize processes and manufacturing even has been referenced in the State of the

Union address [20]. In 2018 the spending on 3D-printing is estimated at 12 Billion $ [20]. By 2022 the market will increase to 20 Billion $ [20]. 3D-printing is primarily applied to manufacturing tasks that have a high degree of complexity and/or customization [7]. Due to its additive approach, 3D-printing is able to produce complex parts at the same price than complex parts. It is even possible to easily cope with complexities making conventional manufacturing difficult or impossible. In [21] typical complexities are identified: features, geometries, parts consolidation and fabric step consolidation.

2.2   Co-creation with 3D-Printing

For a long time, product design and development were driven by a serial approach [22], e.g. waterfall like model. Starting from a collection of requirements, more and more concrete specifications are developed [22]. They are basis for the design of the product. Finally, production starts and is transported to the customer.

This classical design approach [22] is expert-driven, top-down-oriented and uses a strict separation between the role of the product designer and the product user. In this approach, the core-competency for product design is assumed nearly completely at certain experts, that build up their knowledge through own studies and experience. They create a plan how to match the assumed or collected user requirements by a certain design and implementation of the product. Typical for this approach is the strict separation of designer and consumer roles. The consumer is involved only at clearly defined points of developments e.g. he was interviewed or asked to fill out questionnaires.

Nowadays, however, the advantages to integrate the customer more intensively are broadly accepted [23]. First concepts such as open innovation [13] recognized the value provided by inputs of external stakeholders such as the consumer. Co-creation is the active involvement of the consumer into the design, creation and distribution of products [14, 24]. Both terms overlap partially. However there is co-creation outside open innovation if the input of the consumer does not end in a commercialized product [14]. At the same time open innovation may happen with other stakeholders than the consumer, thus not being considered a co-creation [14].

3D-printing is an enabler for co-creation by facilitating to capture ideas, suggestions and feedback of the consumer. Through 3D-printing the customer can be better integrated into the production process. The spectrum ranges from influencing the design of mass products to the individual design of products [14]. By using a co-creation approach design processes can be improved in terms of quality, time and costs [25]. Also the relationship and related business processes with customer can be strengthened [26].

The integrating of customers into business processes is always a challenge in research and practice [27–30] for various different reasons. In example, mostly it is quite difficult for customers to participate in the processes. Besides complicated user interfaces, there is also a lack of knowledge due to the actual structuration of the process and the possible opportunities to improve it. Furthermore, finding a place where customers are able to contribute to a certain process is still quite hard. Customers normally want a comfortable solution. That means in fact, that they want to contribute at a time, place and way determined by themselves and not the supplier. Any restriction like an enhanced booting time of the computer could have a negative impact.

3 Pre-study Design: Benefits from Using 3D-Printing

Unfortunately, there is a lack of research about the benefits from 3D-printing processes based on a structured literature review in leading databases like SpringerLink, AISeL, IEEEXplore recommend by the literature [31]. Nevertheless, production processes are an important area for BPM projects [16–18]. The integration of 3D-printing can improve related business processes and integrate the customer into the production process. Therefore, we designed and implemented an empirical pre-study to discover the benefits from 3D-printing processes. This step is important to prepare future studies as well as ensure that it gain relevant and significant results. The study design, implementation and results are described in the following and is summarized in the following Fig. 1.

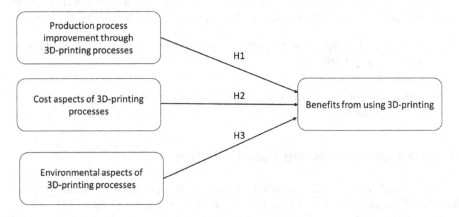

Fig. 1. Research model

The improvement of processes is a very important factor in BPM research and practice (e.g., [25, 32, 33]). The improvement of processes is related with a huge effort and integrates concepts of co-creation as well as is knowledge intensive [32]. To improve co-creation and integrate the individual preferences of a customer, the use of 3D-printing can be useful [12]. Therefore, we suppose that the improvement of production processes via 3D Printing creates benefits by more co-created products and a more flexible production:

H1:   A production process improvement through the use of 3D-printing influences positively the benefits from using 3D-printing

Cost aspect of process and related information systems as well as production systems should be not neglected (e.g., [34–36]). For instance, 3D-printing processes could reduce the costs of the design and production of a product by increasing co-creation with the customer and to be more flexible in the e.g. selection of the production place and related shipping costs. Furthermore, the customer's preferences can be captured more easily and correctly by using a co-creation approach. the necessity of additional queries and adaptions due to wrong interpretations or analysis is reduced. This leads us to:

H2:   Cost aspects of 3D-printing processes are influencing benefits from using 3D-printing

Environmental aspects are getting more and more important in BPM research and practice as well [36]. Customers, (governmental) institutions, the organization itself and many more stakeholders want to optimize their environmental impact and related business processes [36, 37]. Production processes using 3D-printing and capturing the customer needs more exactly by co-creation can be more environmental-friendly. In natural resources can be saved in the production process as well as the waste of e.g. unused products can be avoided. Therefore, we assume a positive influence of the environmental aspects of 3D-printing processes to the benefits from using 3D-printing:

H3:   Environmental aspects of 3D-printing processes influence positively the benefits from using 3D-printing

In the following, we describe our research methods and the data collection to discover our research model.

4   Research Methods and Data Collection

For the investigation of our research model, we used a quantitative research method conducted via an online-based survey like recommended in the literature [38, 39].

Our study was implemented through the open source survey tool Limesurvey [40] and pre-tested in the fourth quarter of the year 2017. After improving the questionnaire based on the pre-test results, we implemented our survey also in the fourth quarter of the year 2017. At the beginning of our questionnaire, we implemented check questions to ensure that we only get answers of 3D-printing experts with related process knowledge. We contacted the experts formally and informally via email, professional social networks (like XING, LinkedIn), Blogs, telephone etc. According to the research model, the main questions of the survey were ranked on a five-point Likert scale [41]. The relevant questions can be found in the appendix section of the paper. Other questions like the years of working experience were collected using an open question format. After cleaning our data because of e.g. missing values or expert level/correct check questions, we got a final sample of n = 111 experts. On average, the participants had 13.3 years of working experience in the relevant field. Most of our experts in the sample currently using 3D-printing (77.27%). The other experts are planning to use, consult or have worked with 3D-printing as well as have the related knowledge. The experts worked for leading enterprises in Germany, Austria and Switzerland. In general, our experts assign high benefits to using 3D-printing processes according to our study.

For analysing our research model with our collected empirical data, we used a structural equation modelling approach (SEM) [42, 43]. The approach connects our causal model (research model) to the empirical data via the use of partial least square regression [42, 43]. Significances were analysed via the recommended bootstrapping algorithms [42, 43]. We used Smart PLS version 3.2 [44] to develop the SEM. This research approach is often used in research (e.g., [44–47]).

The quality metrics of our data are satisfying, therefore we assume that our results are valid and reliable. According to Chin [48] the coefficient of the determination (R^2) is in a good range (0.475 > 0.19). Furthermore, Cronbach's Alpha (>0.70), and the composite reliability (>0.70) are satisfied. All quality metrics of our model are listed in Table 1.

The results are more precisely described in the next section.

5   Results

Regarding the research model and our collected data, we got the following results of our SEM analysis (Fig. 2 as well as Table 1):

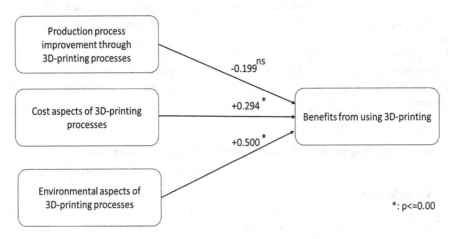

Fig. 2. Results of the SEM

Hypothesis 1 (*A production process improvement through the use of 3D-printing influences positively the benefits from using 3D-printing.*) must be rejected, because of a missing significance (p = 0.113 > 0.05). This might be explained by deeply divided opinions of our experts in this issue. Maybe there are some business case specifics (e.g., current level of production process automation), we have not covered in our survey. Future research should there investigate this aspect more detailed.

Regarding our analysis, we can confirm hypothesis 2 (*Cost aspects of 3D-printing processes are influencing benefits from using 3D-printing*). We discovered a significant positive influence (+0.294) cost aspects have on the perceived benefits of using

3D-printing. Our experts see great potentials in reducing the cost of a business process with related 3D-printing technology in the production environment. For instance, as explained before the cost for stocking raw materials sink strongly, because now the customer himself has to take care for supplying necessary materials. Furthermore, cost related to the execution of the process (e.g., energy, manpower) were transferred to the customer as well.

Finally, our data support the supposition given in hypothesis 3 (*Environmental aspects of 3D-printing processes influence positively the benefits from using 3D-printing*). A significant, positive path coefficient (+0.500) indicates, that the improved environmental aspects of the business process lead to a higher benefit. Our experts see high potential of 3D-printing processes by improving environmental aspects. Integrating 3D-printing in related business processes can improve environmental aspects and also the benefits of using 3D-printing. This is in line with current general research about environmental aspects of BPM (e.g., [37]).

In summary, the important details of the SEM are described below:

Table 1. Quality metrics of the SEM

	Path coefficient	Significance (p-values)	Cronbach's Alpha	Composite reliability
Production process improvement	−0.199	0.113	1 (*1 item*)	1 (*1 item*)
Cost aspects	+0.294	0.00	0.709	0.811
Environmental aspects	+0.500	0.00	0.766	0.894
Benefits from using 3D printing	–	–	0.798	0.845

In the following section we want to conclude based on our results.

6 Conclusion

The use of 3D-printing generates promising potential both for research and practice. We addressed some gaps in the existing research about the benefits of 3D-printing processes. We developed and implemented a first pre-study to get empirical insights. We found that cost as well as environmental aspects of 3D-printing processes are positively influencing the perceived benefits from using 3D-printing.

We contribute to the current literature in different ways. We extend previous work on the use of co-creation in business processes and show the relation to 3D-printing. Furthermore, we add knowledge on the environmental aspects of manufacturing related business processes based on 3D-printing processes. Managers can use our knowledge e.g. for decision support and evaluation of 3D-printing business cases. Regarding their business model, they can reduce costs and can implement more environmental processes.

Limitations can be found in the research method and asked experts. It was not possible to address all possible experts. However, regarding the current literature (e.g., [34, 50]) we collected a satisfying sample. Capturing the arguments given while discussion hypothesis's 1 result, also the composition of the questionnaire might be able to improve in terms of different business case specifics.

Future research projects should start at this point and extend the sample e.g. to other countries like US, Australia, BRIC states and compare as well as extend our results. The use of research methods like Case study research for the evaluation of 3D-printing processes could be a good starting point for future research. Furthermore, a deeper look into the factors influencing 3D-printing benefits, new ways of designing the collaboration network of 3D-printing process partners (e.g., through smart contracts [52]) and a broader case-individual discussion of e.g. environmental aspects should be done. Also implications on information system design [51], enterprise architecture [49] are important to discover.

Appendix

The excerpt of the main pre-study items (Table 2):

Table 2. The excerpt of the main pre-study items

	Items*
Production process improvement	• Improvement of production
Cost aspects	Cost reduction of: • piece production costs • material expenses • storage costs • changeover costs • labor costs
Environmental aspects	• Waste avoidance • Use of recyclable and renewable resources
Benefits from using 3D-printing	• Benefits of 3D-printing in general and resulting from digitization, customer-integration, place, production output & time, etc.
Working experience	• Years of working experience

*: *translated from the German language*

References

1. Byrnes, N.: Why "America first" policies won't keep China from industrial domination. MIT Technology Review. https://www.technologyreview.com/s/603848/competing-with-the-chinese-factory-of-2017/. Accessed 31 May 2018
2. Baldwin, R., Lopez-Gonzalez, J.: Supply-chain trade: a portrait of global patterns and several testable hypotheses. World Econ. **38**(11), 1682–1721 (2013)

3. Wakoya, A.G., Bayiley, Y.T.: The effect of mass customization on competitive strategy. J. Manag. **3**(1), 31–42 (2015)
4. Bogner, E., Löwenb, U., Frankea, J.: Systematic consideration of value chains with respect to the timing of individualization. Procedia CIRP **60**, 368–373 (2017)
5. Suri, R.: It's About Time: The Competitive Advantage of Quick Response Manufacturing. CRC Press, Boca Raton (2016)
6. Audretsch, D.B., Lehmann, E.: The Seven Secrets of Germany: Economic Resilience in an Era of Global Turbulence. Oxford University Press, New York (2016)
7. Conner, B.P., et al.: Making sense of 3-D printing: creating a map of additive manufacturing products and services. Addit. Manuf. **1–4**, 64–76 (2014)
8. Anderson, C.: Makers: The New Industrial Revolution. Cornerstone Digital, New York (2012)
9. Basiliere, P.: '3D-Printing changes Business Models'
10. Müller, A., Karevska, S.: Global 3D Printing Report 2016. Ernst & Young, Mannheim (2016)
11. Ihl, C., Piller, F.: 3D printing as driver of localized manufacturing: expected benefits from producer and consumer perspectives. In: Ferdinand, J.-P., Petschow, U., Dickel, S. (eds.) The Decentralized and Networked Future of Value Creation. PI, pp. 179–204. Springer, Cham (2016). https://doi.org/10.1007/978-3-319-31686-4_10
12. Rayna, T., Striukova, L., Darlington, J.: Co-creation and user innovation: the role of online 3D printing platforms. J. Eng. Technol. Manag. **37**, 90–102 (2015)
13. Chesbrough, H.: Managing open innovation. Res.-Technol. Manag. **47**(1), 23–26 (2004)
14. Rayna, T., Striukova, L.: The impact of 3D printing technologies on business model innovation. In: Benghozi, P., Krob, D., Lonjon, A., Panetto, H. (eds.) Digital Enterprise Design & Management. Advances in Intelligent Systems and Computing, vol. 261, pp. 119–132. Springer, Cham (2014). https://doi.org/10.1007/978-3-319-04313-5_11
15. Rengier, F., et al.: 3D printing based on imaging data: review of medical applications. Int. J. Comput. Assist. Radiol. Surg. **5**(4), 335–341 (2010)
16. Estruch, A., Heredia Álvaro, J.A.: Event-driven manufacturing process management approach. In: Barros, A., Gal, A., Kindler, E. (eds.) BPM 2012. LNCS, vol. 7481, pp. 120–133. Springer, Heidelberg (2012). https://doi.org/10.1007/978-3-642-32885-5_9
17. Hongjun, L., Nan, L.: Process improvement model and it's application for manufacturing industry based on the BPM-ERP integrated framework. In: Dai, M. (ed.) ICCIC 2011, Part I. CCIS, vol. 231, pp. 533–542. Springer, Heidelberg (2011). https://doi.org/10.1007/978-3-642-23993-9_77
18. Barber, K.D., Dewhurst, F.W., Burns, R., Rogers, J.B.B.: Business-process modelling and simulation for manufacturing management: a practical way forward. Bus. Process Manag. J. **9**(4), 527–542 (2003)
19. Scheer, A.-W., Boczanski, M., Muth, M., Schmitz, W.-G., Segelbacher, U.: Prozessorientiertes Product Lifecycle Management. Springer, Heidelberg (2005)
20. The Editors: Obama's 2013 State of the Union Speech: Full Text. The Atlantic, 12 February 2013
21. Atzeni, E., Salmi, A.: Economics of additive manufacturing for end-usable metal parts. Int. J. Adv. Manuf. Technol. **62**(9–12), 1147–1155 (2012)
22. Ulrich, K.T.: Product Design and Development. Tata McGraw-Hill Education, New York (2003)
23. Kambil, A., Friesen, G.B., Sundaram, A.: Co-creation: a new source of value. Outlook Mag. **3**(2), 23–29 (1999)

24. Benkler, Y.: The Wealth of Networks : How Social Production Transforms Markets and Freedom. Yale University Press, London (2006)
25. van der Aalst, W.M.P., Rosa, M.L., Santoro, F.M.: Business process management. Bus. Inf. Syst. Eng. **58**(1), 1–6 (2016)
26. O'Hern, M.S., Rindfleisch, A.: Customer co-creation. Rev. Mark. Res. **6**, 84–106 (2010)
27. Laudon, K.C.: Managing Information Systems: Managing the Digital Firm, 13th edn. Prentice Hall (2013)
28. Chesbrough, H., Spohrer, J.: A research manifesto for services science. Commun. ACM **49**(7), 35–40 (2006)
29. Harrison-Broninski, K.: Human interactions: the heart and soul of business process management: how people really work and how they can be helped to work better (2005)
30. Möhring, M., et al.: Using smart edge devices to integrate consumers into digitized processes: the case of amazon dash-button. In: Business Process Management Workshops, pp. 374–383 (2017)
31. Webster, J., Watson, R.T.: Analyzing the past to prepare for the future: writing A. MIS Q. **26**(2), 494–508 (2002)
32. Seethamraju, R., Marjanovic, O.: Role of process knowledge in business process improvement methodology: a case study. Bus. Process Manag. J. **15**(6), 920–936 (2009)
33. 'Business Case BPM WP'. Oracle Whitepaper
34. Schmidt, R., Möhring, M., Keller, B.: Customer relationship management in a public cloud environment–key influencing factors for european enterprises. In: Proceedings of the 50th Hawaii International Conference on System Sciences, pp. 4241–4250 (2017)
35. Scheer, A.W., Brabänder, E.: The process of business process management. In: vom Brocke, J., Rosemann, M. (eds.) Handbook on Business Process Management 2. International Handbooks on Information Systems, pp. 239–265. Springer, Heidelberg (2010). https://doi.org/10.1007/978-3-642-01982-1_12
36. Zairi, M.: Business process management: a boundaryless approach to modern competitiveness. Bus. Process Manag. J. **3**(1), 64–80 (1997)
37. Nowak, A., Leymann, F., Schumm, D., Wetzstein, B.: An architecture and methodology for a four-phased approach to green business process reengineering. In: Kranzlmüller, D., Toja, A.M. (eds.) ICT-GLOW 2011. LNCS, vol. 6868, pp. 150–164. Springer, Heidelberg (2011). https://doi.org/10.1007/978-3-642-23447-7_14
38. Parasuraman, A., Grewal, D., Krishnan, R.: Marketing Research. Cengage Learning, Boston (2006)
39. Recker, J.: Scientific Research in Information Systems: A Beginner's Guide. Springer, Heidelberg (2013)
40. LimeSurvey - the free and open source survey software tool !, 24 April 2011. http://www.limesurvey.org/de/start. Accessed 23 Nov 2014
41. Likert, R.: A technique for the measurement of attitudes. Arch. Psychol. **140**, 5–55 (1932)
42. Wong, K.K.-K.: Partial least squares structural equation modeling (PLS-SEM) techniques using SmartPLS. Mark. Bull. **24**, 1–32 (2013)
43. Hooper, D., Coughlan, J., Mullen, M.: Structural equation modelling: guidelines for determining model fit. Electr. J. Bus. Res. Methods **6**, 53–60 (2008)
44. Ringle, C.M., Wende, S., Will, A.: SmartPLS 2.0 (beta), Hamburg, Germany (2005)
45. Urbach, N., Ahlemann, F.: Structural equation modeling in information systems research using partial least squares. JITTA J. Inf. Technol. Theory Appl. **11**(2), 5 (2010)
46. Münstermann, B., Eckhardt, A., Weitzel, T.: The performance impact of business process standardization: an empirical evaluation of the recruitment process. Bus. Process Manag. J. **16**(1), 29–56 (2010)

47. Schmidt, R., Möhring, M., Härting, R.-C., Reichstein, C., Neumaier, P., Jozinović, P.: Industry 4.0 - potentials for creating smart products: empirical research results. In: Abramowicz, W. (ed.) BIS 2015. LNBIP, vol. 208, pp. 16–27. Springer, Cham (2015). https://doi.org/10.1007/978-3-319-19027-3_2
48. Chin, W.W.: The partial least squares approach to structural equation modeling. Mod. Methods Bus. Res. **295**(2), 295–336 (1998)
49. Lapalme, J., Gerber, A., Van der Merwe, A., Zachman, J., De Vries, M., Hinkelmann, K.: Exploring the future of enterprise architecture: a Zachman perspective. Comput. Ind. **79**, 103–113 (2016)
50. Lahrmann, G., Marx, F., Winter, R., Wortmann, F.: Business intelligence maturity: development and evaluation of a theoretical model. In: 2011 44th Hawaii International Conference on System Sciences (HICSS), pp. 1–10 (2011)
51. Hevner, A., Chatterjee, S.: Design science research in information systems. In: Design Research in Information Systems. Integrated Series in Information Systems, vol. 22, pp. 9–22. Springer, Boston (2010). https://doi.org/10.1007/978-1-4419-5653-8_2
52. Möhring, M., Keller, B., Schmidt, R., Rippin, A.L., Schulz, J., Brückner, K.: Empirical insights in the current development of smart contracts. In: Proceedings of the 22nd Pacific Asia Conference on Information Systems (PACIS), AIS, Yokohama, Japan (2018)

Evaluation of WfMC Awards for Case Management: Features, Knowledge Workers, Systems

Johannes Tenschert[✉] and Richard Lenz

Institute of Computer Science 6 (Data Management),
Friedrich-Alexander-Universität Erlangen-Nürnberg (FAU), Erlangen, Germany
{Johannes.Tenschert,Richard.Lenz}@fau.de

Abstract. In recent years, many production case management (PCM) and adaptive case management (ACM) systems have been introduced into the daily workflow of knowledge workers. In many research papers and case studies, the claims about the nature and requirements of knowledge work in general seem to vary. While choosing or creating a case management (CM) solution, typically one has the target knowledge workers and their domain-specific requirements in mind. But knowledge work shows a huge variety of modes of operation, complexity, and collaboration. We want to increase transparency on which features are covered by well-known and award-winning systems for different types of knowledge workers and different classes of systems. This may not unveil gaps between requirements and offered solutions, but it can uncover differences in solutions for varying user bases. We performed a literature review of 48 winners of the WfMC Awards for Excellence in Case Management from 2011 to 2016 and analyzed case studies in regard to targeted knowledge workers, advertised features, and type of system. Different types of knowledge workers showed a different bias on certain system types and features in regard to collaboration and variability of processes.

Keywords: Adaptive case management ·
Production case management · Knowledge-intensive business process ·
Knowledge work

1 Introduction

In recent years, many PCM and ACM systems have been introduced with the goal of improving the efficiency and quality of knowledge work. Typically, one may find sentences like "Most knowledge workers spend their time in business applications like Salesforce." [1], "Due to [...] and the high degree of interactivity between knowledge workers [...]" [2], "If knowledge workers can rely on [...] they can provide simplified automated process fragments without worrying about all possible exceptions" [3]. Each of these articles may be right with their assumptions for target users. But is this really true for knowledge work in general?

© Springer Nature Switzerland AG 2019
F. Daniel et al. (Eds.): BPM 2018 Workshops, LNBIP 342, pp. 107–120, 2019.
https://doi.org/10.1007/978-3-030-11641-5_9

According to Davenport [4], knowledge work can be classified into four groups by complexity and level of interdependence (i.e. required collaboration). He distinguishes the transaction, integration, expert, and collaboration model [4, p. 27]. The transaction model ($\downarrow$collaboration, $\downarrow$complexity) covers work reliant on formal procedures and training, i.e. routine work. The expert model ($\downarrow$collab., $\uparrow$compl.) is judgment-oriented and highly reliant on individual expertise and experience. The integration model ($\uparrow$collab., $\downarrow$compl.) covers repeatable work reliant on formal processes but also dependent on integration across functional boundaries. The collaboration model ($\uparrow$collab., $\uparrow$compl.) covers complex improvisational work and relies on deep expertise across functions. This already suggests that different types of knowledge workers may have different requirements in regard to collaboration, variability in processes, and management of case data.

There are many ways to classify CM systems, e.g. by area of operations: CRM, ERP, ECM, issue tracker. We apply a condensed categorization of [5] that introduces seven categories ranging from predictable, repeatable to variable and unique processes. We distinguish BPM, PCM and ACM. BPM systems cover predictable and repeatable work. PCM systems are more flexible and tailored to a particular domain. ACM systems support unexpected workflows and unstructured data, and they are useful not only in one domain. We reviewed winners of the WfMC Awards for Excellence in Case Management that were published as case studies in regard to their classes of targeted knowledge workers, advertised features, and type of system. This analysis is intended to increase transparency on which features are covered by well-known and award-winning systems. It is not intended to unveil gaps between requirements and offered solutions.

In the following sections, we briefly introduce related work outlining features and requirements in CM, and the methods for our analysis. In Sect. 4, we introduce the extracted features, and Sect. 5 presents our findings with a complete matrix and resulting analyses. Afterwards, we discuss our approach and results, and in Sect. 7 we conclude the paper.

2   Related Work

Palmer et al. [6] compared characteristics of ACM, ECM, CRM, and BPM based on typical requirements in ACM and hint which type of system to use based on business problems addressed and expected workflow. However, they do not directly address PCM systems and do not cover the interdependence of different types of knowledge workers. Matthias [7] outlines requirements of ACM, e.g. in regard to decisions, database organization, variability, and access control. Our literature review focuses on the features available in award-winning CM systems to approximate different requirements of different types of knowledge workers.

Motahari-Nezhad and Swenson [8] outline the state of the art in CM with six groups for systems and examples. We also compare classes of systems, although on a more granular level, to find similarities and differences of systems tailored to different types of knowledge workers. Hauder et al. [9] derive ten requirements for ACM based on a literature review. But in the original sources, these requirements were probably formulated with a specific knowledge worker in mind. Hauder et al. did not find one reference implementation fulfilling all extracted requirements, which suggests that requirements of knowledge workers also differ within ACM.

3   Methods

We analyzed 48 winners of the WfMC Awards for Excellence in Case Management from 2011 to 2016 [10–15]. There were 51 award winners, but the case studies Kirtland AFB (2014), Remfry and Sagar (2015), and Texas Office of the Attorney (2015) have not been published and were therefore omitted.

We performed this analysis by evaluating each case study and extracting features elaborated in the text and figures. For each case study, we compiled a list of features with a reference to their source in the text or figure. The case studies did not always show a consistent vocabulary for the names of features and many were similar. Hence, we used more general terms for similar features, e.g. "reporting" and "dashboard" yielded "BI (analytics/reporting)". The approach was iterative. Obviously, this method yields only features that are elaborated and deemed important by the authors, and actual systems may have a different focus. They ranged from support for (semi-)structured processes, (un-)structured artifacts, and collaboration support to non-functional features like "system of record". Due to space constraints, some features that were not characteristic for system types and type of knowledge work are omitted. For the categorization of a case study to ACM, PCM, and BPM, we apply a condensed categorization derived from Swenson [5]. There were no clear differentiators, so many systems fall in more than one category. This also impacts aggregated statistics: Percentages are provided for *all* systems of a group.

The categorization of knowledge work is based on Davenport's classification of knowledge-intensive processes [4]. Knowledge work is divided into four groups by complexity of work and level of interdependence (collaboration). Systems can target multiple stakeholders with a varying degree of complexity and interdependence. These dimensions were classified based on available user descriptions and if necessary on assumptions (e.g. work of lawyers and physicians typically yields the expert model). For each case study, the extracted knowledge workers, features, and type(s) of system were inserted into Table 1. We used RapidMiner[1] to find correlations between types of system and types of knowledge work, between types of knowledge work and features, between system types and features, and conditional probabilities between all attributes (knowledge work, systems, features). Due to space constraints, some analyses had to be omitted here.

4   Extracted Features

We extracted the following features by evaluating all case studies, i.e. their text and figures, for descriptions, portrayals, and occurrences of features. Structured

---

[1] https://rapidminer.com.

Table 1. Extracted features, system types, and knowledge workers of evaluated case studies

Year / Book and Case Study columns (grouped by year):

- **2011 – Taming the Unpredictable:** BAA Heathrow, UK; Los Angeles County Advisory Body, USA; Lakshmi Kumaran & Sridharan, India; Pinellas County Clerk of the Circuit Court, USA; Wellhde Hospital Integrated Care Pathway, UK; DVIT – Financial Service, Netherlands
- **2012 – How Knowledge Workers Get Things Done:** Cognocare, an ACM-based System for Oncology, Netherlands; General Italian Insurance Company S.A.; HSBC Bank, India; New York State Office of Children and Family Services, USA; MATS Norwegian Food Safety Authority, Norway; Pearson GmbH, Austria; Gepper, Australia; UWV, Netherlands; Vision Service Plan (VSP)
- **2013 – Empowering Knowledge Workers:** Axlo Group Holdings Ltd., UK; CargoNet AS Norway; Department of Transport, South Africa; Directorate for the Construction of the Facilities for EURO 2012, Ukraine; Fleet One, USA; Info Edge India Ltd., India; Norwegian Courts Administration, Norway; Texas Office of the Attorney General Crime Victim Services Division, USA; U.S. Department of Housing and Urban Development (HUD), USA; UBS Bank, Worldwide
- **2014 – Thriving on Adaptability:** Cognocare, an ACM-based System for Oncology; Crawford & Company, United States; Department of Human Services, State of Hawaii, USA; Library McCamish Systems, USA; The National Police Immigration Service, Norway; Frambu LLC, a BNV Mellon company, USA; The Antwerp Port Authority, USA; TIAA-CREF, USA; WESTMD DEPI Practice; African Reinsurance, Africa
- **2015 – Best Practices for Knowledge Workers:** Eaton Vance Investment Managers, USA; EDP Renewables, United States; National Institute of Allergy and Infectious Diseases (NIAID), USA; United States Nuclear Regulatory Commission, USA; Pediatric Hospital "Bambino Gesú" Kidney Transplant Integrated Care Pathway, USA; PENSCO Trust Company, USA; Universal Forest Products, USA; Fenn Such Kahn & Shepard, P.C., USA
- **2016 – Intelligent Adaptability:** Grinnel Mutual, USA; Leading European Bank – Banking Correspondence Management System, USA; Molina Healthcare Inc., USA; Univerzit Leading Croatia; WPS Health Solutions, USA

Feature rows (left column labels):

System:
- BPM
- PCM
- ACM
- Other

KW (models):
- Transaction model
- Integration model
- Expert model
- Collaboration model

Features:
- Ad-Hoc Activities
- Artifact templates
- Automatic notification
- BI (Analytics / Reporting)
- Business Rules
- Case Conversations
- Case History
- Case State
- Case Templates
- Case Tags
- Checklists
- Collaboration within a case
- Configurability
- Deadlines
- Declarative Process Modelling
- Document Generation
- Documents
- Domain-specific Information
- Guard Rails
- Integration with external services
- Interactions of a Case
- Meetings of a Case
- Mobile Devices
- Notes on Artifacts
- Queries / Buckets of cases
- Queries / Worklists across cases
- Relations between cases
- Relations between interactions
- Resource Allocation
- Roles
- Search
- Stakeholders of a Case
- Semi-Structured Process (Predefined Tasks)
- Semi-Structured Process (Stages)
- Semi-Structured Process (Aspect)
- Structured Process (Complete)
- Suggestions of Activities
- System of Record
- Task State
- Task Templates
- Tasks
- Typed Interactions
- User Access Control
- Views for Multiple Stakeholders / Roles

(aspect, complete) and semi-structured (tasks, stages) processes originated from "(semi-)structured processes" that were detected in nearly every case study.

Ad-hoc Activities. Tasks or goals that were not predefined in a process model or (task) repository and are nonetheless captured in the system as part of a case. Ad-hoc activities should cover tasks that were not anticipated at design-time.

Artifact Templates. The system supports templates for artifacts that are not only generated at case initiation. Tasks and documents are excluded from this definition. Instead, they are covered in the features *document generation* and *task templates*.

Automatic Notifications. The system provides automated messages, e.g. for warnings, alerts, or reminders.

BI (Analytics/Reporting). The system provides some dashboard or reporting for KPIs of the case. Subsumed in one feature due to similarity and intention.

Business Rules. The system enables either predefined business rules, e.g. creating a case if a specific event is detected, or user-defined rules.

Case Conversations. The system provides threaded or flat discussions for cases.

Case History. The system provides a visible activity stream of a case.

Case State. Different states of a case are emphasized and more sophisticated than *open* and *closed*. Varying from three predefined states to completely user-defined states assigned in an ad-hoc fashion.

Case Tags. The system allows to annotate cases with keywords to facilitate search.

Case Templates. In the system, cases and their contents are initiated with a template containing predefined artifacts or relations. A template may be initiated using some existing artifact, e.g. an event for a new customer or incident. Examples for predefined artifacts are a set of related tasks for a certain type of case, and a copy of an existing case with generated case-specific contents.

Checklists. The system provides a list of tasks that are typically performed or a list of attributes to typically ensure. The lists may be part of a case template/type or specific to a certain state of a case.

Collaboration within a Case. The case study emphasizes case-specific collaboration, e.g. collaborative access, collaborative workflow, chat. May differ from *notes on artifacts* in emphasizing on the collaboration aspect. A user-to-user chat that is only visible to the chatting participants can be *collaboration within a case*, but does not qualify as *case conversations* visible to all stakeholders with access rights (e.g. [16]).

Configurability. The case study emphasizes that the solution is easily configurable to end users or domains.

Deadlines. The system displays for each case pending calendar entries or deadlines for tasks or milestones.

Declarative Process Modeling. The system enables modeling (parts of) processes by stating the dependencies between tasks.

Document Generation. The system provides automatic generation of letters (e.g. MS Word, email) based on templates. This feature covers correspondence between stakeholders of a case, but not system messages indicating certain events or reminders.

Documents. The system covers document management, i.e. documents are stored and related to a case.

Domain-specific Information. The system is tailored to a specific domain, e.g. by considering certain attributes and providing a domain-specific user interface.

Guard Rails. The system prevents users from or actively reminds them to performing certain actions based on regulatory and organizational rules.

Integration with External Services. The system provides integration with existing systems in order to streamline processes and reduce redundant data entry.

Interactions of a Case. The system provides an overview of case-specific correspondence.

Meetings of a Case. The system tracks case-specific meetings and may cover meeting minutes as well.

Table 2. System type by supported knowledge worker

Given ↓ / Then →	ACM	PCM	BPM	Other	Σ
Collaboration model	100.0%	25.0%	0.0%	0.0%	4
Expert model	36.8%	84.2%	21.1%	10.5%	19
Integration model	52.9%	70.6%	5.9%	0.0%	17
Transaction model	18.2%	93.9%	27.3%	0.0%	33

Table 3. Supported knowledge workers by type of system

Then → Given ↓	Collaboration model	Expert model	Integration model	Transaction model	Σ
ACM	28.6%	50.0%	64.3%	42.9%	14
PCM	2.5%	40.0%	30.0%	77.5%	40
BPM	0.0%	36.4%	9.1%	81.8%	11
Other	0.0%	100.0%	0.0%	0.0%	2

Mobile Devices. The system provides a mobile-enabled website or an "App" for a mobile platform that connects to the system.

Notes on Artifacts. The system enables users to annotate cases or artifacts.

Queries/Buckets of Cases. The system provides predefined queries or views to display a selection of cases based on status, type, tags, assigned users or teams, stakeholders, or resource constraints.

Queries/Worklists Across Cases. The system provides predefined queries or views to display a selection of tasks based on status, assigned users or teams, priority and deadlines, or skill-based. These views are provided across cases.

Relations Between Cases. The system enables to link related cases either with or without a parent-child-relationship.

Relations Between Interactions. The system relates interactions of a case to other interactions, e.g. linking two requests. One occurence in [2].

Resource Allocation. The system provides (automated) resource allocation, resource planning, or documentation of resource allocation.

Roles. The system allows to assign roles. If roles are case-specific, this feature overlaps with *stakeholders of a case*.

Search. The system provides some sort of search to find cases or artifacts.

Semi-structured Process (Predefined Tasks). The system provides a task repository covering typical process instances. Users may decide in which order and what tasks of this repository are performed.

Semi-structured Process (Stages). Depending on the case state (stage), the system suggests potential activities.

Stakeholders of a Case. In the system, stakeholders (e.g. clients, assigned personnel) are captured for each case. For example, specific stakeholders are assigned to tasks and artifacts, or cases contain a list of (annotated) stakeholders.

Structured Process (Aspect). The system enables completely structured and optionally automated subprocesses, e.g. predefined and related forms and tasks in a BPMN model of one aspect of the case. One aspect does not cover the whole process, i.e. it is a subgoal of the case.

Structured Process (Complete). In the system, the workflow of the case is modeled a priori and performed as a standardized process. The system might allow certain deviations from the expected workflow, but typically cases adhere to the model.

Suggestions of Activities. There are multiple courses of action to proceed in a case. The system actively suggests certain activities, interactions, or escalations, but ultimately the knowledge worker decides whether to apply the suggestion. The displayed options may be based on prioritized outcomes or restrictions.

System of Record. The article emphasizes a system of record, i.e. the system is used as and users are aware it is the authoritative data source for cases and information.

Tasks. The systems monitors case-related tasks that are either predefined or ad hoc.

Task State. The system enables different states of a task that are more sophisticated than *open* or *closed*.

Task Templates. The system provides a task library with entries that are either plain or adaptable to the context, or reuse of previous tasks as a template for the task at hand. This differs from *semi-structured process (predefined tasks)* in not modeling the tasks of the process. Obviously there is some overlap between these features and many award winners provide both.

Typed Interactions. The system distinguishes different types of requests, classifies correspondence by intention (e.g. comment, complaint, proposal) or provides typed response options to correspondence (e.g. agree, decline, counteroffer).

User Access Control. The system provides access control to cases or certain artifacts, e.g. role-based or by sharing.

Views for Multiple Stakeholders/Roles. The system offers differing views to users depending on their role in the system or case.

5   Findings

Table 1 shows for each case study which system types, types of knowledge work, and features were extracted. We generated Tables 2, 3, 4 and 5 based on this matrix. First, the evaluation confirms that different types of knowledge workers use different system classes (Table 2), and that a certain type of system may indicate the type of knowledge work (Table 3). Since there are overlaps in the user base and systems often cannot be completely attributed to one category, rows in all tables do not add to 100%. The most dominant system class is PCM. There was a high overlap between PCM and BPM systems: Only one system classified as BPMS was not also classified as PCMS. Seven ACMSs were categorized as ACMS only, the other seven also overlap with PCMS. Two case studies (Other) could not be categorized into ACMS, BPMS, or PCMS at all [17,18], but they overlap with PCMS. Obviously, these overlaps have influence on Table 4.

Knowledge workers of the collaboration model are only covered in four case studies, so their results should be taken with a grain of salt. Nonetheless, the low number indicates that they may have been largely ignored, even though they need exactly what ACM aims to offer: Great flexibility and collaboration in one system of record. One environment is usually shared by different types of users with different requirements, but 25 of 48 of the case studies seem to be tailored to only one type of knowledge worker. The transaction model alone is represented by 16 case studies, and 10 systems are classified for users of both the transaction and the integration model. Due to these overlaps, Table 5 cannot completely discern the features different types of knowledge workers seem to require.

The analyzed case studies in the collaboration model all provided document management (*documents*), *case conversations*, a *case state*, an activity stream of a case (*case history*), and three out of four provided support for mobile devices. However, due to the low number of case studies categorized in the collaboration model, those could be outliers. *Documents* seem to be important in every type of knowledge work. Unsurprisingly, *case conversations* are rare in the expert and transaction model. Both *case history* and *case state* appear in half of the case studies classified as expert or integration model. Unlike the feature *case state*, no case study allowed user-defined *task states*. In the integration model, there is a high emphasis on *case templates* and *collaboration within a case*. The expert model shows the highest share of systems that integrate with external services.

Most case studies provide some support for predefined structured or semi-structured processes with a complete process model, a process model of aspects of a process, predefined tasks, or task support based on the current stage of a case. All systems classified as BPMS either model the complete process or aspects of it. Obviously, ACMSs focus less on structure. Of those systems, 35.7% provide predefined tasks, and 28.6% model aspects of a process. PCMSs also model aspects of a process (47.5%) and offer predefined tasks (32.5%). For case studies classified as ACMS only, no instance had support for structured aspects.

In ACMSs, the most prevalent features seem to be *documents, tasks, collaboration within a case*, and surprisingly *roles*. Table 4 suggests that *BI (analytics/reporting)*, would be important for ACMSs as well. However, for the seven case studies classified as ACM-only, only two seem to provide this feature. The high share of ACMSs with reporting capabilities stems from ACM/PCM hybrids. Moreover, providing a *case history, case state, case templates*, notes on certain artifacts, *access control*, and *ad-hoc activities* seems to be characteristic for ACMSs. All system types have a high emphasis on being a system of record.

Ad-hoc activities are available both in ACMS (50%) and PCMS (31%). Unsurprisingly, support for document and task management seems to be important across all types of system and knowledge work. Many systems offer different views for different stakeholders (around 30% per type), but usually every stakeholder has the same view. Support for mobile devices is aligned with the type of the system: 36% of ACMSs, 18% of PCMSs and 9% of BPMSs have some support for mobile devices. Moreover, the support is highest in the collaboration model (75%) and integration model (35%), i.e. the types of knowledge work emphasizing on collaboration. For the collaboration features *case conversations, stakeholders of a case, notes on artifacts*, and *collaboration within a case*, 71% of all ACMSs had support for at least two of these features. For PCMSs, only 33% had at least two of these features. All collaboration features have the highest probability of being present in an ACMS rather than in other system types, but due to the distribution of system types, they usually imply PCMS. Unsurprisingly, they are usually present in the integration and collaboration model.

Table 4. System types and features

Feature	Feature implies system class					System class implies feature			
	ACM	PCM	BPM	Other	Σ	ACM	PCM	BPM	Other
Ad-hoc activities	44%	75%	0%	13%	16	50%	30%	0%	100%
Artifact templates	100%	0%	0%	0%	1	7%	0%	0%	0%
Automatic notifications	28%	80%	20%	8%	25	50%	50%	45%	100%
BI (analytics / reporting)	24%	91%	24%	6%	34	57%	78%	73%	100%
Business rules	24%	95%	29%	5%	21	36%	50%	55%	50%
Case conversations	64%	55%	9%	0%	11	50%	15%	9%	0%
Case history	44%	72%	22%	6%	18	57%	33%	36%	50%
Case state	47%	71%	12%	0%	17	57%	30%	18%	0%
Case tags	67%	33%	0%	0%	3	14%	3%	0%	0%
Case templates	42%	74%	16%	0%	19	57%	35%	27%	0%
Checklists	60%	40%	0%	0%	5	21%	5%	0%	0%
Collaboration within a case	47%	79%	16%	0%	19	64%	38%	27%	0%
Configurability	100%	67%	0%	0%	3	21%	5%	0%	0%
Deadlines	25%	88%	31%	6%	16	29%	35%	45%	50%
Declarative process modeling	100%	100%	0%	0%	3	21%	8%	0%	0%
Document generation	43%	100%	14%	0%	14	43%	35%	18%	0%
Documents	31%	86%	22%	0%	36	79%	78%	73%	0%
Domain-specific information	19%	97%	32%	6%	31	43%	75%	91%	100%
Guard rails	27%	87%	20%	13%	15	29%	33%	27%	100%
Integration with external services	29%	90%	26%	6%	31	64%	70%	73%	100%
Interactions of a case	44%	78%	11%	0%	9	29%	18%	9%	0%
Meetings of a case	50%	100%	0%	0%	2	7%	5%	0%	0%
Mobile devices	50%	70%	10%	0%	10	36%	18%	9%	0%
Notes on artifacts	44%	67%	17%	0%	18	57%	30%	27%	0%
Queries / buckets of cases	29%	71%	43%	0%	7	14%	13%	27%	0%
Queries / worklists across cases	43%	86%	7%	0%	14	43%	30%	9%	0%
Relations between cases	56%	67%	22%	0%	9	36%	15%	18%	0%
Relations between interactions	100%	0%	0%	0%	1	7%	0%	0%	0%
Resource allocation	14%	100%	29%	0%	7	7%	18%	18%	0%
Roles	29%	87%	23%	3%	31	64%	68%	64%	50%
Search	75%	100%	0%	0%	4	21%	10%	0%	0%
Semi-structured process (predefined tasks)	29%	76%	18%	12%	17	36%	33%	27%	100%
Semi-structured process (stages)	33%	67%	0%	0%	3	7%	5%	0%	0%
Stakeholders of a case	43%	79%	7%	14%	14	43%	28%	9%	100%
Structured process (aspect)	20%	95%	30%	0%	20	29%	48%	55%	0%
Structured process (complete)	0%	100%	78%	0%	9	0%	23%	64%	0%
Suggestions of activities	20%	80%	20%	40%	5	7%	10%	9%	100%
System of record	26%	87%	29%	3%	38	71%	83%	100%	50%
Task state	67%	33%	0%	0%	3	14%	3%	0%	0%
Task templates	50%	90%	0%	10%	10	36%	23%	0%	50%
Tasks	30%	84%	19%	5%	37	79%	78%	64%	100%
Typed interactions	60%	80%	0%	0%	5	21%	10%	0%	0%
User acces control	42%	84%	16%	0%	19	57%	40%	27%	0%
Views for multiple stakeholders / roles	36%	93%	29%	7%	14	36%	33%	36%	50%
Σ						14	40	11	2

Finally, a large share (65%) of the analyzed case studies describes some sort of templating: *case templates*, *document generation*, *task templates*, and *artifact templates*. The feature *artifact templates* appears only once [19] as a generic way for content templates in a wiki. All systems providing document generation are classified as PCMS. For *document generation*, the case studies show a clear emphasis on the transaction and expert model. Except for the systems in [16,20], all document templates seem to be predefined. In the future, this could be improved by introducing it for user-specific correspondence as well. *Case templates* are characteristic for ACMSs (57%), but they are provided in PCMSs as well.

Table 5. Knowledge workers and features

Feature	Feature implies types of knowledge work					Type of knowledge work implies feature			
	Collab.	Expert	Integr.	Trans.	Σ	Collab.	Expert	Integr.	Trans.
Ad-hoc activities	13%	31%	38%	63%	16	50%	26%	35%	30%
Artifact templates	0%	0%	100%	0%	1	0%	0%	6%	0%
Automatic notifications	4%	36%	40%	68%	25	25%	47%	59%	52%
BI (analytics / reporting)	3%	47%	32%	68%	34	25%	84%	65%	70%
Business rules	5%	48%	29%	76%	21	25%	53%	35%	48%
Case conversations	36%	27%	64%	45%	11	100%	16%	41%	15%
Case history	22%	56%	44%	44%	18	100%	53%	47%	24%
Case state	24%	53%	47%	53%	17	100%	47%	47%	27%
Case tags	0%	33%	67%	33%	3	0%	5%	12%	3%
Case templates	5%	32%	63%	74%	19	25%	32%	71%	42%
Checklists	20%	20%	80%	60%	5	25%	5%	24%	9%
Collaboration within a case	11%	42%	58%	68%	19	50%	42%	65%	39%
Configurability	0%	33%	100%	100%	3	0%	5%	18%	9%
Deadlines	0%	56%	31%	56%	16	0%	47%	29%	27%
Declarative process modeling	0%	67%	33%	67%	3	0%	11%	6%	6%
Document generation	7%	64%	36%	64%	14	25%	47%	29%	27%
Documents	11%	39%	36%	72%	36	100%	74%	76%	79%
Domain-specific information	3%	55%	26%	68%	31	25%	89%	47%	64%
Guard rails	13%	33%	47%	67%	15	50%	26%	41%	30%
Integration with external services	6%	48%	23%	68%	31	50%	79%	41%	64%
Interactions of a case	11%	56%	67%	67%	9	25%	26%	35%	18%
Meetings of a case	0%	50%	100%	50%	2	0%	5%	12%	3%
Mobile devices	30%	30%	60%	70%	10	75%	16%	35%	21%
Notes on artifacts	17%	33%	50%	67%	18	75%	32%	53%	36%
Queries / buckets of cases	0%	43%	29%	86%	7	0%	16%	12%	18%
Queries / worklists across cases	14%	36%	50%	64%	14	50%	26%	41%	27%
Relations between cases	11%	67%	56%	56%	9	25%	32%	29%	15%
Relations between interactions	100%	0%	100%	0%	1	25%	0%	6%	0%
Resource allocation	0%	29%	29%	71%	7	0%	11%	12%	15%
Roles	10%	35%	29%	81%	31	75%	58%	53%	76%
Search	0%	50%	50%	75%	4	0%	11%	12%	9%
Semi-structured process (predefined tasks)	0%	47%	35%	65%	17	0%	42%	35%	33%
Semi-structured process (stages)	33%	0%	33%	67%	3	25%	0%	6%	6%
Stakeholders of a case	14%	50%	57%	43%	14	50%	37%	47%	18%
Structured process (aspect)	0%	55%	35%	70%	20	0%	58%	41%	42%
Structured process (complete)	0%	22%	11%	100%	9	0%	11%	6%	27%
Suggestions of activities	20%	60%	20%	20%	5	25%	16%	6%	3%
System of record	5%	39%	32%	74%	38	50%	79%	71%	85%
Task state	33%	67%	67%	33%	3	25%	11%	12%	3%
Task templates	10%	60%	40%	50%	10	25%	32%	24%	15%
Tasks	5%	49%	35%	65%	37	50%	95%	76%	73%
Typed interactions	40%	20%	60%	60%	5	50%	5%	18%	9%
User acces control	11%	37%	32%	74%	19	50%	37%	35%	42%
Views for multiple stakeholders / roles	14%	43%	21%	71%	14	50%	32%	18%	30%
Σ						4	19	17	33

6   Discussion

Since the classifications we applied for knowledge work and system types lack clear differentiators, they obviously are subjective to a certain degree. Moreover, the features extracted depend on the authors of that text to actually write about a particular feature or to provide a comprehensive screenshot. Hence, some features of systems that are not described or displayed will be missing. Nonetheless, our analysis shows significant differences between features present in types of knowledge work and types of system. Moreover, knowledge workers of different type focus on different types of systems, i.e. ACMSs seem to be tailored to or required by the collaboration model. These differences are sufficiently clear to not just stem from erroneous or subjective detection.

The features detected in our analysis can stem from users asking for certain features and the provider assuming requirements as well. Hence, they can only pose as an approximation to the requirements of certain types of knowledge worker. However, the results at least suggest, that a requirements analysis for new or existing ACMS and PCMS cannot be limited to a literature review on typical requirements of knowledge workers, but has to consider peculiarities of the target users at least on the granularity of Davenport's classification.

The feature *typed interactions* covers suggested responses like agree, decline, counter-offer or comment in negotiations [2], as well as approve and deny of document templates for change management [21]. Even though the authors did not comment on speech act theory [22,23] and probably are not aware of supporting their processes by emphasizing on the pragmatic intention of the user's interactions, they are providing such a support. Finally, the feature *system of record* was not extracted from all case studies. Nonetheless, if the investments are made to support knowledge workers with a CM system, one main motivation might have been to provide such a system. Hence, providers of these systems could put more emphasis on creating a system of record in order to make this goal visible.

7   Conclusion

In this analysis, we evaluated 48 winners of the WfMC Awards for Excellence in Case Management in regard to their classes of targeted knowledge workers, features, and type of system. We confirmed that different types of knowledge workers use different system classes and that the type of system also indicates the type of knowledge work supported. Different types of knowledge work showed a different emphasis in the provided features. Knowledge workers in the collaboration model seem to be underrepresented, i.e. either in the awards or in the target users of CM systems. A large share of award-winning case studies was classified as PCMS. ACMSs focus on semi-structured processes and ad-hoc activities. Unsurprisingly, ACM systems have a higher emphasis on collaboration than PCMSs and BPMSs. Support for document management seems to be important for CM regardless of system type or type of knowledge work.

Of course, this analysis only covers features of award-winning case studies, not requirements of the systems and knowledge workers. Nonetheless, these features were elaborated by the authors of the case studies and most likely deemed important by them. They can hint at what sort of features is most likely asked for by certain users. In order to gain the actual requirements of different types of knowledge workers, interviews and further analyses are still necessary.

References

1. Franks, P.C.: Integrated ECM solutions: where records managers, knowledge workers converge. Inf. Manag. **50**(4), 18–22, 47 (2016)
2. 4Spires: Fleet one, USA. In: Fischer, L. (ed.) Empowering Knowledge Workers: New Ways to Leverage Case Management, pp. 145–150. Future Strategies Inc. (2014)
3. Tenschert, J., Michelson, G., Lenz, R.: Towards speech-act-based compliance. In: 2016 IEEE 18th Conference on Business Informatics (2016)
4. Davenport, T.H.: How knowledge workers differ, and the difference it makes. In: Thinking for a Living: How to Get Better Performances and Results from Knowledge Workers, pp. 25–38. Harvard Business Press (2005)
5. Swenson, K.D.: Innovative organizations act like systems, not machines. In: Fischer, L. (ed.) Empowering Knowledge Workers: New Ways to Leverage Case Management, pp. 31–42. Future Strategies Inc. (2014)
6. Palmer, N.G., Dugan, L.: Understanding and evaluating case management software. In: Fischer, L. (ed.) Thriving On Adaptability: Best Practices for Knowledge Workers, pp. 31–40. Future Strategies Inc. (2015)
7. Matthias, J.T.: Technology for case management. In: Mastering the Unpredictable: How Adaptive Case Management will Revolutionize the Way that Knowledge Workers Get Things Done, pp. 63–88. Meghan-Kiffer Press (2010)
8. Motahari-Nezhad, H.R., Swenson, K.D.: Adaptive case management: overview and research challenges. In: 2013 IEEE 15th Conference on Business Informatics (CBI), pp. 264–269. IEEE (2013)
9. Hauder, M., Münch, D., Michel, F., Utz, A., Matthes, F.: Examining adaptive case management to support processes for enterprise architecture management. In: IEEE 18th International Enterprise Distributed Object Computing Conference Workshops and Demonstrations, pp. 23–32, September 2014
10. Fischer, L. (ed.): Taming the Unpredictable: Real-World Adaptive Case Management. Future Strategies Inc. (2011)
11. Fischer, L. (ed.): How Knowledge Workers Get Things Done: Real-World Adaptive Case Management. Future Strategies Inc. (2012)
12. Fischer, L. (ed.): Empowering Knowledge Workers: New Ways to Leverage Case Management. Future Strategies Inc. (2014)
13. Fischer, L. (ed.): Thriving On Adaptability: Best Practices for Knowledge Workers. Future Strategies Inc. (2015)
14. Fischer, L. (ed.): Best Practices for Knowledge Workers: Innovation in Adaptive Case Management. Future Strategies Inc. (2016)
15. Fischer, L. (ed.): Intelligent Adaptability. Future Strategies Inc. (2017)
16. ISIS Papyrus: "Paneon GmbH, Austria". In: Fischer, L. (ed.) How Knowledge Workers Get Things Done: Real-World Adaptive Case Management, pp. 155–164. Future Strategies Inc. (2012)
17. Castillo, L.: Cognocare, an ACM-based system for oncology. In: Fischer, L. (ed.) How Knowledge Workers Get Things Done: Real-World Adaptive Case Management, pp. 119–128. Future Strategies Inc. (2012)
18. Castillo, L.: Cognocare, an ACM-based system for oncology. In: Fischer, L. (ed.) Thriving On Adaptability: Best Practices for Knowledge Workers, pp. 119–126. Future Strategies Inc. (2015)
19. Callewaert, F.: The antwerp port authority. In: Fischer, L. (ed.) Thriving On Adaptability: Best Practices for Knowledge Workers, pp. 213–230. Future Strategies Inc. (2015)

20. Hyland "Grinnell mutual, USA". In: Fischer, L. (ed.) Intelligent Adaptability, pp. 99–106. Future Strategies Inc. (2017)
21. ISIS Papyrus Europe AG "Leading european bank - banking correspondence management system". In: Fischer, L. (ed.) Intelligent Adaptability, pp. 107–121. Future Strategies Inc. (2017)
22. Austin, J.L.: How to do Things with Words. Oxford University Press, Oxford (1975)
23. Searle, J.R.: Speech Acts: An Essay in the Philosophy of Language. Cambridge University Press, New York (1969)

Investigating the Trade-off Between the Effectiveness and Efficiency of Process Modeling

Jeroen Bolle and Jan Claes[(✉)]

Department of Business Informatics and Operations Management,
Ghent University, Tweekerkenstraat 2, 9000 Ghent, Belgium
{jeroen.bolle,jan.claes}@ugent.be

Abstract. Despite recent efforts to improve the quality of process models, we still observe a significant dissimilarity in quality between models. This paper focuses on the syntactic condition of process models, and how it is achieved. To this end, a dataset of 121 modeling sessions was investigated. By going through each of these sessions step by step, a separate '*revision*' phase was identified for 81 of them. Next, by cutting the modeling process off at the start of the revision phase, a partial process model was exported for these modeling sessions. Finally, each partial model was compared with its corresponding final model, in terms of time, effort, and the number of syntactic errors made or solved, in search for a possible trade-off between the effectiveness and efficiency of process modeling. Based on the findings, we give a provisional explanation for the difference in syntactic quality of process models.

Keywords: Conceptual modeling · Process modeling · Syntactic quality · Revision phase · Business process management

1 Introduction

Because of the ever-increasing complexity of business processes, the demand for process models of high quality is rising. In order to be competitive, businesses need to be as productive as possible, while using their resources economically. When processes are designed, this implies that the supporting process models should be constructed in an efficient way, without wasting unnecessary time or effort. Unfortunately, the process of creating process models is often not effective and/or not efficient. Therefore, this paper investigates whether the time and effort used for constructing a process model influence the quality of the resulting model.

This research goal was triggered by two observations. First, previous research concluded that different process modeling strategies exist, and that the *efficiency* of modeling can be influenced by applying a strategy that optimally aligns with the cognitive properties of the modeler. The *effectiveness* of process modeling, on the other hand, appeared to be harder to influence [1]. Second, by comparing several process models that represent the same process, but that were made by different modelers, a high dissimilarity in the pragmatic, semantic and syntactic quality was noticed. To explain

© Springer Nature Switzerland AG 2019
F. Daniel et al. (Eds.): BPM 2018 Workshops, LNBIP 342, pp. 121–132, 2019.
https://doi.org/10.1007/978-3-030-11641-5_10

this difference, the construction processes of these models were investigated with PPMCharts. A PPMChart is a convenient visualization tool that represents the various steps used to create a process model as dots on a grid, using different colors for each type of operation and different shapes for each type of element on which the operation is performed [2]. During the inspection of these charts, it was noticed that near the end of the modeling session two things happened fairly often: the modeler executed 'create' operations in a much slower fashion than at the start, and the modeler looped back through the model and made changes to parts created earlier. This raised our suspicion that some people revised their process model before handing in the result, most likely in search for mistakes.

In this light, the question can be asked whether the conceived process model benefits from such a revision phase. In other words, do these people use the changes of the revision phase to actually improve the model, or do they waste time and effort? Do they mainly correct errors or are they working on improving secondary aspects of the model? Is it possible that they introduce more errors than they solve?

In this exploratory paper, we formulate a provisional answer to the questions mentioned above. An existing dataset was used, containing the operational details of 121 modeling sessions. For each session we tried to determine whether there was a revision phase and if so, the process of process modeling was cut into two separate phases: a first phase that was mostly used for building the model, and a second phase during which the modeler predominantly revised the model created throughout the first part. For each of the 81 instances for which two phases could be distinguished, a process model was exported at two points in time: the model created during the first phase only, called the **partial model**, as opposed to the **final model**, which is the result of the modeling session as a whole, including the second phase. The properties of both phases and both models were then compared in order to collect insights related to the above questions.

The research is limited in scope in two ways. First, it focuses on one specific type of conceptual models: sequence flow process models. Moreover, the models in the dataset were restricted to use only 6 constructs. Second, because we prioritized depth of the research over breadth, this study was narrowed to only one quality level, i.e. syntactic quality. Methodologically, it makes sense to first investigate syntactic quality, because in comparison to semantic and pragmatic quality, it can be measured more accurately. On the other hand, syntactic quality may be the least relevant of the three because many tools help to avoid syntactic mistakes completely. Nevertheless, we believe that the insights from the syntax dimension may also shed light on the other two quality dimensions, because the source of cognitive mistakes is probably the same (i.e., cognitive overload due to the complexity of modeling).

This paper is structured as follows. Section 2 presents related research. In Sect. 3, the variables that were selected are presented and discussed. Section 4 explains the applied method to determine the partial models. Section 5 provides the results. Section 6 concludes the paper with a summary of the findings and a brief discussion.

2   Related Work

For the past three decades, business process modeling has been one of the important domains within the information systems research field [3]. The recent developments in research about process models can be classified into three distinct research streams. The first stream focuses on the application of process models (i.e., *why are process models created*). Davies et al. [4] performed this type of study in the broader research domain of conceptual modeling. The second category studies the quality of process models (i.e., *what makes a process model of a high standard*). In the absence of a unanimously agreed quality framework for process models [3, 5, 6], multiple publications have proposed their own version [7, 8].

Recently, a third stream originated within the process modeling research field. This sub-branch investigates the process of process modeling instead of the result of this process, the final model (i.e., *how is a qualitative process model constructed*). The underlying idea is that the quality of the final model is largely based on the quality of the preceding modeling process. This paper is associated with the last stream.

During the last decades, a lot of interesting publications that studied the process of process modeling have been introduced. Hoppenbrouwers et al. [9] laid an important groundwork for future research by presenting fundamental insights into the process of creating conceptual models. Mendling [10] studied the effects of process model complexity on the probability of making errors during the process of process modeling. Pinggera et al. [11] investigated if the structuring of domain knowledge could improve the modeling process of casual modelers. Haisjackl et al. [12] provide valuable insights into the modeler's behavior towards model lay-out and its implications on pragmatic quality of process models.

While some publications propose modeling techniques and tips to boost the effectiveness of process modeling [13], others suggest an approach that would improve the efficiency of the modeling process [14]. Yet other publications present modeling methods that enhance both the effectiveness and efficiency of process modeling [15, 16]. The aforementioned publications resulted in general guidelines and techniques for process modeling, intended for all conceptual/process modelers. As opposed to these works, recent studies began acknowledging the existence of different modeling styles for creating business process models [17]. Finally, Claes et al. [1] developed a differentiating modeling technique that matches the modeling strategy to the modeler's cognitive profile, in order to enhance the modeler's ability to create process models in an efficient way. Furthermore, a series of different tools for visualizing modeling processes were created, which have contributed significantly to the investigation of the process of process modeling [2, 18]. One of these tools – the PPMChart – was used extensively during our research, for the purpose of identifying the two modeling phases (cf. supra).

3   Measurements

In 2015, 146 students of the Business Engineering master program at Ghent University took part in an experiment, in the context of a Business Process Management course. The experiment consisted of three consecutive tasks: a series of cognitive tests, a benchmarking modeling task and an experimental modeling task. In between the second and third task, a subset of the participants received a treatment, with the intention of giving them an advantage over the other students when performing the third task. In this last task, all students were asked to create a process model about a mortgage request case, based on a written description of the process. A more detailed discussion of the experiment is provided by Claes et al. [1].

This experiment generated a dataset [19], containing the operational details of the 146 modeling sessions for the experimental modeling task. For 25 of these sessions, not all variables of interest could be measured adequately, rendering them useless for the research. Hence, 121 modeling sessions were left to analyze.

Due to the lack of a consensus on the definition of high-quality process modeling [3, 5, 6], it was first decided which elements to consider for characterizing a modeling process of high quality. Because of the ever-increasing need for productivity that exists today, we believe that evaluating the modeling session solely by the quality of the result does not suffice, which is why the whole process of process modeling (PPM) was studied for this paper. In order to assess the PPM quality, three quality indicators are proposed, based on the *Devil's Triangle* of Project Management [20]:

- First, **time** is defined as the number of seconds it takes to reach a certain point of the modeling process.
- Second, the **effort** is determined by the number of operations on elements (create, delete, move, ...) that are performed in the modeling tool during the process. These actions are divided into mutually exclusive subgroups, as shown in Table 1.
- Third, the **syntactic quality** of the resulting model consists of two measures: the number of errors made, and the number of errors solved by the modeler. The construction of a list of potential syntax errors and the subsequent detection of these violations was performed by De Bock and Claes [21], who delivered an overview of when errors are made and corrected throughout the modeling process.

Table 1. Classification of modeling actions

Subtype	Definition	Operations
Create	All operations that add new elements to the model	Create Activity, Create AND Gate, Create Edge, Create End Event, Create Start Event, Create XOR Gate
Fix	All (non-create) operations that can correct mistakes	Delete Activity, Delete AND Gate, Delete Edge, Delete End Event, Delete Start Event, Delete XOR Gate, Reconnect Edge, Name Activity, Name Edge, Rename Activity, Rename Edge
Adjust	All operations that alter the model without changing the syntactic quality	Create Edge Bend Point, Delete Edge Bend Point, Move Activity, Move AND Gate, Move Edge, Move Edge Bend Point, Move End Event, Move Start Event, Move XOR Gate

All three quality indicators are of continuous nature, meaning they can be determined for every moment in the modeling process. Time and effort combined express the efficiency of the process, while the correctness is an indicator of the effectiveness of process modeling.

4 The Partial Model

During an initial scan of the dataset, it was noticed that near the end of a modeling process, some subjects exhibited behavior that suggested they were inspecting the model they created thus far, potentially in search for mistakes. It was theorized that, in some cases, a modeling session could be split into two distinct phases: a **model building phase**, during which the biggest part of the model was created (i.e., the *"partial model"*), and a **revision phase**, mainly used for tracking down errors and trying to correct them (i.e., resulting in the *"final model"*). From here on out, we will refer to the moment that separates the two phases as **PT** or **Partial Time**, a name derived from the partial model that is exported at that instant. As a result of this definition of the modeling phases, we used the following guidelines to determine PT in all modeling sessions in a consistent way:

- The biggest part of the model should be produced before PT. Additional *'create'* operations past this moment are allowed but should mainly be used to replace deleted chunks of elements or add a finishing touch to the model (e.g., to close gaps between activities or to create an end event).
- The revision phase should be primarily composed of operations that imply that the subject is inspecting the process model: moving elements around, deleting or renaming elements, extensive scrolling, etc.

The identification of PT was performed by the first author, who was inexperienced in this stream of process modeling research at the time of the identification. This was a deliberate choice, with the intention of avoiding that the decision-making process would be biased in favor of the expected results. In order to determine PT as objectively and as consistently as possible, 5 types were identified:

- Type A: Near the end of the modeling process, the subject clearly tries to correct mistakes using operations from the *'fix'* category (cf. Table 1). PT is fixed at the last *'create'* operation before these correction attempts take place.
- Type B: This type is similar to type A, but the second modeling phase is now identified by a notable time span between the last *'create'* operation and the end of the modeling process. The biggest part of this time span is used to perform *'adjust'* actions and scrolls, operations that suggest the subject is reviewing his model in search for errors. PT is fixed at the last *'create'* operation before this time gap.
- Type C: The subject deletes unused elements from the model. These are elements that were created somewhere during the modeling process but that were never connected to the rest of the model. PT is fixed at the last *'create'* operation before these loose elements are erased.

- Type D: A specific type for subjects who use an aspect-oriented approach. This means that the subject postpones the creation of edges until the (majority of) the nodes of the model are created. Often, the subject first creates all nodes, then connects them with arrows, and typically proceeds with moving elements around to improve the readability of the model, after which they try to identify and solve errors. PT is fixed at the first of this series of '*adjust*' actions.
- Type E: Type E includes all models that don't fit one of the previous descriptions. Typically, these subjects don't seem to have a separate revision phase. Instead, they either make corrections throughout the whole process of modeling, or they don't adjust the model at all. As a result, no partial model was exported for these observations. They were left out for part of the analysis (cf. infra).

Two tools were used for the identification of PT. First, we created two PPMCharts for every observation in the dataset: one version including the '*move*' actions, the other excluding them. Per model, we jointly analyzed its two charts and visually identified provisional time points to be considered for the choice of PT. Next, the Cheetah Experimental Platform [22] was used to inspect the modeling processes, by reviewing a step-by-step replay of each session. Combining insights offered by both tools, we were able to determine a partial model for 81 of the 121 models originally considered. The other 40 were assumed to lack a separate revision phase (i.e., type E).

Figure 1 depicts two examples of PPMCharts [2]. On the left is a clear example of a type A model, where the arrow indicates the first '*delete*' action (red) of a sequence, with only a few '*create*' actions (green) in between. The chart on the right is an example of a type D model, where the subject first created the nodes (squares) and then the edges (triangles). The arrow indicates the first (very light) blue '*move*' action of a series.

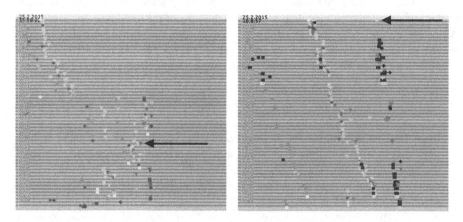

Fig. 1. Example of a type A model (left) and a type D model (right) (Color figure online)

5 Investigation of the Theorized Trade-off

5.1 Analysis of the Total Modeling Process

For this part of the analysis, all 121 models from the dataset are considered. During an initial data inspection, a disparity in the size of the models that were created was noticed: the number of elements varied between 61 and 120. In order to avoid biased results - the bigger a process model, the more opportunities for the subject to make mistakes - the data were normalized by expressing the number of errors as a percentage of the number of elements in the final model.

Figure 2 shows the relationship between the total time it took a subject to create the process model, and the relative number of errors (s)he made/solved during that time. It also displays the relative number of errors remaining at the end of the process, simply defined as 'errors made' minus 'errors solved'. The following correlation coefficients confirm relations that can also be inferred from Fig. 2:

- 'Total time' and 'errors solved': 0,253* (p = 0,050)
- 'Total time' and 'errors made': −0,200 (p = 0,820)
- 'Total time' and 'errors remaining': −0,183* (p = 0,044)

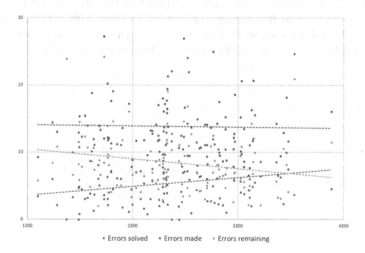

Fig. 2. Total modeling time and the relative number of syntactic errors solved, made and remaining: plots and linear trend lines

Very similar results are obtained when time is replaced with effort, the number of operations performed by the subject during the process. The coefficients are:

- 'Total effort' and 'errors solved': 0,349** (p = 0,000)
- 'Total effort' and 'errors made': −0,003 (p = 0,970)
- 'Total effort' and 'errors remaining' −0,229* (p = 0,012)

This similarity is not really surprising, since time and effort are positively correlated with a coefficient of 0,457** (p = 0,000). Subjects that perform more actions, generally take more time.

Our interpretation of these results is as follows. Making mistakes does not per se take a lot of time, hence the absence of a significant relationship between '*total time*' and '*errors made*'. Solving errors on the other hand is more time consuming and requires extra effort. Therefore, people who use more time and operations to finish a model typically solve more problems, which is why their final model generally contains less syntactic errors. These observations suggest a certain trade-off between efficiency and effectiveness of process modeling.

5.2   Comparison Between the Model Building and the Revision Phase

From this point on, we look at the modeling process as the combination of two phases: a model building phase, during which the subject creates most of the model, followed by a revision phase, where the subject takes time to go over the model in search for mistakes. Dropping the 40 cases for which we weren't able to identify two separate phases (i.e., type E), leaves us with 81 models for this part of the analysis.

Figure 3 contains boxplots that compare the distribution of the total modeling time of people for which a revision phase was identified with the total modeling time of the remaining subjects. Notwithstanding some outliers, people with a separate revision phase generally exhibit a longer total modeling time. This accentuates once more that detecting and trying to correct errors requires additional time (and effort).

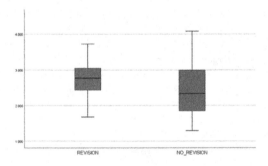

Fig. 3. Total modeling time for modeling sessions with and without revision phase

As a result of our definition of both phases, the first phase is characterized by a prevalence of '*create*' actions, while the second phase mostly contains '*adjust*' and '*fix*' operations. In the next part of the analysis, it is investigated if most errors are in fact made in phase 1 and corrected in phase 2. After all, creating elements in the model building phase does not necessarily introduce mistakes, and deleting elements in the revision phase does not necessarily solve mistakes. Moreover, most people also made adjustments during the model building phase.

This time, the number of errors is normalized according to the size of the model at the end of the corresponding phase, as well as to the duration of that phase, since the revision phase is significantly shorter than the model building phase in most cases. The result is a variable representing the number of errors made/solved per hour, divided by the number of elements the model contains (i.e., *'errors made'* and *'errors solved'*). The graphs in Fig. 4 depict the *difference* in relative number of errors made/solved between both phases (i.e., *'errors made/solved$_{phase2}$'* - *'errors made/solved$_{phase1}$'*). A positive value reflects a higher proportion in the revision phase than in the model building phase, a negative value indicates the opposite.

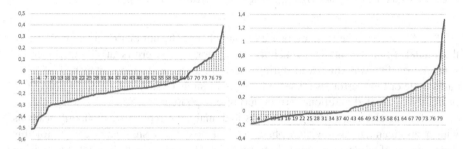

Fig. 4. Difference between the revision phase and the model building phase in terms of relative number of errors made (left) and errors solved (right), arranged in increasing order

The graph shows that people have a tendency to make more mistakes in the first phase – this is the case for 67 of the 81 subjects in this study. The resolution of errors, on the other hand, is spread more evenly across both phases: 40 subjects solved relatively more in the second phase and 38 in the first phase (the remaining 3 didn't fix any issues at all). Errors seem to be resolved in a more efficient way in the second phase, however, since the positive values generally diverge further from zero.

Two considerations are in order here:

- To be able to solve errors, they obviously have to exist in the first place. As a result, the proportion of errors solved will by default be situated more towards the ending of the modeling process than the proportion of errors made.
- Because of the way the emergence of syntactic errors was determined, a counteracting effect could also take place for the *'errors made'*. Since it is sometimes unclear in a partial model whether or not a certain syntactic violation is deliberate, or a consequence of the model not being finished yet, syntactic errors were sometimes detected (too) late in the modeling process. As a result, the proportion of *'errors made'* may have shifted towards the end of the process.

The question can also be asked what would happen with the process models in terms of syntactic quality if the revision phase was to be ignored. In other words: which models generally contain less syntactic errors, the partial or the final models? For this purpose, it was examined how the number of errors evolved during the revision phase of each model. The result is shown below in Fig. 5.

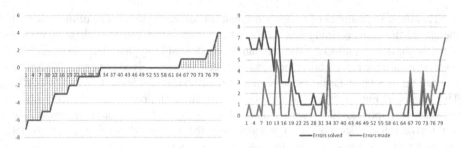

Fig. 5. Net result of the revision phase on the number of errors (left) and the number of errors made/solved during this phase (right), arranged by increasing net result

For 33 subjects, the revision phase did not change the number of errors included in the model, often because they neither made nor solved errors at all. In 31 cases, syntactic quality was improved during the second phase. In more than half of these cases three or more errors were solved. The remaining 17 subjects – a noteworthy 20% of the data - created additional errors during this phase.

Unfortunately, inspecting PPMCharts and analyzing replays of modeling sessions with a disadvantageous revision phase did not reveal any significant causes for this deterioration of syntactic quality. Therefore, the 31 subjects with a beneficial revision phase were compared with the 17 subjects who made additional errors. For most variables, no substantial contrast between both groups was found. However, the number of 'create' operations reflected the most notable difference between the groups, as shown in Fig. 6.

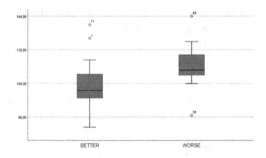

Fig. 6. Number of *create* operations during the total modeling process in case of a beneficial and a deteriorating revision phase

People who created additional errors during the revision phase generally performed significantly more 'create' operations throughout the total modeling process. As a result, they typically also used more effort in general and created a larger process model. A possible explanation is that these subjects created unnecessary elements and thus more complex process models, which caused them to make more errors. More extensive research is required to formulate a substantiated explanation.

6   Conclusion

Using an existing dataset containing data about modeling sessions performed by students, a distinction was made between people who end their modeling process with a separate revision phase and people who don't. Investigating these sessions resulted in findings that – although they are provisional – we believe lay the foundation for interesting future research. First, it was demonstrated that subjects who take more time to create a process model generally produce a model of higher syntactic quality, mostly because extra time was used to detect and correct mistakes. Second, it was shown that subjects tend to make more errors in the first part of the modeling process than near the end, while the opposite is true for the correction of mistakes. For this dataset, 38% of the subjects made good use of their revision phase, 41% had an unimpactful revision phase and 21% created additional errors. No particular general reason was found as to why the last group made the syntactic quality deteriorate at the end of the modeling process.

Being an exploratory study, one should be careful to generalize these findings and take the next three limitations into consideration. First, the use of an arbitrary dataset containing data generated by students may not be representative for all conceptual modelers. Second, the research is limited by the specific choice of process model quality indicators, thus ignoring the semantic and pragmatic aspect of model correctness. Finally, although considerable effort was spent to avoid biases, the subjective nature of the identification of the revision phases may have had an influence on the results. This research can be extended in multiple ways. Alternative ways to identify a revision phase could be used to verify the results. The definition of model quality can be widened to include semantic and pragmatic quality. Finally, future work should include an extension towards a more complete specification of the BPMN language or towards other conceptual modeling languages.

References

1. Claes, J., Vanderfeesten, I., Gailly, F., et al.: The Structured Process Modeling Method (SPMM) - what is the best way for me to construct a process model? Decis. Support Syst. **100**, 57–76 (2017). https://doi.org/10.1016/j.dss.2017.02.004
2. Claes, J., Vanderfeeste, I., Pinggera, J., et al.: A visual analysis of the process of process modeling. Inf. Syst. E-bus. Manag. **13**, 147–190 (2015). https://doi.org/10.1007/s10257-014-0245-4
3. De Oca, I.M.-M., Snoeck, I., Reijers, H.A., et al.: A systematic literature review of studies on business process modeling quality. Inf. Softw. Technol. **58**, 187–205 (2015). https://doi.org/10.1016/j.infsof.2014.07.011
4. Davies, I., Green, P., Rosemann, M., et al.: How do practitioners use conceptual modeling in practice? Data Knowl. Eng. **58**, 358–380 (2006). https://doi.org/10.1016/j.datak.2005.07.007
5. Soffer, P., Kaner, M., Wand, Y.: Towards understanding the process of process modeling: theoretical and empirical considerations. In: Daniel, F., Barkaoui, K., Dustdar, S. (eds.) BPM 2011, Part I. LNBIP, vol. 99, pp. 357–369. Springer, Heidelberg (2012). https://doi.org/10.1007/978-3-642-28108-2_35
6. Recker, J.: A socio-pragmatic constructionist framework for understanding quality in process modelling. Australas. J. Inf. Syst. **14**, 43–63 (2007). https://doi.org/10.3127/ajis.v14i2.23

7. Sánchez-González, L., García, F., Ruiz, F., et al.: Toward a quality framework for business process models. Int. J. Coop. Inf. Syst. **22**, 1–15 (2013). https://doi.org/10.1142/S0218843013500032

8. Vanderfeesten, I., Cardoso, J., Mendling, J., et al.: Quality metrics for business process models. In: Fisher, L. (ed.) BPM and Workflow Handbook, pp. 179–190. Future Strategies (2007)

9. Hoppenbrouwers, S.J.B.A., Proper, H.A., van der Weide, T.P.: A fundamental view on the process of conceptual modeling. In: Delcambre, L., Kop, C., Mayr, H.C., Mylopoulos, J., Pastor, O. (eds.) ER 2005. LNCS, vol. 3716, pp. 128–143. Springer, Heidelberg (2005). https://doi.org/10.1007/11568322_9

10. Mendling, J.: Metrics for process models: Empirical foundations of verification, error prediction and guidelines for correctness. LNBIP, vol. 6. Springer, Heidelberg (2008). https://doi.org/10.1007/978-3-540-89224-3

11. Pinggera, J., Zugal, S., Weber, B., et al.: How the structuring of domain knowledge helps casual process modelers. In: Parsons, J., Saeki, M., Shoval, P., Woo, C., Wand, Y. (eds.) ER 2010. LNCS, vol. 6412, pp. 445–451. Springer, Heidelberg (2010). https://doi.org/10.1007/978-3-642-16373-9_33

12. Haisjackl, C., Burattin, A., Soffer, P., et al.: Visualization of the evolution of layout metrics for business process models. In: Dumas, M., Fantinato, M. (eds.) BPM 2016. LNBIP, vol. 281, pp. 449–460. Springer, Cham (2017). https://doi.org/10.1007/978-3-319-58457-7_33

13. Silver, B.: BPMS Watch: Ten tips for effective process modeling. http://www.bpminstitute.org/resources/articles/bpms-watch-ten-tips-effective-process-modeling

14. Laue, R., Mendling, J.: Structuredness and its significance for correctness of process models. Inf. Syst. E-bus. Manag. **8**, 287–307 (2010). https://doi.org/10.1007/s10257-009-0120-x

15. Xiao, L., Zheng, L.: Business process design: process comparison and integration. Inf. Syst. Front. **14**, 363–374 (2012). https://doi.org/10.1007/s10796-010-9251-3

16. Li, Y., Cao, B., Xu, L., et al.: An efficient recommendation method for improving business process modeling. IEEE Trans. Ind. Inform. **10**, 502–513 (2014). https://doi.org/10.1109/TII.2013.2258677

17. Pinggera, J., Soffer, P., Fahland, D., et al.: Styles in business process modeling: an exploration and a model. Softw. Syst. Model. **14**, 1055–1080 (2013). https://doi.org/10.1007/s10270-013-0349-1

18. Pinggera, J., Zugal, S., Weidlich, M., et al.: Tracing the process of process modeling with modeling phase diagrams. In: Daniel, F., Barkaoui, K., Dustdar, S. (eds.) BPM 2011, Part I. LNBIP, vol. 99, pp. 370–382. Springer, Heidelberg (2012). https://doi.org/10.1007/978-3-642-28108-2_36

19. Bolle, J., De Bock, J., Claes, J.: Data for Investigating the trade-off between the effectiveness and efficiency of process modeling, Mendeley Data v1 (2018). https://doi.org/10.17632/5b8by4k244.1

20. The Project Management Hut: Devil's Triangle of Project Management. https://pmhut.com/devils-triangle-of-project-management

21. De Bock, J., Claes, J.: The origin and evolution of syntax errors in simple sequence flow models in BPMN. In: Matulevičius, R., Dijkman, R. (eds.) CAiSE 2018. LNBIP, vol. 316, pp. 155–166. Springer, Cham (2018). https://doi.org/10.1007/978-3-319-92898-2_13

22. Pinggera, J., Zugal, S., Weber, B.: Investigating the process of process modeling with Cheetah Experimental Platform. In: Mutschler, B. et al. (eds.) Proceedings of ER-POIS 2010. CEUR-WS 6, pp. 13–18. CEUR-WS (2010)

The Repercussions of Business Process Modeling Notations on Mental Load and Mental Effort

Michael Zimoch[1]([✉]), Rüdiger Pryss[1], Thomas Probst[2], Winfried Schlee[3], and Manfred Reichert[1]

[1] Institute of Databases and Information Systems, Ulm University, Ulm, Germany
{michael.zimoch,ruediger.pryss,manfred.reichert}@uni-ulm.de
[2] Department for Psychotherapy and Biopsycho Health,
Danube University Krems, Krems an der Donau, Austria
thomas.probst@donau-uni.ac.at
[3] Department of Psychiatry and Psychotherapy, Regensburg University,
Regensburg, Germany
winfried.schlee@googlemail.com

Abstract. Over the last decade, plenty business process modeling notations emerged for the documentation of business processes in enterprises. During the learning of a modeling notation, an individual is confronted with a cognitive load that has an impact on the comprehension of a notation with its underlying formalisms and concepts. To address the cognitive load, this paper presents the results from an exploratory study, in which a sample of 94 participants, divided into novices, intermediates, and experts, needed to assess process models expressed in terms of eight different process modeling notations, i.e., BPMN 2.0, Declarative Process Modeling, eGantt Charts, EPCs, Flow Charts, IDEF3, Petri Nets, and UML Activity Diagrams. The study focus was set on the subjective comprehensibility and accessibility of process models reflecting participant's cognitive load (i.e., mental load and mental effort). Based on the cognitive load, a factor reflecting the mental difficulty for comprehending process models in different modeling notations was derived. The results indicate that established modeling notations from industry (e.g., BPMN) should be the first choice for enterprises when striving for process management. Moreover, study insights may be used to determine which modeling notations should be taught for an introduction in process modeling or which notation is useful to teach and train process modelers or analysts.

Keywords: Business process modeling notations · Cognitive load · Mental load · Mental effort · Human-centered design

1 Introduction

Business process models specify in terms of textual or graphical artifacts the business processes in an enterprise [1]. In this context, insights on the comprehension of process models demonstrate that *process model comprehension* plays

© Springer Nature Switzerland AG 2019
F. Daniel et al. (Eds.): BPM 2018 Workshops, LNBIP 342, pp. 133–145, 2019.
https://doi.org/10.1007/978-3-030-11641-5_11

an important role when analyzing and optimizing processes [2,3]. As a result, enterprises are confronted with an influx of *process modeling notations* (e.g., *Business Process Model and Notation (BPMN) 2.0* [4], *Event-driven Process Chains (EPCs)* [5], or *Flow Charts* [6]) for the documentation of their business processes within process models. However, for an effective use of process models, the latter must ensure that the processes of an enterprise are comprehended correctly by all involved stakeholders.

In prior research, we investigated process model comprehension in order to reveal factors fostering or thwarting respective comprehension [7,8]. Furthermore, focusing on cognitive neuroscience and psychology, we proposed valuable lessons learned on how to optimize empirical studies for a deeper investigation on process model comprehension [9].

To enhance our previous work on process model comprehension, this work presents the results obtained from an exploratory process model comprehension study. In detail, a sample consisting of $n = 38$ novices, $n = 21$ intermediates, and $n = 35$ experts in the domain of process modeling are confronted with process models expressed in terms of eight different *process modeling notations*. The objective of the study was to evaluate the perceived *cognitive load* (i.e., effort being used in the working memory) of participants caused when comprehending respective *modeling notations*. Based on the results we obtained, we derived for each *process modeling notation* a *mental difficulty level*.

This work contributes to the field of process model comprehension in two ways. First, in research, we want to learn more about the cognitive load and adverse effects when comprehending process models in terms of different *modeling notations* [10]. The obtained insights can foster related empirical investigations in this context. Second, in practice, enterprises can be supported in making decision about the adoption of a particular *process modeling notation* or which modeling tool should be used when adopting process-oriented thinking.

The remainder of the paper is structured as follows: Sect. 2 introduces theoretical backgrounds. Study setting and operation are explained in Sect. 3. In Sect. 4, the obtained results are described empirically and discussed. Finally, Sect. 5 discusses related work, while Sect. 6 summarizes the paper and gives an outlook on future work.

2 Theoretical Background

The *cognitive load* can be defined as a multidimensional construct representing an individuals cognitive capacity used to work on or to solve a task as well as to address a problem [11]. Thereby, *cognitive load* has a causal dimension reflecting the interaction between task- (e.g., inherent difficulty of the task) and subject-specific characteristics (e.g., knowledge about a topic). Particularly, *cognitive load* is comprised of the assessment dimensions describing the measurable aspects *mental load* and *mental effort* [12,13]. The *mental load* relates to a task, which indicates the cognitive capacity needed to cope with the complexity of a task. Juxtaposing *mental load*, the *mental effort* is subject-specific and refers to the invested cognitive capacity of an individual while working on a task [14].

In many fields (e.g., psychology, education), the observation as well as measurement of the *cognitive load* has become crucial. Reasons for this are that a reduced *cognitive load* has a positive impact on the working memory, thus promoting information processing and assuring a greater success in learning processes [15]. In turn, a high *cognitive load* (i.e., overloading the working memory) inhibits information processing leading to confusion and a higher risk of making mistakes. As a consequence, *cognitive load* should be kept at an appropriate level [16]. Therefore, an appropriate level of *cognitive load* can be ensured from the ideal interplay of *mental load* and *mental effort*. Particularly, by designing tasks and presenting information in such way that an individual is not confronted with challenges, demanding more capacity in the working memory [17].

3 Study Setting

Any *process modeling notation* has its own strengths and weaknesses regarding, for example, model conformance checking or expressibility [23]. Considering this fact, enterprises are confronted with the important decision about which *process modeling notation* fulfills their requirements and covers all their needs. Some of the *process modeling notations* are offering an extensive syntax to express business processes in a fine-grained level. On the other, some *notations* provide only a limited set, which is, however, sufficient for correctness verification of process models [24]. For an effective use of *process modeling notations*, their formalisms and methodologies must be comprehended correctly. Thereby, the acquisition of knowledge about a *process modeling notation* represents a cognitive task. In this context, several aspects of a *modeling notation* are learned more easily and quickly, while, on the other, some aspects are difficult to learn, having different impact on the *cognitive load* of an individual. However, this effect cannot be generalized and, hence, is completely different between individuals. Especially novices without any knowledge in process modeling are often confronted with difficulties how to properly comprehend *process modeling notations*. To address this issue, we conduct an exploratory comprehension study in which novices, intermediates, and experts from the ·domain Thereby, we want to investigate the impact of *modeling notations* on the *cognitive load* (i.e., *mental load* and *mental effort*) of individuals. In detail, we agree on using the following *notations* as described in Table 1. Aside well-known *modeling notations* (e.g., *BPMN*), we chose *notations* that are rarely seen in the process repositories of enterprises (e.g., *IDEF3*). Table 1 contains an additional column *specific* (i.e., Spec.), stating whether the use of a *modeling notation* is more focused on particular aspects.

3.1 Study Planning

Participants. All participants have an academic background. In detail, students and research associates as well as professionals, who take a distance e-learning course, are invited for the study at Ulm University. There are no prerequisites for participating in the study and all participants are recruited on a voluntary basis. Further, all participants have given their consent.

Table 1. Process modeling notations used in the study

Notation	Brief Description	Spec.
BPMN 2.0 [4]	*BPMN* is a graphical notation for documenting business processes based on flowcharting techniques. Nowadays, BPMN is an established standard for process modeling	
Declarative [18]	*Declarative Process Modeling* is an approach specifying in an implicit manner through the use of constrains the order of execution for a process model	✓
eGantt [19]	*eGantt Chart* is an extension of the Gantt Chart with an emphasis to capture time characteristics of a process in a process model	✓
EPC [5]	*EPC* is a flow chart type mainly used for the implementation and configuration of enterprise resource planning	
Flow [6]	*Flow Chart* is a diagram for the graphical modeling of processes. *Flow Charts* are widely used and enable the creation of easy-to-understand process models	
IDEF3 [20]	*IDEF3* is a method used for modeling of processes with a focus on the process flow as well as the state of objects and respective conditions	✓
Petri Net [21]	*Petri Net* is a notation mainly used for the description of distributed systems and the only notation with an exact mathematical theory	✓
UML Act. [22]	*UML Activity Diagram* documents activities as well as the control flow in a flowchart manner and are often used for process modeling	

Object. Participants need to assess eight different process models (cf. Table 1) regarding their subjective *comprehensibility* and *accessibility*. With this assessment, we want to draw conclusions about the perceived *mental load* and *mental effort* of the participants. Thereby, the process models reflect three different *levels of complexity* (i.e., *easy*, *medium*, and *hard*). To be more precise, the *easy process model* contains only basic modeling elements. With rising *level of complexity*, the total number of elements is increased and new elements, previously not contained in the process model, are added. For each *level of complexity*, process models are created using the mentioned eight *process modeling notations* respectively (cf. Table 1). As the semantic description of the process models is not relevant in this context, we use abstract labels (i.e., alphabetical letters) for the single modeling elements.[1] Furthermore, it is ensured that all process models are comparable within a specific *level of complexity*. Therefore, experts and novices in the domain of process modeling, who were not participating in the study, ranked and compared the used process models.

---

[1] Material: https://www.dropbox.com/sh/4shyxdy2p4xsf71/AAAfY1EMfwrluYA2B Q-g4z18a?dl=0.

Instrumentation. There are three different questionnaires in the study. First, a *pre-study questionnaire* is used gathering demographic data (e.g., age, gender) and asking about prior knowledge on *process modeling notations*. In addition, participants were asked about their familiarity with specific *modeling notations*.

Second, a *mid-study questionnaire* is used providing four items regarding the *mental load* (i.e., comprehensibility and accessibility of the process models): the ① process model is comprehensible, the ② process model is accessible, the ③ used modeling constructs (e.g., split and join) are comprehensible, and the ④ the process flow is understood properly. All items are rated on a 7-point Likert scale, ranging from 0 (i.e., strongly disagree) to 6 (i.e., strongly agree).

Third, a *post-study questionnaire* is used with four items capturing the *mental effort* of participants: the ① mental demand for performing the task is high, the ② task is complex, the ③ overall performance during the task, and the ④ effort level for performing the task. All items are rated on a 7-point Likert scale, ranging from 0 (i.e., strongly disagree) to 6 (i.e., strongly agree).

In addition, in the *post-questionnaire* participants need to categorize the assessed process models according their subjective preferences regarding the *comprehensibility* as well as *accessibility*, beginning from the simplest to the most difficult process model. Finally, participants are able to leave qualitative feedback.

3.2 Study Design and Procedure

The design and procedure of the study is based on the guidelines set out by [25] on how to systematically plan, conduct, and evaluate studies. Precedent, a pilot study with four students and four research associates was conducted for the improvement of the process models and to ensure their comparability with each other. Moreover, the pilot study was used to eliminate potential ambiguities as well as misunderstandings. Further, the chosen language in the study is German. Within a period of two weeks, several sessions for conducting the study are offered to the participants. Each session took about 20 min and ran as follows: An introduction is given to the participants, in which the procedure of the study is explained and the study materials (i.e., process models, questionnaires) are handed out. Afterwards, the participants need to answer the *pre-study questionnaire*. Following this, they are asked to read and assess eight randomly selected process models. Thereby, it is ensured that participants need to assess each *modeling notation* in a study run. For each assessed process model, the participants have to answer statements regarding subjective *comprehensibility* and *accessibility* (i.e., *mental load*). After completing this step, participants answer the *post-study questionnaire*, providing information about perceived *mental effort* and they need to categorize assessed process models in a ranking system. The study procedure is illustrated in Fig. 1.

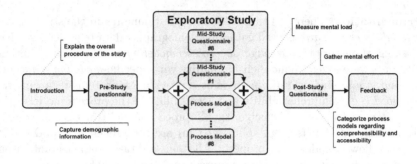

Fig. 1. Study design

4 Data Analysis and Interpretation

In total, data from 94 participants were collected. A median split is performed to categorize the participants into samples of novices, intermediates, and experts. Therefore, we determine the median for the number of process models a participant has analyzed and created during the last 12 months. Consequently, $n = 38$ novices, $n = 21$ intermediates, and $n = 35$ experts participate in the study. Each participant assesses eight randomly selected process models regarding subjective *comprehensibility* and *accessibility* (i.e., *mental load*), resulting in $n = 752$ assessed process models. Table 2 summarizes the detailed distribution of the assessed *process modeling notations* for each *level of complexity*.

Table 2. Number of assessed process models

		Process modeling notations							
		BPMN	Declarative	eGantt	EPC	Flow	IDEF3	Petri	UML Act.
Level	Easy	30	31	33	31	36	28	34	35
	Medium	29	32	33	36	28	35	33	31
	Hard	36	31	28	27	30	31	27	27

4.1 Descriptive Statistics

The obtained data regarding the perceived *mental load* for each *process modeling notations* are presented in Table 3 as means for the entire sample size as well as each sample respectively (i.e., novices, intermediates, and experts). Higher values indicate less *mental load*. As described in Sect. 3.1, *mental load* is determined with four aggregated items. As a prerequisite, all response variables must show a high reliability [26]. For this purpose, *Cronbach's* α (i.e., several items are an accurate estimate of an accumulated item) is calculated.[2] For *mental load*, a Cronbach with $\alpha = 0.79$ was calculated.

---

[2] According to [26], $\alpha > 0.6$ acceptable reliability; $0.7 < \alpha < 0.9$ good reliability.

Table 3. Mental load

		BPMN	Declarative	eGantt	EPC	Flow	IDEF3	Petri	UML Act.
All	Easy	5.77	4.70	5.34	5.60	5.61	4.62	5.66	5.54
	Medium	4.61	3.61	3.70	5.13	4.86	4.34	4.51	4.68
	Hard	3.60	3.27	3.01	4.99	3.99	3.64	3.90	4.46
Nov.	Easy	5.76	5.12	5.03	5.18	5.43	4.47	5.87	5.60
	Medium	4.47	3.45	3.28	4.54	5.14	3.67	4.83	4.69
	Hard	3.22	3.20	2.98	4.63	2.72	2.42	4.10	4.35
Int.	Easy	5.84	4.21	5.43	5.78	5.63	4.88	5.33	5.70
	Medium	4.66	3.66	4.13	5.32	5.17	4.78	4.88	4.50
	Hard	4.53	3.47	3.45	5.55	5.29	4.80	3.95	4.92
Exp.	Easy	5.70	4.77	5.56	5.84	5.78	4.50	5.79	5.33
	Medium	4.71	3.71	3.68	5.54	4.26	4.56	3.81	4.84
	Hard	3.05	3.00	2.58	4.79	3.97	3.69	3.66	4.11

Figures 2, 3 and 4 depict the two-dimensional data (i.e., *process modeling notations, level of complexity*) regarding the *mental load* for the entire sample and each sample respectively (i.e., novices, intermediates, and experts) as radar charts. Thereby, higher values stand for less *mental load* on the used scale.

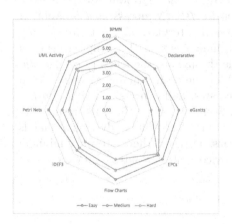

Fig. 2. Mental load - All   **Fig. 3.** Mental load - Novices

Regarding the *easy level of complexity* (cf. Figs. 2, 3 and 4), several *process modeling notations* (e.g., *BPMN*, *Petri Nets*, and *UML Activity Diagrams*) are close to the maximum of the scale (i.e., 6 - reflecting less *mental effort*). However, comparing the values obtained from each sample, there is an indication that *easy process models* expressed in *Declarative* or *IDEF3* appear to be more challenging to comprehend and, consequently, demand a higher *mental load* from the participants. In general, with rising *level of complexity*, an erratic decrease

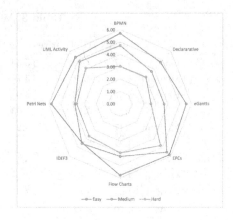

Fig. 4. Mental load - Intermediates   **Fig. 5.** Mental load - Experts

in the *mental load* is observable (cf. Table 2). In detail, instead of an uniform decrease of the values for the *mental load*, for several *modeling notations*, an abrupt decline can be seen in the samples. For example, there is a significant drop in the *mental load* between the *easy* and *medium BPMN process models* in each sample. For novices and intermediates, there is only a little decline in the *mental load* between the *medium* and *hard BPMN model*. However, experts show again a significant drop in the *mental load* for *BPMN process models*. Another example concerns the process models expressed in terms of *EPCs*. There are only slight decreases in the *mental load*, which indicates that the comprehension of *EPC process models* appears to be feasible throughout the *levels of complexity*.

Table 4 shows the *mental effort (ME)* for the entire sample as well as each sample respectively (i.e., novices, intermediates, and experts). *Mental effort* is determined by four aggravated items (cf. Sect. 3.1), here, *Cronbach* resulted in $\alpha = 0.83$. Higher values stand for less *mental effort*. The study task demands a moderate *mental effort* from all samples. Further, there are only minimal differences in the *mental effort* between novices, intermediates, and experts.

Table 4. Mental effort

	All	Novices	Inter.	Experts
ME	2.84	2.76	2.89	2.94

4.2 Discussion

It is obvious that with rising *level of complexity* the *mental load* regarding the *process modeling notations* is decreasing (cf. Fig. 2). Despite the comparability of the process models, however, it is interesting to see how the *mental load* is decreasing in a different manner between novices, intermediates, and experts.

For example, the *mental load* from intermediates regarding the *medium* and *hard process model* is about the same. In turn, for novices and experts, the differences between these two *levels of complexity* are clearly discernible. The same can be observed regarding the *IDEF3 process models*. Further, the results provide a good indication about the *mental load* when comprehending process models in particular *modeling notations*. More complex process models expressed in *Declarative, IDEF3*, or *eGantt* appear to be more challenging to comprehend. Reason might be that these *notations* are not as widespread in practice as other *notations*. Often, amongst others, tertiary educational institutions are teaching more common *notations* widely used in practice such as *BPMN* and *EPCs*.

While there are different characteristics for the *mental load*, however, the *mental effort* needed for process model comprehension is approximately the same between novices, intermediates, and experts. Although the *mental effort* for comprehending a process model is on a moderate level, however, process model comprehension is a complex matter that needs to be taken into account.

Based on the study results, Fig. 6 presents the derived *mental difficulty* for each *process modeling notation* and respective *level of complexity*. Therefore, for each process model, the percentage proportion is calculated based on the categorization the participants indicated in the *post-study questionnaire* (cf. Sect. 3.1). Therefore, we considered respective position of each process model as well as their *level of complexity*. The calculated percentage proportion serves as a factor used for multiplying the aggravated constructs *mental load* and *mental effort* to derive the *mental difficulty* [27].

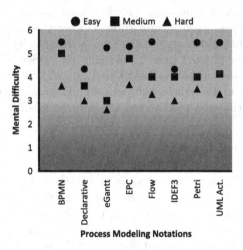

Fig. 6. Mental difficulty

As shown in Fig. 6, regarding an *easy level of complexity*, almost all values, except for *Declarative* and *IDEF3*, are still in the green range. This means that the comprehension of such process models should not be a challenge. With rising *level of complexity*, the values for the comprehension of process models tend towards the orange or red range respectively. However, none of the values is located completely in the red range and, consequently, it seems to be that process models can be comprehended independently of the *modeling notation* and *level of complexity*. In addition, the results confirm the further use of the widespread and established *modeling notations* such as *BPMN* and *EPCs*.

4.3 Threats to Validity

Study results can only be accurately interpreted when limitations are also discussed. *First*, the comparison of the different *process modeling notations* can also be seen as a comparison like the idiom apples and oranges. Particular *modeling notations* are usually not used for the direct modeling of business processes, but for other purposes (e.g., *Petri Nets* are mainly used for the verification of process models) and, hence, a direct comparison is difficult. *Second*, the classification of participants into the samples of novices, intermediates, and experts solely based on a median split by considering the number of process models a participant has analyzed and created during the last 12 months may be oversimplified. Hence, results might differ significantly with experts in process modeling with several years of experience. *Third*, the size of the process models might not be representative. In general, real world process models are usually much larger than the models we used in the study. *Fourth*, in addition, to ensure comparability between the process models, the latter are modeled only with basic modeling elements because some *modeling notations* don't have the expressiveness for the documentation of complex processes. *Fifth*, the respective level of complexity reflected by the process models constitutes another threat. The models might be considerably unbalanced between the level of complexity and, hence, working memory capacity of participants, especially for novices, may be exceeded. *Seventh*, specific *process modeling notations* (e.g., *BPMN*) may be considered as easy to comprehend since these *notations* are more familiar than others (e.g., *IDEF3*), thus resulting in a positive impact on *mental load* and *mental effort*.

5 Related Work

A review of existing business *process modeling notations* in literature is presented in [23]. Further, the authors present a framework for the classification of *modeling notations* according their purpose. [28] discusses an evaluation of *modeling notations* and proposes a meta-model to capture the concepts of the evaluated *notations*. In turn, [29] compares and discusses different process modeling methods based on aspects like process model representation and tool support.

Concerning research on *process model comprehension*, [30] gives insights into subject-specific characteristics (e.g., theoretical knowledge) influencing *process model comprehension*. In turn, [31] evaluates different *process modeling notations* with respect to their comprehensibility. A discussion about factors having an influence on the *comprehension of process models* is presented in [32].

Regarding cognitive aspects, a measure to determine the cognitive complexity for process models is proposed in [33]. Further, [34] shows an approach to reduce the *cognitive load* when comprehending process models by applying patterns for improving the model comprehensibility. The empirical assessment of participants' *mental effort*, while creating or comprehending process models, is demonstrated in [35]. The *mental effort* needed between inexperienced and experienced modelers to create process models is investigated in [36].

Altogether, there are several works regarding the comparison of *process modeling notations*. However, to the best of our knowledge, none of the discussed works deal with such a comparison of *process modeling notations*, while taking the *cognitive load* (i.e., *mental load* and *mental effort*) of participants into account.

6   Summary and Outlook

This paper presented the impact of different *process modeling notations* on the *cognitive load* (i.e., *mental load* and *mental effort*) of $n = 38$ novices, $n = 21$ intermediates, and $n = 35$ experts in the domain of process modeling. Therefore, study participants needed to comprehend and assess process models of different *levels of complexity* (i.e., *easy, medium, hard*) in terms of eight different *modeling notations*, i.e., *BPMN 2.0, Declarative Process Modeling, eGantt Charts, EPCs, Flow Charts, IDEF3, Petri Nets*, and *UML Activity Diagrams*. We gathered information related to the *mental effort* and *mental load* (i.e., *cognitive load*) of participants while performing the study task. Based on the *cognitive load*, we derived a factor representing the *mental difficulty* for each *process modeling notation*. The high availability of *process modeling notations* resulted in a strong demand from enterprises to compare and evaluate these *notations* in order to find an appropriate one covering the needs of an enterprise. The presented *mental difficulty* may support enterprises in making this decision considering the perceived *cognitive load* of an individual, while comprehending process models expressed in terms of different *modeling notations*. In addition, the results from this paper may help in answering questions about which *modeling notations* should be supported with a greater emphasis in modeling tools. Further, the results show that the use of established *modeling notations* (e.g., *BPMN*) is recommendable for enterprises and future process modelers as well as analysts.

Future research is needed in order to determine more precise indications about an individuals *cognitive load* while comprehending process models. Besides an in-depth statistical analysis of the results, for example, instead of using abstract labels the process model elements could be described with concrete labels. Further, participants should be confronted with process models from real projects expressed in terms of the assessed *modeling notations*. Finally, *cognitive load* is only one factor having an effect on the comprehension of process models expressed in different *modeling notations* and, therefore, the consideration of additional factors (e.g., *expressiveness, level of automation*) will be subject of future studies.

References

1. Nowak, A., et al.: Flexible information design for business process visualizations. In: 5th International Conference on Service-Oriented Comp and App (SOCA 2012), pp. 1–8 (2012)
2. Bandara, W., Indulska, M., Chong, S., Sadiq, S.: Major issues in business process management: An Expert Perspective, pp. 1240–1251 (2007)
3. Gulla, J.A., Brasethvik, T.: On the challenges of business modeling in large-scale reengineering projects. In: International Conference on Requirements Engineering, pp. 17–26 (2000)
4. OMG: Business Process Management & Notation 2.0 (2018). www.bpmn.org. Accessed 27 Feb 2018
5. van der Aalst, W.M.P.: Formalization and verification of event-driven process chains. Inf. Soft, Tech., 41(10), 639–650 (1999)
6. Schultheiss, L.A., Heiliger, E.: Techniques of flow-charting. In: Proceedings of the 1963 Clinic on Library Applications of Data Processeing, pp. 62–78 (1963)
7. Zimoch, M., et al.: Cognitive insights into business process model comprehension: preliminary results for experienced and inexperienced individuals. In: Proceedings of the BPMDS 2017, pp. 137–152 (2017)
8. Zimoch, M., et al.: Using insights from cognitive neuroscience to investigate the effects of event-driven process chains on process model comprehension. In: Proceedings of the 1st International Conference on Cognitive Business Process Management, pp. 446–459 (2017)
9. Zimoch, M., et al.: Eye tracking experiments on process model comprehension: lessons learned. In: Proceedings of the BPMDS 2017, pp. 153–168 (2017)
10. Zugal, S., Pinggera, J., Weber, B.: Assessing process models with cognitive psychology. EMISA 190, 177–182 (2011)
11. Paas, F., et al.: Cognitive load measurement as a means to advance cognitive load theory. Educ. Psychol. 38, 63–71 (2003)
12. Wickens, C.D.: Multiple resources and mental workload. Hum. Factors 50, 449–455 (2008)
13. Paas, F.G., Van Merriënboer, J.J.: The efficiency of instructional conditions: an approach to combine mental effort and performance measures. Hum. Factors 35, 737–743 (1993)
14. Krell, M.: Evaluating an instrument to measure mental load and mental effort considering different sources of validity evidence. Cogent Educ. 4, 1280256 (2017)
15. Sweller, J.: Cognitive load during problem solving: effects on learning. Cogn. Sci. 12, 257–285 (1988)
16. Ayres, P.: Impact of reducing intrinsic cognitive load on learning in a mathematical domain. Appl. Cogn. Psychol. 20, 287–298 (2006)
17. Baddeley, A.: Working memory. Science 255, 556–559 (1992)
18. van Der Aalst, W.M., et al.: Declarative workflows: balancing between flexibility and support. Comp. Sci-Res. Dev. 23, 99–113 (2009)
19. Lanz, A., Kolb, J., Reichert, M.: Enabling personalized process schedules with time-aware process views. In: Franch, X., Soffer, P. (eds.) CAiSE 2013. LNBIP, vol. 148, pp. 205–216. Springer, Heidelberg (2013). https://doi.org/10.1007/978-3-642-38490-5_20
20. Kim, C.H., Yim, D.S., Weston, R.: An integrated use of IDEF0, IDEF3 and petri net methods in support of business process modelling. Proc. Inst. Mech. Eng. Part E: J. Process. Mech. Eng. 215(4), 317–329 (2001)

21. Murata, T.: Petri Nets: properties, analysis and applications. Proc. IEEE **77**, 541–580 (1989)
22. Dumas, M., ter Hofstede, A.H.M.: UML activity diagrams as a workflow specification language. In: Gogolla, M., Kobryn, C. (eds.) UML 2001. LNCS, vol. 2185, pp. 76–90. Springer, Heidelberg (2001). https://doi.org/10.1007/3-540-45441-1_7
23. Aguilar-Saven, R.S.: Business process modelling: review and framework. Int. J. Prod. Econ. **90**, 129–149 (2004)
24. White, S.A.: Process modeling notations and workflow patterns. In: Workflow Handbook, pp. 265–294 (2004)
25. Wohlin, C., Runeson, P., Höst, M., Ohlsson, M.C., Regnell, B., Wesslen, A.: Experimentation in Software Engineering - An Introduction. Kluwer, New York (2000)
26. Kline, P.: Handbook of Psychological Testing, vol. 2. Routledge, Abingdon (1999)
27. Sweller, J.: Cognitive load theory, learning difficulty, and instructional design. Learn. Instr. **4**, 295–312 (1994)
28. List, B., Korherr, B.: An evaluation of conceptual business process modelling languages. In: Proceedings of the Symposium on Applied Computing (SAC 2006), pp. 1532–1539 (2006)
29. Wang, W., Ding, H., Dong, J., Ren, C.: A comparison of business process modeling methods. In: International Conference on SOLI 2006, pp. 1136–1141 (2006)
30. Mendling, J., Recker, J., Reijers, H.A., Leopold, H.: An empirical review of the connection between model viewer characteristics and the comprehension of conceptual process models. In: Information Systems Frontiers, pp. 1–25 (2018)
31. Kiepuszewski, B., ter Hofstede, A.H.M., Bussler, C.J.: On structured workflow modelling. In: Wangler, B., Bergman, L. (eds.) CAiSE 2000. LNCS, vol. 1789, pp. 431–445. Springer, Heidelberg (2000). https://doi.org/10.1007/3-540-45140-4_29
32. Mendling, J., Strembeck, M., Recker, J.: Factors of process model comprehension-findings from a series of experiments. Decis. Support Syst. **53**(1), 195–206 (2012)
33. Gruhn, V., Laue, R.: Adopting the cognitive complexity measure for business process models. In: 5th International Conference on Cognitive Informatics (ICCI 2006), pp. 236–241 (2006)
34. Gruhn, V., Laue, R.: Reducing the cognitive complexity of business process models. In: 8th International Conference on Cognitive Informatics (ICCI 2009), pp. 339–345 (2009)
35. Zugal, S., Pinggera, J., Reijers, H., Reichert, M., Weber, B.: Making the case for measuring mental effort. In: EESSMod 2012 (2012)
36. Martini, M., et al.: the impact of working memory and the process of process modelling on model quality: investigating experienced versus inexperienced modellers. In: Scientific Reports, vol. 6 (2016)

First International Workshop on Process-Oriented Data Science for Health Care (PODS4H)

First International Workshop on Process-Oriented Data Science for Health Care (PODS4H)

The world's most valuable resource is no longer oil, but data. The ultimate goal of data science techniques is not to collect more data, but to extract knowledge and insights from existing data in various forms. For analyzing and improving processes, event data is the main source of information. In recent years, a new discipline has emerged combining traditional process analysis and data-centric analysis: process-oriented data science (PODS). The interdisciplinary nature of this new research area has resulted in its application for analyzing processes in different domains such as education, finance, and especially health care.

The International Workshop on Process-Oriented Data Science for Health Care 2018 (PODS4H 2018) aimed at providing a high-quality forum for interdisciplinary researchers and practitioners (both data/process analysts and a medical audience) to exchange research findings and ideas on health-care process analysis techniques and practices. PODS4H research includes a wide range of topics from process mining techniques adapted for health-care processes, to practical issues on implementing PODS methodologies in health-care centers' analysis units. For more information, visit pods4h.com

The first edition of the workshop attracted a remarkable number of high-quality submissions, from which nine regular papers and four success cases were selected for presentation. The papers included a wide range of topics: process-oriented dashboards for operation room analysis, process mining for medical training, conformance checking for melanoma surveillance, drug use patterns, trauma patients transport, interactive analysis of emergency processes, event log improvement based on indoor location system data, care pathway simulation, and emergency room episode analysis. The success cases include a framework for managing data quality issues when performing process mining on health records, the combination of real-time location systems with process analysis, a methodology for applying process mining in the emergency room, and the NETIMIS care pathway simulation tool. The workshop also hosted a panel where several topics were discussed, including new projects and initiatives in the area.

This edition of the workshop included two awards to the best regular paper and the best success case. The PODS4H 2018 Best Paper Award was given to the "Tailored Process Feedback Through Process Mining for Surgical Procedures in Medical Training" by Ricardo Lira, Juan Salas-Morales, Rene de La Fuente, Ricardo Fuentes, Marcos Sepúlveda, Michael Arias, Valeria Herskovic, and Jorge Munoz-Gama. The PODS4H 2018 Best Success Case Award was given to "Applying Value-Based Health Care in Hospital Management with Process Mining and Real Time Location Systems" by Carlos Fernandez-Llatas, Vicente Traver, Salvador Vera, Eduardo Monton, and

Jordi Rovira. The prize included a voucher for a professional Data Scientist Training, provided by the Celonis Academic Alliance.

The workshop was an initiative of the Process-Oriented Data Science for Healthcare Alliance. The goal of this international alliance is to promote the research, development, education, and understanding of process-oriented data science in health care. For more information about the activities and its members, visit pods4h.com/alliance

The organizers would like to thank all the Program Committee members for their valuable work in reviewing the papers, and the BPM 2018 Organizing Committee for supporting this successful event.

November 2018

Jorge Munoz-Gama
Carlos Fernandez-Llatas
Niels Martin
Owen Johnson

Organization

Program Committee

Robert Andrews	Queensland University of Technology, Australia
Joos Buijs	Eindhoven University of Technology, The Netherlands
Andrea Burattin	Technical University of Denmark, Denmark
Daniel Capurro	Pontificia Universidad Católica de Chile, Chile
Josep Carmona	Universitat Politècnica de Catalunya, Spain
Carlos Fernandez-Llatas	Universitat Politècnica de Valencia, Spain
René de la Fuente	Pontificia Universidad Católica de Chile, Chile
Roberto Gatta	Università Cattolica del Sacro Cuore, Italy
Zhengxing Huang	Zhejiang University, China
Owen Johnson	University of Leeds, UK
Felix Mannhardt	SINTEF, Norway
Ronny Mans	VitalHealth Software, USA
Niels Martin	Hasselt University, Belgium
Renata Medeiros de Carvalho	Eindhoven University of Technology, The Netherlands
Jorge Munoz-Gama	Pontificia Universidad Católica de Chile, Chile
Ricardo Quintano	Pontifícia Universidade Católica do Rio de Janeiro, Brazil/Philips Research
David Riaño	Universitat Rovira i Virgili, Spain
Eric Rojas	Pontificia Universidad Católica de Chile, Chile

Lucia Sacchi   University of Pavia, Italy
Fernando Seoane   Karolinska Institutet, Sweden
Marcos Sepúlveda   Pontificia Universidad Católica de Chile, Chile
Minseok Song   Pohang University of Science and Technology,
                                   South Korea
Vicente Traver   Universitat Politècnica de Valencia, Spain
Wil van der Aalst   RWTH Aachen University, Germany
Rob Vanwersch   Maastricht University Medical Center,
                                   The Netherlands
Chuck Webster   EHR Workflow Inc., USA

Expectations from a Process Mining Dashboard in Operating Rooms with Analytic Hierarchy Process

Antonio Martinez-Millana[1]([✉]), Aroa Lizondo[1], Roberto Gatta[2],
Vicente Traver[1,3], and Carlos Fernandez-Llatas[1,3]

[1] ITACA, Universitat Politècnica de València,
Camino de Vera S/N, 46022 Valencia, Spain
anmarmil@itaca.upv.es
[2] Polo Scienze Oncologiche ed Ematologiche,
Universit Cattolica del Sacro Cuore, Fondazione Policlinico Universitario Agostino
Gemelli, Largo Francesco Vito 1, 00168 Rome, Italy
[3] Unidad Mixta de Reingeniería de Procesos Sociosanitarios,
Instituto de Investigación Sanitaria del Hospital Universitario y Politecnico La Fe,
Bulevar Sur S/N, 46026 Valencia, Spain

Abstract. The wide-spread adoption of real-time location system is
boosting the development of software applications to track persons and
assets on real-time and perform analytics. Among the vast amount of
data analysis techniques, process mining allows to conform work-flows
with heterogeneous multivariate data, enhancing the model understandability and usefulness in clinical environments. However, such applications still find entrance barriers in the clinical context. In this paper we
have identified the preferred features of a process mining based dashboard deployed in the operating rooms of a hospital equipped with a
real-time location system. Work-flows are inferred and enhanced using
process discovery on location data of patients undergoing an intervention,
drawing nodes (states in the process) and transitions across the entire
process. Analytic Hierarchy Process has been applied to quantify the
prioritization of the features contained in the process mining dashboard
(filtering data, enhancement, node selection, statistics, etc..), distinguishing on the priorities that each of the different roles in the operating room
service assigned to each feature. The staff in the operating rooms (N=10)
was classified into three groups: Technical, Clinical and Managerial staff
according to their responsibilities. Results show different weights for the
features in the process mining dashboard for each group, suggesting that
a flexible process mining dashboard is needed to boost its potential in
the management of clinical interventions in operating rooms.

1 Introduction

Operating Rooms (ORs) is an essential and central element in modern hospitals
[1]. Cost estimations of ORs are around 16% and 20% of total hospital budget, due to the amount of human resources mobilized, use of high technology,

© Springer Nature Switzerland AG 2019
F. Daniel et al. (Eds.): BPM 2018 Workshops, LNBIP 342, pp. 151–162, 2019.
https://doi.org/10.1007/978-3-030-11641-5_12

the heterogeneity of roles involved the importance it plays in the image of the institution [2,3]. Efficiency in the management and interventions in ORs aims to reduce the non-occupation time of the surgical blocks and the way to organize the different types of interventions to optimize medical teams (surgeons and nurses). To achieve a good efficiency rates in OR it is crucial to have a protocol and a high skilled staff to take over daily-basis decision making for scheduled and unscheduled interventions [3].

Recent reviews conclude that managerial surgeons make decisions to increase the clinical work per time-unit in individual ORs and that command displays may be an effective way to gain efficiency [4]. This behavior is well-known as reactive scheduling, which involves schedule definition and posterior assessment [5]. The schedule definition entails to foresee starting, duration and ending times for the regular operations sequences (preparation, anesthesia induction, intervention, wake-up and turnover) in terms of time and resources. The posterior assessment monitors the schedule execution and adapts the planned scheduled to deal with unexpected events [6]. Reactive scheduling process occurs when unexpected events or disruptions occur along the process [7]. Nevertheless, the main disadvantage in the ORs planning and management is that processes are often recorded manually by nurses. This issue has been already identified as a bottleneck in the performance assessment of ORs [8]. Moreover, manual notes can also lead to bias: unnecessary delays, under-use of the operation rooms, unnecessary transfers, etc. In addition it could cause an increase in the probability of the adverse effects in the surgical process which, according to [9] stand for the 40% of all the adverse effects in hospitals.

Information and Communications Technology (ICT) can provide tools and systems to support both programming and assessment of operations. To perform this tracking process a pragmatic task *Gonzalez et altres* proposed a semi-automatic information collector [10]. Nowadays, some hospitals are equipping them with Real-Time Location System (RTLS) to manage the location of patients and assets that could help to optimize the management of ORs by applying process mining (PM) techniques.

An example of the use of the extracted data from the RTL Systems to manage the patient locations is introduced by *Fernandez-Llatas et altres* [11], in which PM is applied to perform an analysis of operation sequences and locations in the ORs showing the most common paths and insights of the entire process in ORs based on RTLS data. In this study, researchers developed a front-end application to analyze ORs process providing a complete suite of tools to discover, compare and enhance surgical processes.

Technologies should be presented in a meaningful way to the ORs staff to ensure a successful deployment. The application of computer decision systems in an interactive way will not only increase their effectiveness and efficacy [12], but also involve the staff of ORs in the process of knowledge extraction, avoiding frustrations using technology for managing complex processes [13].

Process mining has multiple unknowns when landing to real applications when the intended user is the clinical staff. In this paper we report a study based on the Analytic Hierarchy Process (AHP) to identify which are the preferred features of a web-based process mining dashboard. This front-end has been presented in [11]. To distinguish the preferences of each role, we have grouped the users in three groups: manager, hospital staff and technical staff of the ORs. We have obtained feature prioritization of 10 subjects including the three roles. The AHP is particularly effective for quantifying experts' opinions that are based on personal experience and knowledge to design a consistent framework for the application of process mining in ORs.

2   Related Work

Real-Time Location Systems (RTLS) monitor with the position of a moving element with a given sampling frequency. In a hospital, moving elements are equipped with an active or passive element (tag) which identifies it when is nearby a beacon. In our study patients are the moving elements who bear a wristband with the tag before entering the operating room service.

Lean principles present a condensed primer of RTLS in health care environment [14]. Throughput is a key performance indicator for a patient pathway across a facility, which linked to RTLS could provide valuable information such as waiting times and resources utilization [15].

RTLS systems in 23 hospitals in the US have been analyzed from a qualitative perspective in [16]. In this work, researchers observed the systems in use and conducted 80 semi-structured interviews with hospital personnel and vendors. Authors find asset tracking as the best feature and identify several obstacles related to the technical set-up and organizational context.

Specifically, the operating room service (Fig. 1) consist of several spaces, each of them equipped with a beacon to identify a patient whenever he/she goes to that specific area.

The application of process mining techniques in combination with RTLS systems provided an easy to use and unobtrusive way to achieve a view of the deployed process. In this paper we analyze the web-based dashboard to perform process mining discovery and enhancement analytics presented by Fernandez-Llatas et altres [11] (Fig. 2).

Analyzing RTLS data from a discovery perspective and enhancing these work flows with information related to the average time and overload it is possible to create pre-programmed contingency plans for the management and allocation of resources of the operating rooms. But, why this promising applications are still not widely used in the clinical context? Instead of using a semistructured interview we have used the Analytic Hierarchy Process [17] to quantify the features of a dashboard for discovering and enhancing processes based on RTLS data in a ORs service.

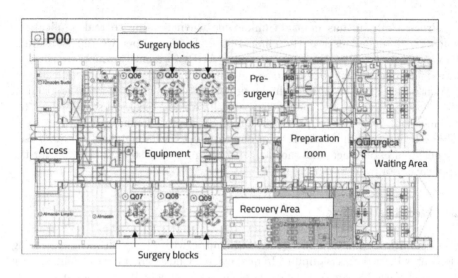

Fig. 1. Composition of the operating room service

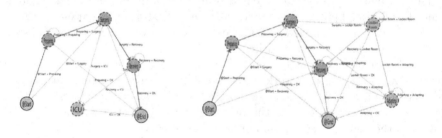

Fig. 2. Example of the inferred and enhanced work flows of patients across the operating room service [11].

3   Materials and Method

3.1   Analyzing ORs Processes with PALIA

PALIA consists on a web-based dashboard which allows to perform process mining analysis on a given dataset. The software (Fig. 3) is composed of three major areas: Filters, for the selection of the data; Mining, for the configuration of the visualization of work flows and Information, to show the information about the operation tool and the selected tracks. These three major areas are divided into five functional areas:

1. **Filters (1–2)**, for the selection of the input data in each analysis. There are several types of filters depending of the type of data and the required information: dates, times, durations, type of intervention, etc. This component shows the percentages of the samples meeting the filtering characteristics, so the user can have an idea on the extension of the selected subgroup of data.

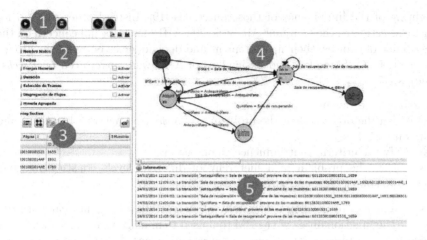

Fig. 3. Areas of the analyzed PM Tool

2. **Miner (3–4)**, for the configuration of the work flow visualization. Graphical representations of the inferred processes are depicted in the central part of the web-tool(4) by means of ORs states (nodes) and transitions (arrows). The visualization component allows adding meta-information to the work flow, for instance, rendering heats maps to discover frequencies or occupation in the ORs. There is also a list with all the samples selected with the filters (1–2)

3. **Information (5)**, which shows details about how the process mining algorithm was applied and also features of the selected samples: Information on the number of merged branches, a log on how the process mining algorithm analyzed the events and infers work-flows, errors on data selection and statistics on the transitions and states in the work-flow.

PALIA works in the following way: First a Comma-separated Value file containing the RTLS data is loaded using a default file dialog window. Then, the user is able to filter the data and apply a discovery topological algorithm. The inferred work-flow appears in the screen with nodes and arrows, which represent the track followed by the patients across the surgical process (Preparation - Surgery - Recovery - Intensive Care Unit - Locker room - Adaptation).

3.2 Analytic Hierarchy Process to Determine Priorities

AHP is a methodology for decision-making which aims at solving complex problems. It allows quantifying opinions and transforming them into a coherent decision model. The process is derived from a pairwise comparison using a numerical scale. AHP has found its widest applications in multi-factors decision-making, planning, and resource allocation and in conflict resolution [18]. The AHP is a method which incorporates benefits and risks, explicitly by combining the importance of differences in probabilities of outcomes related to alternatives and the weighting of the importance of those outcomes [19]. Instead other method for feature selection as the Conjoint Analysis [20], AHP has a special concern on the consistency of choices, their measurement and dependencies between the groups of elements [21]. Some key and basic steps of AHP were introduced by *Pecchia et altres* [17]:

- Define the problem.
- Broaden the objectives of the problem or consider all actors, objectives, and its outcome.
- Identify the criteria that influence the behavior.
- Structure the problem in a hierarchy of different levels constituting goal, criteria, sub-criteria and alternatives.

Our aim was to develop a hierarchy of elements grouped into categories to describe the functionalities of the PM tool used to manage ORs processes. These categories were ranked using questionnaires to extract the relative importance of each need per category (local weights, LW), the relative importance of each category (category weights, CW), and the importance of each need compared to all the others (Global weights, GW) [22].

3.3   Applying AHP to PALIA

The AHP was applied to this study, mainly because of its inherent capability of handling qualitative and quantitative criteria used in reclamation method selection problems. Six factors for applying AHP [18] were considered and fulfilled in this study. The hospital was motivated to adopt PALIA to support ORs processes management and was committed to implement the decision and involved staff from the ORs department. Stakeholders were active participants in the entire decision process from development to implementation. In order to identify the elements and the categories of the hierarchy we have used the different functionalities of PALIA. The hierarchy is composed of four levels which have a 1:n relationship with on the three functional areas described in Subsect. 3.1. The first hierarchy level is generic and only describes the functional area (Filters, Miner and Log). The second hierarchy level describes the main features of the functional area (for Filters it contains Dates, Times, Duration, Type of Intervention, Type of OR, etc...). The third hierarchy level contains details of the main features within the functional area (in Filters, for the Type of Intervention it contains details of the medical service, the type of program, the surgical process, the surgeon in chief, etc..). The fourth and final hierarchy level contains low granularity details. For creating the hierarchy tree and collecting the answers we have used the application web BPMSG AHP Online System (https://bpmsg. com/academic/ahp.php). In this study we focus on two hierarchy priority Levels:

1. **Hierarchy Level 1**, Functional Area: Filter, Miner and Information
2. **Hierarchy Level 2**, Features of the functional level:
    - **Filter:** Dates, Time, Duration, Level, Node Name, Features, Stretch, Disgregation, Statistics

- **Miner:** Frequency, Occupation, Transitions, List of samples
- **Information:** Sample cleaning, Wrong Selection, Evolution of process, Error messages and extra information.

3.4 Participants

Table 1 shows the profiles of the involved participants and their specific role into the ORs service. Only one of the initial 11 participants (complete ORs Staff) was unable to fill the AHP questionnaire successfully and was discarded for the analysis. Responders, who signed the informed consent of this study, were employers of the *Hospital General de Valencia*, one of the four hospitals of reference at the city covering a population of 350,000 inhabitants. It has 27 operating rooms and in the 2014 it registered 26,497 surgeries [23]. We organized a meeting in the Hospital General with all the participants which lasted 2 h. The session consisted of an introduction to the dashboard (Fig. 3) and a walk through with real data to showcase examples.

Table 1. Profiles of the participants in the AHP study

Variable	Type	Distribution
Role	Manager	20%
	Hospital staff	60%
	Technical	20%
Age		46.2 ± 10.3
Gender	Male	40%
	Female	60%
Years of expertise		21.2 ± 10.7
Computer literacy	Low	0%
	Medium	70%
	High	30%

4   Results

A total of 10 questionnaires were collected after the session held in the University General Hospital of Valencia with professionals working in the ORs service. For the analysis of the responses, participants were grouped into three categories depending of their roles within the ORs service.

The AHP questionnaire allows to assign priorities for each of the features contained in the defined hierarchy levels, each of which correspond to a particular functional area of the PALIA web-tool.

The overall analysis of priorities shows a 57% level of consensus (relative homogeneity $\beta = 77\%$), in which the Miner component achieves the higher priority rates for the features of Visualization (heat maps) and views selection (maximum occupation and current occupation). Collecting the relative priorities within each of the three main categories, we could extract the importance that each group of users assign to each of the PALIA functional areas.

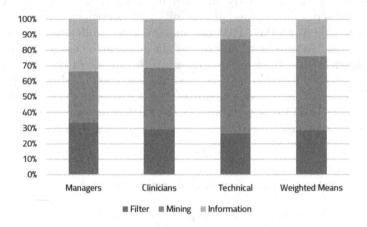

Fig. 4. Assigned priorities for the functional areas

Figure 4 depicts the relative weights (%) assigned to each of the modules. Manager and Clinical user groups show a similar consensus on the prioritization of functional areas, whereas the group composed by the Technical staff provides more priority to the Mining component. The stack in the right part of Fig. 4 is a weighted mean of the relative priority assigned by participants, in which we can see that the Mining component is still the most important, and the two other components (Filters and Information) share a lower similar importance.

Figures 5–7 shows the spider-web diagram containing the assigned priorities within the Hierarchy Level 2 for each of the features, splitting responses by user groups.

Regarding the Filters (Fig. 5) we can see a similar consensus between the Managers and the Clinicians, whereas the Technical staff is weighting two features which were not that relevant for the former groups: Date and Level selection. The answers for the Filter component achieved a 65.7% group consensus.

Regarding the Miner (Fig. 6), which was the most weighted component overall, we can see that Clinicians are more interested on knowing the Occupation of the rooms flow, but with respect to the Frequency, there are similar priorities with the Manager's choice. This figure shows also that only the Technical staff is interested on having the list of samples which composed the work-flow. The answers for the Miner component achieved a 78.5% group consensus.

Regarding the Information (Fig. 7), we can see that the priorities assigned by the Clinicians are the same for each feature, which could indicate a strong

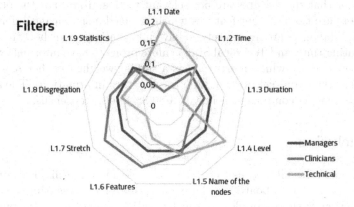

Fig. 5. Priorities for filter functional module features

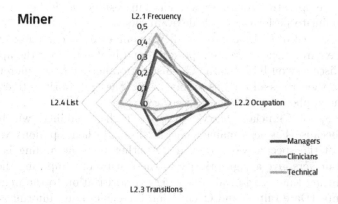

Fig. 6. Priorities for miner functional module features

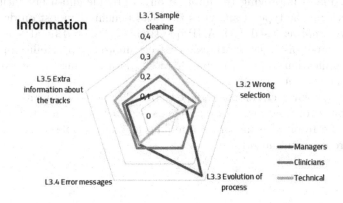

Fig. 7. Priorities for information functional module features

consensus or that the features are not relevant to this group. For the other two groups there are two different features which received a significant different prioritization. Managers are more likely to have information about how the process evolved during time, and Technical Staff is more prone to have information about the sample cleaning (which moreover has a similar weight to other related features such as the wrong selection of samples and extra information). The answers for the Information component achieved a 80.8% group consensus.

5   Conclusions

AHP questionnaires allowed us to extract valuable and quantifiable information about the use of a dashboard for the exploitation of Process Mining on Real-Time Location System samples in the Operating Rooms of a Hospital. The sample size of the questionnaire could be considered small (n=10) to provide significant findings. Nevertheless, this population contains all professional staff working in the operating room services and our results should be considered as a starting point to perform large scale evaluations.

The assessment of PALIA features allow to enhance the communication with the clinical environment to create a powerful and usable tool for the application of process discovery on RTLS data. The distinction between the different groups according to their roles allowed us to analyze how assign priorities to each of the stages of the application of process discovery.

The group of Clinicians shows always a high variability, which can be explained because has less management tasks and their opinions vary more depending of the specific work they do. Another relevant finding is how the group of Managers give a high priority to the feature of comparing the process evolution during time. Technical staff assigns prioritization to the management of timestamps (Date Filters and Occupation Frequency) and information of the data cleaning process.

Moreover, knowing the priorities each role assigned to the dashboard features we are capable of improving the application to provide end-users with specific tools to perform the type of analytics they can benefit the most on daily basis. New features (not assessed with AHP questionnaire) should be also assessed in a semi-structured discussion with experts to evaluate the possibility of creating a specific application to perform specific tasks (data filtering, process discovery, process enhancement, etc...).

Future work will focus on using the dashboard to perform advanced tasks, for instance how errors and high noise problems (e.g. patient safety and medication issues) can be resolved using process mining as one possible application in an operating room environment.

References

1. Agnoletti, V., et al.: Operating room data management: improving efficiency and safety in a surgical block. BMC Surg., **13**(1), 7 (2013)
2. Marques, I., Captivo, M.E., Pato, M.V.: An integer programming approach to elective surgery scheduling. OR Spectr., **34**(2), 407–427 (2011)
3. Haynes, A.B., et al.: A surgical safety checklist to reduce morbidity and mortality in a global population. N. Engl. J. Med., **360**(5), 491–499 (2009)
4. Dexter, F., Epstein, R.H., Traub, R.D., Xiao, Y.: Making management decisions on the day of surgery based on operating room efficiency and patient waiting times. J. Am. Soc. Anesth. **101**(6), 1444–1453 (2004)
5. Smith, S.F.: Reactive scheduling systems. In: Intelligent scheduling systems. Springer, Boston, pp. 155–192 (1995). https://doi.org/10.1007/978-1-4615-2263-8_7
6. Raheja, A.S., Subramaniam, V.: Reactive schedule repair of job shops. Int. J. Adv. Manuf. Technol., **19**(10), 756-763 (2002)
7. Tanimizu, Y., Komatsu, Y., Ozawa, C., Iwamura, K., Sugimura, N.: Co-evolutionary genetic algorithms for reactive scheduling. J. Adv. Mech. Des. Syst. Manuf. **4**(3), 569–577 (2010)
8. Redondi, A., Chirico, M., Borsani, L., Cesana, M., Tagliasacchi, M.: An integrated system based on wireless sensor networks for patient monitoring, localization and tracking. Ad Hoc Netw. **11**(1), 39–53 (2013)
9. de Vries, E.N., Ramrattan, M.A., Smorenburg, S.M., Gouma, D.J., Boermeester, M.A.: The incidence and nature of in-hospital adverse events: a systematic review. Qual. Saf. Health Care **17**(3), 216–223 (2008)
10. González-Arévalo, A., Gómez-Arnau, J., García del Valle, S.: Coordinación y gestión de las áreas quirúrgicas. Tratado de Anestesia y Reanimación. Torres. Arán Ediciones, **1**, 221–238 (2001)
11. Fernandez-Llatas, C., Lizondo, A., Monton, E., Benedi, J.M., Traver, V.: Process mining methodology for health process tracking using real-time indoor location systems. Sensors, **15**(12), 29821–29840 (2015)
12. Fernández-Llatas, C., Meneu, T., Traver, V., Benedi, J.M.: Applying evidence-based medicine in telehealth: an interactive pattern recognition approximation. Int. J. Environ. Res. Public Health **10**(11), 5671–5682 (2013)
13. Westbrook, J.I., Braithwaite, J.: Will information and communication technology disrupt the health system and deliver on its promise? Med. J. Aust. **193**(7), 399 (2010)
14. Drazen, E., Rhoads, J.: Using tracking tools to improve patient flow in hospitals, issue brief. Calif. HealthCare Found., **4**(1) (2011)
15. Miclo, R., Fontanili, F., Marquès, G., Bomert, P., Lauras, M.: RTLS-based process mining: towards an automatic process diagnosis in healthcare. In: IEEE International Conference on Automation Science and Engineering (CASE), pp. 1397–1402. IEEE (2015)
16. Fisher, J.A., Monahan, T.: Evaluation of real-time location systems in their hospital contexts. Int. J. Med. Inform. **81**(10), 705–712 (2012)
17. Pecchia, L., et al.: Analytic hierarchy process (AHP) for examining healthcare professionals assessments of risk factors. Methods Inf. Med. **50**(5), 435–444 (2011)
18. Sloane, E.B., Liberatore, M.J., Nydick, R.L., Luo, W., Chung, Q.: Using the analytic hierarchy process as a clinical engineering tool to facilitate an iterative, multidisciplinary, microeconomic health technology assessment. Comput. Oper. Res. **30**(10), 1447–1465 (2003)

19. Saaty, T.L.: A scaling method for priorities in hierarchical structures. J. Math. Psychol. **15**(3), 234–281 (1977)
20. Bridges, J.F., et al.: Conjoint analysis applications in health–a checklist: a report of the ISPOR good research practices for conjoint analysis task force. Value Health **14**(4), 403–413 (2011)
21. World Health Organization: Safe Surgery Saves Lives (2009)
22. Saaty, T.: How to structure and make choices in complex problems. Hum. Syst. Manag. **3**(4), 255–261 (1982)
23. CHGUV: Anual report 2014. http://chguv.san.gva.es/documents/10184/81032/Informe_anual2014.pdf/713c6559-0e29-4838-967c-93380c24eff9

Tailored Process Feedback Through Process Mining for Surgical Procedures in Medical Training: The Central Venous Catheter Case

Ricardo Lira[1], Juan Salas-Morales[1], Rene de la Fuente[2], Ricardo Fuentes[2], Marcos Sepúlveda[1], Michael Arias[3], Valeria Herskovic[1], and Jorge Munoz-Gama[1(✉)]

[1] Computer Science Department, School of Engineering, Santiago, Chile
{rlira2,jisalas1,jmun}@uc.cl, {marcos,vherskov}@ing.puc.cl
[2] Department of Anesthesiology, School of Medicine,
Pontificia Universidad Católica de Chile, Santiago, Chile
{rdelafue,rfuente}@med.puc.cl
[3] Department of Business Computer Science, Universidad de Costa Rica,
San Jose, Costa Rica
michael.arias_c@ucr.ac.cr

Abstract. In healthcare, developing high procedural skill levels through training is a key factor for obtaining good clinical results on surgical procedures. Providing feedback to each student tailored to how the student has performed the procedure each time, improves the effectiveness of the training. Current state-of-the-art feedback relies on Checklists and Global Rating Scales to indicate whether all process steps have been performed and the quality of each execution step. However, there is a process perspective not successfully captured by those instruments, e.g., steps performed but in an undesired order, part of the process repeated an unnecessary number of times, or excessive transition time between steps. In this work, we propose a novel use of process mining techniques to effectively identify desired and undesired process patterns regarding rework, order, and performance, in order to complement the tailored feedback of surgical procedures using a process perspective. The approach has been effectively applied to analyze a real Central Venous Catheter installation training case. In the future, it is necessary to measure the actual impact of feedback on learning.

Keywords: Process mining · Healthcare · Feedback · Medical training · Surgical procedures

1 Introduction

The development of procedural skills is critical for physicians of any specialty. Better technical skills are associated with better clinical outcomes [8] and the

© Springer Nature Switzerland AG 2019
F. Daniel et al. (Eds.): BPM 2018 Workshops, LNBIP 342, pp. 163–174, 2019.
https://doi.org/10.1007/978-3-030-11641-5_13

absence of them is the most important factor in errors derived from the operator in healthcare [13]. Historically, procedural skills have been taught in daily clinical work, in a master-apprentice model, assuming sufficient exposure time to obtain them. However, there are a number of complexities that increasingly make this model more difficult: health system efficiency considerations, time constraints on clinical training activities, and patient safety [19]. In particular, [19] mentions some examples of efficiency issues: resident work hour restrictions have reduced the exposure of residents to their surgical mentors; and changes in reimbursement and insurance and legal issues have introduced productivity constraints on the surgical procedures. Therefore, training in simulation environments prior to contact with patients has extended as an effective practice to achieve positive effects on the learning process [5]. However, this teaching methodology has a high cost [20] and presents aspects that are not fully resolved, including how to give effective feedback to the students.

Feedback in clinical education of procedural skills is defined as the delivery of specific information on the comparison between the student's performance and a standard [15]; it ensures that certain standards are met, and promotes learning [4]. Feedback is effective when it is used to promote a positive and desirable development [2]. In the case of procedural skills taught in a simulation environment, feedback can be delivered by an instructor, partner or computer, either during or after the simulation activity. However, standard oral feedback has some drawbacks, since it depends on the availability of a person who is trained in the procedure, which is usually a difficult to obtain and expensive resource. In addition, it is an essentially subjective opinion of the evaluator.

To establish when an apprentice has reached an acceptable level of competence that guarantees patient safety, various tools have been developed. The two most commonly used are Checklists and Global Rating Scales (GRS) [11]. Checklists break down the procedure into a series of steps, and check whether the student has performed each step. Meanwhile, GRS consider the evaluation of the student's performance in different areas. In both cases, at the end of each training session, they allow to provide feedback to the student about which steps/areas deserve a greater attention. Checklists have the limitation of giving similar weights to different errors in the execution of a procedure, even though some of these have greater implications for the clinical outcome and patient safety [14]. On the other hand, GRS provide a more qualitative evaluation, but have the limitation that their reliability is dependent on the characteristics and training of the evaluators [3].

The performance of a surgical procedure can be seen as a process [16], i.e., a set of activities (procedure steps) and events that are executed in a specific order so as to achieve a certain goal. A process-oriented feedback seeks to emphasize the relevance of following this orderly sequence of activities, identifying deviations such as: rework, execution of activities in a different order than desired, slow execution of activities, or slow transition times between activities. Process Mining [1] is an emerging discipline that allows analyzing the execution of a process based on the knowledge extracted from event logs created from the data stored in information systems. The event logs record the execution of the different activities in which a process can be broken down.

In this article, we propose the novel use of process mining in order to complement the tailored feedback of surgical procedures using a process perspective. In particular, we believe it is possible to identify when the student repeats some activities (rework), when the student performs some activities in an incorrect order, or when it takes too long to perform an activity or a transition between two consecutive activities. To the best of our knowledge, this is the first application of process mining in medical procedures training.

The proposed method has been validated by applying it to a course, taught by the simulation center of the Pontificia Universidad Católica de Chile, where students learn the procedure for installing the Central Venous Catheter with ultrasonography. In this course, the traditional method of feedback delivery is immediate oral feedback by the instructor, along with an evaluation based on Global Rating Scales and Checklists.

2   Objectives and Context

This article has two main contributions. First, we propose a novel method for applying process mining techniques to effectively identify desired and undesired process patterns regarding rework, order, and performance, in order to complement the tailored feedback of surgical procedures using a process perspective. Second, we illustrate how this method was applied to a real Central Venous Catheter installation training case.

2.1   Objectives

Our main research objective (**O**) is to propose how *process mining* can be used to identify desired/undesired process patterns as part of the tailored feedback on medical procedural training. It can be broken down into specific objectives:

O1: To identify desired/undesired process patterns regarding *rework*.
O2: To identify desired/undesired process patterns regarding the *order* in which activities are performed.
O3: To identify desired/undesired process patterns regarding *performance*.

2.2   Central Venous Catheter Installation Training Case

The simulation center at the School of Medicine of the Pontificia Universidad Católica de Chile, developed a training program for 42 first-year residents of anesthesiology, emergency medicine, cardiology, intensive medicine and nephrology, in the context of the research "Simulation-based training program with deliberate practice for ultrasound-guided jugular central venous catheter placement" [6]. This program is developed in three stages:

I. Online Instruction and PRE recording: three online classes are available through a web platform, each with mandatory and complementary readings. At the end of this stage, a written evaluation is taken and a recording of a first procedure execution is made (identified as PRE video).

II. Demonstration Session: a demonstration of the entire procedure of installation of a Central Venous Catheter (CVC) in a "Blue Phantom Torso"[1] is given by an expert to the entire group of residents. In addition, the 4 stations of deliberate practice are presented: (a) preparation of the ultrasonography equipment, the patient and the work tools; (b) handling ultrasonography equipment; (c) venous puncture with ultrasound guidance; (d) catheter installation and fixation.

III. Deliberate Practice: residents must complete four deliberate practice sessions accompanied by an instructor who supervises and delivers immediate feedback in the development of stations described in stage II.

After the end of the course, a second video (identified as POST video) of the procedure of installation of a CVC with ultrasonography is recorded for each resident, which is used to evaluate the training program.

Parallel to this training session, we recorded videos (identified as EXP videos) of the execution of the same procedure under the same conditions by different professionals from the anesthesiology division, with at least 5 years of clinical practice and experience in the installation of CVC with ultrasonography.

3   Method

Unlike more classical process mining methodologies, in our approach event logs are not generated automatically from the execution of the procedure. Instead, our method uses an observer-based approach [12], i.e., observation performed by a human observer. In this case, event logs were generated based on the off-line observation, by medical specialists, of the aforementioned recorded videos.

The proposed method is inspired by the process mining PM2 methodology [7], which has been previously used in the healthcare domain [17,18] and it is suitable for the analysis of both structured and unstructured processes. The proposed method is decomposed into five stages (see Fig. 1): (1) video recording - data are extracted from the videos recorded for both students and experts; (2) video tagging - the videos were tagged by two medical doctors, identifying for each case (each execution of the CVC installation), activities (procedure steps) and timestamps (time elapsed since the beginning of the procedure); (3) event log generation - tagging information is used to generate an event log containing the data of all executions; (4) model discovery - process mining discovery algorithms are applied to the event log in order to describe the observed behaviour of the procedure; and, (5) model analysis - the discovered process models are analyzed in order to generate feedback for each student.

---

[1] http://www.bluephantom.com.

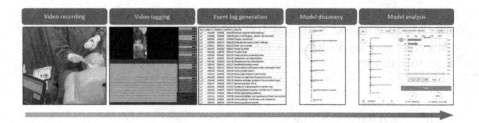

Fig. 1. Stages of the proposed method.

3.1   Video Recording Stage

In this stage, we extracted trace data from different recorded videos that register how the procedure was executed by each of the students. The videos were organized into three categories: PRE: executions previous to the training program, POST: executions after the training program, and EXP: executions by experts.

3.2   Video Tagging Stage

In this stage, each video is tagged by two medical doctors. VCode Vdata is used as tagging software [10]. A set of activities that are characteristic of the CVC procedure are used as possible labels for each of the activities observed in the videos. The result of the tagging process is a plain text file for each video; each of its rows displays information on: activity name, start time and duration time in hundredths of a second, and finally an optional field of notes.

3.3   Event Log Generation Stage

In this stage, we created the event log that is used by process mining algorithms. The event log is a comma separated value file that included a row for each of the events tagged, grouping in a single file the data gathered from all the videos. It contains the following columns: case id (each video), event (activity name), start and end timestamps (both with a granularity at the second level), and an observation field. In addition, for each video, the following fields are recorded: performer (participant id), type of performer (student or expert), category (PRE, POST or EXP), and success or failure in the execution of the procedure.

3.4   Model Discovery Stage

We processed the event log with a discovery algorithm to obtain a process model representing the behaviour of each student when performing the procedure. In the PM literature, there is a wide range of discovery algorithms [1]. We selected the Celonis algorithm and its implementation in the Celonis commercial tool[2].

---

[2] Celonis tool: http://www.celonis.com/en/product.

This algorithm is based on the Fuzzy algorithm concept [9] combined with some characteristics from the family of Heuristic algorithms [1], providing process models that are easy to interpret for an interdisciplinary audience. Moreover, Celonis tool also integrates a set of filtering options to explore the process data interactively and to address the specific objectives we have.

3.5   Model Analysis Stage

Celonis was used to identify aspects of the execution of the process that can be delivered as feedback to the student, to guide their learning. Specifically: (1) identify rework, i.e., when the student repeats one or more activities in the execution of the procedure; (2) identify the execution of activities in an incorrect order of execution, comparing it with the order in which the experts perform it; (3) analyze the student's performance, including duration of activities and transition time between them, comparing it with the performance of the experts.

We consider two key features to create custom views that can be used to provide process-oriented feedback to the students:

Filter: filters can be applied to the event log in order to obtain more specific process models. We use three kind of filters:

Activity selection filters: They allow to exclude some activities from a process model. For example, we distinguished two type of activities: action activities are those that are performed in order to install the CVC, and checking activities are those that are performed in order to verify whether some critical activities produce the desired outcome. In some models, we exclude the checking activities.

Case selection filters: They allow to create a process model using only some process instances. For example, we use these filters to create process models that display only the activities performed by the student we want to give feedback to, either in the PRE or POST scenario, and to create a process model that display only the activities performed by the group of experts (EXP).

Collapsing filters: They allow to group some activities in a process model, so that they are represented by a single element in the model. For example, some phases of the CVC installation are regularly performed well by most participants. Therefore, all activities corresponding to one of those phases can be clustered in a single cluster element.

Compose View: We compose views that include different models of a process based on the data loaded from an event log. Each of those models can be created by applying different filters to the event log in order to obtain more specific process models.For example, Fig. 2 shows a composition of three views, displaying the execution of the procedure by a student before/after the training, compared to the execution of the procedure by an expert.

4   Results

The process-oriented feedback for a student who is learning about a surgical procedure is delivered through different process models that show patterns in which their performance is compared to the desired behavior. This information is useful for students because it helps them to focus their attention and effort, so as to avoid making mistakes in future executions. It can be a guide for future sessions of deliberate practice, which allows focusing efforts on simpler and independent tasks, e.g., puncturing the vein with ultrasonography or passing the guidewire, thus simplifying the training to some steps that are difficult due to rework, lack of order, or slowness in execution. Different views and filters were applied to address the three specific objectives:

O1. To identify desired/undesired process patterns regarding rework: For this objective, we compose a view with three models (as shown in Fig. 2). The first two models display how the student performed the procedure before and after the training (PRE and POST). The third model displays how an expert would perform the procedure. Since the purpose is to illustrate the occurrence of reworks in the action activities, checking activities are excluded. This view allows to provide feedback about which activities the student repeated (**O1.1**); to comment on the performance's evolution, by comparing the reworks observed in PRE versus POST (**O1.2**); finally, by including the execution of an expert, it is possible to compare the student's performance against the desired outcome (**O1.3**). In the example shown in Fig. 2 a rework in the trocar installation can be observed.

O2. To identify desired/undesired process patterns regarding the order in which activities are performed: To achieve this objective, we compose a view with three models (Fig. 3). The first two models display how the student performed the procedure before and after the training (PRE and POST). The third model displays how an expert would perform the procedure. Since the purpose is to illustrate problems in the order in which activities are performed, those phases that are correctly performed by most participants are excluded. This view allows to provide feedback about which activities the student performed in the wrong order, or activities that were not performed at all (**O2.1**); to comment on the performance's evolution, by comparing the performance PRE versus POST (**O2.2**); finally, it is also possible to compare the student's performance against the desired order as performed by an expert (**O2.3**). In the example shown in Fig. 3, *Remove trocar* and *Check guidewire* activities are performed in the opposite order to the desired one.

O3. To identify desired/undesired process patterns regarding performance: To achieve this objective, we compose a view with three models (as shown in Fig. 4). The first two models display how the student performed the procedure before and after the training (PRE and POST) including the time required to perform each activity and the time elapsed during the transition between two consecutive activities. The third model displays the average time it takes the group of experts (EXP) to perform the procedure. This view allows to provide feedback about

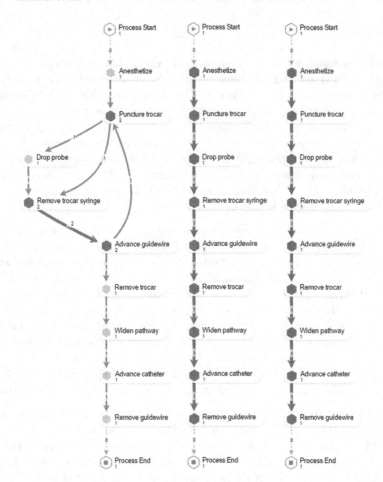

Fig. 2. Process model of student *16F69F* in PRE (left) and POST (center), and a generic expert EXP (right). Numbers show activity frequency per case. Activity names are shortened and only action activities are shown for readability reasons. Regarding **O1.1**, the PRE model shows rework during the phase of venous puncture with trocar: the student unsuccessfully performs a first trocar installation, then perform a successful one on the second iteration. Moreover, during the second iteration, the probe is not properly dropped (in the sterile zone) closing the door for a possible third iteration without sterilizing everything again. Notice that, the exact path followed in each iteration can be analyzed using the case animation feature of the tool. Regarding **O1.2**, POST model is able to capture that the same student does not perform any rework, and the process match exactly the reference process performed by the EXP (**O1.3**).

which activities were performed too slow (**O3.1**), or when the student hesitated taking too much time between activities (**O3.2**). In this case, it is always useful to have as a reference the performance of the experts. In Fig. 4, it can be observed

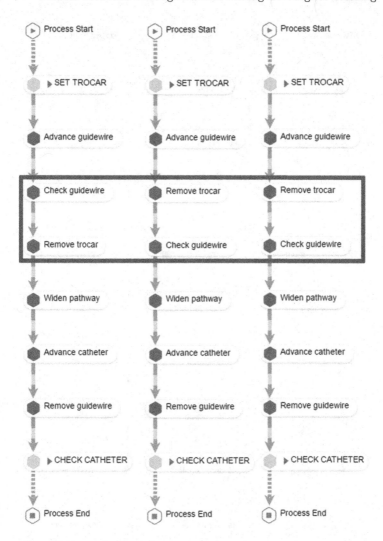

Fig. 3. Process model of student *12F67F* in PRE (left) and POST (center), and a generic expert EXP (right). Activities in *set trocar* and *check catheter* phases are clustered for the sake of readability (they show no difference in order between PRE, POST and EXP). Activity names are shortened and both action and checking activities are shown. Regarding **O2.1**, the PRE model shows the student checked the guidewire and then removed the trocar; however, it is desirable to do these activities in the opposite order, because removing the trocar may affect the position of the guidewire. Regarding **O2.2**, POST model is able to capture that the student learned to perform the activities in the right order, and the process match exactly the reference process for the procedure performed by the EXP (**O2.3**).

in the PRE model that *Puncture trocar* was performed too slow and that the transition time between *Remove trocar* and *Widen pathway* took too long.

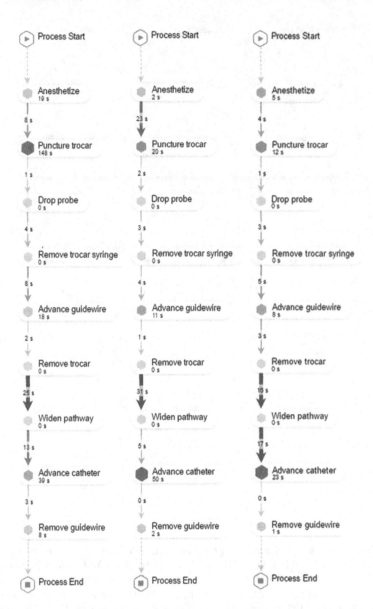

Fig. 4. Process model of student *45D59F* in PRE (left) and POST (center), and the average of the 8 experts in EXP (right). Activity names are shortened and both action and checking activities are shown. Regarding **O3.1**, it is possible to highlight *Puncture trocar* was performed too slow in the PRE model (notice the symbol of this activity is larger and darker). Regarding **O3.2**, it can be observed that the transition time between *Remove trocar* and *Widen pathway* is too long (notice the arrow between these activities is wider and darker). Regarding **O3.3**, it is possible to highlight the duration of *Puncture trocar* in the POST model got close to the EXP model. However, the time between *Remove trocar* and *Widen pathway* is still too long (**O3.4**).

It is also possible to provide feedback about student's evolution from PRE to POST regarding the duration of activities (**O3.3**) and the time between transitions (**O3.4**). It could happen that some activities (or times between activities) have similar execution times to the average time of the experts, but others are still not close to the average time of the experts. It is then possible to highlight positive aspects, and others where there is still room for improvement. In the example shown in Fig. 4, it is possible to highlight in the POST model that the duration of *Puncture trocar* got close to the EXP model, but the transition time between *Remove trocar* and *Widen pathway* is still too long (**O3.4**).

5 Conclusions and Future Work

In this article, it has been shown that process mining can be used to provide a process-oriented feedback to students who are learning procedural skills, by effectively identifying desired and undesired process patterns regarding rework, order, and performance, in order to complement the tailored feedback of surgical procedures. The approach has been effectively applied to analyze a real CVC installation training case. To the best of our knowledge, this is the first application of process mining in medical procedures training. It opens a novel approach to the analysis of training programs by generating tailored feedback to students. This approach is generic and therefore can be replicated in other medical training programs. The main limitation of this approach is its scalability when a high amount of videos needs to be tagged. However, in specialized medical procedure training, as CVC installation, it is a viable option. Another potential limitation is when the medical procedure has more complex patterns, e.g., when the order among some activities is not relevant. In such a case, more advanced conformance checking techniques should be considered.

In the future, it is necessary to measure the actual impact of feedback on learning. To this end, a feedback methodology will be applied that includes the tagging of videos by the own students, and the delivery of a report based on the results of the process mining analysis after the PRE evaluation. To measure whether the process-oriented feedback has a statistically significant impact on students' learning, their performance on the PRE and POST scenarios will be evaluated using GRS, comparing an experimental group with a control group.

Acknowledgments. This work is partially supported by CONICYT FONDECYT 181162, CONICYT FONDECYT 11170092, CONICYT REDI 170136, VRI-UC Interdisciplinary 2017, and FOND-DCC 2017-0001. We thank Jerome Geyer-Klingeberg and Celonis Academic Alliance for their support and material.

References

1. van der Aalst, W.M.P.: Process Mining - Data Science in Action. Springer, Heidelberg (2016). https://doi.org/10.1007/978-3-662-49851-4
2. Archer, J.C.: State of the science in health professional education: effective feedback. Med. Educ. **44**(1), 101–108 (2010)

3. Bould, M., Crabtree, N., Naik, V.: Assessment of procedural skills in anaesthesia. Br. J. Anaesth. **103**(4), 472–483 (2009)
4. Chowdhury, R.R., Kalu, G.: Learning to give feedback in medical education. Obstet. Gynaecol. **6**(4), 243–247 (2004)
5. Cook, D.A., et al.: Technology-enhanced simulation for health professions education: a systematic review and meta-analysis. Jama **306**(9), 978–988 (2011)
6. Corvetto, M., Pedemonte, J., Varas, D., Fuentes, C., Altermatt, F.: Simulation-based training program with deliberate practice for ultrasound-guided jugular central venous catheter placement. Acta Anaesthesiologica Scandinavica **61**(9), 1184–1191 (2017)
7. van Eck, M.L., Lu, X., Leemans, S.J.J., van der Aalst, W.M.P.: PM^2: a process mining project methodology. In: Zdravkovic, J., Kirikova, M., Johannesson, P. (eds.) CAiSE 2015. LNCS, vol. 9097, pp. 297–313. Springer, Cham (2015). https://doi.org/10.1007/978-3-319-19069-3_19
8. Fecso, A.B., Szasz, P., Kerezov, G., Grantcharov, T.P.: The effect of technical performance on patient outcomes in surgery: a systematic review. Ann. Surg. **265**(3), 492–501 (2017)
9. Günther, C.W., van der Aalst, W.M.P.: Fuzzy mining – adaptive process simplification based on multi-perspective metrics. In: Alonso, G., Dadam, P., Rosemann, M. (eds.) BPM 2007. LNCS, vol. 4714, pp. 328–343. Springer, Heidelberg (2007). https://doi.org/10.1007/978-3-540-75183-0_24
10. Hagedorn, J., Hailpern, J., Karahalios, K.G.: Vcode and vdata: illustrating a new framework for supporting the video annotation workflow. In: Proceedings of the Working Conference on Advanced Visual Interfaces, pp. 317–321. ACM (2008)
11. Ilgen, J.S., Ma, I.W., Hatala, R., Cook, D.A.: A systematic review of validity evidence for checklists versus global rating scales in simulation-based assessment. Med. Educ. **49**(2), 161–173 (2015)
12. Lalys, F., Jannin, P.: Surgical process modelling: a review. Int. J. Comput. Assist. Radiol. Surg. **9**(3), 495–511 (2014)
13. Leape, L.L., et al.: The nature of adverse events in hospitalized patients: results of the harvard medical practice study ii. N. Engl. J. Med. **324**(6), 377–384 (1991)
14. Ma, I.W., et al.: Comparing the use of global rating scale with checklists for the assessment of central venous catheterization skills using simulation. Adv. Health Sci. Educ. **17**(4), 457–470 (2012)
15. Nesbitt, C.I., Phillips, A.W., Searle, R.F., Stansby, G.: Randomized trial to assess the effect of supervised and unsupervised video feedback on teaching practical skills. J. Surg. Educ. **72**(4), 697–703 (2015)
16. Neumuth, D., Loebe, F., Herre, H., Neumuth, T.: Modeling surgical processes: a four-level translational approach. AI Med. **51**(3), 147–161 (2011)
17. Rojas, E., Munoz-Gama, J., Sepúlveda, M., Capurro, D.: Process mining in healthcare: a literature review. J. Biomed. Inform. **61**, 224–236 (2016)
18. Rojas, E., Sepúlveda, M., Munoz-Gama, J., Capurro, D., Traver, V., Fernandez-Llatas, C.: Question-driven methodology for analyzing emergency room processes using process mining. Appl. Sci. **7**(3), 302 (2017)
19. Walter, A.J.: Surgical education for the twenty-first century: beyond the apprentice model. Obstet. Gynecol. Clin. **33**(2), 233–236 (2006)
20. Zendejas, B., Wang, A.T., Brydges, R., Hamstra, S.J., Cook, D.A.: Cost: the missing outcome in simulation-based medical education research: a systematic review. Surgery **153**(2), 160–176 (2013)

An Application of Process Mining in the Context of Melanoma Surveillance Using Time Boxing

Christoph Rinner[1]📷, Emmanuel Helm[2]([✉])📷, Reinhold Dunkl[3],
Harald Kittler[4]📷, and Stefanie Rinderle-Ma[3]📷

[1] Center for Medical Statistics, Informatics, and Intelligent Systems (CeMSIIS),
Medical University of Vienna, Spitalgasse 23, Vienna, Austria
[2] Research Department of Advanced Information Systems and Technology,
University of Applied Sciences Upper Austria, Softwarepark 13, Hagenberg, Austria
emmanuel.helm@fh-hagenberg.at
[3] Faculty of Computer Science, University of Vienna,
Währinger Strasse 29, Vienna, Austria
[4] Department of Dermatology, Medical University of Vienna,
Währinger Gürtel 18-20, Vienna, Austria

Abstract. Background: Process mining is a relatively new discipline that helps to discover and analyze actual process executions based on log data. In this paper we apply conformance checking techniques to the process of surveillance of melanoma patients. This process consists of recurring events with time constraints between the events. Objectives: The goal of this work is to show how existing clinical data collected during melanoma surveillance can be prepared and pre-processed to be reused for process mining. Methods: We describe an approach based on time boxing to create process models from medical guidelines and the corresponding event logs from clinical data of patient visits. Results: Event logs were extracted for 1,023 patients starting melanoma surveillance at the Department of Dermatology at the Medical University of Vienna between January 2010 and June 2017. Conformance checking techniques available in the ProM framework were applied. Conclusions: The presented time boxing enables the direct use of existing process mining frameworks like ProM to perform process-oriented analysis also with respect to time constraints between events.

Keywords: Health care processes · Process mining ·
Electronic health records · Medical guidelines

1 Introduction

1.1 Motivation

The availability of electronic data in the early 1980s changed the business models of many companies as well as the medical domain. The proliferation of electronic health data from a variety of sources including medical treatment records,

© Springer Nature Switzerland AG 2019
F. Daniel et al. (Eds.): BPM 2018 Workshops, LNBIP 342, pp. 175–186, 2019.
https://doi.org/10.1007/978-3-030-11641-5_14

administrative health data and public health information systems supports the integration of all available evidence to feed back into medical research and practice. These health data collected during routine care can be used for secondary purposes such as identifying trends, predicting outcomes, influencing patient care, drug development and therapy choices [23]. By analyzing these data it is possible to gain new insights which can be used for the optimization of health care processes, i.e., the insights might enable the transformation of health care processes instead of simply monitoring them.

Skin malignancies are recognized as a major and global health problem. Accounting for about 5% of all skin cancer cases melanoma is the most dangerous form of skin malignancy and causes about 90% of skin cancer mortalities [9]. Incidence rates in many European countries are actually ranging between 12 and 15 cases per 100.000 inhabitants. Currently, the increasing rates are levelling off in some countries. In contrast, however, for distinct subpopulations such as elderly men the rates are still increasing [11]. Early detection of melanoma is of utmost importance and is leading to a favourable prognosis. Since melanoma may appear years after the excision of the primary tumor, patients with melanoma are monitored closely, usually following a predefined protocol, to allow timely detection of recurrent disease [7].

1.2   Problem Statement

In order to improve the surveillance of patients with melanoma traditional studies depended on manual data acquisition. The goal of this work is to show how existing data from routine care can be combined from different sources and reused for process mining to automatically detect processes [1] and compare them to medical guidelines using conformance checking [14,22].

Process mining is a relatively new discipline that helps to discover and analyze actual process executions based on log data. Log data stores events that are produced during process execution, for example, the execution of a process activity "excision". Process mining techniques are particularly promising to be applied in the healthcare domain facing challenges to (a) learn the process of interest; (b) understand deviations, (c) analyze bottlenecks, and (d) monitor organizational behaviour [16].

1.3   Research Questions

This paper focuses on challenges (a) and (b) and aims at learning about the applicability and results of process mining and conformance checking for the treatment and surveillance of melanomas. Previous studies have indicated the potential of process mining in this area [3,6]. In these two studies, however, only a maximum of 10 process instances were analyzed. Moreover, several data challenges were pointed out, specifically that the time granularity of the logged data was too coarse [3]. This problem is also referred to by [16] as imprecise data. Hence, this paper addresses the following research questions:

1. How can existing clinical data be reused for the application of process mining?
2. How can data of recurring events with time constraints that span a long period of time be prepared to apply process mining?
3. How can we apply process mining to check guideline compliance?
4. What can we learn from process mining in the context of surveillance of melanoma patients?

1.4   Contribution

This study provides conceptual extensions towards data preparation for imprecise log data. In particular, we describe a method for data preparation using a specific naming convention to model the time aspects used in medical guidelines (e.g. Check up in 6 months). The methods were tested using follow-up guidelines for melanoma patients and anonymous patient data from the Department of Dermatology at the Medical University Vienna (DDMUV).

In Sect. 2 we give a brief overview of the process mining discipline and the positioning of this paper in the field of process mining in healthcare. In Sect. 3 we describe the data preparation and time boxing steps as well as the conformance checking applied in this paper. In Sect. 4 the study population at the DDMUV is described and time boxing and conformance checking are applied to melanoma surveillance data. In Sect. 5 the generalized applicability of our methods and the medical implications are discussed.

2   Related Work

2.1   Process-Oriented Analysis

Process mining [1] offers techniques for different analysis tasks such as *discovery* of process models, *conformance checking* between process execution logs and process models or guidelines, and *enhancement* of existing models using information about the recorded reality in process execution logs (i.e. event logs). Event logs store events that are produced by the execution of the process tasks during runtime. Existing standardized formats for event logs are MXML and XES [25]. In this article we mainly focus on the task of *conformance checking*.

There are different techniques and algorithms to measure conformance (e.g. [2,22]). The degree of conformance is measured in four orthogonal dimensions. (1) Fitness is the most agreed upon measure to determine if the model reflects the recorded behaviour in the event log. Approaches to measure fitness are listed and evaluated by [4]. (2) Precision aims to identify overly general models and penalize unwanted behaviour. A recent study, however, shows that existing precision metrics do not provide the desired properties to reliably recognize underfitting [24]. (3) Generalization measures the degree of overfitting, i.e. if the model only allows for what has actually been observed. (4) Structural appropriateness, derived from the simplicity quality dimension described in [1], aims to find an easy to understand and not overly complex model.

2.2 Positioning in the Field

Rojas et al. provide a recent survey on existing literature and case studies on process mining in healthcare, collecting 74 publications in this area [20]. Following their terminology, our paper analyzes *organizational* processes based on data from a *clinical support system* for *oncology* in *Austria*. It poses a *specific question* (i.e. guideline compliance) and utilizes the *conformance* perspective. This paper presents a *semi-automated* implementation strategy, providing a novel data preparation approach facilitating the use of the *tool* ProM. The analysis follows the *basic* approach, using the ProM plugin Multi-perspective Process Explorer (MPE) [14].

3   Methods

The process of surveillance of melanoma patients starts with the detection and the excision of the primary tumor (i.e. the baseline visit). Melanoma patients are staged according to the American Joint Committee on Cancer (AJCC) staging system (i.e. stage I to IV). After excision of the primary tumor patients start a 10 year surveillance period. Depending on AJCC staging, follow-up visits have different surveillance intervals and include different types of examinations (i.e. clinical examination, analyzing tumor markers, lymph node sonography, computed tomography of the abdomen, PET-CT etc.). In AJCC stage I for example the interval of the follow-up visits is 6 months in the first 5 years and one year between the 5th and the 10th year. In the other AJCC stages intervals of 3 months in the first five years and 6 month between the 5th and the 10th year are scheduled. The higher the AJCC stage the more often examinations are performed as part of a follow-up program. During surveillance, the AJCC stages are re-evaluated and patients can be assigned a higher AJCC stage and start the corresponding follow-up surveillance from the beginning. In this paper we refer to this upgrading as stage change. Since the observation period of this study only covers 7 years (January 2010 to June 2017) patients are still compliant to the guideline if they have not missed the next-to-last or last follow-up visit before the end of the study (i.e. June 2017). Patients are considered lost to follow-up when the surveillance is terminated prematurely at the DDMUV (e.g. patient changed clinic). The events occurring during melanoma surveillance are depicted in Fig. 1.

We took the clinical data needed to perform the process mining from a local melanoma registry stored in the Research and Analysis (RDA) Platform of the Medical University of Vienna [5]. Additional information about the transfer of patients between different clinics within the hospital, laboratory results as well as treatment information is obtained from the local Hospital Information System (HIS). Since the melanoma registry is maintained manually the information from the HIS is used to detect additional follow-up visits re-using information from routine care. The event logs used in this study are created using the JAVA programming language, a JDBC driver to access the data in the Oracle Database and the OpenXES library to create event logs in the MXML format.

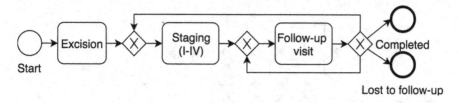

Fig. 1. A BPMN representation of the process of melanoma surveillance. The start event of the surveillance is the excision of the melanoma followed by the AJCC stage classification and the follow-up visits. Depending on the AJCC stage the number of follow-up visits can vary and the AJCC stage can be re-assessed after each follow-up visit. A patient can be lost to follow-up or complete the surveillance successfully.

Conformance checking was performed in the ProM framework. The study was approved by the ethics committee of the Medical University of Vienna (EK Nr.: 1297/2014).

3.1   Data Preparation and Time Boxing

According to the guideline, depending on the AJCC stage of the patient, the follow-up treatment takes place at certain time intervals (i.e. every three, six or twelve months) in a repeated fashion for ten years. The existing process execution logs record the same event for each occurrence (e.g. *follow-up visit* for each follow-up visit), so the process mining algorithms are not able to distinguish between these events depending on the fixed time period specified in the guideline (e.g. second follow-up visit after one year).

Using the simplified process model depicted in Fig. 1, it is not possible to distinguish the different follow-up visits automatically and as a consequence conformance to the guideline cannot be checked using current process mining algorithms. To overcome this problem we propose a naming convention based on *time boxing* for recurring events commonly described in medical guidelines. During the time boxing, each activity (e.g. each follow-up visit) is allocated (i.e. aligned) to a predefined fixed time period it matches in, called a time box. Each time box corresponds to an event in the medical guideline and the events in each time box are named according to the name of the time box (e.g. *I_F_01_1Q* corresponds to an AJCC stage I follow-up visit in the first quarter of the first year). In Fig. 2 the process model with applied time boxing corresponding to the melanoma surveillance guideline used at the DDMUV is shown.

The event log follows the same naming convention as the process model. To generate the event log, all follow-up visits are assigned to the corresponding (i.e. temporally closest) time boxes. All follow-up visits in one time box are merged and represented as one. In order to analyze over-compliance, multiple events could be assigned to the same time box (without merging) and the resulting self-loops considered during the analysis.

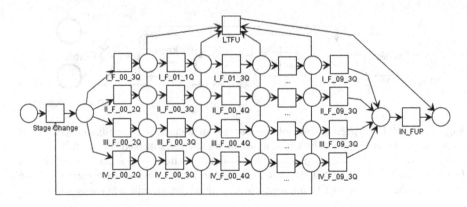

Fig. 2. A simplified petri net model with applied time boxing corresponding to the guideline used at the DDMUV. For each AJCC stage (i.e. I, II, III and IV) the follow-up visits after three, six or twelve month (i.e. 2Q is 2nd quarter, 3Q is 3rd quarter 4Q is 4th quarter) for ten years are shown. After each follow-up visit a patient can proceed to any later follow-up visit, have a state change, be lost to follow-up (LTFU) or complete the surveillance (i.e. IN_FUP meaning *still in follow-up*).

3.2   Conformance Checking with ProM

The conformance checking was done using the process mining framework ProM in version 6.6 and the respective plug-in *Multi-perspective Process Explorer* (MPE) [14]. It allows for fitness and precision calculation and provides different views on the data, including (1) a model view, depicting the base model petri net, (2) a trace view, making it possible to investigate individual traces and (3) a chart view, showing the distribution of attribute values in the log for certain parts of the model.

The MPE is an advanced tool that integrates state of the art algorithms described in [15], that are also able to integrate different perspectives i.e. data, resource and time. The configuration for penalties on log and event moves was adapted to the specific use case. A valid configuration for the alignment parameters (penalties for moves on the log/model) had to be identified. Due to the pre-processing there are no wrong events (events present in the log but not in the model) save for the LTFU (i.e. Lost to follow-up) event, so the alignment algorithm must always identify the missing events (events in the model that are missing in the log). The parameters are described in the results section.

4   Results

The DDMUV is a tertiary referral centre that offers a long-term surveillance program for melanoma patients based on the European guideline on melanoma treatment [9]. An example for the follow-up sub process in the European guideline on melanoma treatment modelled in the Business Process Modeling and Notation (BPMN) can be found in [3]. The melanoma registry at the DDMUV contains data of baseline and follow-up visits of melanoma patients. Excisions are documented way back to the early 1990s, a continuous documentation of the follow-up visits started 2010. In 2017, the melanoma registry covered about 2,200 patients. In this study we included all 1023 patients (43% females, mean age 59 ± 17.5 years) with baseline visit (i.e. excisions) after January 2010 and at least one follow-up visit. Besides the demographic data, different characteristics of the identified melanoma are documented. For the baseline visit this includes among others, (1) melanoma subtype (superficial spreading melanoma, nodular melanoma, lentigo maligna melanoma, acral lentiginous melanoma and others), (2) anatomic site (i.e. abdomen, hand, foot, head etc.), (3) depth of invasion, (4) date of surgery for the primary tumor and (5) staging information. More than one primary tumor can be documented. Only melanoma staging is used for conformance checking. We extracted five different event logs from our real world data, one including all patients (i.e. I–IV), and four for each AJCC stage separately (i.e. I, II, III, IV) based on the highest AJCC stage of the patient. If a patient initially started with AJCC stage I and then moved to AJCC stage II the patient is represented in the AJCC II log file. Table 1 lists the number of patients in each log.

Table 1. Number of patients in log files split by maximal AJCC stage and separated by LTFU (lost to follow-up) and IN_FUP (in follow-up). The distribution of the outcome indicators LTFU and IN_FUP show significant differences between the AJCC stages.

		I–IV	I	II	III	IV
Female	LTFU	286	146	95	20	25
	IN_FUP	153	45	50	17	41
	Total	439	191	145	37	66
Male	LTFU	379	167	124	42	46
	IN_FUP	205	43	70	33	59
	Total	584	210	194	75	105
Total	LTFU	665	313	219	62	71
	IN_FUP	358	88	120	50	100
	Total	1023	401	339	112	171

The number of patients per AJCC stage decreases with higher AJCC stage, which corresponds to the fact that most melanomas in Austria are diagnosed in early stages [10]. Most patients (n = 401) were in AJCC stage I. This group also had the highest number of patients lost to follow-up (n = 313, 78%). The ratio of patients IN_FUP (i.e. in follow-up) was the highest in AJCC stage IV with 58% (n = 100). There is no difference between proportion of individuals lost to follow-up (LTFU) between men and women. Men were generally older than women and there was no significant difference between the LTFU and IN_FUP in respect to the age. Patients in lower AJCC stages were generally younger

(I: mean age 57 ± 17 years; II: mean age 59 ± 18 years; III: mean age 60 ± 18 years; IV: mean age 63 ± 16 years)

4.1 Conformance Checking of Melanoma Surveillance

To check the conformance of our guideline models in regard to the recorded event logs, we replayed the logs on the models using the MPE. For the alignment, the default costs of the MPE for missing events in the log (value: 2) and missing activities in the model (value: 3) leads to undesired behaviour. The alignment algorithm identifies follow-up visits after a long period of time as wrong events. When the penalty for a sequence of missing events exceeds the penalty for a wrong event, the alignment algorithm will declare the current event wrong. In order to ensure a correct alignment, the maximum number of skipped follow-up visits in all traces was identified and the penalties adopted respectively. Since the maximum number of consecutive skipped events for one trace is 19 in our data, we chose a penalty of 1 for missing events and 20 for wrong events. For the LTFU event we reduce the wrong event penalty to 0, thus only penalizing the missing IN_FUP event at the end and not overvalue the outcome indicator for the fitness calculation. The results in the form of fitness and precision indicators can be seen in Table 2.

Table 2. Using the MPE, for each stage log and the combined log (I–IV) the average fitness and precision were calculated in regard to the respective guideline models.

AJCC stage	No. of patients	Avg. fitness % (min - max)	Avg. precision % (#observed / #possible behaviour)
I	401	98,6% (91% - 100%)	75,1%
II	339	98,0% (82% - 100%)	71,4%
III	112	98,2% (85% - 100%)	65,0%
IV	171	98,7% (88% - 100%)	63,1%
I–IV	1.023	98,4% (53% - 100%)	87,0%

Our measurements show that the guideline models have an overall comparable and good fitness value, i.e. the model generally explains the behaviour seen in the log. This originates from three facts. (1) The renaming and clustering of activities was done based on the terminology that was also used for the guideline model. (2) The time boxing method presented in 3.1 leads to an ordered sequence of events, where loops and duplicates cannot occur. (3) The only *wrong* events (i.e. events present only in the log, not in the model) are the LTFU events.

The precision of the model for stage I is 75.1% and declines to 63.1% for stage IV. The ratio between observed and possible behaviour indicates underfitting for low values. The explanation for the generally lower precision values is that the guideline models include the whole time period of ten years of follow-up visits, while the event logs only cover a maximum of seven and a half years. Thus, modelled events like I_F_08_1Q (i.e. stage I, eighth year, first quarter) will never be reached during replay, leading to a lower precision. The explanation for the declining values of precision is that the guideline models for higher stages allow for all the lower stages' events too, since a patient can start in stage I and be re-evaluated to stages II, III or IV during his follow-up visits. The amount of possible behaviour is thus higher while the number of actual patients in the stages is similar (II) or significantly lower (III and IV) than in stage I.

Figure 3 shows the most frequent trace recorded in the complete log. 148 of the 1023 patients follow this trace where they (1) start with the excision (Start), (2) are staged in AJCC I (StageChange), (3) go to their first follow-up (I_F_00_3Q) and (4) are afterwards lost to follow-up (wrong event LTFU). The following missing event (IN_FUP) is in the guideline model but was not present for those traces in the log. Finally, the End event concludes the trace.

Fig. 3. The most frequent trace in the complete log (I–IV) visualized via the MPE's trace view. (fitness 98.8%)

Figure 4 shows a patient that started in stage I and was re-evaluated to stage II and later to stages III and IV. All in all just 1 follow-up visit during stage II was missed and the fitness is very high. The trace spans over the whole observation period, with the start in 2010 and the last follow-up in late 2015. Thus, the patient was identified as in follow-up (IN_FUP).

Fig. 4. A trace comprising all four stages and only one missing follow-up visit. (99.8%)

In Fig. 5 the patient classified in stage II skipped multiple follow-up visits and left the monitoring entirely after four years. The low fitness value correlates with the low guideline compliance.

Fig. 5. A trace with multiple skipped events and thus relatively low fitness. (89.6%)

5   Discussion and Lessons Learnt

5.1   Reuse of Clinical Data for Process Mining

We reused existing patient data available from a local EHR system in the context of melanoma surveillance. In combination with the local melanoma registry additional follow-up visits and laboratory data to the event log were identified. Creating the log file using a procedural programming approach allowed us to add pre-processing steps. For example we tagged patients that successfully terminate the process (i.e. still in the surveillance program (IN_FUP)) during the creation of the log file based on the time they did not show up before the end (i.e. time boxes missed). Beside the melanoma registry, more than 70 other registries are documented in the RDA platform. In a current master thesis a mapping of the melanoma registry data from the RDA data model [19] to the i2b2 star schema [17] and the OMOP common data model [18] is performed. By adapting our approach to these two widely used data models a greater variety of data could be made available to process mining and conformance checking in particular.

5.2   Guideline Compliance Checking

In our approach we pushed the time dimension into the process structure to be able to use the conformance checking capabilities of ProM on imperative models. However, there are two viable alternative approaches: (1) Using a data-aware alignment algorithm would allow to keep the time dimension hidden in the data, thus naming the follow-up visits just *follow-up*, avoiding the initially confusing time boxing notation (e.g. I_F_00_3Q). However, we decided to use our naming approach to make all (missing) steps easily visible in the model. (2) The current version of the ProM framework also includes a declarative mining module that derives sets of constraints in form of a declarative model from log files and offers also conformance checking [13,21]. This needs further investigation, especially in the preparation of a correct declarative constraint set based on the guideline as well as an adapted real log to be replayed.

5.3   Medical Implications

In [12] the prognosis among patients with thin melanomas depending on the surveillance compliance was analyzed. Patients were considered to be compliant with the follow-up regimen if they had at least one annual follow-up examination and non-compliant if they had follow-up intervals of more than one year. They showed that compliant patients before the onset of recurrence had a significantly better prognosis than non-compliant patients.

When using our calculated fitness instead of the fixed time-intervals to evaluate the survival, the same effect can be observed in our data as seen in Fig. 6. We sampled all 246 patients that stayed in follow-up for more than two years based on their fitness value into three equal-sized groups and used the Kaplan–Meier estimator for survival analysis. The survival probability of patients with

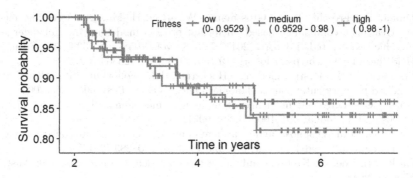

Fig. 6. Survival analysis for all 246 patients that stayed in follow-up for more than two years, sampled into three equal-sized groups depending on their fitness.

a high guideline compliance after five years is about 5% higher compared to the least compliant group. However, adding the patients that stayed for less than 2 years to the estimator, looking at all 358 patients in follow-up, showed a reversed effect. The main reason was that higher fitness is easier to achieve with a shorter stay and many with a short stay died early, e.g. after being staged in IV and the first follow-up visit.

The compliance calculated and formalized using the fitness using MPE is very promising. Yet it has to be further analyzed under which circumstances it correlates to the outcome of the patients. Further we plan to analyze how the compliance affects the tumor progression of the patients, i.e. if patients with a higher compliance are less likely to progress to a higher AJCC stage.

References

1. van der Aalst, W.M.: Process Mining: Discovery, Conformance and Enhancement of Business Processes. Springer, Heidelberg (2011). https://doi.org/10.1007/978-3-662-49851-4
2. Adriansyah, A., van Dongen, B.F., van der Aalst, W.M.P.: Towards robust conformance checking. In: zur Muehlen, M., Su, J. (eds.) BPM 2010. LNBIP, vol. 66, pp. 122–133. Springer, Heidelberg (2011). https://doi.org/10.1007/978-3-642-20511-8_11
3. Binder, M., et al.: On analyzing process compliance in skin cancer treatment: an experience report from the evidence-based medical compliance cluster (EBMC²). In: Ralyté, J., Franch, X., Brinkkemper, S., Wrycza, S. (eds.) CAiSE 2012. LNCS, vol. 7328, pp. 398–413. Springer, Heidelberg (2012). https://doi.org/10.1007/978-3-642-31095-9_26
4. De Weerdt, J., De Backer, M., Vanthienen, J., Baesens, B.: A critical evaluation study of model-log metrics in process discovery. In: zur Muehlen, M., Su, J. (eds.) BPM 2010. LNBIP, vol. 66, pp. 158–169. Springer, Heidelberg (2011). https://doi.org/10.1007/978-3-642-20511-8_14
5. Dorda, W., et al.: ArchiMed: a medical information and retrieval system. Methods Inf. Med. **38**(1), 16–24 (1999)

6. Dunkl, R., Fröschl, K.A., Grossmann, W., Rinderle-Ma, S.: Assessing medical treatment compliance based on formal process modeling. In: Holzinger, A., Simonic, K.-M. (eds.) USAB 2011. LNCS, vol. 7058, pp. 533–546. Springer, Heidelberg (2011). https://doi.org/10.1007/978-3-642-25364-5_37

7. Francken, A.B., Bastiaannet, E., Hoekstra, H.J.: Follow-up in patients with localised primary cutaneous melanoma. Lancet. Oncol. 6(8), 608–621 (2005)

8. Garbe, C.: Increasing incidence of malignant melanoma. Hautarzt 51(7), 518 (2000). https://doi.org/10.1007/s001050051166

9. Garbe, C., et al.: Diagnosis and treatment of melanoma: European consensus-based interdisciplinary guideline. Eur. J. Cancer 46(2), 270–283 (2010)

10. Hackl, M., Ihle, P.: Krebserkrankungen in Österreich, 6th edn. Statistik Austria, Vienna (2018)

11. Jemal, A., et al.: Recent trends in cutaneous melanoma incidence and death rates in the United States 1992–2006. J. Am. Acad. Dermatol. 65(5), 17 (2011)

12. Kittler, H., Weitzdorfer, R., Pehamberger, H., Wolff, K., Binder, M.: Compliance with follow-up and prognosis among patients with thin melanomas. Eur. J. Cancer 37(12), 1504–1509 (2001)

13. Maggi, F.M.: Declarative process mining with the declare component of ProM. In: BPM (Demos), Beijing (2013)

14. Mannhardt, F., De Leoni, M., Reijers, H.A.: The Multi-perspective process explorer. In: BPM (Demos), Innsbruck (2015)

15. Mannhardt, F., De Leoni, M., Reijers, H.A., Van Der Aalst, W.M.: Balanced multi-perspective checking of process conformance. Computing 98(4), 407–437 (2016)

16. Mans, R.S., van der Aalst, W.M., Vanwersch, R.J.: Process Mining in Healthcare: Evaluating and Exploiting Operational Healthcare Processes. Springer, publisher location (2015). https://doi.org/10.1007/978-3-319-16071-9

17. Murphy, S.N., et al.: Serving the enterprise and beyond with informatics for integrating biology and the bedside (i2b2). J. Am. Med. Inform. Assoc. 17(2), 124–130 (2010)

18. Overhage, J.M., Ryan, P.B., Reich, C.G., Hartzema, A.G., Stang, P.E.: Validation of a common data model for active safety surveillance research. J. Am. Med. Inform. Assoc. 19(1), 54–60 (2011)

19. Rinner, C., Duftschmid, G., Wrba, T., Gall, W.: Making the complex data model of a clinical research platform accessible for teaching. J. Innov. Health Inform. 24(1) (2017)

20. Rojas, E., Munoz-Gama, J., Sepúlveda, M., Capurro, D.: Process mining in healthcare: a literature review. J. Biomed. Inform. 61, 224–236 (2016)

21. Rovani, M., Maggi, F.M., de Leoni, M., van der Aalst, W.M.: Declarative process mining in healthcare. Expert Syst. Appl. 42(23), 9236–9251 (2015)

22. Rozinat, A., Van der Aalst, W.M.: Conformance checking of processes based on monitoring real behavior. Inf. Syst. 33(1), 64–95 (2008)

23. Safran, C., et al.: Toward a national framework for the secondary use of health data: an american medical informatics association white paper. J. Am. Med. Inform. Assoc. 14(1), 1–9 (2007)

24. Tax, N., Lu, X., Sidorova, N., Fahland, D., van der Aalst, W.M.: The Imprecisions of Precision Measures in Process Mining. arXiv preprint arXiv:1705.03303 (2017)

25. Verbeek, H.M.W., Buijs, J.C.A.M., van Dongen, B.F., van der Aalst, W.M.P.: XES, XESame, and ProM 6. In: Soffer, P., Proper, E. (eds.) CAiSE Forum 2010. LNBIP, vol. 72, pp. 60–75. Springer, Heidelberg (2011). https://doi.org/10.1007/978-3-642-17722-4_5

Characterization of Drug Use Patterns Using Process Mining and Temporal Abstraction Digital Phenotyping

Eric Rojas[✉] and Daniel Capurro

Internal Medicine Department, School of Medicine,
Pontificia Universidad Católica de Chile, Santiago, Chile
{eric.rojas,dcapurro}@uc.cl

Abstract. Understanding and identifying executed patterns, activities and processes for patients of different characteristics provides medical experts a deep understanding of which tasks are critical in the provided care, and may help identify ways to improve them. However, extracting these events and data for patients with complex clinical phenotypes is not a trivial task. This paper provides an approach to identifying specific patient cohorts based on complex digital phenotypes as a starting point to apply process mining tools and techniques and identify patterns or process models. Using temporal abstraction-based digital phenotyping and pattern matching, we identified a cohort of patients with sepsis from the MIMIC II database, and then apply process mining techniques to discover medication use patterns. In the case study we present, the use of temporal abstraction digital phenotyping helped us discover a relevant patient cohort, aiding in the extraction of the data required to generate drug use patterns for medications of different types such as vasopressors, vasodilators and systemic antibacterial antibiotics. For sepsis patients, combining the use of temporal abstraction digital phenotyping and process mining tools and techniques, was proven to help extract accurate cohorts of patients for health care process mining.

Keywords: Process mining · Digital phenotyping · Drug use patterns

1 Introduction

Currently, medical centers execute a large number of medical activities and procedures, which constitute an essential part of the provided care. To successfully administer the most suitable care to patients, processes must be executed in the most effective and efficient way possible.

Medications are an essential component of clinical care. The pattern at which they are administered can reflect the clinical pathways followed by patients during their episodes of care. Also, in general, following a proper medication process or pathway should also impact the healthcare system, reducing hospitalization

© Springer Nature Switzerland AG 2019
F. Daniel et al. (Eds.): BPM 2018 Workshops, LNBIP 342, pp. 187–198, 2019.
https://doi.org/10.1007/978-3-030-11641-5_15

rates, improving quality of life, reducing health system costs, and, improving quality of care [11].

Nowadays, hospital information systems store large amounts of medical information generated during routine patient care; this includes abundant medication administration data. Using this information, it is possible to conduct several types of analysis. Data analytic tools and techniques, specifically process mining, provide the ability to discover process models, understand the interaction between resources and analyze their performance [1]. The use of process mining techniques does not only facilitate an understanding of the natural complexity of hospital processes and what these genuinely entail, but it also generates improvement opportunities in relation to care services. With the use of process mining, drug administration patterns may be studied to understand which are the drugs administered and in which patterns.

In the past, process mining in healthcare has been applied, generating complex and unreadable spaghetti process models [1]. To reduce this complexity, multiple efforts have been done. For example, applying simple filters guided by the desired objectives [23] or clustering techniques to organize patients [20]. These efforts required multiple tasks to identify the correct cohorts of patients, increasing the possibility of leaving significant patients outside of the analysis. Our study proposes the use of temporal abstraction digital phenotyping to identify more specific cohorts of patients with exact medication and condition types, to use process oriented data analytic techniques to generate detailed patterns.

The structure of the paper is as follows: Sect. 2 includes the background of the study. Section 3 describes the followed method. Section 4 describes a case study using the proposed method, including the results and discussion. Finally, conclusions an future work are included at the end of this paper.

2   Background

The use of clinical data is relevant in the big data era. Each time more and more data are being produced in clinical environments. In order to study the executed processes, the data must be accessed, selected and extracted faster and faster to not loose track of details included in it. The selection process of the adequate data is vital to obtain the desired results. In the past, filtering [23] and clustering [20] techniques have been good approaches to define the exact data to extract for process analysis, but can be time consuming and generate cohorts of patient's data that may not be the desired ones.

The need to generate good techniques to identify and extract data for patients satisfying very punctual conditions is required in process mining. In our approach we focus on presenting a method that combines the use of temporal abstraction digital phenotyping with the process mining tools and techniques available.

2.1   Temporal Abstraction Digital Phenotyping

Patient phenotyping has become a key component of complex data analytics in healthcare. The identification of predictor variables or risk factors using large volumes of routinely collected clinical data between two patient types—for example, comparing older adults with and without dementia, understanding complex processes in emergency rooms and analyzing clinical data to identify heralding attributes—has become a prevailing study design and it requires establishing clear and robust patient phenotypes and methods to retrieve patients that match that phenotype. However, establishing robust clinical phenotypes in databases of routinely collected clinical data is not a trivial task. Current approaches include the use of billing information, diagnostic codes, developing complex algorithms, natural language processing and, ultimately, manual abstraction of clinical records [5, 13,27]. Our research group has advanced the use of temporal abstraction and pattern matching to design digital phenotypes and query clinical databases to retrieve patients that meet the designed phenotype [4]. Briefly, there are two main approaches when attempting to identify patient cohorts based on temporal patterns of clinical data in raw medical records.

The first and probably more cumbersome one is developing a complex query using database query languages or specific query languages with temporal capacities, usually built over an existing one, such as SQL. Such an effort has been made in the past not only by clinical and health informatics researchers, but by database and data storage researchers too: temporal querying and temporal database capacities is a long dated problem. Starting in 1992, TSQL2 [26] was a significant attempt on making temporal databases a reality, until its death on 2001 after hard criticism. A decade later, SQL:2011 [25] included temporal behavior on its core definition, but with little real functionality added. This approach has emerged in some health informatics researchers as well. CHRONUS introduced TimeLineSQL (TLSQL) [7], and later in CHRONUS II an extension to the SQL query language [16], greatly inspired in TSQL2 [25], was also developed as a means to directly execute temporal based queries on databases. The most significant hurdle of these methods involve developing very complex database queries that require advance programming knowledge with limited reusability.

The other main way of dealing with temporal querying difficulties is by abstracting the underlying raw data onto higher level models suitable for time expressiveness. Typically, this involves the ability to represent time concepts such as instants, intervals and bound times, along with temporal relations among them. DXtractor [15] introduced a hybrid model where simple plain SQL queries were produced to retrieve patient sets from some precompiled options, and then allowed to perform some basic Boolean and temporal operations over these sets in a chained way. IDAN [2] presents a temporal-abstraction mediation approach, implemented in a modular architecture. In a way, it decouples the effort of temporal reasoning on clinical data in several small components, each with a specific task. PROTEMPA [19] aimed to offer a system for specifying temporal and mathematical relationships between data elements to retrieve cohorts of patients that meet the given requirements. It is a framework like system, with clearly distinguishable modules for different tasks, such as data extraction, knowledge and abstractions definition, temporal trends detection, etc. A later and more novel system, Eureka [18], was constructed to address some of PROTEMPA's issues and extend its functionality. It is commonly used to integrate with the I2B2 framework, and offers many useful tools for other integrations too. Opposed to its predecessor, it carries one general ETL (Extract, Transform and Load) system capable of extracting data from different sources, but in practice this means not only configuring some metadata XMLs (eXtensible Markup Language) to describe the underlying model, but also coding a specific data extractor to fully determine the data backend. Eureka is highly configurable and offers a rich API to integrate the researcher's database to the system, but programming knowledge and a deep understanding of how Eureka's ETL works is needed in order to achieve this. ClinicalTime [8] aims to facilitate the task of describing the database by imposing on the user no programming knowledge, only data knowledge. This means that the researcher only needs to know how his data looks like in order to define the desired mapping to the program's model. Using this knowledge the researcher is able to design a clinical phenotype and retrieve patient cohorts with detailed attributes and conditions.

2.2   Process Mining in Healthcare

Process mining is a research discipline that focuses on the extraction of information from data generated and stored in the databases of information systems. The data is extracted to create events logs, which can be viewed as a set of cases in which each one contains all the activities executed for a process instance [1].

Process-Aware Information Systems (PAIS) [1] are systems that should be readily able to produce event logs. Specific examples of such applications include Enterprise Resource Planning systems and Hospital Information Systems (HIS [22]). Event log data are not limited simply to the data from these tools, as many other systems can also provide useful data about process execution. Moreover, data regarding an specific complex process can come from multiple information systems or data sources.

There are three main types of process mining analysis that can be performed: process discovery, conformance checking, and enhancement. Process discovery allows process models to be extracted from an event log; conformance checking allows monitoring deviations by comparing a given model with the event log; and how enhancement allows extending or improving an existing process model using information about the actual process recorded in the event log [1].

Process mining has been successfully employed for analysis and study purposes across different industries, including the education [3], marketing [17], among others. The healthcare domain is not the exception [10,14,20,22]. Normally, any activity executed in a hospital by a physician, nurse, technician or any other resource to give care to a patient is stored in a HIS (compound of databases, systems, protocols, events, etc.). Activities are recorded in event logs for support, control and further analysis. Process models are created to specify the order in which different health workers are supposed to perform their activities within a given process, or to analyze critically the process design. Moreover, process models are also used to support the development of HIS, for example, to understand how the system is expected to support the process execution [22].

3   Method

Figure 1 presents an overview of the methods used for the analysis. It consists of four phases that combine temporal abstraction-based digital phenotyping and process mining to discover drug patterns. The phases are the following:

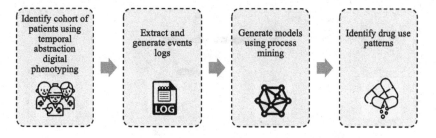

Fig. 1. Method

1. *Identify specific cohort of patients using temporal abstraction digital phenotyping from specific data sources.* Through ClinicalTime [8] a tool that includes temporal abstraction digital phenotyping, precise conditions and time intervals can be defined to extract specific patients, allowing clinicians and researchers to study more accurate groups of patients. An example of the way temporal queries are visually constructed using ClinicalTime can be seen in Fig. 2.
2. *Extract and generate events logs for the identified cohorts of patients.* Having identified an specific cohort of patients, an event log for an research purpose can be generated. An event log must include at least a case unique identifier, an executed activity, and a timestamp.
3. *Generate models using process mining tools and techniques.* With the extracted event log, several tools (such as Disco [12], or PALIA [10]) and techniques (such as heuristic miner [28] or the Palia algorithm [21]), can be used to generate process models and complementary information for analysis.
4. *Identify drug use patterns based on the models discovered in previous steps.* Using the models previously obtained, the medication data can be used to identify drug use patterns on the extracted set of identified patients.

4   Results

We validated the methods presented in Sect. 3 using a case study of patients with Sepsis. For each of the established steps of the method, the executed tasks are described in the next subsections.

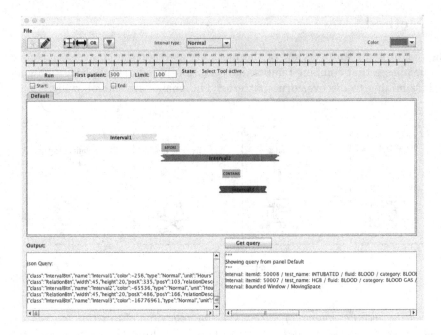

Fig. 2. Temporal query example in ClinicalTime

4.1 Phase 1: Define Cohorts of Patients Using Temporal Abstraction Digital Phenotyping

For phase 1 of the method, we retrieved the patient cohort from the MIMIC II database. MIMIC II is an anonymized database containing more than 30,000 intensive care unit episodes [24]. Several tables were used to identify the cohort of patients and the data for the event log: *icustay_events, ioevents, labevents, medevents, physicianorderentry_events, procedure_events, ioevents, microbiology_events, labevents, medevents, icustay_events,*. A full description of the MIMIC II data model can be found in [6].

This database was accessed through ClinicalTime, an application to build and execute complex temporal abstraction digital phenotyping queries [8]. Clinical-Time is an application that enables researchers to define clinical phenotypes using a graphical interface. In that interface, the users define clinical temporal instants and intervals (instantaneous, bounded), as well as their temporal and mathematical relationships. Additional conditions can be defined for every interval such as: duration, number of repeated instants within an interval, etc. [8].

In addition, temporal relations can be defined between intervals and instants; ClinicalTime implements the full set of temporal relations. Researchers can define other relations between intervals such as temporal distance, an increase or decrease of a certain magnitude, a percent change in values, etc. Once an interval pattern is defined, it can be saved and combined with other interval patterns. This allows the creation of arbitrarily complex interval patterns to describe clinical phenotypes. ClinicalTime then uses its search algorithms to abstract the temporal intervals from a clinical relational database (in our case MIMIC II), and identifies and returns a list of patients matching the pattern. This patient set becomes the patient cohort. To ensure precision, the phenotyping algorithm was validated against a manually annotated subset of MIMIC II.

For our case study we decided to extract drug use patterns for Sepsis. Sepsis is one of the main causes of admission to the intensive care unit world-wide and has significant associated morbidity and mortality. To meet the definition, a sepsis patient must meet two of the following criteria: (a) temperature $<36\,^{\circ}C$ or $>38\,^{\circ}C$, (b) respiratory rate $>20/min$ or $PaCO2$ $<32\,mmHg$, (c) heart rate $>90/min$, (d) white blood cell (WBC) count $<4,000$ or $>12,000$ or $>10\%$ bands (immature white blood cells). In addition to these criteria, the condition must be a response to an *active infection*. An active infection was defined as a combination of clinical and laboratory results. We created the sepsis phenotype using the above criteria, and extracted from the MIMIC II a cohort of patients that meet these criteria.

4.2 Phase 2: Create Event Log from Identified Cohort of Patients

Based on the cohort of patients generated using Clinical Time, we proceeded to extract and generate the event log from the database MIMIC II. Based on an exploratory analysis of the data, the pharmacy provider order entry (POE) records were necessary to execute the drug use analysis.

In the database, multiple categories for medications and drugs were discovered. A drug may be classified by the chemical type of the active ingredient or by the way it is used to treat a particular condition. In our case we centered our analysis in vasodilators, vasopressors, and systemic antibacterial antibiotics. These three categories were selected to analyze. Vasodilators are medicines that dilate (widen) blood vessels, allowing blood to flow more easily through. Some act directly on the smooth muscle cells lining the blood vessels [9]. Some examples of Vasodilators are nitroglycerin and desmopressin. Vasopressors are medicines that constrict (narrow) blood vessels, increasing blood pressure. They are used in the treatment of extremely low blood pressure, especially in critically ill patients [9]. Some examples of Vasopressors are phenylephrine and dopamine. Antibiotics are drugs that can either kill an infectious bacteria or inhibit its growth. Different antibiotics work by different mechanisms and are used to treat infections caused by bacteria that are sensitive to that particular antibiotic [9]. Some examples of antibiotics are vancomycin and ampicillin.

After identifying the categories of the drugs or medications, we proceeded to extract the event logs, including all patients that have two conditions:

(i) were identified as having Sepsis as the main cause of admission to the intensive care unit (results of phase 1), and;

(ii) have been medicated with either vasodilators, vasopressors, or systemic antibacterial antibiotics.

Three specific cohort of patients were extracted into three event logs. For the drug use pattern of sepsis and vasodilators, 20 cases were identified with 6 start and end medication activities. For sepsis and vasopressors, 33 cases were identified with 12 start and end medication activities. And, finally, for sepsis and systemic antibacterial antibiotics, 60 cases were identified with 21.

4.3   Phase 3: Generate Models from Event Logs

After extracting each event log from the previous phase, process mining tools were used to generate process models. In our case, we used Disco [12] to generate the models. Following are the three resulting drug use patterns discovered.

First, the process model for patients with sepsis and vasodilators medications is presented in Fig. 3. This model includes the starting and ending medication activities for nitroglycerin, and nitroprusside sodium, which were the identified vasodilators. The arcs correspond to the sequential order for all the different 20 cases. Nitroglycerin was the most frequently used vasolidator.

Second, the process model reflecting the drug use for patients with sepsis and vasopressors medications is presented in Fig. 4. This model includes the starting and ending medication activities for norepinephrine, phenylephrine, epinephrine, vasopressin, dopamine and dobutamine, which were the identified vasopressors. The arcs correspond to the sequential order for all the different 33 cases. Norepinephrine was the vassopressor medicated the most.

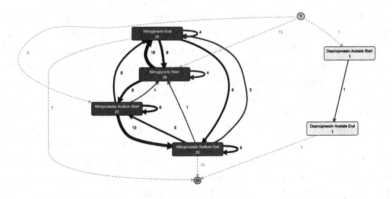

Fig. 3. Process for vasodilators

And, finally, the process model reflecting the drug use for patients with sepsis and systemic antibacterial antibiotics is presented partially in Fig. 5. This model includes only the starting medication activities for vancomycin, levofloxacin, metronidazole, piperacillin-tazobactam, ceftriaxone, aztreonam, meropenem, ampicillin, cefazolin, azithromycin, linezolid, ciprofloxacin, clindamycin, unasyn, sulfameth/trimethoprim, penicillin G, potassium, erythromycin, nafcillin, oxacillin, dicloxacillin, and, imipenem-cilastatin. These were the identified antibiotics.

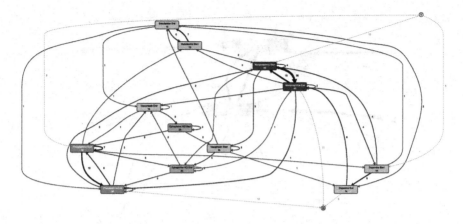

Fig. 4. Process for vasopressors

In this case to improve the visualization of the process model, only the starting activities were selected and not the ending of the treatment. The arcs correspond to the sequential order for all the different 60 cases. Vancomycin and levofloxacin were the systemic antibacterial antibiotics most frequently prescribed.

4.4 Phase 4: Identify Drug Use Patterns

Finally, we identified the drug use patterns using the information and models acquired through process mining in the previous phases.

First, for the vasodilators, the main pattern is featured in Fig. 3. Although a full clinical analysis is beyond the scope of this paper, the observed pattern, its sequences and frequencies, is consistent to what clinicians use in the real-world. It is frequent, in most of the cases, to either only prescribe nitroglycerine (35%), or, prescribe it combined with nitroprusside sodium (35%), being this drug medicated in 70% of cases. Also, 65% of cases begin with nitroglycerine, being the first option as a vasodilator to be medicated.

Secondly, for the vasopressors, the main pattern is featured in Fig. 4. Twenty-four percent of all cases only prescribe Norepinephrine, while Epinephrine is never prescribed alone. The most prescribed are norepinephrine, which is present in 70% of the cases and Phenylephrine, which is present in 42% of them. Additional to this, 27% of the cases include two different vasopressors, and only 12% include 4 or more different ones. The trend in these cases is to start either with Norepinephrine in 42% of the cases, or start with Phenylephrine in 30%.

And finally, for the systemic antibacterial antibiotics, part of the main pattern is shown in Fig. 5, where some characteristics were identified. There is no predominant antibiotic that is prescribed alone, the top antibiotics prescribed alone are levofloxacin (only 8.3%) and cefazolin (only 5%). More than 50% of the cases prescribe 3 or more antibiotics (53.3%), while 20% prescribe only two. Fifty-three percent of the cases tend to start the antibiotic medication pattern

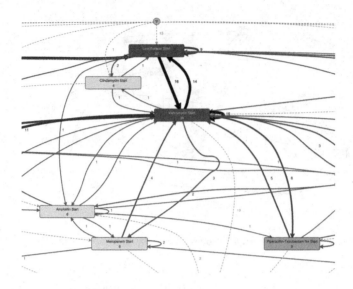

Fig. 5. Process for antibiotics

with either vancomycin (28.3%) or levofloxacin (25%), being these two the initials and the most prescribed antibiotics. While no cases start directly with aztreonam, linezolid, nafcillin, oxacillin or dicloxacillin. When clinicians are presented with these patterns, they are concordant with current clinical practices, highlighting the potential of the presented methods for clinical data analytics.

5   Discussion

The discussion will be addressed both from a process mining and a clinical perspective. From the process mining point of view our method and case study contributed with an easier and more accurate method to identify cohorts of patients when multiple and temporally-related clinical conditions must be met. This facilitated the extraction and generation of event logs, because only the necessary data was included for the analysis, eliminating or reducing the filtering/clustering phases utilized by most process-mining methodologies. In addition, this simplified the complexity involved in generating accurate queries to extract data from electronic health records. Finally, this approach generated better, more understandable and readable models, that are based on cohorts of patients with very specific conditions, making them easier to analyze.

From the clinical point of view, analyzing drug use patterns may help in measuring conformance of clinical practice with guideline recommendations, identify changes in prescription pattern over time, for example, when new resources or drugs are incorporated to care and whether those changes are clinically explained or not. This method also generates ways to verify, monitor and control drug prescription on specific groups of patients, opening avenues to improving the provided care and the quality of the outcome.

6   Conclusions and Future Work

In this case study, we applied a method based on the combination of process mining and temporal abstraction-based digital phenotyping to help discover drug use patterns among patients with sepsis. Each step of the proposed method has been executed: identifying cohorts of patients using temporal abstraction digital phenotyping, creating the event logs, generating models, and finally identifying drug use patterns. Although beyond the scope of this demonstration, the ability to quickly discover drug utilization patterns should be a useful tool to study healthcare resource utilization, its patterns and how they might change over time. Future work will include expanding the case study to include additional data sources and phenotypes (such as additional drug classes and clinical outcomes), the inclusion of additional process mining techniques to improve pattern discovery, and, finally, formal clinical validation of the identified patterns.

References

1. Van der Aalst, W.M.P.: Process Mining: Data Science in Action. Springer, Heidelberg (2016). https://doi.org/10.1007/978-3-662-49851-4
2. Boaz, D., Shahar, Y.: Idan: a distributed temporal-abstraction mediator for medical databases. In: Dojat, M., Keravnou, E.T., Barahona, P. (eds.) AIME 2003. LNCS (LNAI), vol. 2780, pp. 21–30. Springer, Heidelberg (2003). https://doi.org/10.1007/978-3-540-39907-0_3
3. Bogarín, A., Romero, C., Cerezo, R., Sánchez-Santillán, M.: Clustering for improving educational process mining. In: Proceedings of the Fourth International Conference on Learning Analytics and Knowledge, pp. 11–15. ACM (2014)
4. Capurro, D., Barbe, M., Daza, C., Santa María, J., Trincado, J., Gomez, I.: ClinicalTime: identification of patients with acute kidney injury using temporal abstractions and temporal pattern matching. AMIA Summits Transl. Sci. Proc. **2015**, 46 (2015)
5. Che, Z., Kale, D., Li, W., Bahadori, M.T., Liu, Y.: Deep computational phenotyping. In: Proceedings of the 21st ACM SIGKDD International Conference on Knowledge Discovery and Data Mining, pp. 507–516. ACM (2015)
6. Clifford, G.D., Scott, D.J., Villarroel, M., et al.: User guide and documentation for the MIMIC II database. MIMIC-II database version 2(95) (2009)
7. Das, A.K., Tu, S.W., Purcell, G.P., Musen, M.A.: An extended SQL for temporal data management in clinical decision-support systems. In: Proceedings of the Annual Symposium on Computer Application in Medical Care, p. 128. American Medical Informatics Association (1992)
8. Daza, C., Santa Maria, J., Gomez, I., Barbe, M., Trincado, J., Capurro, D.: Phenotyping intensive care unit patients using temporal abstractions and temporal pattern matching. In: Proceedings of the 7th ACM International Conference on Bioinformatics, Computational Biology, and Health Informatics, pp. 508–509. ACM (2016)
9. Drugs.com: Date source of drug information online (2018). https://www.drugs.com. 1 May 2018
10. Fernandez-Llatas, C., Lizondo, A., Monton, E., Benedi, J.M., Traver, V.: Process mining methodology for health process tracking using real-time indoor location systems. Sensors **15**(12), 29821–29840 (2015)

11. Gilberg, K., Laouri, M., Wade, S., Isonaka, S.: Analysis of medication use patterns: apparent overuse of antibiotics and underuse of prescription drugs for asthma, depression, and CHF. J. Manag. Care Pharm. **9**(3), 232–237 (2003)
12. Günther, C., Rozinat, A.: Disco: discover your processes. BPM **940**, 40–44 (2012)
13. Liao, K.P., et al.: Development of phenotype algorithms using electronic medical records and incorporating natural language processing. BMJ **350**, h1885 (2015)
14. Mans, R.S., van der Aalst, W.M.P., Vanwersch, R.J.B.: Process Mining in Healthcare: Evaluating and Exploiting Operational Healthcare Processes. SBPM. Springer, Cham (2015). https://doi.org/10.1007/978-3-319-16071-9
15. Nigrin, D.J., Kohane, I.S.: Temporal expressiveness in querying a time-stamp-based clinical database. JAMIA **7**(2), 152–163 (2000)
16. O'Connor, M.J., Tu, S.W., Musen, M.A.: The Chronus II temporal database mediator. In: Proceedings of the AMIA Symposium, p. 567 (2002)
17. Osses, A.S., et al.: Business process analysis in advertising: an extension to a methodology based on process mining projects. In: 2016 35th International Conference of the Chilean Computer Science Society (SCCC), pp. 1–12. IEEE (2016)
18. Post, A.R., et al.: Temporal abstraction-based clinical phenotyping with Eureka! In: AMIA Annual Symposium Proceedings, vol. 2013, p. 1160. American Medical Informatics Association (2013)
19. Post, A., Harrison Jr., J.: Protempa: a method for specifying and identifying temporal sequences in retrospective data for patient selection. J. Am. Med. Inform. Assoc. **14**(5), 674–683 (2007)
20. Rebuge, Á., Ferreira, D.R.: Business process analysis in healthcare environments: a methodology based on process mining. Inf. Syst. **37**(2), 99–116 (2012)
21. Rojas, E., et al.: Palia-er: bringing question-driven process mining closer to the emergency room. In: Proceedings of the Business Process Management Conference 2018 (BPM 2018) Demo Track, Barcelona, Spain (2018)
22. Rojas, E., Munoz-Gama, J., Sepúlveda, M., Capurro, D.: Process mining in healthcare: a literature review. J. Biomed. Inform. **61**, 224–236 (2016)
23. Rojas, E., Sepúlveda, M., Munoz-Gama, J., Capurro, D., Traver, V., Fernandez-Llatas, C.: Question-driven methodology for analyzing emergency room processes using process mining. Appl. Sci. **7**(3), 302 (2017)
24. Saeed, M., et al.: Multiparameter intelligent monitoring in intensive care II (MIMIC-II): a public-access intensive care unit database. Crit. Care Med. **39**(5), 952 (2011)
25. Snodgrass, R.T.: The TSQL2 Temporal Query Language, vol. 330. Springer, New York (2012)
26. Snodgrass, R.: The TSQL2 Temporal Query Language. Springer, London (1995). https://doi.org/10.1007/978-1-4615-2289-8
27. Wei, W.Q., Teixeira, P.L., Mo, H., Cronin, R.M., Warner, J.L., Denny, J.C.: Combining billing codes, clinical notes, and medications from electronic health records provides superior phenotyping performance. JAMIA **23**(e1), e20–e27 (2015)
28. Weijters, A.J.M.M., Ribeiro, J.T.S.: Flexible heuristics miner (FHM). In: 2011 IEEE Symposium on Computational Intelligence and Data Mining (CIDM) (2011)

Pre-hospital Retrieval and Transport of Road Trauma Patients in Queensland
A Process Mining Analysis

Robert Andrews[1]([⊠]), Moe T. Wynn[1], Kirsten Vallmuur[1],
Arthur H. M. ter Hofstede[1], Emma Bosley[2], Mark Elcock[3],
and Stephen Rashford[2]

[1] Queensland University of Technology (QUT), GPO Box 2434, Brisbane, Australia
r.andrews@qut.edu.au
[2] Queensland Ambulance Service (QAS), GPO Box 1425, Brisbane, Australia
[3] Retrieval Services Queensland (RSQ), GPO Box 1425, Brisbane, Australia

Abstract. Existing process mining methodologies, while noting the importance of data quality, do not provide details on how to assess the quality of event data and how the identification of data quality issues can be exploited in the planning, data extraction and log building phases of any process mining analysis. To this end we adapt CRISP-DM [15] to supplement the Planning phase of the PM² [6] process mining methodology to specifically include data understanding and quality assessment. We illustrate our approach in a case study describing the detailed preparation for a process mining analysis of ground and aero-medical pre-hospital transport processes involving the Queensland Ambulance Service (QAS) and Retrieval Services Queensland (RSQ). We utilise QAS and RSQ sample data to show how the use of data models and some quality metrics can be used to (i) identify data quality issues, (ii) anticipate and explain certain observable features in process mining analyses, (iii) distinguish between systemic and occasional quality issues, and, (iv) reason about the mechanisms by which identified quality issues may have arisen in the event log. We contend that this knowledge can be used to guide the extraction, preprocessing stages of a process mining case study.

Keywords: Process mining · Data quality · Pre-hospital · GEMS · HEMS

1 Introduction

Pre-hospital care and transport can be supplied by road services, aero-medical services or a combination of these two services. Comparing the various transport modes and escort levels etc. may lead to a better understanding of factors contributing to patient outcomes. However, there is limited research internationally examining the retrieval processes for patients from roadside to definitive care, and there has been no research conducted in the Queensland context.

© Springer Nature Switzerland AG 2019
F. Daniel et al. (Eds.): BPM 2018 Workshops, LNBIP 342, pp. 199–213, 2019.
https://doi.org/10.1007/978-3-030-11641-5_16

This project, being conducted in collaboration with Queensland Ambulance Services (QAS) as providers of ground-based transport services, and Retrieval Services Queensland (RSQ) as coordinators of aero-medical transport services, aims to provide insights into a key question of interest to QAS and RSQ, i.e. "are we getting the right level of care to the right patients"? Specific questions to be considered include:

- Are we making the correct decisions about the dispatch of assets?
- Is it possible to validate the existing "45 min" guideline for transport to the nearest facility (and the by-pass to major trauma centre guideline)?
- Can the existing aero-medical request for launch protocol be validated?

It is proposed to address these questions using process mining techniques to:

- Discover the range of different care and transport processes undertaken for road trauma patients from roadside to definitive care
- Conduct conformance (to guidelines) and comparative performance analyses
- Identify key factors influencing deviance from standard care and delivery processes as given in the guidelines

The impact of data quality on process mining analyses is well recognised [5,12] and existing process mining methodologies [6,10] refer to taking into consideration characteristics and quality of the input data in the early stages of a process mining study. However, there is little guidance on how to actually assess process-data quality. We adapt the CRISP-DM (Cross Industry Standard Process for Data Mining as outlined in [15]) and apply it in our case study to (Fig. 1):

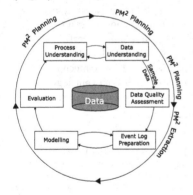

- Gain an understanding of the overall QAS and RSQ dispatch-retrieval-transport processes.
- Gain an understanding of QAS and RSQ data through development of data models and examination of sample data extracts.
- Conduct data quality assessments.
- Prepare event logs.
- Use the sample data to discover models of the individual QAS and RSQ retrieval/transport processes.
- Evaluate/conformance check the models.

Fig. 1. Our approach adapted from the CRISP-DM methodology outlined in [15].

The major contributions of this paper include:

- conceptual data models (Object-Role Models (ORM)) of data held by (i) a ground based ambulance service provider (QAS) and (ii) a coordinator of aero-medical retrieval and transport service provider (RSQ);
- an assessment of the quality (fitness for purpose) of the QAS and RSQ data for process mining analysis;

- a contribution to the knowledge on how to conduct a process mining study through a demonstration of the value of systematically identifying data-related issues prior to carrying out a process mining analysis; and
- a contribution to the knowledge-base in relation to Ground and Helicopter EMS dispatch processes in an Australian context.

2 Related Literature

We considered literature relating to process mining in the healthcare domain (and in pre-hospital transport in particular). We look also at process mining methodology and find that existing methodologies do not highlight the value of data quality assessment in the early stages of a process mining exercise. We found that little work has been done in applying process mining techniques to analyse pre-hospital processes.

2.1 Process Mining and Healthcare

Rebuge and Ferreira [10] propose a Business Process Analysis methodology for healthcare based on process mining. The methodology comprises: (1) the preparation of an event log; (2) log inspection; (3) control-flow analysis; (4) performance analysis; (5) organizational analysis; (6) transfer of results. While the methodology steps (1) and (2) are data-focused, consideration of the quality of data and its suitability for process mining is not considered. In [6] the authors propose PM^2 as a comprehensive, 6 step (Planning, Extraction, Data processing, Mining & Analysis, Evaluation, Process Improvement & Support) process mining methodology. In the description of PM^2, the authors do not mention event data-quality from the point of view of (i) informing the Extraction and Data Processing stages, (ii) possible impacts on Mining & Analysis. Mans et al. [9] discuss event data recorded in Hospital information Systems (HIS) and introduce the Healthcare Reference Model, a comprehensive data model designed to allow analysts to locate event data easily and to support data extraction. Rojas et al. [11] review 74 articles describing applications of process mining in the healthcare domain. Papers were characterised according to 11 points of relevance including process type, data type, frequently asked questions, analysis perspectives, tools, methodologies. The authors conclude that future work should focus on the implementation of process-aware hospital information systems along with improved visualisation and visual analytics techniques and an increased focus on conformance checking in case studies. Andrews et al. [2] discuss the application of process mining techniques in the analysis of healthcare process-related data focussing on data extraction, pre-processing and data quality assessment before considering challenges facing analysts in dealing with the semi-structured nature of healthcare processes when conducting discovery, conformance (comparative) performance analysis before providing some novel visualisation options. Little work has been done in applying process mining techniques to analyse pre-hospital processes. Lamine et al. [8] apply process mining and discrete event simulation to assess the efficiency of emergency call centre operations in France and [3] apply process discovery, conformance checking and performance analysis in a case study involving ambulance services in Iran.

3 Case Study Description

3.1 High-Level Patients Retrieval/Transport Process in Queensland

Figure 2 is a high-level retrieval/transport model derived from operating guideline documents provided by QAS and RSQ. In Queensland, all emergency calls (calls to 000) are routed to a single, statewide call centre operated by QAS. The emergency centre operators gather as much information about the incident as they can from the caller reporting the incident. Usually, QAS will dispatch one or more ground-based ambulances to the scene of the incident, but may directly request aero-medical evacuation of injured person(s). Once on-scene, QAS paramedics (i) will provide first-level support to injured patients, (ii) may contact a senior on-call paramedic or QAS Medical Coordinator (an experienced emergency doctor) for treatment advice, and (iii) where the situation fits guidelines, may request aero-medical evacuation of injured person(s).

Where aero-medical retrieval/transport is required, the QAS Communications Centre Supervisor (CCS) calls the RSQ Communication Centre. The call is picked up by a QAS Emergency Medical Dispatcher (EMD) stationed at the dedicated Rotary Wing Desk within the RSQ Communication Centre. The EMD has access to the statewide QAS Computer Aided Dispatch (CAD) which shows

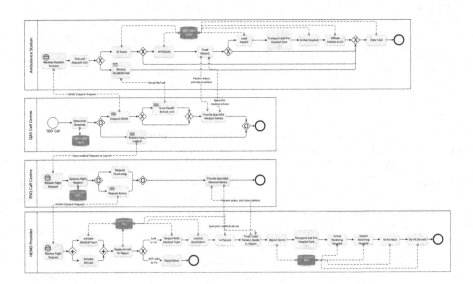

Fig. 2. BPMN model of emergency incident management - ground and aero-medical call centre, asset deployment and patient transport.

the Incident record. The EMD links the QAS CCS with the RSQ Medical Coordinator who discuss the incident and determine the optimal response. If the decision is made to dispatch an aircraft the EMD tasks the aircraft while the RSQ Medical Coordinator contacts the retrieval team to fly on the respective aircraft and provides the patient's clinical details. On arrival at the scene of the incident, or following contact with the patient, the retrieval team contacts the RSQ Medical Coordinator (via satellite phone or mobile phone). The RSQ Medical Coordinator provides specialist advice to, and oversight of, the retrieval team. They then determine the receiving hospital based on the patient's clinical needs and informs the receiving, on-duty Emergency Department Specialist of the incoming patient, their estimated time of arrival and their clinical condition and requirements. On arrival at the Receiving Hospital, the retrieval team hands-over to the Emergency Department Specialist.

3.2 Scenarios

From the process description, high level BPMN model, data models, and discussions with domain experts, it is possible to derive some scenarios which may play out in response to any incident:

1. Road-based response/s with treatment and no transport.
2. Road-based response/s with treatment and at least one primary transport.
3. Road-based response/s with treatment and a rotary wing primary transport.
4. Rotary wing inter-hospital (secondary) transfer.
5. Fixed wing primary transport.
6. Multileg primary transport (road + rotary or fixed wing).

In this preliminary study, only scenarios 1–3 are considered. The full study will consider all scenarios.

3.3 Data Models - Ground and Aero-medical Retrieval/Transport

From our understanding of emergency incident reporting-to-retrieval/transport (developed through interviews with domain experts, documentation describing QAS and RSQ data and informed by our literature review) we identified data relevant to the study that allows end-to-end traceability (notification to delivery to definitive care) and which allows segmentation of the data into cohorts of retrieval/transport cases of interest to the process stakeholders. The Object-Role Model [7] in Fig. 3 depict the main data attributes necessary to allow end-to-end traceability and case segmentation for QAS. A simlar model (not shown) was developed for RSQ. The main categories of data are as follows:

1. **Incident data** such as location of the incident, notification datetime the incident was reported to the emergency call centre and the priority of the incident.
2. **Patient data** including patient name, age, gender, pre-existing conditions, allergies, current medications and indigenous status.

3. **Transport data** which includes timestamped way-point data representing key case milestones, details of assessment of the scene, patient and injury by the paramedics, observations of the patient, management activities and procedures carried out by the ground-based paramedics or aircraft medical team, the destination hospital, and the patient outcome.

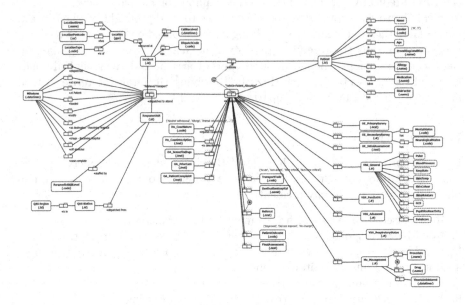

Fig. 3. ORM model of QAS data

4   Data Quality Assessment

Data quality is described as a multi-dimensional concept [13] with each dimension representing some (quantifiable) characteristic. For this study, we use 3 (Completeness, Precision, Uniqueness) of the 20 quality dimensions frequently mentioned in [14] and their associated metrics. The metrics were chosen as they provide insights into not only the state of the data in a particular column, but also into some possible impacts on process modeling. For instance, low values for the Precision metric [13] for datetime columns indicates coarse granularity (e.g. some values in the column may be at day level granularity). From a process mining perspective, this presents some issues in sequencing the events properly (day level granularity events will always appear to occur before milli-second level granularity events for events that have the same date). The Completeness metric [1] measures the fraction of the rows of the data set that have a value in the column. The Completeness metric then gives an indication of the suitability of the column for inclusion in an event log. For instance, if the column values are intended to be used to differentiate between cohorts of cases and the column is only 25% complete, it will not be possible to properly segment the set of cases. Lastly, the Uniqueness metric [4] provides a measure of the similarity of values in the column. For datetime columns that represent event log times, it is often good to have high uniqueness, while for columns that represent activity labels, a certain degree of sameness is desirable. To conduct the assessment we (i) loaded the sample data into a relational database, then (ii) applied the column level quality checks, and (iii) checked for the presence of event log imperfection patterns as described in [12].

4.0.1 QAS Sample Data comprised a de-identified sample of 500 incidents attended by QAS between 01-July-2016 and 09-Jul-2016. The data set was compiled from two separate information systems maintained by QAS. The Computer Aided Dispatch (CAD) system records the datetime of incident notification, (first) vehicle assignment, vehicle arrival on scene, departure from the scene, arrival at destination (hospital) and finally completion of the assignment. Not all waypopint times are recorded for vehicles not involved in a patient transport. The Electronic Ambulance Report Form (eARF) records waypoint times for individual patients including vehicle en route, arrival at the scene, paramedics at the patient, patient loaded (for transport) and patient off-load (at hospital). Again, as not all attendances result in a transport, not all fields are populated.

The data was provided in tabular (Excel) format where each of the 15 columns represented an attribute of the attendance/transport (incident identifier, patient identifier and patient and vehicle waypoint times). The data set contained 12 datetime type columns with 2 waypoint times, one from the CAD system and one from the eARF, that likely represent the same event ('At Scene'). From the QAS process description, we note that there can be multiple units attending a single incident and multiple patients involved in a single incident. It is therfore possible to consider the data from at least three different "case" perspectives, i.e. an **incident** may be considered as a case, each **patient** may be considered as a case, or each **response unit** may be considered as a case. After consulting with the domain experts, it was determined that each patient should be the subject of the case. For the purposes of this part of the study, it was decided that eARF could be treated as surrogate patients, i.e. the eARF number would be the case identifier. Some of the time stamps are standardised across all records relating to a given incident to reflect the 'First Assigned' time (that is, all vehicles attending an incident will have the same value for the FIRST_ASSIGNED_CAD waypoint time). Others, such as On Scene/Depart Scene/Destination/Clear reflect the times for that specific unit. Not all timestamps are relevant to all attending units, hence some are empty e.g. D_LOADED_VACIS time isn't recorded for units not transporting a patient. After considerable cleaning and de-duping of the data it was possible to match 723 eARF (VACIS) records with response unit (CAD) records.

Table 1 provides values for three column-level metrics useful in assessing the quality of the de-duplicated data. Here we note that:

- the **Completeness** metric shows that only 3 of the date time columns are 100% complete which indicates that in any incident, not all patient and vehicle waypoints are completed. In particular, the 50% complete value for OFF_STRETCHER_VACIS indicates that only half the patients involved in incidents required road transport to hospital.
- the **Precision** metric (for datetime) values gives an indication of mixed granularity among the various timestamps.
- the **Uniqueness** metric gives an indication of the degree of distinct values found in the column. The FIRST_ASSIGNED_CAD value shows low Uniqueness indicating many repeated values. This reflects the QAS policy of assigning to all vehicles involved in an incident, the timestamp of the first vehicle assigned to attend the incident.

Table 1. QAS - Column-level data quality summary

Column name	Data type	Completeness	Precision	Uniqueness
T_INCIDENT	int	100%		100%
D_RECEIVED_CAD	datetime	100%	83%	100%
FIRST_ASSIGNED_CAD	datetime	100%	83%	58%
CAD (Vehicle) waypoints				
ON_SCENE_CAD	datetime	97%	83%	83%
DEPART_SCENE_CAD	datetime	57%	83%	82%
AT_DEST_CAD	datetime	57%	83%	82%
CLEAR_CAD	datetime	100%	83%	83%
eARF (Patient) waypoints				
EN_ROUTE_VACIS	datetime	94%	66%	84%
AT_SCENE_VACIS	datetime	95%	66%	86%
AT_PAT_VACIS	datetime	90%	66%	90%
LOADED_VACIS	datetime	52%	66%	89%
NOTIFY_VACIS	datetime	7%	66%	53%
OFF_STRETCHER_VACIS	datetime	50%	66%	93%

The distinctly different values of the Precision metric between the _CAD timestamps and _VACIS timestamps suggests a difference in granularity between the sets of timestamps. Investigation revealed that all the _VACIS timestamps were recorded at minute-level granularity while the _CAD timestamps were recorded at second-level granularity. The immediate effect of the mixed granularity on event ordering can be seen when considering two events that must, in reality, occur in a particular order, but which appear to happen in a different order (according to their timestamps). For instance, D_RECEIVED_CAD is the time when QAS Call Centre is notified of an incident and D_EN_ROUTE_VACIS is the time a response unit is recorded as travelling to the incident scene. There are 52 (out of 723) cases where the D_EN_ROUTE_VACIS time is earlier than the D_RECEIVED_CAD time, and in 49 of these cases, the two timestamps are the same to minute-level granularity (as one example, for N_EARF = 76507098, D_RECEIVED_CAD = 2016-07-05 07:34:08 and D_EN_ROUTE_VACIS = 2016-07-05 07:34:00). This fits the description of the 'Inadvertent Time Travel' log imperfection pattern described in [12] and as such, if left unaddressed, has likely impact on process mining in terms of temporal ordering of events no longer matching reality, and incorrect activity/case durations and will likely result in discovered process models showing these two events in parallel rather than, as expected, in sequence. We note that there are also 15 cases where the value of D_CLEAR_CAD is earlier than the D_OFF_STRETCHER_VACIS time, however, this discrepancy appears to be due to some other mechanism (the difference between the two times is up to 1 h).

4.0.2 RSQ Data RSQ provided a de-identified sample of 500 aero-medical transports with case dates between 01-Mar-2017 and 28-Apr-2017 comprising 419 Inter-hospital Transfers, 78 Primary Response missions and 3 Search and Rescue missions. The data set was provided in tabular (Excel) format where each row represented a separate mission and each of the 128 columns represented an attribute of the mission. The data included 62 mission records where the Mechanism of Injury value was 'Vehicle accident' (comprising 35 Inter-hospital Transfers and 27 Primary Response missions). The data set contained only 12 datetime type columns. From a process mining perspective, this gives, at most, 12 different activities that can be extracted from the data. Table 2 provides values for some column-level metrics useful in assessing the quality of the data.

Table 2. RSQ - Column-level data quality summary for a sample of columns

Column name	Data type	Completeness	Precision	Uniqueness
SOURCE_ID	int	100%		100%
DATE_RETRIEVAL_REQUESTED	datetime	100%	33%	7%
TEAM_ACTIVATED	datetime	100%	65%	95%
READY_TO_DEPART	datetime	100%	65%	96%
DEPART_WITH_MEDICAL_TEAM	datetime	100%	65%	95%
LAND_AT_DESTINATION	datetime	100%	65%	96%
AT_SCENE_PATIENT	datetime	100%	65%	97%
DEPARTURE_READY	datetime	100%	65%	96%
ACTUAL_TIME_DEPART	datetime	100%	65%	96%
ARRIVE_AT_RECIEVING_HOSPITAL	datetime	100%	65%	96%
DEPART_RECIEVING_HOSPITAL	datetime	100%	65%	96%
ARRIVE_BACK_AT_BASE	datetime	100%	66%	93%
AVAILABLE_FOR_NEXT_TASKING	datetime	100%	63%	92%
MECHANISM_OF_INJURY	string	25%	100%	27%

Here we note that:

- the **Completeness** metric shows that all values are populated for the date time columns, while only 25% of the records in the log have a value for the MECHANISM_OF_INJURY column;
- the **Precision** metric (for datetime) values gives an indication of coarse granularity among the various timestamps. For instance, all values of the DATE_RETRIEVAL_REQUESTED column are day-level granularity, while all other datetime columns are at minute-level granularity.
- the **Uniqueness** metric gives an indication of the degree of distinct values found in the column. The DATE_RETRIEVAL_REQUESTED value shows low Uniqueness indicating many repeated values. This is not surprising given the narrow range of case dates (many cases on any given day). The SOURCE_ID column shows perfect uniqueness (every value different from all others), while Uniqueness value of 27% for the MECHANISM_OF_INJURY column is reflective of the value being populated from a limited set of allowed values (e.g. a pull-down on a form).

The datetime columns represent milestone events in a mission and are expected to be sequential. We note that there are several violations of such ordering apparent in the sample data. For instance:

Table 3. RSQ - Milestone activity ordering violations

Milestone activities	a before b	$a = b$	a after b
a = DEPARTURE_READY b = ACTUAL_TIME_DEPART	434	66	0
a = ARRIVE_AT_RECEIVING_HOSPITAL b = DEPART_RECEIVING_HOSPITAL	470	29	1
a = ARRIVE_BACK_AT_BASE b = AVAILABLE_FOR_NEXT_TASKING	315	105	80

4.1   Preliminary Process Mining Analysis

In this section we complete the quality analysis by (i) generating event logs from the sample respective data sets, and (ii) using PromLite 1.2 to perform basic process discovery (Inductive Visual Miner plugin) and conformance analysis (Multi-perspective Process Explorer plugin) to check that the event logs are suitable for process mining.

4.1.1   QAS Process Discovery and Conformance

An event log was generated from the de-duplicated QAS sample data by (i) treating each eARF in the sample data as a case, (ii) mapping the N_EARF column to the event log case identifier attribute, (ii) creating an event from each datetime column in the dataset by mapping the column name to the activity label and the row value of the column to the timestamp value. A process model was discovered and conformance checking (see Fig. 4) showed the model had 94.3% fitness (490 wrong and missing events out of 5,595 events in total). The discovered model highlighted some variations from the expected process behaviour as described by the QAS domain expert (illustrated in Fig. 2 and also highlighted some of the data quality issues discussed in 4.0.1. For instance, the expected behaviour is sequential execution of milestone tasks while the discovered model shows parallelism, (e.g. D_EN_ROUTE_VACIS and D_AT_SCENE_VACIS occur in a parallel block). Investigation showed that while there no cases where D_EN_ROUTE_VACIS preceded D_AT_SCENE_VACIS, there were 14 cases where the timestamp values for these two activities were the same. As observed earlier, the data quality analysis precision metric for the _VACIS times indicated only minute-level granularity. This may represent a "field dispatch" (i.e. non-tasked ambulance encounters an accident and notifies EMD it is on-scene). As such, the milestone events actually occurred in the expected sequence, but very close together (i.e. within the same minute) such that the recorded values were identical. In a similar vein, investigating the parallelism exhibited around the D_ON_SCENE_CAD and D_AT_PAT_VACIS activities showed that for the 581 cases where both activities occurred, in 198 cases the D_AT_PAT_VACIS activity occurred before the D_ON_SCENE_CAD activity. However, 174 of these cases had timestamps within 1 min of each other. Taking into account the minute-level granularity of the _VACIS times, it is again possible that these milestone events, in reality, occurred in the expected sequence, but very close together (i.e. within the same minute) but that the mixed granularity of the _VACIS and _CAD times results in incorrect event ordering. (We note that there were in fact 24 cases where there was 'real' deviation from the expected event ordering.)

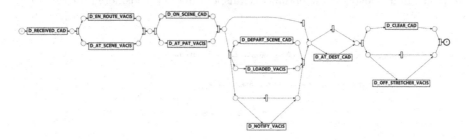

Fig. 4. QAS conformance model derived from sample data

Lastly, we note that the discovered model reflects the nature of the various types of attendance. For instance, (i) the 359 cases which skip the D_LOADED_VACIS and D_DEPART_SCENE_CAD steps reflect that not all attendances required the transport of a patient to hospital, and (ii) the 53 cases where a D_AT_DEST_CAD event occurs without a corresponding

D_OFF_STRETCHER_VACIS event which may reflect a non-transporting unit (e.g. Critical Care Paramedic backup) has proceeded to the hospital to accompany the transporting unit.

4.1.2 RSQ Process Discovery and Conformance

An event log was generated from the RSQ sample data by (i) treating each row in the sample data as an individual case, (ii) mapping the SOURCE_ID column to the event log case identifier attribute, and (iii) creating an event from each datetime column in the dataset by mapping the column name to the activity label and the row value of the column to the timestamp value. A process model was discovered and conformance checking (see Fig. 5) showed the model had 98.7% fitness (156 wrong and missing events out of 5,922 events in total). The discovered model highlighted some variations from the expected process behaviour as described by the RSQ domain expert (illustrated in Fig. 2 and also highlighted some of the data quality issues discussed in 4.0.2. The model shows parallelism for activities following AT_SCENE_PATIENT where the expected behaviour is sequential. The data quality assessment (see Table 3) and the model identified the activities and the extent of the deviation from expected behaviour. The conformance analysis revealed other event ordering issues including 10 cases where the first activity was not DATE_RETRIEVAL_REQUESTED.

5 Discussion and Lessons Learned

Data modelling prior to process mining informs the data extraction phase of the case study. The data models and relationship cardinalities show there are many possible *case* perspectives that are relevant (i.e. an incident may be considered a case, an individual patient experience may be considered a case, an individual response unit's dispatch/attendance/transport may be considered a case, etc.).

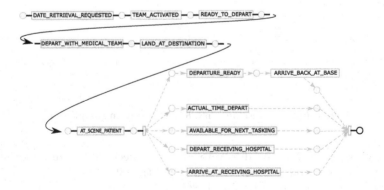

Fig. 5. RSQ conformance model derived from sample data

An important consideration in extracting the final dataset will be ensuring that, as well as the stakeholder's view that the patient experience is the case perspective, it will be possible to investigate other case perspectives.

The quality assessment of the (sample) data, conducted **prior** to the discovery and conformance analyses, adds value to the overall process mining exercise in at least **four** ways.

1. **Identifying event-data quality issues allows for the anticipation of certain observable features in subsequent process mining analysis.** For instance, the mixed granularity in timestamps led the analysts to anticipate incorrect event ordering (subsequently confirmed in the parallelism apparent in the discovered models). Further, the fact that the (RSQ) DATE_RETRIEVAL_REQUESTED values are all at day-level granularity precludes the possibility to properly assess performance aspects of various phases of aero-medical retrieval (for instance, how long does it take to activate a medical team following a retrieval request?). For the ground-based retrieval/transport data, the quality analysis showed duplication in the FIRST_ASSIGNED_CAD values. After discussion with QAS it emerged that it is QAS practice to include, for all response units dispatched to attend an incident, the same value for FIRST_ASSIGNED_CAD. This can be taken into account by making this a case attribute. Identifying this issue through quality assessment headed-off issues that may have arisen in the process mining analysis had the FIRST_ASSIGNED_CAD milestone been included as an activity for all eARFs and response units involved in the incident.

2. **Quantifying quality issues means that it is possible to separate systemic from occasional quality 'breaches'.** For instance, the fact that all (QAS) _VACIS timestamps were at a low level of precision (i.e. minute-level granularity) points to a systemic cause.

3. **Identifying quality issues allows for reasoning about the mechanisms that may have caused the event data quality issue.** For instance, it is unlikely that **all** (QAS) _VACIS events happened exactly on the minute, but, it is likely that, either the system recording the event had only minute-level precision, or that in extracting the data for analysis, seconds and milli-seconds were 'masked'. The fact that some (RSQ) cases have ARRIVE_AT_RECEIVING_HOSPITAL and DEPART_RECEIVING_HOSPITAL occurring at the same times may indicate a combination of human and system issues, i.e a human omission to record the ARRIVE time when the aircraft arrives (possibly due to patient care needs), and a system requirement that an ARRIVE time needs to be entered before a DEPART time can be entered.

4. **An understanding of 2 and 3 above facilitates informed engagement with process stakeholders and decisions about data quality remediation actions.** For instance, if the _VACIS granularity issues were as a result of incorrect data extraction, this quality issue can be resolved by simply extracting the data at the appropriate granularity.

Limitations associated with this current work include the fact that the approach has been trialled on only two, small data sets. Future work will focus on applications to larger datasets.

Acknowledgement. The work in this paper was funded from a grant from the Queensland Motor Accident Insurance Commission (MAIC).

References

1. van der Aalst, W.M.: Extracting event data from databases to unleash process mining. In: vom Brocke, J., Schmiedel, T. (eds.) BPM - Driving Innovation in a Digital World. MP, pp. 105–128. Springer, Cham (2015). https://doi.org/10.1007/978-3-319-14430-6_8
2. Andrews, R., Suriadi, S., Wynn, M., ter Hofstede, A.H.: Healthcare process analysis. In: Process Modelling and Management for HealthCare. CRC Press (2017)
3. Badakhshan, P., Alibabaei, A.: Using process mining for process analysis improvement in pre-hospital emergency. In: Middle East North Africa Conference for Information Systems, Paris, March 2018 (2018)
4. Batini, C., Cappiello, C., et al.: Methodologies for data quality assessment and improvement. ACM Comput. Surv. (CSUR) **41**(3), 16 (2009)
5. Bose, R.J.C., Mans, R.S., van der Aalst, W.M.: Wanna improve process mining results? In: CIDM 2013, pp. 127–134 (2013)
6. van Eck, M.L., Lu, X., Leemans, S.J.J., van der Aalst, W.M.P.: PM^2: a process mining project methodology. In: Zdravkovic, J., Kirikova, M., Johannesson, P. (eds.) CAiSE 2015. LNCS, vol. 9097, pp. 297–313. Springer, Cham (2015). https://doi.org/10.1007/978-3-319-19069-3_19
7. Halpin, T., Morgan, T.: Information Modeling and Relational Databases. Morgan Kaufmann, San Francisco (2010)
8. Lamine, E., Fontanili, F., Di Mascolo, M., Pingaud, H.: Improving the management of an emergency call service by combining process mining and discrete event simulation approaches. In: Camarinha-Matos, L.M., Bénaben, F., Picard, W. (eds.) PRO-VE 2015. IAICT, vol. 463, pp. 535–546. Springer, Cham (2015). https://doi.org/10.1007/978-3-319-24141-8_50
9. Mans, R.S., van der Aalst, W.M.P., Vanwersch, R.J.B., Moleman, A.J.: Process mining in healthcare: data challenges when answering frequently posed questions. In: Lenz, R., Miksch, S., Peleg, M., Reichert, M., Riaño, D., ten Teije, A. (eds.) KR4HC/ProHealth 2012. LNCS (LNAI), vol. 7738, pp. 140–153. Springer, Heidelberg (2013). https://doi.org/10.1007/978-3-642-36438-9_10
10. Rebuge, Á., Ferreira, D.R.: Business process analysis in healthcare environments: a methodology based on process mining. Inf. Syst. **37**(2), 99–116 (2012)
11. Rojas, E., Munoz-Gama, J., Sepúlveda, M., Capurro, D.: Process mining in healthcare: a literature review. J. Biomed. Inform. **61**, 224–236 (2016)
12. Suriadi, S., Andrews, R., ter Hofstede, A., Wynn, M.: Event log imperfection patterns for process mining: towards a systematic approach to cleaning event logs. Inf. Syst. **64**, 132–150 (2017)

13. Wand, Y., Wang, R.: Anchoring data quality dimensions in ontological foundations. Commun. ACM **39**(11), 86–95 (1996)
14. Wang, R.Y., Strong, D.M.: Beyond accuracy: what data quality means to data consumers. J. Manag. Inf. Syst. **12**(4), 5–33 (1996)
15. Wirth, R., Hipp, J.: CRISP-DM: towards a standard process model for data mining. In: PAKDDM, pp. 29–39 (2000)

Analyzing Medical Emergency Processes with Process Mining: The Stroke Case

Carlos Fernandez-Llatas[1](✉), Gema Ibanez-Sanchez[1], Angeles Celda[2], Jesus Mandingorra[2], Lucia Aparici-Tortajada[1], Antonio Martinez-Millana[1], Jorge Munoz-Gama[3], Marcos Sepúlveda[3], Eric Rojas[3], Víctor Gálvez[3], Daniel Capurro[4], and Vicente Traver[1]

[1] ITACA, Universitat Politècnica de València, Valencia, Spain
cfllatas@itaca.upv.es
[2] Hospital General de Valencia, Valencia, Spain
[3] School of Engineering, Pontificia Universidad Católica de Chile, Santiago, Chile
[4] School of Medicine, Pontificia Universidad Católica de Chile, Santiago, Chile

Abstract. Medical emergencies are one of the most critical processes that occurs in a hospital. The creation of adequate and timely triage protocols, can make the difference between the life and death of the patient. One of the most critical emergency care protocols is the stroke case. This disease demands an accurate and quick diagnosis for ensuring an immediate treatment in order to limit or even, avoid, the undesired cognitive decline. The aim of this paper is perform an analysis of how Process Mining techniques can support health professionals in the interactive analysis of emergency processes considering critical timing of Stroke, using a Question Driven methodology. To demonstrate the possibilities of Process Mining in the characterization of the emergency process, we have used a real log with 9046 emergency episodes from 2145 stroke patients that occurred from January of 2010 to June of 2017. Our results demonstrate how Process Mining technology can highlight the differences of the stroke patient flow in emergency, supporting professionals in the better understanding and improvement of quality of care.

Keywords: Process mining · Stroke · Emergency · Healthcare

1 Introduction

Since the arrival of Data Science, the application of its advantages to health is a growing desire. The analysis of large amount of data available in hospitals to support the optimization of clinical processes is one of current challenges [19]. These technologies are very useful to support health professionals in the creation of better care processes that allows the improvement of the Quality of Care to patients and the effectiveness of the treatments, and also it will allow a better management of the cost of patients, making the health system sustainable [4]. Improving the management of the clinical processes will not only save lives, but can also bring a better and personalized care to more patients.

© Springer Nature Switzerland AG 2019
F. Daniel et al. (Eds.): BPM 2018 Workshops, LNBIP 342, pp. 214–225, 2019.
https://doi.org/10.1007/978-3-030-11641-5_17

One of the cases where the accurate coordination of clinicians is crucial, is the case of stroke. Stroke is one of the illnesses that has a higher morbidity and mortality impact. It is specially significant because of the high sociosanitary and associated disability cost that it can cause [9,15]. That supposition not only has a strong impact in the quality life of patients, but can also increase costs for the health system [14]. The adequate diagnosis, and the timely and coordinated action of health professionals is decisive to save the live of the patient, and also stops the cognitive decline of the patients [2]. The prognosis of stroke depends largely of the possibility of reducing, to the maximum, the brain injury. One of the success keys is the creation and optimization of primary care and emergency protocols for quick diagnosis and treatment, to shorten the time between the stroke event and the application of the adequate treatment. In this way, the creation of Data Science tools for the continuous analysis of the primary care and emergency protocols will be a clear advantage for the improvement of clinical processes for critical diseases like stroke.

Process Mining Technology can be a good option for supporting health professionals in the understanding of the clinical process of emergencies. Process Mining [1] is a relatively new technology that have been used successfully in different fields. Process Mining use machine learning technologies for infer and analyze flows in a human understandable way. This can be used by health professionals for a better understanding of the clinical process, enabling the application of interactive models [12] that have natural application in the medical domain [10].

The aim of this paper is to evaluate the capabilities of Process Mining to analyze the hospital flow of emergencies via the analysis of the stroke processes. In order to do that we have applied a Question Driven methodology [24] based in two main questions:

- Q1: Can Process Mining detect and measure the special characteristics of the stroke emergency processes?
- Q2: Is Process Mining able for measuring organizational changes that affects the emergency process?

The objective of this paper is to show how Process Mining can offer solutions to these questions in the medical domain offering information about the statistical significance of the processes. In medical domain, it is not enough to provide information about the processes differences to demonstrate findings. To discover medical knowledge it is mandatory to evaluate the statistical significance, in other case the findings are not conclusive [6]. In that way, we have analyzed a real log of 9046 Emergency episodes of 2145 patients that suffer at least one stroke event between January of 2010 and June of 2017. This log was used to evaluate the questions using process mining technologies and measure their statistical significance.

The paper is organized as follows. First, a related work section to analyze the field and following the emergency flow is presented in more detail. After that, the results proposed and the selected experiments performed are explained. Finally, a discussion part concludes the paper.

2   Related Work

Despite the great advantages of the application of Big Data to healthcare, health professionals have some suspicions about how the current expert clinical knowledge should be integrated into the automatically learned clinical models [16]. So, it is needed to use new models to incorporate, in a better way, this working knowledge into data science models. In this line, Interactive Pattern Recognition (IPR) [12] is a formal framework that introduces the medical expert in the middle of the learning process allowing them to correct the hypothesis model in each iteration avoiding undesirable errors, and to converge to a solution in an iterative way. However, this framework requires human understandable machine learning frameworks to take advantage of this possibilities. The application of Process Mining within this framework can be a good solution to solve this gap.

Process Mining [1] is a research discipline area that uses existing information in clinical databases and Hospital Information Systems (HIS) to create human understandable views that support healthcare stakeholders in the better understanding of the clinical process. In last years Process Mining has been applied in the medical domain [18,23]. There are some applications where the applications of Process Mining has successfully demonstrated how the medical experts can discover the clinical protocols in different disciplines like dental treatments [17]; surgery flow [11]; or chemotherapy [3]. Also, Process Mining Interactive methodologies has been proposed for supporting the application of these technologies in the medical domain. This is the case of the Question Driven methodology [24]. This methodology propose the formulation of research questions, based on daily problems of physicians, and use Process Mining technologies to solve it.

In the case of medical emergencies there are some recent studies that apply Process Mining. In [8,21] the authors apply Process Mining control-flow discovery and clustering techniques for inferring the most common emergency unit flows. In [20], different hospitals has been compared attending to triage protocols measuring the patients flows using discovery techniques. In [22], the emergency flow is analyzed based on a Question Driven methodology, which is an interactive [12] methodology that are intended to support health professionals via solving their specific questions in an iterative way.

In addition to discovering the flows, in order to use these techniques in the case of stroke where the time is crucial, it is also necessary to analyze the time spent in each one of the emergency stages, to properly characterize and compare the processes.

In addition, the concept of statistical significance has a critical importance to create new medical knowledge. To evaluate and measure medical processes there is a need to show the differences between them, besides is mandatory to show the statistical significance of the findings. Although there are some suspicions with the interpretation and use of statistical significance indexes like P-Value [5,13], it is clear that most of clinical literature is focused on measuring the statistical significance using P-Value [6]. For that, in order to provide trustable information to healthcare professionals, it is desirable to provide a measure of statistical significance within the flow, to allow acceptance of the results achieved by Process Mining algorithms in the medical domain.

3 The Emergency Room Treatment Process

Figure 1 illustrates the Emergency Process in a hospital. The process starts when a patient arrives to the hospital and being administrative *admitted*. Then, the patient waits until the clinical staff performs the *triage*. The triage step is guided by a software that provides questions to be asked to the patient in order to determine a level of priority (classified into one of the 5 existing levels, from less (5) to more (1) priority according to the classical Manchester codification [25]). Next, the patient *waits* until the system assigns a physician to the case based on the physician discipline, the patient priority, and the availability of the resources). Then, the patient receives medical *attention* and the case is discharged. Given that this work focuses on the stroke emergency process, we distinguish three possible discharges: *Ordinary Discharge* (the patient is sent home), *Stroke* (the patient goes directly to the special unit that treats this cases), and *Hospital Admission* (to treat other complex cases out of the scope of this work).

In this moment, everything is ready to receive medical attention, depending on the seriousness priority. When a physician is free, he selects a patient in the computer system depending on his specialty and the patient's priority. This starts the Medical *Attention* process that finishes with the discharge of the patient. Depending on the final assessment, the patient is identified as a: *Stroke* case, *Ordinary Discharge*, or a *Hospital Admission*.

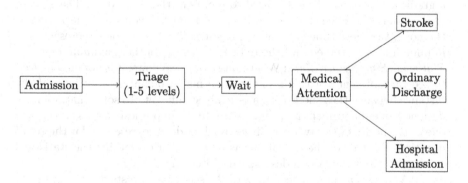

Fig. 1. General flow of medical emergencies.

In this work, we have used real data from 2145 patients that have suffered a stroke episode between the period since January 2010 to June 2017. An emergency episode is related to the process followed by a patient in the emergency area. The log information is acquired from the Hospital Information System (HIS) with a time stamp granularity in seconds. In this log, we have a set of 9046 emergency episodes that can be divided in three kind of episodes depending on the discharge destiny (available in the HIS), 5536 (54%) are Ordinary episodes, 2265 (35%) corresponds to Stroke episodes, and 1222 (11%) are episodes with a hospital admission non directly related to stroke.

4   Statistical Significance and Time Maps

For the Process Mining analysis we have used a Desktop version of PALIA Suite called PMCode [11]. The main characteristic of PALIA Suite/PMCode is that is focused on the creation of custom Process Mining dashboards for their use in the medical domain. This framework has been successfully tested in the application of Question Driven methodology over emergency data [22] and other medical domains like surgery [11] or diabetes [7]. As the discovery algorithm, we have selected PALIA Light algorithm, implemented on PMCode, a version of the PALIA algorithm [11] for Non-Parallel Logs, because is implemented on PMCode and more efficient for the problem than the complete version of PALIA.

In PMCode it is possible to create render maps for customizing the colors and features of the nodes of discovered model. This allows to apply specific enhancement maps over the process previously discovered in order to highlight the nodes depending on a customized formulation. In this way, for the experiments performed in this paper we have used two main custom maps: Time and Statistical Significance maps.

– *Time Maps*: These maps provide a gradient color view representing the time spent in each one of the nodes. The time spent is represented by the average of the duration time spent in each one of the stages of the Emergency flow. Figure 3 is an example of how the stages duration is presented. In this view, a gradient from green to red is used to represent the time spent. The greener is the color, the quicker is the activity and, on the contrary, the redder is the node, the more time is spent in the stage. The green color represents the minimum time observed and the red color represents the maximum one.
– *Statistical Significance Maps*: When comparing two flows, it is not only needed to see the difference of colors in the time maps, but also to evaluate if the distance between these nodes is statistically significant. To solve that, we have designed an enhancement map that compute the hypothesis test between each one of the nodes of the two workflows. Each node is represented for the set of durations for each of the executions associated. For evaluating the statistical significance between two nodes, we calculated the P-Value.
  The P-Value is calculated according to the sequence of tests defined in Fig. 2. A Kolmogorov-Smirnov Test has been used for evaluating the normality of the distribution of the time values of the nodes of the two flows. If the set of values pass the normality test, we applied a classical T-Student Test for the computation of P-Value. On the contrary, if the result is not normal we apply a Mann-Whitney-Wilcoxon Test. The P-Value threshold for statistical significance is fixed at 0.05.

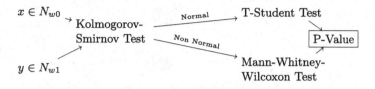

Fig. 2. Calculation of P-Value in the statistical significance map.

Figure 4 shows an example of the applications of time and statistical significance Maps. While the colors represents the time spent in each one of the stages, the nodes which statistical significance are below the threshold are highlighted in with a yellow line. Using this view, the health professionals not only can evaluate the changes in the flow, but also distinguish which of those changes are statistically significant. This measure will support health professionals in the detection of differences with a strong evidences. A comparison between two nodes can have a high median/average differences, but a high P-Value. This means that there are not real evidences that the behavior of patients at this point are actually different. On the contrary, if the P-Value is lower than the threshold (typically 0.05) it is assumed that there are strong reasons to think that the behavior of the patients in that point of the flow is different. This information is crucial to discover and demonstrate medical evidence.

5   Experiments

In this section, we evaluated the proposed questions using Process Mining Techniques.

5.1   Q1: Can Process Mining Detect and Measure the Special Characteristics of the Stroke Emergency Processes?

The aim of this question is to evaluate how Process Mining can show the differences in the stroke emergency process. Although topologically, the stages followed by stroke episodes are the same than ordinary or other hospital admissions, it is expected that it should show differences in the time spent in some of the stages due to the special characteristics of the problem. To show the differences among the time spent, depending on the level of emergency, we have labeled the nodes of wait and attention time with the level of triage selected. Also, we have labeled the events with the most common discharge destinies (Exitus (Death), Home, Primary Care,...).

Figure 3 shows the ordinary episodes flow after applying discovery and time maps. In this map, it is possible to see the time spent in each one of the stages in a qualitative way. As expected, the time of waiting is inversely proportional to the emergency priority, while the time of attention is directly proportional. That means most complex emergencies have lower waiting time but take more time to be treated.

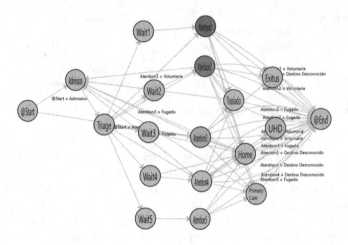

Fig. 3. Flow of the ordinary discharge episodes for Q1. The colours in nodes represents the average time spent in each one of the activities. (Color figure online)

Figure 4 shows the comparison between the Brain Stroke Unit, the Ordinary flow and Table 1 show the numerical stats. The differences are more significant than in the hospital admission care. The Admission and the Triage times are significantly lower, as well as the Attention times for high priority emergencies. Also, Low priority emergencies have a higher time to attention in stroke case, increasing significantly level 4 episodes time stay in 212 min (8,67 times worst than ordinary episodes).

Table 1. Analysis of statistical significance between ordinary and stroke unit admission nodes (Interquartile range in minutes). Bold rows are the one that have statistical significance.

Ordinary emergency			Stroke emergency		
Activity	N	IQ range	N	IQ range	P-Value
Admission	**5630**	**9,27 [4,62, 18,53]**	**1475**	**7,12 [4,13, 13,80]**	**0,00**
Triage	**5630**	**1,00 [0,00, 2,00]**	**1475**	**1,00 [0,00, 2,00]**	**0,05**
Wait1	41	7,97 [2,97, 15,47]	126	4,97 [2,72, 8,97]	0,13
Attention1	**53**	**389,50 [208,82, 667,47]**	**180**	**110,82 [74,60, 213,07]**	**0,00**
Wait3	2960	53,47 [21,97, 110,97]	555	36,97 [12,97, 83,97]	0,07
Attention3	3016	220,56 [128,48, 355,56]	576	247,07 [152,45, 373,81]	0,72
Wait2	829	7,97 [4,97, 15,97]	613	7,97 [3,97, 16,97]	0,83
Wait4	1571	51,97 [22,97, 103,97]	43	61,97 [23,97, 121,97]	0,62
Attention4	**1590**	**27,68 [10,54, 99,55]**	**43**	**240,15 [116,78, 384,62]**	**0,00**
Attention2	**866**	**305,53 [195,90, 568,57]**	**673**	**210,62 [133,84, 304,88]**	**0,00**
Wait5	105	73,97 [34,97, 116,47]	3	3,97 [0,97, 185,97]	0,77
Attention5	105	25,17 [9,22, 68,44]	3	86,30 [54,05, 563,27]	0,45

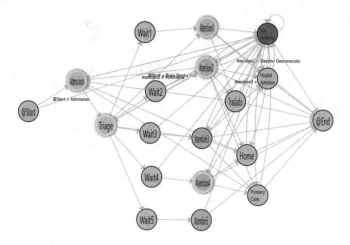

Fig. 4. Flow of Stroke episodes with statistical significance map. The colours in nodes represents the average time spent in each one of the activities. (Color figure online)

5.2 Q2: Process Mining Is Able to Measure Organizational Changes in the Stroke Emergency Process?

In March of 2017 an organizational change in the Emergency protocol of the Hospital was deployed, enabling a second place for triage. The aim of modification was to improve the time of admission of, at least, most complex emergencies.

Table 2. Analysis of statistical significance between single and double triage Nodes (Interquartile range in minutes). Bold rows are the one that have statistical significance.

Single triage			Double triage		
Activity	N	IQ range	N	IQ range	P-Value
Admission	**284**	**11,01 [4,71, 24,27]**	**425**	**7,75 [3,53, 17,96]**	**0,00**
Triage	284	2,00 [2,00, 4,00]	425	2,00 [1,00, 4,00]	0,29
Wait5	3	26,97 [24,97, 151,97]	7	73,97 [20,97, 157,97]	0,66
Attention5	3	25,13 [9,62, 141,52]	7	56,68 [33,78, 254,33]	0,27
Wait2	85	6,97 [3,97, 12,47]	108	6,97 [4,97, 13,97]	0,44
Attention2	88	242,48 [170,38, 399,66]	119	275,58 [195,95, 497,25]	0,48
Wait3	**142**	**56,47 [21,97, 128,22]**	**210**	**40,47 [15,97, 79,47]**	**0,00**
Attention3	142	222,09 [123,20, 410,48]	216	209,33 [124,15, 344,74]	0,35
Wait4	43	51,97 [22,97, 110,97]	74	59,97 [25,72, 107,22]	0,92
Attention4	43	63,42 [22,18, 255,33]	76	36,48 [13,51, 190,50]	0,19
Stroke	63	8640 [5760, 14400]	90	8640 [5760, 13305]	0,56
Wait1	7	7,97 [2,97, 10,97]	4	6,97 [4,22, 10,47]	0,69
Attention1	8	112,41 [47,95, 314,73]	7	149,57 [63,35, 563,50]	0,33

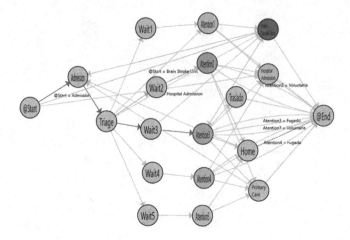

Fig. 5. Flow for single triage (January to March 2017). The colours in nodes represents the average time spent in each one of the activities, and colours in edges represents the quantity of patients that follows this path .

This question is oriented to evaluate if Process Mining is able to detect this organizational change and quantify how affects to the stroke emergency process. In this way, we have performed a study from January to June of 2017 over stroke episodes, splitting the log in single triage (before March) and double triage (after March). In this log we have 284 (40%) episodes of single triage and 425 (60%) of double triage.

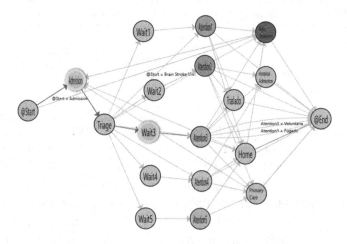

Fig. 6. Flow for double triage (March to June 2017). The colours in nodes represents the average time spent in each one of the activities, and colours in edges represents the quantity of patients that follows this path

Figure 5 shows the flow inferred with double triage. The arrows colors quantify the number of patients over the flow. This highlights the most common paths. Figure 6 shows the comparison between the single and double triage flow and Table 2 show the numerical stats. According to that, there is a significant decrease of admission times in 3,26 min (30% of improvement). Also, there is a significant decreasing of time in Wait3 node in 16 min (28% of improvement), that is the most populated waiting stage.

6   Discussion and Conclusions

In this paper, we have analyzed how Process Mining can support health Professionals in the analysis of emergency processes taking as example the stroke problem. We have and compared different process and we have provide a tool to state the statistical significance of these differences using P-Value method. The measure of the statistical significance is crucial for achieving medical trust ability. In clinical world, if the results are not supported by an evaluation of the statistical evidence, we can't trust on them as an actual medical evidence.

On one hand, we have evaluated if Process Mining technologies can discover the characteristics of processes followed for specific diseases. In this way, we have stated the differences between the ordinary emergency process with the stroke emergency process. We have observed there is a clear difference in the time of triage, admission and attention with the stroke emergency process. The stroke process requires a specific treatment that should be covered by the stroke unit of the hospital and the emergency physicians should stabilize and derive the patient to the unit as soon as possible. In this process, the triage is crucial, the selection of a correct emergency level decreases significantly the time of stay in emergencies. In our study we have detected a set of under-triaged stroke patients that were incorrectly classified as level 4 according to the emergency classification. This is probably caused by a confusion with a typical level 4 disease. This can dramatically increase the time of stay in a 867% and this should be considered. This increase of time can be decisive for the survival or cognitive decline of the patient.

On the other hand, we have analyzed how Process Mining can measure the impact of organizational changes in the triage process. Specifically, we have compared the differences between a triage with two nurses and the triage with just one nurse. As expected, we have demonstrated that there is a significant change in the time of admission of brain stroke patients. Also, we have discovered that this change affects positively the waiting time of level 3 patients, that, in fact, are the most common. In addition, there is no evidence that this change affects the quality of the triage, because there is not significant changes observed in the time of attention. Our findings demonstrate that the use of Process Mining, not only allows health professionals to understand the clinical processes, but also can support the optimization of the process in an interactive way by measuring the impact of the organizational changes in critical diseases like stroke. The statistical significance maps provides a layer of trustability to health professionals enabling them in paying attention to specially significant differences.

In that way, Process Mining can be an exceptional tool for analyzing widely the processes, to detect special circumstances that experts should pay attention, and, after that, can support them in the impact analysis in the posterior correctional actions in an iterative and interactive way.

References

1. van der Aalst, W.M.P.: Process Mining: Data Science in Action, 2nd edn. Springer, Heidelberg (2016). https://doi.org/10.1007/978-3-662-49851-4
2. Alberts, M.J., et al.: Recommendations for the establishment of primary stroke centers. JAMA **283**(23), 3102–3109 (2000)
3. Baker, K., et al.: Process mining routinely collected electronic health records to define real-life clinical pathways during chemotherapy. Int. J. Med. Inform. **103**, 32–41 (2017)
4. Bates, D.W., Saria, S., Ohno-Machado, L., Shah, A., Escobar, G.: Big data in health care: using analytics to identify and manage high-risk and high-cost patients. Health Aff. **33**(7), 1123–1131 (2014)
5. Calster, B.V., Steyerberg, E.W., Collins, G.S., Smits, T.: Consequences of relying on statistical significance: some illustrations. Eur. J. Clin. Invest. **48**(5), e12912 (2018)
6. Chavalarias, D., Wallach, J.D., Li, A.H.T., Ioannidis, J.P.A.: Evolution of reporting P values in the biomedical literature, 1990–2015. JAMA **315**(11), 1141–1148 (2016)
7. Conca, T., et al.: Multidisciplinary collaboration in the treatment of patients with type 2 diabetes in primary care: Analysis using process mining. J. Med. Internet Res. **20**(4), e127 (2018)
8. Delias, P., Manolitzas, P., Grigoroudis, E., Matsatsinis, N.: Applying process mining to the emergency department. In: Encyclopedia of Business Analytics and Optimization, pp. 168–178. IGI Global (2014)
9. Feigin, V.L., et al.: Global burden of diseases, injuries and risk factors study 2013 and stroke experts writing group: global burden of stroke and risk factors in 188 countries, during 1990–2013: a systematic analysis for the global burden of disease study 2013. Lancet. Neurol. **15**(9), 913–924 (2016)
10. Fernandez-Llatas, C., Bayo, J.L., Martinez-Romero, A., Benedi, J.M., Traver, V.: Interactive pattern recognition in cardiovascular diseases management. a process mining approach. In: Proceedings of the IEEE International Conference on Biomedical and Health Informatics 2016, Las Vegas, EEUU (2016)
11. Fernandez-Llatas, C., Lizondo, A., Monton, E., Benedi, J.M., Traver, V.: Process mining methodology for health process tracking using real-time indoor location systems. Sensors (Basel, Switzerland) **15**(12), 29821–29840 (2015)
12. Fernandez-Llatas, C., Meneu, T., Traver, V., Benedi, J.M.: Applying evidence-based medicine in telehealth: an interactive pattern recognition approximation. Int. J. Environ. Res. Publ. Health **10**(11), 5671–5682 (2013)
13. Goodman, S.N.: Toward evidence-based medical statistics. 2: the Bayes factor. Ann. Intern. Med. **130**(12), 1005–1013 (1999)
14. Gustavsson, A., et al.: Cost of disorders of the brain in Europe 2010. Eur. Neuropsychopharmacol. J. Eur. Coll. Neuropsychopharmacol. **21**(10), 718–779 (2011)
15. Howard, G., Goff, D.C.: Population shifts and the future of stroke: forecasts of the future burden of stroke. Ann. N. Y. Acad. Sci. **1268**, 14–20 (2012)

16. Lazer, D., Kennedy, R., King, G., Vespignani, A.: The parable of Google flu: traps in big data analysis. Science **343**(6176), 1203–1205 (2014)
17. Mans, R., Reijers, H., van Genuchten, M., Wismeijer, D.: Mining processes in dentistry. In: Proceedings of the 2nd ACM SIGHIT International Health Informatics Symposium, IHI 2012, pp. 379–388. ACM, New York (2012)
18. Mans, R.S., van der Aalst, W.M.P., Vanwersch, R.J.B.: Process Mining in Healthcare: Evaluating and Exploiting Operational Healthcare Processes. SBPM. Springer, Cham (2015). https://doi.org/10.1007/978-3-319-16071-9
19. Murdoch, T.B., Detsky, A.S.: The inevitable application of big data to health care. JAMA **309**(13), 1351–1352 (2013)
20. Partington, A., Wynn, M., Suriadi, S., Ouyang, C., Karnon, J.: Process mining for clinical processes: a comparative analysis of four Australian hospitals. ACM Trans. Manage. Inf. Syst. **5**(4), 19:1–19:18 (2015)
21. Rebuge, l., Ferreira, D.R.: Business process analysis in healthcare environments: a methodology based on process mining. Inf. Syst. **37**(2), 99–116 (2012)
22. Rojas, E., Fernández-Llatas, C., Traver, V., Munoz-Gama, J., Sepúlveda, M., Herskovic, V., Capurro, D.: PALIA-ER: bringing question-driven process mining closer to the emergency room. In: 15th International Conference on Business Process Management (BPM 2017) (2017)
23. Rojas, E., Munoz-Gama, J., Sepúlveda, M., Capurro, D.: Process mining in healthcare: a literature review. J. Biomed. Inform. **61**, 224–236 (2016)
24. Rojas, E., Sepúlveda, M., Munoz-Gama, J., Capurro, D., Traver, V., Fernandez-Llatas, C.: Question-driven methodology for analyzing emergency room processes using process mining. Appl. Sci. **7**(3), 302 (2017)
25. Storm-Versloot, M.N., Ubbink, D.T., Kappelhof, J., Luitse, J.S.K.: Comparison of an informally structured triage system, the emergency severity index, and the manchester triage system to distinguish patient priority in the emergency department. Acad. Emerg. Med. **18**(8), 822–829 (2011)

Using Indoor Location System Data to Enhance the Quality of Healthcare Event Logs: Opportunities and Challenges

Niels Martin[✉] [ID]

Hasselt University, Agoralaan Building D, 3590 Diepenbeek, Belgium
niels.martin@uhasselt.be

Abstract. Hospitals are becoming more and more aware of the need to manage their business processes. In this respect, process mining is increasingly used to gain insight in healthcare processes, requiring the analysis of event logs originating from the hospital information system. Process mining research mainly focuses on the development of new techniques or the application of existing methods, but the quality of all analyses ultimately depends on the quality of the event log. However, limited research has been done on the improvement of data quality in the process mining field, which is the topic of this paper. In particular, this paper discusses, from a conceptual angle, the opportunities that indoor location system data provides to tackle event log data quality issues. Moreover, the paper reflects upon the associated challenges. In this way, it provides the conceptualization for a new area of research, focusing on the systematic integration of an event log with indoor location system data.

Keywords: Event log · Data quality · Hospital information systems · Indoor location systems data · Process mining

1 Introduction

The healthcare sector in general, and hospitals in particular, are confronted with challenges such as tightening budgets contrasted to increased care needs due to the aging population [4,15,19]. To face these challenges, hospitals are becoming increasingly aware of the need to manage their processes in order to improve them [15]. In this respect, process mining is gaining more attention as a way to gain insights in healthcare processes. Process mining is the extraction of knowledge from an event log containing process execution information from a process-aware information system such as a hospital information system (HIS).

Process mining research mainly focuses on the development of new techniques to extract knowledge from an event log or the innovative application of existing techniques [5]. However, consistent with the "garbage in - garbage out" principle, the quality of all process mining analyses ultimately depends on the quality of

© Springer Nature Switzerland AG 2019
F. Daniel et al. (Eds.): BPM 2018 Workshops, LNBIP 342, pp. 226–238, 2019.
https://doi.org/10.1007/978-3-030-11641-5_18

the event log used as an input [19]. Bose et al. [5] state that most real-life event logs struggle with issues such as incompleteness, noisiness, and imprecision. This also holds in healthcare, where it is not always possible to extract high-quality data from a HIS [8]. Despite such observations, which are broadly shared given their inclusion in the Process Mining Manifesto [2], limited research has been done on the improvement of data quality within the process mining field. One research direction, which did not yet receive explicit attention, is considering the enrichment and improvement of event logs using other process-related data sources such as location data.

This paper discusses, from a conceptual angle, how indoor location system (ILS) data can be used to alleviate data quality issues present in an event log originating from a HIS. In healthcare, ILS systems are increasingly used for e.g. patient flow and staff workflow management [4]. Data generated by such systems provides information on the location of process participants such as patients, staff members, and medical equipment at a particular moment. This paper outlines which opportunities ILS data provides to tackle event log quality issues, but also reflects upon the associated challenges. In this way, it provides the conceptualization for a new research area, focusing on a systematic integration of an event log with ILS data. The resulting enhanced event log will contribute towards exploiting the full potential of process mining in healthcare practice.

This paper is structured as follows. Section 2 introduces the notions of an event log and ILS data. In Sect. 3, an overview of related work is provided. Section 4 details the potential of ILS data to alleviate event log quality issues. Despite these opportunities, several challenges are still ahead, as discussed in Sect. 5. The paper ends with a conclusion in Sect. 6.

2   Preliminaries

This section outlines the notions of an event log (Sect. 2.1) and ILS data (Sect. 2.2), which are the key data sources considered in this paper.

2.1   Event Log

An event log is a data file containing process execution information. It consists of a collection of events, e.g. the completion of patient registration at the reception desk, associated to a case such as a patient. It minimally consists of an ordered set of events for each case, but typically also includes information such as a timestamp, and the resource associated to the event [1].

Table 1 illustrates the structure of an event log, where each line represents an event. For instance: the first line indicates that the registration of patient 103 by resource Mike started on April, 25th at 10:59:41. He completed this registration at 11:04:04, as shown in the second row of Table 1.

Table 1. Illustration of event log structure

Case id	Timestamp	Activity	Transaction type	Resource	...
...	...	...	...	...	...
103	25/04/2018 10:59:41	Register patient	start	Mike	...
103	25/04/2018 11:04:04	Register patient	complete	Mike	...
104	25/04/2018 11:04:04	Register patient	start	Mike	...
109	25/04/2018 11:06:22	Clinical examination	start	Judy	...
...	...	...	...	...	...

2.2 Indoor Location System Data

ILS data originates from an indoor location system (ILS), also referred to as a real-time location system. From a technical perspective, wireless Radio Frequency Identification (RFID) technology is used. Locations are determined using RFID tags, which can e.g. be integrated in patient identification bracelets or staff cards, and antennas which are deployed in a particular department [4]. For more technical details on an ILS, the reader is referred to [4].

Raw ILS data records the location of a process participant at a particular point in time. Technology provider's software often summarizes this raw data such that ILS data expresses time periods during which a process participant resided at a particular location [8,23]. When this is not the case, preprocessing is required to obtain a dataset in the format exemplified in Table 2. For instance: the first line shows that the triage nurse with RFID-tag 1044 was present in the second triage room on April 25th from 12:58:06 until 13:22:14.

Table 2. Illustration of ILS data structure

Tag id	Type	Location	Start	End
...	...	...	...	...
1044	triage nurse	Triage room 2	25/04/2018 12:58:06	25/04/2018 13:22:14
7862	patient	Triage room 2	25/04/2018 13:12:14	25/04/2018 13:16:44
7809	patient	Box 12	25/04/2018 13:12:24	25/04/2018 13:52:03
1003	nurse	Box 12	25/04/2018 13:18:22	25/04/2018 13:26:02
...	...	...	...	...

3   Related Work

Given the potential of process mining to gain profound insights in processes, it is increasingly studied in a healthcare context. Process mining methods are, amongst others, used to retrieve the activity order in healthcare processes [6,9], to mine social networks [6,18], and to check the conformance between an event log and a process model [15,22]. A recent literature review on process mining in healthcare can be found in [21].

As for other data-oriented research domains, data quality should be one of the process mining community's prime concerns. Data quality is a multi-dimensional concept [24] which is studied in fields such as statistics, management, and computer science [3]. Literature provides several general frameworks to classify data quality issues [3,14]. Moreover, dedicated data quality research has been done for specific types of data. For instance, Gschwandtner et al. [11] focus on time-related data and distinguish between quality issues present in a single dataset and problems stemming from the combination of several datasets. Identified data quality issues include the start of a time interval being later than its end, and different timestamp structures in multiple datasets [11].

Even though insights from general data quality works are conceptually relevant for event logs, the particularities of process-related data warrant dedicated research efforts [24]. In this respect, the Process Mining Manifesto [2] defines five event log maturity levels, with increasing maturity levels implying improved process mining potential. While the maturity levels are rather generic, Bose et al. [5] identify 27 specific event log quality issues, which are grouped in four categories: (i) missing data, (ii) incorrect data, (iii) imprecise data, and (iv) irrelevant data. Examples of issues are missing events, missing case attributes, and incorrect timestamps. Mans et al. [19] use the taxonomy by Bose et al. [5] to evaluate data quality at the Maastricht University Medical Centre. Taking an interview-based approach, they assign an occurrence frequency to each quality issue, with the three most frequently occurring issues being missing events, imprecise timestamps, and imprecise resource information. In the same line of research, Mans et al. [20] discuss how data quality issues influence the potential of process mining to answer frequently asked questions by healthcare practitioners. They mainly focus on timestamp-related data quality issues: incorrect timestamps and timestamps recorded at an insufficiently granular level.

While Bose et al. [5] focus on the identification of event log quality issues, Suriadi et al. [24] both specify 11 data quality issues based on their experience and describe semi-automatic methods to rectify them. The proposed fixes often require domain knowledge to e.g. specify a minimal activity ordering. Moreover, the provided solutions are confined by the boundaries of the event log as this is the only data source considered. For example: a common issue involves data inserted into the HIS using electronic forms, implying that all events recorded when submitting the form share the same timestamp. To tackle this issue, Suriadi et al. [24] suggest merging all these events into a single event. While this approach can be defended when the event log is the sole data source, it needs to be recognized that information can be lost for analysis purposes. Hence, existing event log improvement literature can be extended by considering the use of other sources of process-related information. This paper considers ILS data, which has not been considered for this purpose yet.

In recent years, ILS data has been used for process mining purposes. It is either (i) used directly to perform process mining on sequences of locations, or (ii) converted to an event log using domain knowledge to apply process mining afterwards. ILS data has e.g. been directly used for process mining to retrieve patients' movement processes [7], to mine the workflow of medical devices [16], and to study a surgical process [8]. In these papers, a process instance consists of a sequence of locations and not a sequence of activities. Both only coincide when each activity takes place in a dedicated location. While this might be reasonable for highly specialized hospital units, this assumption will often not hold, e.g. when several activities are executed in a box at the emergency department.

In an effort to link ILS data to activities, Senderovich et al. [23] aim to convert ILS data to an event log. To this end, the interaction concept is introduced, which expresses a period of time during which e.g. a patient and a staff member are simultaneously present at a location. The detected interactions are mapped to activity labels using an integer linear program in which domain knowledge is encoded [23].

Despite the recent uptake in the use of ILS data for process mining purposes, the integration of an event log with its accompanying ILS data has not been considered in literature. Nevertheless, as will be argued in Sect. 4, ILS data can be helpful to alleviate data quality issues associated to HIS data.

4   Using ILS Data to Tackle Event Log Quality Issues

This section outlines, from a conceptual angle, the opportunities that ILS data offers to tackle event log quality issues. ILS data contains location patterns of process participants such as patients, medical staff and potentially even medical equipment. It enables to determine e.g. when a patient visited the room in which MRI-scans are made, even when this is recorded in the HIS at another moment. Besides location patterns, co-locations between process participants can also be identified by matching location pattern. A co-location, consistent with an interaction in [23], is a period of time during which multiple process participants are present at the same location. Co-location patterns convey valuable information for event log improvement as it typically reflects the execution of an activity.

To structure the remainder of this section, the 27 event log quality problems identified in [5] are used. When tackling these issues, ILS data will provide more solid support for some of them compared to others. To this end, a distinction is made between level 1 (Sect. 4.1) and level 2 support (Sect. 4.2). Quality issues for which level 1 support is provided require extensive domain knowledge to solve. However, ILS data can generate useful insights to facilitate the consultation of domain experts. Level 2 support means that ILS data provides a stronger foundation to directly enrich the event log or correct data errors. However, this does not imply that domain knowledge becomes redundant. A last group of issues are not considered in a healthcare context, as discussed in Sect. 4.3.

4.1   Level 1 Support from ILS Data

Missing Activity Names. This quality problem refers to the absence of activity names for particular events. A missing activity name can be retrieved from ILS data by looking for similar location or co-location patterns. However, many activities might have e.g. the same co-location pattern such as the co-location between a patient and a nurse. Consequently, domain experts will play an important role in mapping an ILS pattern to the appropriate activity.

Missing Timestamps. A timestamp is missing when it is absent for an event. When events are recorded automatically by a HIS, e.g. after a click action, a timestamp is automatically generated and is, hence, unlikely to be missing. This is consistent with the case study in [19], indicating that this is a quality issue with low occurrence frequency. When a timestamp is absent, ILS data can complement domain knowledge in an effort to insert a proxy for the missing timestamp, e.g. when the activity should be executed at a particular location.

Incorrect Cases. This quality issue reflects the presence of cases in the event log which are related to another process. When ILS data centers around patients of a particular process, cases included in the ILS data can be compared to the cases included in the event log. Even when this is not the case, patients visiting a different zone in the hospital could be incorrect cases. However, domain knowledge is required to define particular filtering rules.

Incorrect Events. Incorrect events are events which are recorded in the HIS, but did not occur in reality. ILS data can support the detection of such events by checking whether e.g. a co-location between a patient and a resource took place at or around the event's timestamp. In case of an incorrect event, no such pattern should be found in ILS data, implying that the event should be deleted.

Incorrect Activity Names. An incorrect activity name occurs when the name of the activity is registered incorrectly when e.g. clicking a drop-down menu value or entering it manually. ILS data can be helpful by detecting inconsistencies between the activity label of an event and the location or co-location of a patient in ILS data. Such inconsistencies can serve as an input for domain expert consultation. It should be noted that Mans et al. [19] mark this as a low frequent issue.

Imprecise Relationships. This quality issue occurs when events cannot be linked to a case because of the case definition that is used. When studying patient-related healthcare processes, a patient will often be considered as a case. This is confirmed by the case study in [19], where this issue did not occur. When multimorbidity prevails, similar events might be associated to different conditions that a patient suffers and it might not be clear which events relate to the process under study. When such a connection is absent, ILS data can be used to study the locations which are visited or the medical staff that is involved.

Imprecise Activity Names. Imprecise activity names are activity names which are defined rather coarsely, causing them to occur multiple times for a particular patient, even though they refer to different actions. ILS data can support the domain expert by conveying insights in potential differences in location or co-location patterns between several occurrences of the same activity name.

Irrelevant Cases. Irrelevant cases are present when the event log contains cases which are not relevant for a particular analysis. ILS data can support judging whether a particular patient is relevant when e.g. movement patterns or sequences of visited locations play a role. Hence, ILS data can support filtering operations in close consultation with domain experts.

Irrelevant Events. Irrelevant events are events which are not relevant for a particular analysis question. When actions occurring at a particular location are not deemed relevant, ILS data can support filtering operations. Similar to irrelevant cases, this will require close interaction with domain experts.

4.2 Level 2 Support from ILS Data

Missing Events. Missing events are events that have not been recorded for a patient. This can occur e.g. when particular events need to be recorded in the HIS manually. For instance: intermediate checkups by a physician or a nurse might not be recorded in the patient's file. This is, according to the case study in [19], one of the most frequently occurring data quality issues. ILS data can be used to detect those missing events as they will e.g. generate a co-location pattern between a patient and medical staff. Based on contextual information from domain experts, missing events can be imputed in the event log.

When only a single event, e.g. the start of a treatment, is recorded for each activity execution in the HIS, the corresponding complete event can also be seen as a missing event. In this respect, ILS data is a rich source of information to add events with other transaction types related to a particular activity execution. This does not only hold for start and complete events, but also for e.g. suspend and resume events defined by the XES lifecycle extension [12].

Missing/Incorrect Relationships. Missing and incorrect relationships are events which are not associated to a patient or associated to a wrong patient, respectively. When the activity under consideration requires a particular location or co-location pattern, ILS can be used to determine the associated patient or to rectify incorrect relationships. For instance: when co-location is required, it can be determined whether the resource associated to the event is co-located with a patient at a particular point in time. In the case study of Mans et al. [19], both missing and incorrect relationships were marked as a low frequency issue. The fact that missing relationships are infrequent can be attributed to the fact that all actions in a HIS are typically related to a specific patient.

Missing Case Attributes. A case attribute such as a patient's physical condition is missing when its value is not recorded for particular patients. ILS data is unlikely to support the specification of e.g. the patient's weight. However, it can be used to impute new location-related case attributes in the event log, which can be considered as missing case attributes from the event log perspective. Examples are the distance traveled or the number of locations visited.

Missing Event Attributes. When an event attribute value is absent, the missing event attributes issue occurs. Similar to missing case attributes, ILS data will probably not enable the specification of attributes which are completely unrelated to the patient's location. However, a location attribute can be added to the event log. Adding this information enables studying the use of particular hospital areas, determining the relationship between activities and locations, etc.

Incorrect/Imprecise Case/Event Attributes. This quality problem occurs when a wrong or inaccurate value for a case/event attribute is entered. For location-related attributes, ILS data can be leveraged to e.g. correct values which are recorded manually in the HIS or to provide a more detailed value. In [19], imprecise case/event attributes are absent and the occurrence frequency of incorrect case/event attributes is marked as low.

Incorrect Timestamps. A recorded timestamp is incorrect when it does not correspond with the actual time of activity execution. Even though it is marked as a low frequent issue in [19], it should not be ignored as making registrations in the HIS is sometimes postponed. For instance: a physician might record his/her findings after visiting several patients. When such behavior is present, recorded timestamps will not coincide with actual activity execution. ILS data can be used to correct these timestamps as activity execution will be characterized by particular location or co-location patterns. For instance: a checkup by a physician is characterized by a co-location between a patient and a physician.

Imprecise Timestamps. An imprecise timestamp is not recorded at a sufficiently detailed level but, e.g., at the date level. This is marked as a relatively frequently occurring issue in [19]. Similar to incorrect timestamps, ILS data can be leveraged to impute more detailed timestamps in the event log.

Missing Resources. This quality issue implies that resource information is not recorded for a particular event. ILS data can be used to retrieve missing resource information by detecting a co-location of the patient associated to the event and a resource at that particular point in time. However, activity execution does not, by definition, require a co-location between a patient and resource (e.g. when fulfilling an administrative task). When no co-location is required for an activity and it is known at which location it is executed, ILS data can still be helpful.

Incorrect Resources. Incorrect resources imply that the resource associated to an event is not the one actually executing the activity. Even though it is indicated as non-occurring in [19], it can be quite common in healthcare when, e.g., all registrations on a particular computer take place under the account of one staff member. In this respect, ILS data can be used to determine which staff member was co-located with the patient at the moment the activity is executed.

Imprecise Resources. Resource information is imprecise when it does not refer to a specific staff member, but e.g. to a staff category such as nurse or physician. It is one of the more frequent quality issues in the case study of Mans et al. [19]. ILS data can be used to impute more detailed resource information in the event log by detecting the execution of the activity in terms of a location or co-location pattern. When multiple staff members are involved, the HIS will probably only record the resource entering the activity in the system. In that sense, resource information can still be imprecise, even when it refers to a specific staff member. ILS data is highly relevant here as the co-location between multiple staff members indicates that several resources are responsible for activity execution.

4.3   Other Event Log Quality Issues

Missing/Incorrect/Imprecise Position. Data quality issues related to the event's position in a trace are not taken into consideration. This is due to the fact that they relate to event logs without timestamps, which is not considered relevant within the context of a HIS.

Missing Cases. Missing cases refer to patients for which no data is recorded in the HIS. As no file is recorded for these patients, they are not registered upon arrival. When ILS data is recorded by e.g. integrating an RFID tag in a patient identification wristband, it is likely that no ILS data will be recorded for these patients. However, this quality issue seems to be less relevant in healthcare, as is also supported by the case study in [19].

5   Challenges

From Sect. 4, it follows that ILS data can play an important role in improving the quality of a healthcare event log. However, to operationalize this data integration, several challenges need to be taken into account. In this section, four challenges are discussed, demonstrating the need for future research.

5.1   Presence of Data Quality Issues in ILS Data

While this paper focuses on event log quality issues, it should be recognized that ILS data can also suffer from data quality issues. Gal et al. [10] discuss ILS data quality challenges in queue mining, which is a subfield of process mining.

In particular, they highlight the absence of a case identifier for some instances, the difficulty to determine the start and end point of activity execution, and issues related to reaching an appropriate level of data granularity for the analysis.

It should be noted that Gal et al. [10] and other research using ILS data for process mining use ILS data in isolation. This paper advocates the use of both ILS and HIS data. Consequently, HIS data can also be used to contextualize patterns observed in ILS data, which can be helpful to e.g. determine a missing case identifier. Nevertheless, data quality assessment of ILS data is still required prior to its use to alleviate event log quality issues. Data quality should also be a prime concern when the ILS is installed e.g. by performing data accuracy and data completeness tests [25]. Moreover, technology provider's middleware often automatically filters out some inaccuracies present in the data [13].

5.2 Simultaneous Presence of Event Log Quality Issues

Section 4 outlined how ILS data can be helpful to alleviate a series of event log quality issues. In doing so, the perspective of one specific data quality problem is taken. However, in reality, multiple issues can be present simultaneously. Consider, for instance, that a nurse checks up on multiple patients and afterwards records it under the account of a colleague in the HIS. This constitutes a combination of the quality issues incorrect timestamps and incorrect resources.

To know which event log issues are present, systematic data quality assessment needs to be performed. Event log quality assessment is currently often carried out on an ad-hoc basis. Hence, developing and implementing a systematic and generic way to perform data quality assessment on an event log is an important research challenge. The data quality assessment tool should follow a 'signaling' approach in which potential issues are highlighted. Whether these latter issues actually constitute data quality problems requires domain knowledge as this might be context-dependent.

5.3 Need for a Systematic Way to Capture Domain Knowledge

From the prior challenge and Sect. 4, the importance of domain knowledge becomes apparent. In order to use ILS data to improve healthcare event log quality, a relationship must be established between events from HIS data and location/co-location patterns in ILS data. Domain experts play a critical role in defining this relationship given the wide diversity of healthcare processes and HIS implementations. Consequently, there is a need for a systematic way to capture domain knowledge, marking an important research challenge.

To operationalize this, the basic idea of activity patters, introduced in [17] to map low-level events to high-level activities, can be leveraged. An activity pattern is a labeled process model containing the events registered during activity execution, and e.g. conditions related to resource use and timing restrictions.

Activity patterns can also be used to specify the relationship between HIS data and location/co-location patterns in ILS data. This implies that a set of intuitive activity pattern building blocks (with executable semantics) needs to be developed, relating to both HIS and ILS data. In contrast to [17], this enables hospital data specialists to create activity patterns themselves. A simplified example of an activity pattern for activity 'Follow-up patient treatment' is provided in Fig. 1. It shows that the execution of the activity involves an ordering of an event (in the event log) and one or more co-locations between a patient and a nurse (in ILS data). Moreover, the constraint indicates that the follow-up of patient treatments can only take place in rooms C1 to C4. Future work will define more complex building blocks to e.g. express choice or optional event registration.

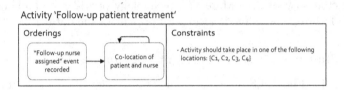

Fig. 1. Illustration of an activity pattern.

5.4   Need to Perform the Integration in a Semi-automated Way

The integration of HIS data and ILS data should be conducted in a semi-automated way. Using domain knowledge, captured in the form of activity patterns, an enhanced event log should be automatically created in which event log quality issues are tackled and ILS patterns are contextualized. During the data integration process, the data specialist can be asked for additional inputs when issues appear. This close interaction with the data specialist and the domain expert ensures that context-specific information is taken into account.

While shaping the semi-automated integration process already poses a research challenge, a related challenge is that the resulting enhanced event log will be location-aware. As the current XES-format [12] does not explicitly include location-related information, a novel location XES extension needs to be defined. This enables conducting location-aware process analyses, which will become more important in process mining given the increasing interest in ILS.

6   Conclusion

This paper discussed, from a conceptual angle, the opportunities that ILS data provides to tackle event log quality issues. This is a novel perspective as prior work on event log quality improvement did not consider the use of other sources of process-related data. ILS data can play an important role to alleviate quality issues considered important in healthcare such as incorrect/imprecise timestamps and imprecise resource information. While prior process mining research centers around the use of either HIS data or ILS data, this paper showed the benefits of integrating both to obtain an enhanced event log. As outlined above, ILS data can be used to tackle common event log quality issues. Conversely, the event log provides rich contextualization of patterns observed in ILS data.

Besides the potential benefits of ILS data to create an enhanced event log, this paper also outlined some challenges. Hence, further research is still required to systematically integrate HIS and ILS data. However, these efforts are worthwhile as a richer and more accurate enhanced event log will make an important contribution towards exploiting the full potential of process mining in practice. Moreover, the enhanced event log will enable the development of new techniques related to e.g. resource behavior analysis and patient waiting time analysis.

References

1. van der Aalst, W.M.P.: Process Mining: Data Science in Action. Springer, Heidelberg (2016). https://doi.org/10.1007/978-3-662-49851-4
2. van der Aalst, W., et al.: Process mining manifesto. In: Daniel, F., Barkaoui, K., Dustdar, S. (eds.) BPM 2011. LNBIP, vol. 99, pp. 169–194. Springer, Heidelberg (2012). https://doi.org/10.1007/978-3-642-28108-2_19
3. Batini, C., Scannapieco, M.: Data Quality: Concepts. Methodologies and Techniques. Springer, Heidelberg (2006). https://doi.org/10.1007/3-540-33173-5
4. Bendavid, Y.: RTLS in hospitals: technologies and applications. In: Encyclopedia of E-Commerce Development, Implementation, and Management, pp. 1868–1883. IGI Global, Hershey (2016). https://doi.org/10.4018/978-1-4666-9787-4.ch132
5. Bose, R.J.C.P., Mans, R.S., van der Aalst, W.M.P.: Wanna improve process mining results? It's high time we consider data quality issues seriously. Technical report BPM Center Report BPM-13-02 (2013)
6. Caron, F., Vanthienen, J., Vanhaecht, K., Van Limbergen, E., De Weerdt, J., Baesens, B.: Monitoring care processes in the gynecologic oncology department. Comput. Biol. Med. **44**, 88–96 (2014). https://doi.org/10.1016/j.compbiomed.2013.10.015
7. Fernandez-Llatas, C., Benedi, J.M., Garcia-Gomez, J.M., Traver, V.: Process mining for individualized behavior modeling using wireless tracking in nursing homes. Sensors **13**(11), 15434–15451 (2013). https://doi.org/10.3390/s131115434
8. Fernandez-Llatas, C., Lizondo, A., Monton, E., Benedi, J.M., Traver, V.: Process mining methodology for health process tracking using real-time indoor location systems. Sensors **15**(12), 29821–29840 (2015). https://doi.org/10.3390/s151229769
9. Forsberg, D., Rosipko, B., Sunshine, J.L.: Analyzing PACS usage patterns by means of process mining: steps toward a more detailed workflow analysis in radiology. J. Digit. Imaging **29**(1), 47–58 (2016). https://doi.org/10.1007/s10278-015-9824-2
10. Gal, A., Senderovich, A., Weidlich, M.: Challenge paper: data quality issues in queue mining. J. Data Inf. Qual. **9**(4), 1–5 (2018). paper 18
11. Gschwandtner, T., Gärtner, J., Aigner, W., Miksch, S.: A taxonomy of dirty time-oriented data. In: Quirchmayr, G., Basl, J., You, I., Xu, L., Weippl, E. (eds.) CD-ARES 2012. LNCS, vol. 7465, pp. 58–72. Springer, Heidelberg (2012). https://doi.org/10.1007/978-3-642-32498-7_5
12. Gunther, C.W., Verbeek, H.M.W.: XES standard definition. Technical report, Eindhoven Unversity of Technology, Eindhoven, The Netherlands (2014)

13. Jeffery, S.R., Garofalakis, M., Franklin, M.J.: Adaptive cleaning for RFID data streams. In: Proceedings of the 32nd International Conference on Very Large Databases, pp. 163–174 (2006)

14. Kim, W., Choi, B.J., Hong, E.K., Kim, S.K., Lee, D.: A taxonomy of dirty data. Data Mining Knowl. Discov. **7**(1), 81–99 (2003)

15. Kirchner, K., Herzberg, N., Rogge-Solti, A., Weske, M.: Embedding conformance checking in a process intelligence system in hospital environments. In: Lenz, R., Miksch, S., Peleg, M., Reichert, M., Riaño, D., ten Teije, A. (eds.) KR4HC/ProHealth 2012. LNCS (LNAI), vol. 7738, pp. 126–139. Springer, Heidelberg (2013). https://doi.org/10.1007/978-3-642-36438-9_9

16. Liu, C., Ge, Y., Xiong, H., Xiao, K., Geng, W., Perkins, M.: Proactive workflow modeling by stochastic processes with application to healthcare operation and management. In: Proceedings of the 20th ACM International Conference on Knowledge Discovery and Data Mining, pp. 1593–1602 (2014)

17. Mannhardt, F., de Leoni, M., Reijers, H.A., van der Aalst, W.M.P., Toussaint, P.J.: Guided process discovery - a pattern-based approach. Inf. Syst. **76**, 1–18 (2018). https://doi.org/10.1016/j.is.2018.01.009

18. Mans, R., Reijers, H., van Genuchten, M., Wismeijer, D.: Mining processes in dentistry. In: Proceedings of the 2nd ACM International Health Informatics Symposium, pp. 379–388 (2012)

19. Mans, R.S., van der Aalst, W.M.P., Vanwersch, R.J.B.: Process Mining in Healthcare: Evaluating and Exploiting Operational Healthcare Processes. Springer, Heidelberg (2015). https://doi.org/10.1007/978-3-319-16071-9

20. Mans, R.S., van der Aalst, W.M.P., Vanwersch, R.J.B., Moleman, A.J.: Process mining in healthcare: data challenges when answering frequently posed questions. In: Lenz, R., Miksch, S., Peleg, M., Reichert, M., Riaño, D., ten Teije, A. (eds.) KR4HC/ProHealth 2012. LNCS (LNAI), vol. 7738, pp. 140–153. Springer, Heidelberg (2013). https://doi.org/10.1007/978-3-642-36438-9_10

21. Rojas, E., Munoz-Gama, J., Sepúlveda, M., Capurro, D.: Process mining in healthcare: a literature review. J. Biomed. Inform. **61**, 224–236 (2016). https://doi.org/10.1016/j.jbi.2016.04.007

22. Rovani, M., Maggi, F.M., de Leoni, M., van der Aalst, W.M.P.: Declarative process mining in healthcare. Expert Syst. Appl. **42**(23), 9236–9251 (2015). https://doi.org/10.1016/j.eswa.2015.07.040

23. Senderovich, A., Rogge-Solti, A., Gal, A., Mendling, J., Mandelbaum, A.: The ROAD from sensor data to process instances via interaction mining. In: Nurcan, S., Soffer, P., Bajec, M., Eder, J. (eds.) CAiSE 2016. LNCS, vol. 9694, pp. 257–273. Springer, Cham (2016). https://doi.org/10.1007/978-3-319-39696-5_16

24. Suriadi, S., Andrews, R., ter Hofstede, A.H., Wynn, M.T.: Event log imperfection patterns for process mining: towards a systematic approach to cleaning event logs. Inf. Syst. **64**, 132–150 (2017). https://doi.org/10.1016/j.is.2016.07.011

25. van der Togt, R., Bakker, P.J., Jaspers, M.W.: A framework for performance and data quality assessment of Radio Frequency IDentification (RFID) systems in health care settings. J. Biomed. Inform. **44**(2), 372–383 (2011). https://doi.org/10.1016/j.jbi.2010.12.004

The ClearPath Method for Care Pathway Process Mining and Simulation

Owen A. Johnson[1,2(✉)], Thamer Ba Dhafari[1], Angelina Kurniati[1,3], Frank Fox[1], and Eric Rojas[4]

[1] University of Leeds, Leeds LS2 9JT, UK
o.a.johnson@leeds.ac.uk
[2] X-Lab Ltd., Leeds LS3 1AB, UK
[3] Telkom University, Bandung, Indonesia
[4] Pontificia Universidad Católica de Chile, Santiago, Chile

Abstract. Process mining of routine electronic healthcare records can help inform the management of care pathways. Combining process mining with simulation creates a rich set of tools for care pathway improvement. Healthcare process mining creates insight into the reality of patients' journeys through care pathways while healthcare process simulation can help communicate those insights and explore "what if" options for improvement. In this paper, we outline the ClearPath method, which extends the PM^2 process mining method with a process simulation approach that address issues of poor quality and missing data and supports rich stakeholder engagement. We review the literature that informed the development of ClearPath and illustrate the method with case studies of pathways for alcohol-related illness, giant-cell arteritis and functional neurological symptoms. We designed an *evidence template* that we use to underpin the fidelity of our simulation models by tracing each model element back to literature sources, data and process mining outputs and insights from qualitative research. Our approach may be of benefit to others using process-oriented data science to improve healthcare.

Keywords: Healthcare · Care pathways · Process mining · Process simulation

1 Introduction

The provision of high quality healthcare involves such complex systems that even those involved in their organization and delivery can feel it is impossible to comprehend. The *care pathway* is one well established and useful concept for bringing much needed clarity [1]. A care pathway describes the sequence of care that is recommended for patients with similar conditions requiring similar treatment [2] and is analogous to a *de jure* business process. Process mining of routine electronic healthcare records (EHR) can provide insight into the *de facto* compliance with care pathways including measuring performance and outcomes [3]. Although EHRs are a rich data source they present significant challenges of data quality, veracity and complexity [4]. In reality, care providers support multiple, simultaneous, diverse pathways for patients with

© Springer Nature Switzerland AG 2019
F. Daniel et al. (Eds.): BPM 2018 Workshops, LNBIP 342, pp. 239–250, 2019.
https://doi.org/10.1007/978-3-030-11641-5_19

highly variable personal needs and many of the interactions, events and decisions occur "off the radar" of the electronic systems.

To have utility, the outputs of process mining efforts need to be iteratively refined with the assistance of domain experts and then presented in a form that makes them accessible to wider stakeholders. In previous work [5] we have found agent based models with discrete event simulation presented through an interactive graphical representation to have been effective in stakeholder engagement. We developed the NETIMIS software tool (www.netimis.com) to support healthcare process simulation and this is now a commercial product and available for academic research. Healthcare process mining presents opportunities for understanding some of the reality of real patients' journeys through care pathways while healthcare process simulation can help communicate these discoveries and explore "what if" options for improvement [6]. In our approach, we extend simulation models to *fill in the gaps* by adding process steps missing from the health record data, adding information such as costs and incorporate insights from domain experts and stakeholder feedback.

In this paper, we present the ClearPath method as an extension of the established PM^2 process mining method [11] to incorporate healthcare process simulation modeling. We illustrate the ClearPath method through three case studies within UK hospitals, which show the discovered pathways for alcohol-related illness, giant-cell arteritis and functional neurological symptoms. In each case, the disease pathway needs to fit within busy hospitals where pathways of care for many diseases are taking place simultaneously.

2   Background

2.1   Process Mining in Healthcare

There is growing interest in process mining in healthcare [7]. Process mining can help answer *frequently posed questions* from clinicians and medical specialists [8] from control-flow, performance, conformance, and organizational perspectives [9]. Frameworks for process mining include the L* life-cycle model [10] which describes the life-cycle of a typical process mining project and more recently the Process Mining Project Methodology (PM^2) which incorporates iterations and gives detailed descriptions for six project stages [11]. Bozkaya et al. [12] propose a methodology called Process Diagnostics Method (PDM) which has been adapted to Business Process Analysis in Healthcare environments (BPA-H) [13]. Mans et al. [3] provides a comprehensive guide to process mining in healthcare including health reference models and pathways. Finally, a question driven methodology to answer frequently asked questions was developed for healthcare [14]. The methodologies share similar steps including extracting event data, applying tools and techniques, analyzing the resulting models and improving based on stakeholder feedback. Process mining has been combined with process simulation [15] including to discover models for simulation [16] and at least once in healthcare [6]. In our approach, we have also combined process mining with traditional business process analysis methods to build a richer model than could be achieved by process mining alone.

2.2   Process Simulation in Healthcare

Brailsford *et al.* (2017) reports on 50 years of healthcare simulation [17] and there are recognized frameworks for good practice in developing simulation models for healthcare [18]. In [6], we described the use of NETIMIS, a discrete event simulation tool for care pathway models which includes aspects of agent based approaches (patient characteristics affecting probabilities), and notions of time, cost and simplicity. We linked this to process mining and found challenges of EHR data quality (veracity, missing events, and missing data) and process complexity which suggested a mixed methods approach was required. There are many sophisticated tools for simulation but in this paper, we report on the use of NETIMIS.

NETIMIS is a cloud-based online service accessed using a standard browser and used to draw, share, evaluate and refine models of care pathways as runnable simulations. A NETIMIS model (see Fig. 1) consists of a network of directed edges and nodes. The edges represent activities that take place over a period of time. The nodes represent events such as a decision point or the start or end of an activity. Pathways are animated with multiple moving tokens representing patients (shown as colored dots that move along the edges at a speed consistent with the time of the activity).

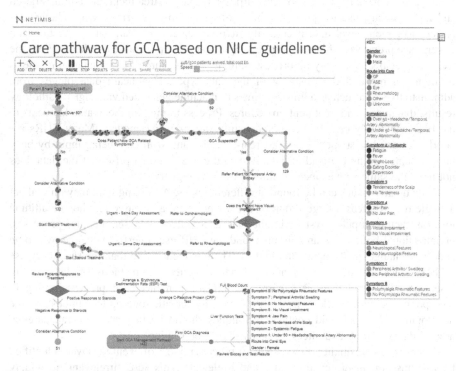

Fig. 1. NETIMIS example showing a run of the Giant Cell Arteritis (GCA) care pathway derived from national guidelines. The simulation model can be run online at www.netimis.com/shared/5ad5fe6f7775761d4c5fd5ec

No patient-level data is needed for NETIMIS as the model is a simulation based on data from population-level analysis. Following agent-based approaches, patient tokens are randomized with attributes that reflect those of the base population and pathway junctions are given probabilities that are dependent on those attributes. Each patient token can be colored with a mini pie chart representing its attributes. The tool supports constraints that can lead to bottlenecks and probabilities that are affected by repetitions. Health outcomes are represented by pathway end nodes and each simulation run calculates total health economic costs and times based on the sum of individual costs and times assigned to each activity completed for all of the patients' care. Unlike Petri Nets, each token represents a single patient that cannot be "split" so there is no support for parallelism. Following the analogy of cars on roads, multiple patient tokens flow through care pathways to create a highly visual and engaging model. Users interact with the visualization through features including accelerate, pause, zoom, inspect, change, share and compare. "As is" and "to be" models can be run side by side so that differences can be explored visually.

2.3 Challenges Using EHR Data for Process Mining Care Pathways

EHR data is normally created for the purposes of patient treatment and administration and its secondary use for process mining of care pathways brings many challenges. Access to patient level data necessarily involves careful ethical, data protection and governance processes which can prove a significant administrative overhead. From the technical perspective, applying process mining to healthcare data is challenging due to its high volume and the diversity of the data types. Healthcare data ranges from administrative data such as admission times to machine generated vital signs, pathology results, diagnoses, and treatment procedures. Process mining all the available events in the EHR inevitably creates incomprehensible *spaghetti-like* models. Many EHRs are poorly designed to support easy, fast real-time use and with data being input by busy human beings doing demanding jobs it should come as no surprise that the data does not have the same provenance as clinical trials or registry data.

Data quality issues can be found at different levels. A missing field may only affect a single row whereas a large group of users who share a negative and hostile attitude towards their computer system might bias a complete data set. People, processes, organizational boundaries and cultures (and the EHR user interface) change over time and these changes will impact on the data. There is recognition that, the secondary use of EHR data for research demands validated, systematic methods of data quality assessment [19] and there is a correspondingly urgent need for process mining to incorporate techniques addressing these issues. Systematic logging techniques and the development of repair and analysis techniques should be in place and transparency around data cleaning and checking steps should be routine.

Four broad data quality dimensions for process mining of event logs were identified by [3]: missing, incorrect, imprecise and irrelevant. This adds 'irrelevant' to widely cited dimensions of EHR data quality that form the basis for the data quality assessment method proposed by Weiskopf & Weng [19] and the valuable harmonized terminology produced by Kahn [20]. These dimensions were further detailed as 27 types of quality issues relating to the case, event and attribute levels of the data in an event log.

The Process Mining Manifesto proposes a useful rating system for data quality ranging from 1-star to 5-star [10] and also emphasizes challenges of incompleteness, noise, granularity, event log complexity and concept drift. In [21] we describe the development of our *data quality management framework* to support the discovery, root-cause investigation, mitigation and careful documentation of these issues using software version control tools that are directly linked to the lines of software code in the Extract and Transform programs used to build the event log. The framework supports a close link between the design of individual process mining experiments and assessment of *fit-enough-for-use* quality.

3   The ClearPath Method

3.1   Rationale for an Agile Approach

Healthcare is a complex business and process mining and simulation in healthcare has some unique requirements. Clinicians work together across organizational and functional boundaries to meet the often highly individual needs of patients with complex conditions. We have found that domain experts such as clinical specialists can have quite limited views on the patient pathways beyond their specialism. Even a simple structured discussion with a number of specialists gathered around a whiteboard or process model has proved beneficial in improving pathways. We have used NETIMIS on multiple projects to structure these pathway discussions, elicit tacit knowledge and generate actionable insights including "what if" scenarios (www.netimis.co.uk/case-studies). Including patients in these discussions has proved incredibly powerful.

There is however a tension in healthcare improvement projects between the desire to drive radical change quickly and the demands of *evidence-based medicine* for detailed and careful reviews, particularly where adverse outcomes can be harmful and even fatal. Our approach has therefore been to adopt *agile methods* with time-boxed iterations which produce process simulation models of increasing fidelity that are backed by strong tooling (ProM, NETIMIS, data mining), traditional academic research methods (literature reviews, qualitative methods) and traditional business process analysis (observation, interviews, sample documents). We use the simulation model as the key output, and an evidence template to underpin the fidelity of model and present both to a Clinical Review Board at the end of each iteration.

3.2   Extending PM²

The ClearPath method follows PM² with the following extensions.

Stage 1: Planning – *research questions* are often simply "what is the care pathway?" or "what does it look like?" and *composing project team* includes identifying a Clinical Review Board and pre-booking meetings so that iterations become time-boxed.

Stage 2: Extraction – the ethics of *extracting event data* when it is sensitive health data often mean long lead times so we make a data request and produce early iterations based on *transferring process knowledge* but with meticulous record keeping of artifacts (interview transcripts, whiteboard photos, journal references) from the investigation. In PM^2 business experts may be part of the project team but in healthcare these are often busy clinicians (or busy managers) so we engage them as interviewees within an iteration and/or in the Clinical Review Board at the end of each iteration.

Stage 3: Data Processing includes *filtering logs* to just include the patients of interest (those involved in the care pathway) and sometimes to *slice and dice* to examine sub-groups of patients (e.g. frailty) and the pathway under different conditions (time of day, day of week) and in different locations. Healthcare reference models and coding systems such as SNOMED-CT and ICD-10 are used for *aggregating events*. We use our data quality framework to document data issues and software code solutions.

Stage 4: Mining and Analysis produces process models which are recreated in NETIMIS (currently by hand) with performance and compliance data added (e.g. mean durations, decision point probabilities) and documented in the evidence template. In Stage 4 we also add in details of the care pathway from our business process analysis, for example where activities are not recorded in the EPR and may also construct multiple models to examine different scenarios (e.g. weekends vs weekdays).

Stage 5: Evaluation includes *verify and validate* results against process insights from multiple reliable sources and root-cause investigations to *diagnose* anomalies. Stage 5 marks the end of each analysis iteration and takes the form of a Clinical Review Board (CRB) meeting where the evidence base, data quality management framework (assumptions, root cause analysis and mitigation decisions) are reviewed together with the latest "as is" and candidate "to be" NETIMIS model as runnable simulations. The CRB meetings are interactive and generally highly productive. The outcome of the meeting is to plan objectives for the next iteration.

Stage 6: Process Improvement and Support is marked by acceptance of the models and evidence by the CRB for *implementing improvements*. Models are published on NETIMIS and can be shared by other organizations and calibrated to local situations.

3.3 The Evidence Template

The ClearPath method focuses attention on the construction of simulation process models and the evidence template plays a key role in supporting early, low-fidelity models and an agile evidence building process. In the evidence template, each model element (patient agent, activity, decision point) and each model attribute (e.g. disease incidence, cost and duration, probability of next activity) are listed with references to the source material so that audit trail can be traced back to the literature sources, process mining outputs or investigation artifacts that were used. The modeler sets a Confidence Indicator (CI) to document their confidence in the evidence base for each element and attribute on a score of 1–5 with 5 being highest.

0 = No confidence/not applicable/system defaults
1 = Guess by Modeler

2 = Estimate from observation or domain expert interview
3 = Empirical evidence from process mining or published literature
4 = Confirmed from multiple reliable sources
5 = Confirmed through Clinical Review Board.

It is evident that a modeler can very quickly create a low fidelity model of the care pathway but will have to record CIs of mostly 0s and 1s. Conversely, a Clinical Review Board could review the evidence for every element in detail recording scores of 5 where they agree and leaving other elements as 3s or 4s where there is still uncertainty. The overall average CI therefore gives a rough indicator of overall confidence in the model and crucially, the modeling can stop when the Clinical Review Board believe they have enough evidence to make a process improvement decisions.

4 Case Studies

4.1 Use of the ClearPath Method

The ClearPath project aims to generate a method for combining process mining and simulation that is suitable for widespread use understanding and improving care pathways in the UK National Health Service. The case studies illustrate aspects of the method in use, some of the achievements and some of the unsolved challenges.

4.2 Case Study 1 Alcohol-Related Emergency Admission Pathway at Liverpool

In some parts of the UK hospital admissions for alcohol related illnesses are rising by 11% per year leading to chronic diseases such as ARLD (Alcohol Related Liver Disease) which has lower survival rates than most common cancers. In busy emergency departments, alcohol-related disruptive behavior may obscure a patient's serious advanced illness and also hamper treatment attempts [22]. There is growing recognition that clinicians need guidance on appropriate care pathways that can help them identifying and deal with alcohol related illness. For this case study, our project team worked with a data and pathway profiling team at the University of Liverpool who had created a data linkage framework based on EHR event data from emergency admissions from hospitals in the North West of England. Our approach consisted of embedding a member of our ClearPath team within the Liverpool team for up to two days per week over a three-month period. We resolved data governance issues by providing tool and analysis advice to the local team. In return, our analyst received sequence, aggregate and conformance data to populate a NETIMIS simulation model (see extract in Fig. 2) and an evidence template (see extract in Fig. 3).

Figure 3 illustrates how the evidence template was used to document the link between the percentages derived from the process data for ICU Disposal to the probability settings in the simulation model. Five iterations of the pathway model were developed starting from simple models from an initial workshop and enriched through investigation and reviews. In the final model these included age-banded probabilities extracted from the routine data and cost data sourced from standard activity tariffs.

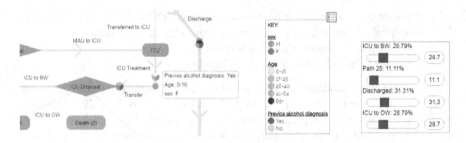

Fig. 2. Close-up view of pathways in and out of the Intensive Care Unit (ICU) and (right) the probability settings for paths exiting ICU Disposal

Pathway Setting ICU Route				
Activity	Time (hours)	Cost (£)	Next Activity (Probability)	CI
ICU Treatment	36.2 [31]	2828.1 [32]	Transfer	3
Transfer	1 [33]		ICU Disposal	1
ICU Disposal			Other Ward 0 28.79%	2
			Specialist Ward – 28.79%	2
			Discharge – 31.31% [34]	3
			Death 11.11% [34]	3

Evidence references

...

31 Mean time from data mining exercise
32 Reference cost from list provided by domain expert
33 Unknown. This is a guess to ensure the model runs OK
34 From process mining of hospital data

Fig. 3. Extract of the evidence template (left) and references (right) for the corresponding ICU elements

Several qualitative researchers had investigated the pathway through patient and clinician interviews and their deep insights helped fill in the gaps and add paths that were not evident from the data. The model was calibrated so that the outcomes reflected published clinical outcomes for the region. The resulting model is being presented as a regional exemplar of data-driven care pathway improvement.

4.3   Case Study 2 - Giant Cell Arteritis (GCA) Care Pathway at Leeds

Giant cell arteritis (GCA) is a rare chronic inflammatory condition of blood vessels (vasculitis) that affects large and medium sized arteries. Symptoms include headaches, tenderness of the scalp, jaw aches and chewing problems and visual impairment. The symptoms displayed can often be mistaken for normal age-related symptoms or other diseases however if GCA is not diagnosed and treated quickly it can lead to visual loss, blindness, or in worst cases a stroke [23]. For this reason, patients are treated with steroids as soon as the diagnosis is considered but this creates other challenges as the steroids impact on the sensitivity of diagnostic tests. Our project team worked with the clinical specialists at Leeds Teaching Hospitals Trust (LTHT) with access to the national MRC-TARGET (Treatment According to Response in Giant cEll arteritis) consortium (https://lida.leeds.ac.uk/target). Figure 1 illustrates the *de jure* pathway for GCA drawn from the National Institute for Health and Care Excellence (NICE) repository (https://pathways.nice.org.uk).

Our objective here was to map the *de facto* pathways in a large and very busy teaching hospital. Our initial approach was to request anonymized data extracts from the hospital EHR for patients with suspected GCA which included time stamped information of the patient's journey starting from their original route of entry into care, through to discharge, or firm diagnosis of GCA. Data quality issues proved insurmountable. It was not possible to accurately identify patients of interest or enough relevant events within the hospital EPR to complete the envisaged process mining exercise. However, we were able to complete a detailed process model through both traditional business analysis investigation and produced five iterations with a Clinical Review Board consisting on the local TARGET consortium leads. We gathered sufficient data from clinical expert interviews and volume and time figures from the EPR and cost figures from hospital tariffs to build a robust working model (mean CI = 4.1). Through documentary sources and interviews with other hospitals were able to produce models for other hospitals and develop a generic model that was a fit with the *de jure* guidelines and could be used to model *de facto* GCA pathways from small district hospitals to the complexity of LTHT. The study concluded with a costed model of the planned future model which identified points at which clinical pathways could be improved by recommending alternate diagnostic approaches. The simulation (Fig. 4) indicates both significant improvements in patient outcomes and simultaneously reduced costs. The models have been presented to the national group. A second phase using process mining of the Leeds EHR data is planned.

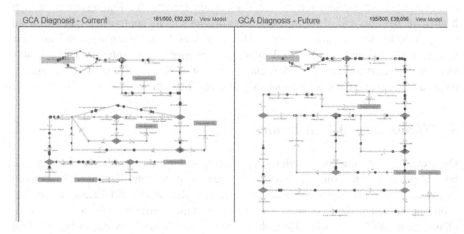

Fig. 4. NETIMIS screenshot showing a side-by-side run of the current Leeds' GCA Diagnosis pathway against the proposed future pathway

4.4   Case Study 3 – Functional Neurological Symptoms (FNS) Care at Leeds

Functional Neurological Symptoms (FNS) are a group of neurological symptoms which include weakness, abnormal movements and blackouts, they cause distress and dysfunction [24]. The symptoms are shared with neurological diseases such as epilepsy, multiple sclerosis or stroke but, in FNS patients, are caused by a brain disfunction. As with GCA, diagnosis is challenging but studies have shown that 31% of patients attending Neurology outpatient clinics had FNS [25]. Many healthcare providers lack specific pathways or services for FNS patients, there are no NICE guidelines and there is little data on acute FNS to inform service improvement. Therefore, there was a need to see the current pathways for FNS patients, and to know how they pass through the different healthcare services over time. Our first step was to generate visual models of the process from first presentation to diagnosis and referral to either psychology or psychiatry therapy.

Our aim with this project was to see whether process mining of the routine data was possible and we worked with a team of neurologists to understand the *de facto* care pathways at LTHT. Clinical inspection of the data quality of the EHR revealed data that was considered too unreliable to use for process mining. Issues included unrecorded events and observations, recording on letters and paper records rather than the EHR, mis-diagnosis and inappropriate referrals. Our alternative approach was to conduct a full audit of all the EHR data, clinical letters and discharge notes for each patient using all available sources (including phone interviews with treating clinicians to complete missing data). These resulting activities include diagnoses, emergency attendances, outpatient clinics, inpatient admissions and psychological/psychiatric referrals. The audit data was collated in the form of an event log which was used with a process mining tool (ProM). The initial results appeared disappointing - a spaghetti diagram with every single patient (n = 205) having a unique and complex variant. It was however a shocking result for the clinical domain experts, the findings show a high healthcare burden, and slow or incomplete movement to appropriate care, with an urgent need for service improvements. The mean time from Presentation to Diagnosis was 22.1 months and a further 7.2 months to an appropriate Referral. These results were presented as a video at an Association of British Neurologist conference and are being used to make the case for clearer pathways for FNS.

5   Discussion and Conclusions

Our experience working with both process mining and process simulation has been that both approaches are complex, challenging and require considerable skill, domain knowledge and perseverance. Healthcare is a complex world and EHRs are not yet capturing sufficient detailed workflow for deep clinical insights. Given this state of affairs, combining as many valid techniques as possible would seem to be the most pragmatic approach for quickly generating insights. However healthcare also demands strong evidence and rigorous methods and the ClearPath method has helped us structure data-driven care pathway investigations that have yielded good results and maintained an audit trail of evidence that ensures the models are defensible.

All three of the case studies here are examples of special case pathways within more general processes such as emergency admissions. Many patients genuinely require variants that differ from the norm. Case Study 1 used a simulation model to combine process mining outputs and other sources to build a useful model backed by evidence that can be traced to its source. Both Case Study 2 and 3 illustrate how difficult it can be to obtain robust EHR data. In Case Study 2 we made do with other sources and in Case Study 3 a manually constructed event log revealed alarming variability in care.

Our methods are evolving through use but we expect to formalize the approach on PM^2, Clinical Review Boards, evidence templates and audit trails. Currently we use NETIMIS for care pathway simulation and there are many alternatives. With NETIMIS we expect to improve tool integration so that it can generate and consume event logs and learn branching probabilities and probability density functions for activity duration from the log. Parallel to this we plan to develop rules to automate the visualization layout, to simplify the task of analyzing complex processes and communicating the results to diverse stakeholders. Our approach has been well received in the UK and may be of benefit to the wider academic and healthcare community seeking to use process-oriented data science to improve healthcare.

Acknowledgment. This work was supported by the cYorkshire Connected Health Cities (CHC) project. The case studies were developed by Luke Naylor, Sahar Salimi Avval Bejestani, Clea Southall and Samantha Haley at the University of Leeds. Case study 1 was supported by Anna Jenkins and the University of Liverpool CHC. Case study 2 was supported by Prof Ann Morgan and the TARGET Consortium for GCA. Case study 3 was supported by Dr Stefan Williams. The third author would also like to thank the Indonesia Endowment Fund for Education (LPDP) for the support given during this research.

References

1. Vanhaecht, K., Panella, M., Van Zelm, R.: An overview on the history and concept of care pathways as complex interventions. Int. J. Care Pathways **14**(3), 117–123 (2010)
2. European Pathway Association, "Care Pathways". http://e-p-a.org/care-pathways/. Accessed 30 May 2018
3. Mans, R.S., van der Aalst, W.M.P., Vanwersch, R.J.B.: Process Mining in Healthcare: Evaluating and Exploiting Operational Healthcare Processes. Springer, Heidelberg (2015). https://doi.org/10.1007/978-3-319-16071-9
4. Weiskopf, N.G., Bakken, S., Hripcsak, G., Weng, C.: A data quality assessment guideline for electronic health record data reuse. eGems (Generating Evid. Methods to Improv. Patient Outcomes) **5**(1), 14 (2017)
5. Johnson, O., Hall, P.S., Hulme, C.: NETIMIS: dynamic simulation of health economics outcomes using big data. PharmacoEconomics **34**(2), 107–114 (2015)
6. Mans, R., Reijers, H., Wismeijer, D., Van Genuchten, M.: A process-oriented methodology for evaluating the impact of IT: a proposal and an application in healthcare. Inf. Syst. **38**(8), 1097–1115 (2013)
7. Rojas, E., Munoz-Gama, J.: Process mining in healthcare: a literature review. J. Biomed. Inform. **61**, 224–236 (2016)
8. Mans, R.S., van der Aalst, W.M.P., Vanwersch, R.J.B., Moleman, A.J.: Process mining in healthcare: data challenges when answering frequently posed questions. In: Lenz, R., Miksch, S., Peleg, M., Reichert, M., Riaño, D., ten Teije, A. (eds.) KR4HC/ProHealth - 2012. LNCS (LNAI), vol. 7738, pp. 140–153. Springer, Heidelberg (2013). https://doi.org/10.1007/978-3-642-36438-9_10
9. van der Aalst, W.: Data science in action. Process Mining, pp. 3–23. Springer, Heidelberg (2016). https://doi.org/10.1007/978-3-662-49851-4_1
10. van der Aalst, W.M.P., et al.: Process mining manifesto. Bus. Process Manag. Work. **99**, 169–194 (2011)

11. van Eck, M.L., Lu, X., Leemans, S.J.J., van der Aalst, W.M.P.: PM^2: a process mining project methodology. In: Zdravkovic, J., Kirikova, M., Johannesson, P. (eds.) CAiSE 2015. LNCS, vol. 9097, pp. 297–313. Springer, Cham (2015). https://doi.org/10.1007/978-3-319-19069-3_19

12. Bozkaya, M., Gabriels, J., van der Werf, J.M.: Process diagnostics: a method based on process mining. In: International Conference on Information, Process, and Knowledge Management, eKNOW 2009, pp. 22–27. IEEE, February 2009

13. Rebuge, A., Ferreira, D.: Business process analysis in healthcare environments: a methodology based on process mining. Inf. Syst. **37**, 99–116 (2012)

14. Rojas, E., Sepúlveda, M., Munoz-Gama, J., Capurro, D., Traver, V., Fernandez-Llatas, C., et al.: Question-driven methodology for analyzing emergency room processes using process mining. Appl. Sci. **7**(3), 302 (2017)

15. van der Aalst, W.M.P.: Business process simulation revisited. Enterp. Organ. Model. Simul. **63**, 1–14 (2010)

16. Rozinat, A., Mans, R.S., Song, M., van der Aalst, W.M.P.: Discovering simulation models. Inf. Syst. **34**, 305–327 (2009)

17. Brailsford, S., Carter, M.W., Jacobson, S.H.: Five decades of healthcare simulation. In: Winter Simulation Conference, vol. 9, no. 62, pp. 365–384 (2017)

18. Marshall, D.A., et al.: Selecting a dynamic simulation modeling method for health care delivery research - part 2: report of the ISPOR dynamic simulation modeling emerging good practices task force. Value Health **18**(2), 147–160 (2015)

19. Weiskopf, N.G., Weng, C.: Methods and dimensions of electronic health record data quality assessment: enabling reuse for clinical research. JAMIA **20**(1), 144–151 (2013)

20. Kahn, M.G., et al.: A harmonized data quality assessment terminology and framework for the secondary use of electronic health record data. EGEMS **4**(1), 1244 (2016)

21. Fox, F., Aggarwal, V., Whelton, H., Johnson, O.: A data quality framework for process mining of electronic health record data. In: Proceedings of the Sixth IEEE ICHI (2018)

22. Lekharaju, P., Thompson, E., Shawihdi, M., Pearson, M., Hood, S., Bodger, K., et al.: PTH-062 emergency admissions for alcohol related conditions: making sense of routine data. Gut **63**(Suppl 1), A236–A236 (2014)

23. Laskou, F., Fiona, C., Aung, T., Benerjee, S., Dasgupta, B.: 074 Fast track giant cell arteritis clinic and pathway for early management of suspected giant cell arteritis: an audit. Rheumatology **57**(suppl_3), key075–298 (2018)

24. Mobini, S.: Psychology of medically unexplained symptoms: a practical review. Cogent Psychol. **2** (2015). Article no. 1033876

25. Stone, J., Carson, A., et al.: Symptoms 'unexplained by organic disease' in 1144 new neurology out-patients. Brain **132**, 2878–2888 (2009)

Analysis of Emergency Room Episodes Duration Through Process Mining

Eric Rojas[1,2(✉)], Andres Cifuentes[1], Andrea Burattin[3], Jorge Munoz-Gama[1], Marcos Sepúlveda[1], and Daniel Capurro[2]

[1] Department of Computer Science, School of Engineering,
Pontificia Universidad Católica de Chile, Santiago, Chile
{eric.rojas,alcifuen,jmun}@uc.cl,marcos@ing.puc.cl
[2] Department of Internal Medicine, School of Medicine,
Pontificia Universidad Católica de Chile, Santiago, Chile
dcapurro@med.puc.cl
[3] Technical University of Denmark, Kgs. Lyngby, Denmark

Abstract. This study presents the proposal of a performance analysis method for ER Processes through Process Mining. This method helps to determine which activities, sub-processes, interactions and characteristics of episodes explain why the process has long episode duration, besides providing decision makers with additional information that will help to decrease waiting times, reduce patient congestion and increment quality of provided care. By applying the exposed method to a case study, it was discovered that when a loop is formed between the Examination and Treatment sub-processes, the episode duration lengthens. Moreover, the relationship between case severity and the number of repetitions of the Examination-Treatment loop was also studied. As the case severity increases, the number of repetitions increases as well.

Keywords: Process mining · Healthcare · Emergency Room

1 Introduction

Performance measurements in Emergency Room (ER) processes are highly important, because of the information they can provide to identify behaviour of episodes with extended waiting times for patients expecting attention. Generally, patients are categorized into triage categories, normally by a nurse, to determine their attention waiting times in the ER. In this ER, the Manchester triage [1] is used to classify patients in five color categories: red, orange, yellow, green and blue. Red being the most critical patients with lowest waiting times, while blue being the least severe patients with highest waiting times. Identifying improvement opportunities in the ER processes can help reduce waiting times, improve the quality of provided services and reduce overcrowding [2].

F. Daniel et al. (Eds.): BPM 2018 Workshops, LNBIP 342, pp. 251–263, 2019.
https://doi.org/10.1007/978-3-030-11641-5_20

Our approach to study the performance of ER processes is based on process mining [3] as the main component to identify, discover and analyze activities executed during ER process episodes. Process mining is a research discipline that focuses on extracting process knowledge from data generated and stored in the databases of (corporate) information systems, in this case Hospital Information Systems (HIS) [3–5]. Process execution data is extracted as event logs, which are sets of episodes, each containing all the activities executed for a particular process instance. Process mining tools and techniques can be applied to discover process models, verify conformance, analyze organizational patterns and check the performance of a process in any hospital [3]. This paper proposes a process mining methodology to study the performance of ER processes. These processes have been analyzed using simulation, data mining along with process mining [6,7], among others. Data mining has been used to find patterns and understanding the causes of certain process behavior while in the other hand process mining describes how processes are currently performed.

Previous studies describe how process mining has been applied to analyze the executed processes within ER [7–9]. These studies have given insight about the process and the flow of activities during episodes (for example, medication or discharge activities). The first case is a study in a Portugal hospital where a software suite was defined to extract data, build an event log and discover any process in the medical center [7]. An specific case study was done using ER data, but the solution is general. The solution includes clustering techniques and Markov chain models. The second study is exploratory and was conducted in four Australian hospitals [8], where data was extracted, an event log was built, and discovery techniques of process mining were applied. The third one proposes a methodology based on frequently posed questions, and provides a case study, but no performance analysis was executed [9]. Performance analysis of ERs has been previously researched [10,11], but did not included a method for performance analysis of the ER episodes or any ER metric in general using process mining techniques.

The objectives of this paper are to analyze the ER episodes behavior using a process mining methodology comprehensively described, to apply it to a case study, and to determine which activities, sub-processes and their interactions in the ER process explain why the process get stalled and has a longer duration, and, to identify any existing relationships between some characteristics of the process activities or sub-processes and the process performance. This paper is an exploratory study of performance analysis in ER using process mining. Further studies must be done in order to complete more in depth quantitative analysis.

The structure of the paper is as follows: Sect. 2 describes the proposed methodology. Section 3 describes a case study where the methodology has been applied including results and discussion. Finally, conclusions and future work are highlighted.

2   Method

A 6 phases methodology for the process performance analysis of ER episodes is proposed, as depicted in Fig. 1[1]. To generate this methodology previous studies from the authors were considered [9].

Phase 1: Extraction and Transformation. ER processes are supported by clinical and non-clinical activities that are executed by different types of resources (physicians, nurses, administrative staff). Each of these series of activities correspond to an episode, which is registered in any Hospital Information System (HIS) [4]. HIS are computer systems designed to ease management of the hospital's medical and administrative information and to improve quality of healthcare [12]. HIS store records as events (or activities) that include all the necessary data to create an event log to perform process mining analysis. An event log is a file record that provides an audit trail that can be used to understand the

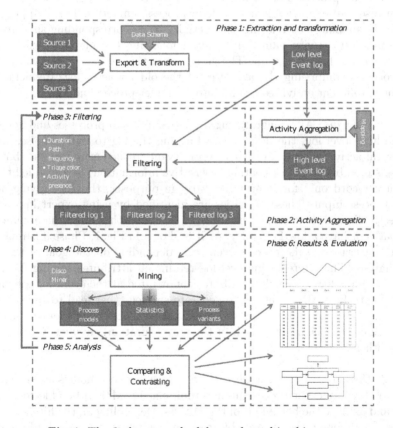

Fig. 1. The 6 phases methodology adopted in this paper.

---

[1] All figures presented in this paper can be seen with more details in the following link https://wp.me/p9ZlAl-1g.

system activities. In this phase, HIS records are extracted from different sources (databases, repositories, etc.) and transformed into a standard event log format that is readable by process mining tools like CSV, XLS, XES, among others.

Phase 2: Activity Aggregation. The complexity of the ER process is defined by the amount of different activities involved in one process instance and the lack of process structure: a classic spaghetti process [3]. Spaghetti processes are unstructured models with several activities connected to each other and are difficult to visualize because of the amount of information that is shown when represented as a graphical model. Thus, it becomes necessary to bring the log to a higher abstraction level, with a general view of the process in terms of flow and structure, by shifting from a perspective which is data-oriented to a process-centric one. The result is an event log that contains sub-processes rather than activities as its basic unit, which makes inspection and analysis more accessible. The followed approach aggregates activities by mapping them to the sub-process they belong to. Sub-processes are defined by some activities that work as separators and, in case the separation is not evident, process semantics. This mapping can be done by manually correlating each activity to its corresponding sub-process. Afterwards, an activity name substitution following the mapping will have a high-level event log as a result. The activities of this new event log will be the sub-processes comprising the activities of the old low-level log. Reducing the amount of different activities results into a simpler process model.

Phase 3: Filtering. After aggregating activities into sub-processes and generating the high-level log, the next step is filtering the log to reduce noise. There are several activities that are either non relevant to the entire process (but are still registered because they contain descriptive information) or are added to the historical record only for information storage purposes, thus not giving additional process input. These activities are identified by using expert knowledge to directly discard them and by filtering out activities that follow highly uncommon paths or have a low frequency. As it is shown in Fig. 1, Phase 3 is revisited later to generate specific process event logs depending on the goal of the process analysis. The filtered high-level log retains all attributes coming from the low-level source log. Using the attributes enumerated at the event log schema it is possible to filter the event log to generate sub-logs that lead to specific process models that can answer questions related to process behaviour depending on Triage color, Diagnosis, length of the stay (duration), frequency of a path, among others.

Phase 4: Discovery. Using the filtered event logs, process models are generated in Disco[2] (or any other process mining tool, e.g. ProM[3]) to be able to look in depth and analyze the behaviour of the process. Depending on the filters applied at Phase 3, different models are obtained. In this phase, numerical information

---

[2] See http://www.fluxicon.com/disco.

[3] See http://www.promtools.org.

can be collected for different process variants. The goal is to collect data of the process variants to elaborate a comprehensive analysis during the next phase. To build process models event logs, process mining discovery algorithms go through the log and maps it onto a process model such that the model is "representative" of the behavior seen in the log [3].

Phase 5: Analysis. Inspecting the process model makes it easier to identify sub-processes by finding exit activities (i.e., activities that are frequently the end of most paths of a section of a process). Before analyzing it is necessary to get the correct model by filtering the log according to what is going to be analyzed and then discovering a new process model, which means that there is a need to go back to Phase 3 and 4 to filter the log and generate the corresponding model. Figure 1 shows this procedure with an arrow starting at Phase 5 and going back to Phase 3, representing the possibility to iterate through Phases 3, 4 and 5 to perform a comprehensive analysis of the process behavior.

Phase 6: Results and Evaluation. This last step considers the delivery of process models, summarizing the data extracted from process models and comparison charts for process model duration, and performing validation and evaluation of these results and conclusions through expert opinion.

3   Case Study

Using the methodology described in the previous section, a case study was conducted at the Clinical Hospital of Red de Salud UC CHRISTUS, using historical data collected at its ER during July 2014. The hospital is located in Santiago, Chile. Figure 2 shows a BPMN diagram[4] of the expected high-level flow of the process according to experts. A Triage sub-process and a Treatment sub-process were identified. Once a Triage sub-process was found, a deeper analysis using the triage color attribute was performed. After the Triage sub-process finishes, the rest of the process corresponds to Treatment and Discharge sub-processes. Identifying the Discharge sub-process is simple since discharging a patient is a straightforward action. The Treatment sub-process involves most of the activities for each case. The conversion from low-level to high-level log will collapse most activities depending on the nature of the tasks performed by the resources of the ER. There are examination activities that figure out what is the problem of the patient based on physical examination and tests, and to confirm if the patient is ready to be discharged. There are also treatment activities that are performed to lead the patient to a stable condition. These collapsed activities represent the Examination for Prediction, Treatment and Examination for Validation sub-processes. Whenever the treatment is not successful, another instance of Examination for Prediction starts, followed by Treatment. In some cases, Examination for Validation is performed by medical order. Analyzing the duration of these sub-processes provides insight about the behaviour of the process

---

[4] See http://www.bpmn.org.

and how performance of these sub-processes affect the whole process. Analyzing the Treatment models through filtering by duration defines differences between short stays and long stays. Finally, considering referrals to specialized physicians provides information about duration of the process depending on the complexity of the diagnosis. A detailed description of the executed tasks performed at each Phase of the study is provided next.

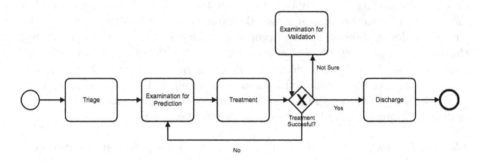

Fig. 2. ER process: high-level BPMN model with the expected high-level process flow.

Phase 1: Extraction and Transformation. The data used to construct the event log is historical data collected by the research team from the Hospital Information Systems Alert ADW Phase I, which corresponds to the system used to store the data in the ER Process. Historical data was built as an enumeration of 309,796 activities registered, composing 7,160 different episodes. There are 64 different activity types registered. Each register includes its timestamp, its resource and the episode id. In addition to these attributes, the analysis considers diagnosis, type of resource (nurse, doctor, technician, auxiliary nurse) and triage color. It is worth noting that the activity timestamps are registered at the hour level. However, timestamps of the first and the last activity are registered considering minutes and seconds. By selecting corresponding attributes using Disco import tool, an event log is built based on this historical data.

Phase 2: Activity Aggregation. Disco was used to filter the event log according to medical expert opinion, who explained that the activities could be classified in 3 sub-processes: Triage, Examination-Treatment and Discharge activities. These groups are initially considered as sub-processes of the ER process. Each of the original activities is mapped to a collapsed activity by enriching the event log with this information. Table 1 shows the relationships created to achieve this representation for the Treatment sub-process. Activities that are not included in this table are disregarded because they are not providing relevant information to the process model, according to the expert.

Concerning the Examination sub-process, activities could be of two types: "for prediction" or "for validation". Examination for prediction is the first examination, performed by a doctor, that provides enough information *to figure out* what problem the patient has. Examination for validation is an activity, also performed by a doctor, that is done with the goal *to check* that everything is fine with the patient and that there is no need to continue staying at the ER. To make a difference between both types it is necessary to check the whole current episode. If current examination is the last one – in the specific episode –, and it is not the first one, it is considered an examination for validation, else it is an examination for prediction. Identifying types of examination was performed using a script implemented for this purpose. The described procedure is formalized in Algorithm 1.

Phase 3: Filtering. At the first filtering phase, events that were not mapped to any sub-processes are removed using Disco filtering capabilities. This filter removes noise and paths that are not required to analyze due to their low frequency.

Table 1. Mapping of activities into sub-processes for the Treatment process.

Original Activities	Sub-process
Nurse task and physician task	Examination
Prescribe Medication, Perform Procedures, Physical Examination, Give Medication, Prescribed Procedures, Prescribed Medication (internal), Required Laboratory Tests, Required Imagenology Tests, Biometry, Other Required Tests, and Cancelled External Medication Prescription	Treatment
Clinic discharge and last discharge	Discharge

Algorithm 1. Identify examination type

```
 1  begin ExaminationType(log)
 2      forall the episode ∈ log do
 3          for i = 0 up to |episode| do
 4              if episode[i].name is "Examination" then
 5                  lastExamination ← TRUE
 6                  for j = i + 1 up to |episode| do        · // |episode| is ep. length
 7                      if episode[j].name is "Treatment" then
 8                          lastExamination ← FALSE
 9                          break

10                  if lastExamination then
11                      set examination type for episode[i] to "validation"
12                  else
13                      set examination type for episode[i] to "prediction"
```

Phase 4: Discovery. Once the filter is applied, Disco generates a process model that represents the portion of the log resulting from the event log aggregation and filtering [13]. Disco also provides performance analysis, but this did not provide significant results for the analysis. From the aggregated event log, we obtain the process model illustrated in Fig. 3. The generated model is similar to the one previously shown in Fig. 2. The main difference between the models is that Fig. 3 includes infrequent paths (e.g., episodes where the patient is taken back from Discharge to Treatment).

The Triage sub-process normally starts with a "Nurse task" activity. After that, the process continues to a "Physician task" and then to an "Intake task" activity. In a third of the cases, a doctor does not take part at the triage sub-process and it goes directly to "Intake task". About 10% of the cases that go through the doctor are then taken by a technician ("Technician task"), and then an intake note is generated. After the generation of the intake note, a physical examination could occur or vital signs could be registered (these activities could be skipped). Finally the first triage is done, and the triage sub-process is finished.

The Treatment sub-process begins with "Examination for Prediction". This activity includes all nurse and doctor activities that occur before any actual treatment is performed. After this examination, a consultation to a specialized doctor could happen. Then, the process continues with the actual treatment. After the treatment is finished, the patient is discharged. Also, a loop between examination for prediction and treatment is identified. And, finally, after the patient is treated, a doctor could request a final examination to validate whether the patient could be discharged or not. It is worth noting that the difference

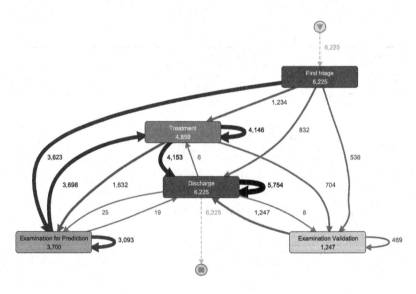

Fig. 3. Process model for the enriched event log. Arrows are decorated with the episode frequency of each path.

between short and long episodes depends almost exclusively of this sub-process. Long episodes of the complete process are the ones that include several of these "examination for prediction - treatment" loops.

Finally, patient discharge is straightforward. After aggregating the activities into one general discharge activity, the sub-process was collapsed into only one activity. The process structure is simple and sequential. Each activity is directly followed by the previous one with no alternative paths.

Phase 5: Analysis. The analysis of the Triage sub-process was done separately since it is the starting point of the ER process, and the examination, treatment and discharge sub-processes were analyzed together to identify loops and possible relationships between them, given that these three sub-processes happen at the attention box and are performed by the same resources.

Triage Sub-process. The initial sub-process in each ER episode consists in determining the severity of each episode according to the patient conditions. Usually, it consists in categorizing each patient to a Manchester triage category (red, orange, yellow, green and blue) [1]. Several activities were selected by the ER specialist from the event log to see the process model followed in the analyzed episodes. Figure 4 shows a process model with the activities executed during the Triage sub-process. For a more detailed analysis of the process, the event log was split in two groups: one half with the fastest episodes and the other half with the slowest episodes. No relevant differences in activities presence or their sequence were discovered.

Examination, Treatment and Discharge Sub-process (ETD). The second sub-process includes three main subsets of activities. The first subset contains the activities related to the examination of the patient. The second subset is the treatment, including treatment activities, medications and exams. Finally, discharge activities, including paperwork and the actual patient discharge, are grouped into a third subset. Figure 5 shows a process model with the activities executed during the ETD sub-process.

For a more detailed analysis of the process, the event log was split into two groups. The first group corresponds to the half with the fastest episodes. In Fig. 6 the process model for these episodes is shown. The other group includes the half with the slowest episodes, and its process model can be seen in Fig. 7. A loop between examination for prediction and treatment was identified on both Figs. 6 and 7. This loop could occur several times in one instance of the process. Every time this loop is followed, the process time increases (considering a discrete average of one hour), impacting on the episode duration. 20% of the cases have a examination for validation before discharging a patient. By inspecting the average duration (discrete average of one hour at treatment-examination validation transition, and another hour for examination validation-discharge transition) and considering that activity times are registered by the hour, it is safe to assume that examination for validation extends from one to two hours,

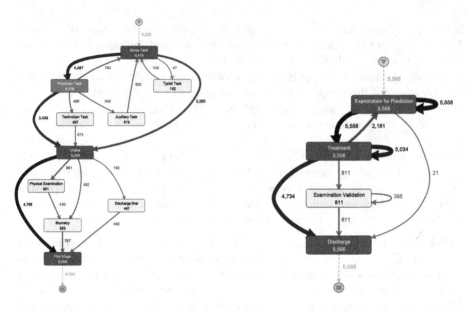

Fig. 4. Triage sub-process model   **Fig. 5.** ETD sub-process

and most of the time it is found at long process instances. Although both groups have the same activity structure, there is a great duration difference between both of them. The reason is that most of fastest cases don't loop back to examination for prediction, while most of slowest cases do.

Analysis by Triage Color. A relevant case attribute for this process is triage color since it informs resources about the severity of the patient case [1,14]. Table 2 shows a detailed description of the characteristics of the loop mentioned in the previous analysis. Higher priority episodes have a higher average of repetitions (red and blue episodes are not considered because they correspond to the least amount in the sample). Orange episodes involve more complex diagnosis and treatments which explains that in average these cases have more than two repetitions of the loop. While the severity of the episode becomes higher, the average number of repetitions increases as well. Green episodes, in average, have half the number of repetitions (1.27 average repetitions) of orange episodes (2.34 average repetitions). Figure 8 show these results. The chart has a value mark at the maximum number of repetitions. The chart is also divided in quartiles. Figure 8 includes all episodes of the event log and the number of repetitions of each episode plotted in black. By simply inspecting the chart, it is clear that more than one third of total episodes have at least two repetitions of the Examination-Treatment loop. When analyzing by color the proportion of episodes with two or more repetitions, this proportion increases as the severity of the case gets higher. This trend is shown in Fig. 9. The trend of higher severity cases is more steep, which reinforces previously shown results. For example, the trend for green cases moves between one and two repetitions, while the trend for orange cases is between one and six repetitions.

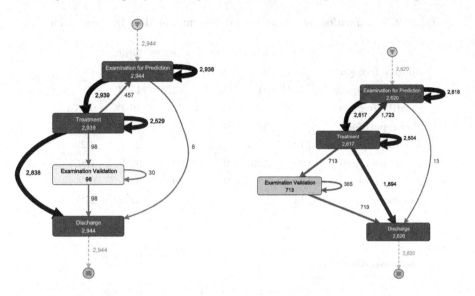

Fig. 6. ETD - fastest cases   **Fig. 7.** ETD - slowest cases

Phase 6: Results and Discussion. By analyzing the generated process models with ER experts, there are two places where the ER process slows its pace, increasing waiting times. The first situation is at the examination-treatment sub-process. Depending on how many times the patient is examined to detect what his/her problem is, the process extends per each examination-treatment loop, impacting total duration. The other section where the process takes longer than expected is when examination for validation takes place, extending the duration by one up to two hours (on average). There is a relationship between triage color and the number of examination-treatment loops. As the severity of the case increases, the number of repetitions increases as well, showing that blue and green cases (low severity) have a below average number of repetitions while yellow, orange and red cases (high severity) have a number of repetitions above the average. Triage color does not affect any other sub-process of the ER process.

Discussion. In this paper, several sub-processes and their relationships were identified. From the ER point of view of the expert, this helps to see which activities implicate an increment on the episode duration. The increment of episode duration directly increases the waiting times for all the episodes, and increments the ER overcrowding, which has become one of the most relevant issues nowadays [15].

Four sub-processes were identified, but two main sub-processes are critical. First, the Examination sub-process, that includes activities where the resources try to determine the diagnostic of the patient to take actions to bring care to the patients. And secondly, the Treatment sub-process where any exam, procedure or medication is provided. In this research, through process mining, a loop between the two sub-processes was identified. Every time it happens, time is added to the

Table 2. Examination for prediction - treatment loop data by triage color

	All	Blue	Green	Yellow	Orange	Red
Episode frequency	5538	44	2347	2374	755	18
Absolute frequency	9184	51	2292	4328	1773	40
Relative frequency	100%	0.56%	32.58%	47.1%	19.3%	0.44%
Max repetitions	13	3	9	10	13	7
Average repetitions	1.66	1.16	1.27	1.82	2.34	2.22
1 or more repetitions	99%	100%	99%	99%	99%	100%
2 or more repetitions	39%	13%	21%	29%	64%	61%
3 or more repetitions	14%	2%	4%	19%	32%	27%

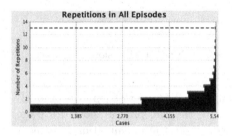

Fig. 8. All episodes

Fig. 9. Episode repetitions trends by triage color (Color figure online)

episode duration. It was also analyzed according to the different triage colors, giving a more detailed description on when this loop happens more often, so as to pay more attention in future episodes.

From the process mining perspective, the methodology was significant to the field. Two levels of analysis were executed, low level detailed analysis was performed with the domain expert to identify the executed activities and the sub-processes to which they belong, and high level analysis was carried out to see the relationships between the different sub-processes to identify which are the causes of slowest episodes.

4   Conclusion and Future Work

By analyzing the ER process, it was discovered that the loop between Treatment and Examination sub-processes increments the duration of ER episodes. This will help with the identification of the episodes where this happens (mainly by triage color), to make any necessary improvements. Identifying these loop as the main cause of the increment in the episode duration time is significant to help reduce it and with this reduction, lower the episode times, free boxes in the

ER, give a faster attention to more patients, shorten waiting times and finally reduce the overcrowding of the ER. Further work will include applying additional process mining and more statistical techniques to analyze the ER process and complete more in depth quantitative analysis. Besides more advanced research of the evidences provided by these models should be conducted in the ER.

References

1. Manchester Triage Group. Emergency Triage. BMJ Publishing (1997)
2. Konrad, R., et al.: Modeling the impact of changing patient flow processes in an emergency department: insights from a computer simulation study. Oper. Res. Health Care **2**(4), 66–74 (2013)
3. van der Aalst, W.M.P.: Data science in action. In: van der Aalst, W.M.P. (ed.) Process Mining, pp. 3–23. Springer, Heidelberg (2016). https://doi.org/10.1007/978-3-662-49851-4_1
4. Mans, R.S., van der Aalst, W.M.P., Vanwersch, R.J.B.: Process Mining in Healthcare: Evaluating and Exploiting Operational Healthcare Processes. SBPM. Springer, Cham (2015). https://doi.org/10.1007/978-3-319-16071-9
5. Rojas, E., Munoz-Gama, J., Sepúlveda, M., Capurro, D.: Process mining in healthcare: a literature review. J. Biomed. Inform. **61**, 224–236 (2016)
6. Kolker, A.: Process modeling of emergency department patient flow: effect of patient length of stay on ED diversion. J. Med. Syst. **32**(5), 389–401 (2008)
7. Rebuge, Á., Ferreira, D.R.: Business process analysis in healthcare environments: a methodology based on process mining. Inf. Syst. **37**(2), 99–116 (2012)
8. Partington, A., Wynn, M., Suriadi, S., Ouyang, C., Karnon, J.: Process mining for clinical processes: a comparative analysis of four Australian hospitals. ACM Trans. Manage. Inf. Syst. (TMIS) **5**(4), 1–19 (2015)
9. Rojas, E., Sepúlveda, M., Munoz-Gama, J., Capurro, D., Traver, V., Fernandez-Llatas, C.: Question-driven methodology for analyzing emergency room processes using process mining. Appl. Sci. **7**(3), 302 (2017)
10. Welch, S.J., Asplin, B.R., Stone-Griffith, S., Davidson, S.J., Augustine, J., Schuur, J., Emergency Department Benchmarking Alliance: Emergency department operational metrics, measures and definitions: results of the second performance measures and benchmarking summit. Ann. Emerg. Med. **58**(1), 33–40 (2011)
11. Michel Sørup, C., Jacobsen, P., Forberg, J.L.: Evaluation of emergency department performance-a systematic review on recommended performance and quality-in-care measures. Scand. J. Trauma Resusc. Emerg. Med. **21**(62), 1–14 (2013)
12. Degoulet, P.: Hospital information systems. In: Venot, A., Burgun, A., Quantin, C. (eds.) Medical Informatics, e-Health. Health Informatics, pp. 289–313. Springer, Paris (2014). https://doi.org/10.1007/978-2-8178-0478-1_12
13. Günther, C.W., Rozinat, A.: Disco: discover your process. BPM (Demos) (2012)
14. Mackway-Jones, K., Robertson, C.: Emergency triage. BMJ Br. Med. J. Int. Ed. **314**(7086), 1056 (1997)
15. Hoot, N., Aronsky, D.: Systematic review of emergency department crowding: causes, effects, and solutions. Ann. Emerg. Med. **52**(2), 126–136 (2008)

First International Workshop on Artificial Intelligence for Business Process Management (AI4BPM)

First International Workshop on Artificial Intelligence for Business Process Management (AI4BPM)

The use of AI in BPM has been discussed as the next disruptive technology that will touch almost all of the business operations and processing activities being performed by humans. In some cases, AI will dramatically simplify human interaction with a process, in other cases it will extensively support humans in the execution of tasks, and in yet other cases it will enable full automation of tasks that have traditionally required manual contributions. Over time, AI may lead to entirely new paradigms for business operations and processes. For example, instead of BPM models centered on process and/or case management, we anticipate models that are based fundamentally on goal achievement, as well as models that fully enable continuous improvement and adaptation based on experiential learning.

AI examples are:

- Machine learning and genetic algorithms for data analysis, e.g., predictive monitoring, workflow discovery, and reachability
- Fuzzy reasoning for supporting business process modeling
- Neural networks for placement detection in assembly workflows

These are only a few examples of what AI technology can do to improve and reengineer business processes. Recently, major IT companies have developed cognitive services to make AI technology ready to use for developing applications. Many large companies started projects to plug these services into their processes or to develop their own AI solutions based on these services. At the same time, AI researchers are discussing safety issues and identifying important sources of risks in AI solutions.

The workshop identified many potential sources for synergies between AI and BPM. On the one hand, several AI solutions can be used in the context of BPM, e.g., planning for adapting or composing business processes, machine learning for process mining and analysis, constraint reasoning for process transformation, verification, and compliance checking.

On the other hand, AI will influence the role of humans within a business process and solutions should be developed to address questions such as novel requirements on employee qualification, shared responsibilities between AI and humans, control and impact of automated decision-making.

Four full and two short papers were presented at the workshop. The focus of the papers ranged from leveraging machine learning approaches for addressing BPM problems to applying planning approaches in BPM scenarios on the one hand, and from analyzing event logs using AI techniques to finding optimal paths in business processes on the other hand.

In particular, Hinkka and colleagues addressed the problem of predicting sequences of activities in business process instances using recurrent neural networks. Ponnalagu and colleagues propose an approach leveraging process context, state, and goals to predict process performance. Høgnason and Debois leverage genetic algorithms for solving the problem of event reachability for DCR (dynamic condition response) graphs. Shing and colleagues present a pipeline for discovering workflows from unstructured natural language texts. Eshuis and Firat propose the use of fuzzy logic for modeling uncertainty in guard stage milestone (GSM) declarative artifact-centric process models. Finally, Knoch and colleagues leverage AI technologies and techniques to automatically detect material picking and placement in the assembly workflow to gather accurate data about human behavior.

The importance of the synergy between the AI and BPM fields also emerged in the keynote by Tijs Slaats. He showed how techniques for temporal logic verification can be applied for discovering declarative process models. Moreover, he also showed how these techniques have recently been used in combination with Petri nets discovery approaches in order to mine hybrid process models, which contain both a declarative and a procedural part.

Each talk was followed by an interesting discussion with the audience. Everybody agreed to continue this experience at next year's BPM conference and to take the chance to further foster the interaction between AI and BPM by targeting with the call for paper also other research areas at the intersection between the two fields.

To sum up, the workshop received a significant number of submissions and a very good attendance. It clearly showed the strong interest of the BPM community in how the AI and BPM fields can potentially fertilize each other. Of course, many of the questions raised in the call for papers remained unaddressed by this year's submissions, e.g., assessing the business risks of AI technologies, understanding the interplay of humans and robots in business processes, using AI technologies such as AI planning to create dynamic business processes. However, we believe that this only goes to show how much more can be done in the future in this research direction.

November 2018

Riccardo De Masellis
Chiara Di Francescomarino
Richard Hull
Krzysztof Kluza
Fabrizio Maria Maggi

Organization

Steering Committee

Boualem Benatallah   University of New South Wales, Australia
Chiara Ghidini   Fondazione Bruno Kessler, Italy

Marco Montali   Free University of Bolzano, Italy
Hamid R. Motahari Nezhad   Ernst and Young, USA
Arik Senderovich   Technion - Israel Institute of Technology, Israel
Biplav Srivastava   IBM T.J. Watson Research Center, USA
Heiner Stuckenschmidt   University of Mannheim, Germany

Program Committee

Fernanda Araujo Baiao   UNIRIO, Brazil
Amin Beheshti   Macquarie University, Australia
Ralph Bergmann   University of Trier, Germany
Andrea Burattin   DTU, Copenhagen, Denmark
Fabio Casati   University of Trento, Italy
Federico Chesani   University of Bologna, Italy
Raffaele Conforti   University of Melbourne, Australia
Giuseppe De Giacomo   Sapienza Università di Roma, Italy
Claudio Di Ciccio   Vienna University of Economics and Business,
                          Austria
Schahram Dustdar   TU Wien, Austria
Paolo Felli   Free University of Bolzano, Italy
Avigdor Gal   Technion, Israel
Aditya Ghose   University of Wollongong, Australia
Monika Gupta   IBM India Research Lab, India
Anup Kalia   IBM T.J. Watson Research Center, USA
Anna Leontjeva   CSIRO61, Sydney, Australia
Henrik Leopold   VU University Amsterdam, The Netherlands
Andrea Marrella   Sapienza Università di Roma, Italy
Paola Mello   University of Bologna, Italy
Fabio Patrizi   Sapienza Università di Roma, Italy
Giulio Petrucci   Google, Switzerland
Manfred Reichert   University of Ulm, Germany
Niek Tax   TU/e Eindhoven, The Netherlands
Irene Teinemaa   University of Tartu, Estonia
Han van der Aa   VU University Amsterdam, The Netherlands
Hagen Voelzer   IBM Zurich Research Lab, Switzerland

Enhancing Process Data in Manual Assembly Workflows

Sönke Knoch$^{(\boxtimes)}$, Nico Herbig, Shreeraman Ponpathirkoottam, Felix Kosmalla, Philipp Staudt, Peter Fettke, and Peter Loos

German Research Center for Artificial Intelligence (DFKI),
Saarland Informatics Campus, Saarland University, Saarbrücken, Germany
soenke.knoch@dfki.de

Abstract. The rise of Industry 4.0 and the convergence with BPM provide new potential for the automatic gathering of process-related sensor information. In manufacturing, information about human behavior in manual assembly tasks is rare when no interaction with machines is involved. We suggest technologies to automatically detect material picking and placement in the assembly workflow to gather accurate data about human behavior. For material picking, we use background subtraction; for placement detection image classification with neural networks is applied. The detected fine-grained worker activities are then correlated to a BPMN model of the assembly workflow, enabling the measurement of production time (time per state) and quality (frequency of error) on the shop floor as an entry point for conformance checking and process optimization. The approach has been evaluated in a quantitative case study recording the assembly process 30 times in a laboratory within 4 h. Under these conditions, the classification of assembly states with a neural network provides a test accuracy of 99.25% on 38 possible assembly states. Material picking based on background subtraction has been evaluated in an informal user study with 6 participants performing 16 picks, each providing an accuracy of 99.48%. The suggested method is promising to easily detect fine-grained steps in manufacturing augmenting and checking the assembly workflow.

Keywords: Manual assembly · Computer vision · BPM · Industry 4.0

1 Introduction

The current trend of automation and data exchange known under the term Industry 4.0 [6,9] addresses the convergence of the real physical and the virtual digital world in manufacturing. It comprises the introduction of cyber-physical systems (CPS), Internet of Things (IoT) and cloud computing in a fourth industrial revolution, where manufacturing companies face volatile markets, cost reduction pressure, shorter product lifecycles, increasing product variability, mass customization leading to batch size 1 and, with rising amounts of data, developments towards a smart factory [1].

© Springer Nature Switzerland AG 2019
F. Daniel et al. (Eds.): BPM 2018 Workshops, LNBIP 342, pp. 269–280, 2019.
https://doi.org/10.1007/978-3-030-11641-5_21

To plan, construct, run, monitor and improve a flexible assembly work station or CPS tackling the challenges of Industry 4.0, engineers and managers require detailed information about the assembly workflow in the life-cycle phases of a CPS, e.g. [16]. In workflows with manual tasks, this information contains data on human behavior, such as grasp distance, assembly time and the effect the workplace design has on the assembly workflow. It can be used to plan and construct efficient assembly work stations and receive information on the executed workflows to establish a continuous improvement process (CIP/Kaizen) within the organization. The data has to be approximated or gathered manually consuming time and money.

We see a large potential in the technical support and automatization of data gathering processes providing information about a workflow's execution regarding time and quality. The accurate detection of human behavior during the assembly workflow, and as a consequence, the generation of meaningful process logs about that workflow, is a challenging task. On the hardware side, the selection of appropriate sensors, their integration into the assembly system and the robustness against the conditions in the manufacturing context have to be considered. On the software side, these sensors have to be used to detect activities reliably, deliver results fast and provide complete information in a homogeneous data format. If the physical setup, the software configuration, and the operation of such a system support workflow execution efficiently, this will add value to the organizations deploying them: (1) Process models will be enriched with detailed information about the assembly steps. (2) Live workflow tracking enables conformance checking online or offline. (3) Further analysis of workflow traces can be used to adapt and optimize the workflow execution.

In a first iteration, an artifact, fast and easy to set up in terms of configuration and instrumentation, was developed integrating multiple sensors to analyze hand and body posture, as well as material picking. For this we used ultrasonic sensors, infrared, RGB+Depth and RGB cameras [7]. This proof-of-concept implementation was then extended and evaluated in a case study with 12 participants in a second iteration [8]. In this work, we reduce the sensor setting to the most promising sensors (2 RGB cameras and 1 hand sensor) applying new methods from computer vision and machine learning to achieve a high resolution in recognizing assembly tasks. Further on, events about task-related events are correlated to process tasks in a BPMN model. Now, manual assembly workflows are controlled by a model that can be created, configured and adapted in a graphical BPMN editor to deploy new workflows, facilitate improvements and support the worker appropriately, thereby achieving a new level of flexibility.

2   Related Work

Related work can be found at the intersection of BPM with cognitive computing, context-awareness and human activity recognition, here mainly vision-based.

Cognitive BPM comprises the challenges and benefits of cognitive computing in relation to traditional BPM. Hull and Motahari Nezhad [4] suggest a cognitive process management system (CPMS) to support cyber-physical processes enabled by artificial intelligence (AI). Marrella and Mecella [11] propose the concept of such a CPMS which automatically adapts processes at run-time, taking advantage of the AI's knowledge representation and reasoning. Similar to our system the alignment between real physical and expected modeled behavior can be measured. In *context-aware BPM*, the user behavior is set in relation to situations affecting the behavior. Transferred to our approach, user behavior corresponds to detected worker activities in the context of a concrete assembly task from a process model instance executed at a concrete assembly work station. Jaroucheh, Liu, and Smith [5] apply linear temporal logic and conformance checking from process mining to compare real with expected user behavior. Since the collection of high resolution data for the aforementioned systems is not a trivial task, augmenting the process with information from cognitive computing platforms is the focus within this research activity.

Vision-based human activity recognition (HAR) mainly focuses on 2D video data and applies various machine learning methods (for an overview see [12]). HAR applications in manufacturing are sparse but occur more frequently since the rise of IoT, industrial internet and Industry 4.0. Within the field of human robot collaboration, [10] use 3D video data to determine the hand position of a worker sitting in front of an assembly work table and collaborating with an assistive robotic system applying Hidden Markov Models (HMM). Two cameras are mounted to the assembly workstation and are calibrated to each other. High calibration effort facilitates the application of the method only in restricted set-ups. Similarly, [13] apply hierarchical HMMs in a multimodal sensor setting within the same domain. They combine RGB-D (Kinect 2) and IR (Leap Motion) cameras to detect fine-grained activities (e.g. assembly, picking an object, fixing with a tool) and gestures (e.g. pointing, thumb up) and analyze the results of the sensors respectively and in combination. Combination of sensors show the best average recognition results for activities in most cases. Unlike these approaches, focusing on gesture detection, we focus on the detection of assembly states in form of image classes allowing the connection to state transitions in process models. It is an example how AI can be applied to ensure process quality and adherence to time constraints.

In the field of *HAR using wearable sensors*, the authors in [2] use convolutional neural networks (CNN) on sequential data of multiple inertial measurement units (IMU). Three IMUs are worn by workers on both wrists and the torso. This way, an order picking process in warehouses can be analyzed, fusing data from all sensors to classify relevant human activities such as walking, searching, picking and scanning. In a similar fashion [15] apply different on-body sensors (RFID, force-sensitive resistor (FSR) strap, IMU) in a "motion jacket" worn by workers, and environmental sensors (magnetic switches and FSR sensors) to detect activities in the automotive industry, such as inserting a lamp, mounting a bar using screws and screwdriver, and verifying the lamp's adjustment. We decided to use contactless activity detection and avoid the instrumentation of workers, because we expect limited user acceptance when wearing sensor equip-

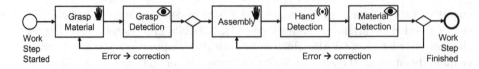

Fig. 1. Flow chart representing the activity detection process within one work step.

ment. In addition, using wearables the reloading of batteries and mechanical signs of fatigue during usage may be issues the operator wants to avoid. Thus, an easy and light set-up is suggested applying state-of-the-art AI technology to the problem of workflow tracking in manual assembly enabling conformance checking and optimization.

3 Concept

BPM has a high potential to support the implementation and adaption of process models which control and sense the assembly workflow. We provide a concept showing how events from the shop floor are correlated to tasks in the model, allowing the supervision of time and quality constraints defined ex-ante in the form of reference times and material states. In addition, the collection of timing information and error frequency supports the optimization of assembly workflows, e.g. through conformance checking. To enable this kind of supervision and optimization, the detection of critical activities within one work step is necessary.

3.1 Activity Detection

Figure 1 shows the three-tired activity detection: first, grasps to container boxes are detected to analyze which part the worker assembles next (*Grasp Detection*). After performing the actual assembly step, the worker places his hands next to the workpiece, where they are detected by the *Hand Detection* module. At this point in time, an image is captured, as there cannot be any occlusions or motion blur. This image is then used for *Material Detection*, analyzing which workstep the assembly led to, i.e. if the step was correct and the next assembly step can be taken or whether one of several possible failure states is reached and a recovery strategy needs to be applied.

Grasp Detection involves the appearance of a foreground object (user's hand) over a stationary background (formed by the image of the container box) seen by the camera observing the container box from the top. It is assumed that this kind of activity occurs when a part is removed from its designated container box. Therefore, activity zones have to be marked within the image to detect the worker's hand. In the process model, material grasping and grasp detection form the first step after receiving the instruction for the current assembly task. Such a system facilitates the supervision of assembly workstations equipped with a lot of similar-looking materials without the effort of hard-wiring the system with the workstation, as in case of classical pick-supervision.

Hand Detection allows the active confirmation of work steps by the worker without reaching to distant touch screens or buttons. With correct and complete execution, this enables positive feedback and forms the point in time when image data for material detection is gathered. This approach ensures that no hands occlude the workpiece and that the workpiece does not move, thereby avoiding motion blur. For the worker, this procedure offers more safety, since subsequent work steps are only started after successful confirmation of the previous ones. In the case of a mistake, the operator is supported with fine-grained assistance.

Material Detection and State Classification is then used on the acquired image to confirm whether the material parts being grasped are correctly assembled in the desired way. This involves verifying if each part is correctly located in the expected orientation. A camera is used to monitor the assembly region (where assembly of product is being performed) which facilitates verification and quality assurance based on the image information. Each material state in the process model, including all error states, is considered as a separate class. For each of these classes, a set of images is acquired to train a classifier. Whenever an image is captured after hand detection, this image is classified to the corresponding state in the BPM, which allows jumping to the next step in the worker guidance system (WGS), a system showing instructions about the current assembly task on a screen, or intervening in case of error states. The training images can be acquired either in a separate gathering process (as done for the presented evaluation), or confirmations/correction input to the WGS can be used to label images with corresponding classes, thus supporting cost-neutral data capture in the wild.

3.2   Process Model

The process model controls the assembly workflow, communicating with activity detection software and the worker guidance system. The model contains the logic and is configured with a set of properties in a graphical BPMN editor, which allows the coupling of events to the model without the specification of concrete devices and due to the generic implementation without further software implementation effort. The binding between model and components in the assembly workflow is based on a topology describing the value range of the property variables necessary to run a concrete workflow.

Messages are divided into *instructions* published by the model controlling software components and *activities* published by software components detecting material picking and placing or user input from the WGS. Every message contains the business key identifying the process instance and a time stamp. Instructions are subdivided according to the destination (*WGS/ActivityDetection*).

A *WGS instruction* contains the mandatory property work step index (used to gather descriptive information and media to explain and present the current assembly task to the user), a list of expected materials, and a list of expected orientations. Optionally, an error is set when the list of found materials or found orientations does not reflect expected ones, e.g. when *WrongMaterial* or *WrongOrientation* occurs. Then, a correction instruction is presented on the WGS.

An *ActivityDetection instruction* contains the expected materials, orientations and the assembly activity, e.g., $grasp_t$, $insert_t$, or $rest_t$, where t is the reference time for one activity. The time t can be used to measure the deviation from the manufacturing time planned. To control software components, we introduce the action state that contains the target state the respective component should occupy, e.g. *start* or *stop* the detection of an activity.

An *ActivityDetection event* is sent when the activity detection was started by the controlling process instance and an activity was detected. Then, the activity detection module delivers the properties material (expected & detected), orientation (expected & detected), activity and a type, which indicates if the source of the detection was from an *automatic* (activity detection) or *manual* (confirm button) origin. Since activity detection might fail, as a fallback it is always possible to confirm an activity by pressing a button on the WGS screen.

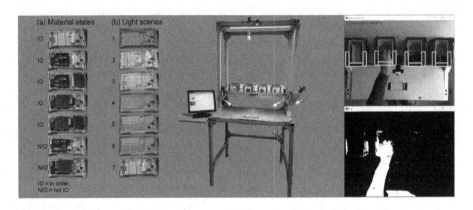

Fig. 2. Assembly workstation equipped with 1 touch screen, 2 cameras and 6 light sources (middle); example material states and 7 light scenes for one part (left); grasp detection using background subtraction with the 'U'-shaped activity zones (right).

4   Implementation

The implementation of the concept presented in the previous section implies an example assembly workflow set up in the lab (Subsect. 4.1), an implemented BPMN 2.0 process model controlling and tracking this workflow (Subsect. 4.2), and three activity detection modules detecting material grasping, hands, and material positioning (Subsects. 4.3–4.5).

4.1 Assembly Workflow and Apparatus

To test and iteratively improve our system, we designed an assembly workflow consisting of four assembly steps. The product to be assembled consists of three printed circuit boards (PCB) to be connected and placed in one 3D printed case. The four materials are provided in small load carriers (SLC), boxes common in manufacturing when dealing with small parts. In steps 1 to 3, material is removed from the SLCs and assembled in the work area on top of the workbench in front of the worker. In step 1 the case is removed and placed with the open side up in the work area. In step 2 one PCB is removed and inserted into the case. In step 3 the two remaining PCBs are removed, connected and inserted into the case. The removal of two parts in parallel is common practice to improve the efficiency in assembly workflows. Finally, within step 4 the assembled construct is removed from the work area and inserted into a slide heading to the back of the assembly workstation. The whole workflow is supported by a basic worker guidance system (WGS) running on a touch screen, showing textual instructions and photos of the relevant materials and of the target state in the respective assembly step.

The assembly workstation, shown in Fig. 2, is made from cardboard prototyping material and instrumented with two consumer-electronics RGB cameras: (1) Logitech C920 HD Pro and (2) Logitech BRIO 4K Ultra HD. The first camera is mounted on top of the assembly station and pointed at the four SLCs loaded with material. The second camera is mounted to the shelf carrying the SLCs and is aimed at the work area.

4.2 Process Model

A process model, modeled in the BPMN language and shown in Fig. 3, controls the activity detection modules and the WGS. It consists of service, send and receive tasks. Service tasks initialize a new work step and can be used to set the relevant parameters necessary for the subsequent work step. Here, we set the index i of the step. Send tasks control components necessary to execute the work step and generate *instruction* messages sent to a target component c. Receive tasks wait for acknowledge events confirming an *activity* a in the current work step.

Send tasks that initialize the activity detection for one concrete material from graph 1 trigger the initialization of the process modeled in graph 2. There, per material, the grasp, hand and material detection are started and stopped in successive order. Within a send task of type start, the activity to observe is defined. When an activity occurs two options exist: (1) the correlated event occurs as expected and the activity is stopped by the subsequent send task, or (2) a correlated activity occurs that detects a behavior that does not match the expected state. Then, the exception handling is started, instructing the worker to correct his activity. Finally, when the process modeled in graph 2 ends, the receive task in graph 1 is informed and continues.

Error handling is important and allows the system to intervene when states are detected that have been classified as erroneous. 38 classes including error states have been generated with the existing material parts by rotating parts on their positions. The BPMN model was re-designed to instruct the worker during the arrangement of materials such that images can be taken in the training phase of the system covering a variety of lighting conditions.

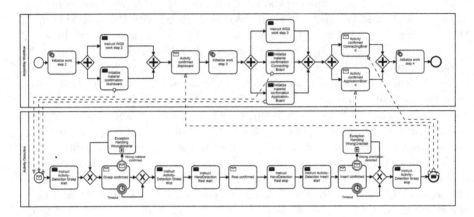

Fig. 3. Assembly workflow snippet (top) and activity detection process (bottom) modeled in BPMN.

4.3   Grasp Detection

The activity zones, shown in Fig. 2, are marked outside the interior of the container boxes because a background subtraction algorithm adapts to a changing background, e.g. when a part is removed from its container box. The activity zones are 'U'-shaped since the direction of approach of the hand is not always perpendicular to the breadth of the container box. The red regions are marked during system setup and the yellow activity zones are automatically identified as a corollary.

The procedure of detecting the start and end of a grasping activity is as follows: When the number of foreground pixels exceeds the total number of pixels in the activity zone by a certain user-defined percentage (40%), activity is said to be detected. To filter noisy indications, the start and end of an activity is detected once there is a considerable number of video frames. Hence a recent history of frames is always observed: Only when a certain user-defined percentage of frames in the recent history of frames contain activity/inactivity is grasp start/end indicated.

4.4 Hand Detection

After the user has assembled the grasped material, he or she actively confirms the work step by putting his hands on designated areas of the workstation close to the assembly area, which have electrically conductive metal plates integrated into the assembly worktop. Those plates are connected to a capacitive sensor. The sensor measures the capacitance of the capacitor, which is formed by the electrodes "metal plate" and "operator". By laying down both hands, the measured capacity reaches a characteristic value, which initializes the manual confirmation of the operator's work step and forms the point in time when the photo for material detection is taken. As these metal plates are integrated directly next to the assembly area, there is no need to reach distant displays or buttons.

4.5 Material Detection and State Classification

As described in Sect. 3, to detect whether materials are assembled correctly, *State Classification* has been implemented based on the image captured at hand detection. Images for every state of the BPM are gathered and a classifier is trained, predicting the state in the BPM based on the image captured during hand detection. Vast improvements in image classification results using deep convolutional neural networks have been achieved in recent years; therefore, we decided to also use convolutional neural nets for our task. Since our task is comparatively simple, we do not apply complex and extremely deep convolutional networks like ResNet [3], but instead design our own very simple network: we use 5 (convolutional, convolutional, max pooling) blocks with filter size of 3 by 3, pool size of 2 by 2, and relu activation, followed by a dense layer of 512 (relu activation), and the final classification dense layer performing softmax. Dropout is used to reduce overfitting.

   To avoid long data gathering phases, we perform an initial training procedure based on parts of the ImageNet dataset [14]. For this, we downloaded a total of 419 classes of the set, including 178 classes we considered roughly related to PCB assembly, such as 'electrical circuit', 'printed circuit', and 'circuit board'. We resize all images to 224×224 pixels and train the network for 300 epochs using a batch size of 64, RMSProp optimization, and a learning rate of 0.0001. These weights are stored and used as a basis for the training process on the images specific to our assembly task. This pre-training should reduce the amount of images required, as basic features like edge, shape, and color detection can already be learned from the ImageNet data. The dense and classification layers at the end are randomly initialized when fine-training on our dataset, as these can be considered specific to the ImageNet data.

5 Evaluation

We conduct an evaluation, testing the individual parts of the activity detection module and analyzing to what extent these can be used to automatically gather insights into the assembly process to optimize it in later steps.

5.1   Grasp and Hand Detection

The grasp and hand detection is evaluated using a series of experimental runs involving 6 users (1 f, 5 m) with different hand sizes (circumference: $\bar{x} = 21.58$, $\sigma = 1.64$ cm; length: $\bar{x} = 18.83$, $\sigma = 1.77$ cm; span: $\bar{x} = 21.67$, $\sigma = 2.29$ cm). During the experiment, the user grasps a part, places it on the worktable and confirms by resting the hands. The user then returns the part and confirms by resting the hands. This procedure is repeated for each part and box with right and left hand alternating 2 times per part in four repetitions comprising all parts. This leads to a 2 (hands) $*$ 2 (remove and return part) $*$ 4 (parts) $*$ 4 (repetitions) = 64 grasp start, stop and rest events. The activity start and stop are monitored by a supervisor with an annotating application to generate ground truth information. The rest detection failed only in 2 cases where two users curved their palms, leading to no sensor response. In total, an accuracy of 99.22% was achieved. For grasp detection, 2 pairs of start and stop events failed where one user approached from angles where the activity zone is least disturbed, such that the threshold of 40% foreground pixels was not reached in the activity zone. In total, an accuracy of 99.48% was achieved.

5.2   Material Detection and State Classification

Training of a neural network usually requires manually labeling a large amount of data samples. For the system described in this paper, we used the BPMN model to automatically label the assembly steps: a custom model which, in combination with the worker guidance system, not only instructed the participant to perform the usual assembly steps, but also directed her to generate erroneous states (i.e. wrong part placement or orientation). This made it possible to semi-automatically (the participant still had to place her hands next to the workpiece after each step) take a picture of every state of the BPMN-model and label it at the same time. Since different lighting conditions easily occur in a real life setting, we also took these into account during the training data acquisition.

To simulate various lighting condition, six Philips Hue lights were placed in different positions around and on the assembly workstation. Using the Philips Hue API allowed us to pragmatically change the lighting scenarios, which results in different shadows on the pictures of the workpieces. During data gathering, the remaining lighting conditions were kept stable (shutter closed, room and workstation lights on). We used seven different simulated lighting conditions (including all lights off, i.e. room lighting) per assembly state. The rationale behind this was to allow for different lighting preferences of the individual workers. Each iteration consisted of 19 states, generating 19 $*$ 7 images and with a duration of approximately 8 min (25 s per assembly step). Two types of iterations were considered, orienting all parts in two directions (left and right) leading to 38 classes in total.

With this setting, we were able to generate 3990 images in 30 runs within 4 h. To avoid irrelevant parts of the surroundings, such as hands or tools, being present in the image, the camera has to be fixed. This allows focusing on the rectangular area that contains the assembled parts which are relevant for classification and detection, which is a requirement to run our system. Overall the goal is to quantify how well the approach works to determine whether material was assembled correctly. For this, we use 50% of the labeled image data for training, 25% for validation, and 25% for testing.

As described in Sect. 4.5, the neural network used to classify the various assembly states was pre-trained on a subset of ImageNet, and then all except for the dense layers are fine-tuned on the training data specific to our task. For this, we again resize all input images to 224 by 224 pixels and train for 25 epochs using a batch size of 8. As expected due to the simple nature of the problem (in comparison to competitions like the ImageNet Large Scale Visual Recognition Challenge (ILSVRC)), the classification performance is very good: a test accuracy of 99.25% was achieved on this 38-class problem.

6   Conclusion

Methods from AI, such as computer vision and machine learning, were tailored to an Industry 4.0 use case and may increase an organization's competitiveness through awareness of error rates and timepass enabling backtracking and intermediate intervention. We have shown that an easy-to-set-up tool set of 2 cameras, 1 capacitive sensor, 6 light sources and activity detection software trained within 4 h has the ability to generate accurate sensor events. Viewing these data through a "process lens," the measurement of time and quality in manual assembly workflows, and thus the optimization of processes, is enabled.

Since it was shown how an Industry 4.0 organization can keep track of its manual assembly workflows, it will be interesting to investigate the potential of unsupervised methods in combination with our approach to discover workflows automatically towards a novel concept of 'video-to-model'. New methods to discover assembly workflows in planning and construction phases, and to monitor or check their conformance online and offline, will provide valuable input for process mining.

Acknowledgments. This research was funded in part by the German Federal Ministry of Education and Research (BMBF) under grant number 01IS16022E (project BaSys4.0). The responsibility for this publication lies with the authors.

References

1. Cavanillas, J.M., Curry, E., Wahlster, W. (eds.): New Horizons for a Data-Driven Economy: A Roadmap for Usage and Exploitation of Big Data in Europe. Springer, Cham (2016). https://doi.org/10.1007/978-3-319-21569-3
2. Grzeszick, R., et al.: Deep neural network based human activity recognition for the order picking process. In: Proceedings of the 4th International Workshop on Sensor-based Activity Recognition and Interaction. iWOAR 2017, pp. 14:1–14:6. ACM Rostock (2017)

3. He, K., et al.: Deep residual learning for image recognition. In: Proceedings of the IEEE Conference on Computer Vision and Pattern Recognition, pp. 770–778 (2016)
4. Hull, R., Motahari Nezhad, H.R.: Rethinking BPM in a cognitive world: transforming how we learn and perform business processes. In: La Rosa, M., Loos, P., Pastor, O. (eds.) BPM 2016. LNCS, vol. 9850, pp. 3–19. Springer, Cham (2016). https://doi.org/10.1007/978-3-319-45348-4_1
5. Jaroucheh, Z., Liu, X., Smith, S.: Recognize contextual situation in pervasive environments using process mining techniques. J. Ambient Intell. Humaniz. Comput. **2**(1), 53–69 (2011)
6. Kagermann, H., et al.: Recommendations for implementing the strategic initiative INDUSTRIE 4.0: Securing the Future of German Manufacturing Industry; Final Report of the Industrie 4.0 Working Group. Forschungsunion (2013)
7. Knoch, S., et al.: Automatic capturing and analysis of manual manufacturing processes with minimal setup effort. In: International Joint Conference on Pervasive and Ubiquitous Computing. UbiComp, pp. 305–308. ACM, Heidelberg, September 2016
8. Knoch, S., Ponpathirkoottam, S., Fettke, P., Loos, P.: Technology-enhanced process elicitation of worker activities in manufacturing. In: Teniente, E., Weidlich, M. (eds.) BPM 2017. LNBIP, vol. 308, pp. 273–284. Springer, Cham (2018). https://doi.org/10.1007/978-3-319-74030-0_20
9. Lasi, H., et al.: Industrie 4.0. Wirtschaftsinformatik **56**(4), 261–264 (2014)
10. Lenz, C., et al.: Human workflow analysis using 3D occupancy grid hand tracking in a human-robot collaboration scenario. In: IEEE/RSJ International Conference on Intelligent Robots and Systems, pp. 3375–3380, September 2011
11. Marrella, A., Mecella, M.: Cognitive business process management for adaptive cyber-physical processes. In: Teniente, E., Weidlich, M. (eds.) BPM 2017. LNBIP, vol. 308, pp. 429–439. Springer, Cham (2018). https://doi.org/10.1007/978-3-319-74030-0_33
12. Poppe, R.: A survey on vision-based human action recognition. Image Vis. Comput. **28**(6), 976–990 (2010)
13. Roitberg, A., et al.: Multimodal human activity recognition for industrial manufacturing processes in robotic workcells. In: Proceedings of the 2015 ACM on International Conference on Multimodal Interaction. ICMI 2015. pp. 259–266. ACM, Seattle (2015)
14. Russakovsky, O., et al.: ImageNet large scale visual recognition challenge. Int. J. Comput. Vis. (IJCV) **115**(3), 211–252 (2015)
15. Stiefmeier, T., et al.: Wearable activity tracking in car manufacturing. IEEE Pervasive Comput. **7**(2), 42–50 (2008)
16. Thoben, K.-D., Pöppelbuß, J., Wellsandt, S., Teucke, M., Werthmann, D.: Considerations on a lifecycle model for cyber-physical system platforms. In: Grabot, B., Vallespir, B., Gomes, S., Bouras, A., Kiritsis, D. (eds.) APMS 2014, Part I. IAICT, vol. 438, pp. 85–92. Springer, Heidelberg (2014). https://doi.org/10.1007/978-3-662-44739-0_11

Modeling Uncertainty in Declarative Artifact-Centric Process Models

Rik Eshuis[✉][iD] and Murat Firat[iD]

School of Industrial Engineering, Eindhoven University of Technology,
Eindhoven, Netherlands
{h.eshuis,m.firat}@tue.nl

Abstract. Many knowledge-intensive processes are driven by business entities about which knowledge workers make decisions and to which they add information. Artifact-centric process models have been proposed to represent such knowledge-intensive processes. Declarative artifact-centric process models use business rules that define how knowledge experts can make progress in a process. However, in many business situations knowledge experts have to deal with uncertainty and vagueness. Currently, how to deal with such situations cannot be expressed in declarative artifact-centric process models. We propose the use of fuzzy logic to model uncertainty. We use Guard-Stage-Milestone schemas as declarative artifact-centric process notation and we extend them with fuzzy sentries. We explain how the resulting fuzzy GSM schemas can be evaluated by extending an existing GSM engine with a tool for fuzzy evaluation of rules. We evaluate fuzzy GSM schemas by applying them to an existing fragment of regulations for handling a mortgage contract.

1 Introduction

Many business processes are performed by knowledge workers, who are responsible for making decisions and adding information, for instance by analyzing data. These knowledge-intensive processes [6] are centered around certain key business entities to which information is added and about which decisions are made, for instance a production order or a request for quotes. Artifact-centric process models have been proposed to model such knowledge-intensive processes [18]. Business artifacts combine data and process aspects of key entities in a holistic way [14]. To allow flexibility, such knowledge-intensive processes are often modeled in a declarative way with rules [9], rather than a procedural way.

Business artifacts contain business rules that specify conditions under which actions are taken [5]. However, these rules are modeled in a classical way, assuming clear-cut boundaries between different states of the world. The variables in the rules may in reality have some degree of uncertainty or may not be described precisely. This is caused by either having insufficient data or lacking concrete definitions of the considered concepts. For example, is there a unique loan amount that distinguishes small loans and big loans? Or in case we only have a small sampling of the requested loan amounts, then how reliable will it be to use a crisp number as the boundary for small and big loans?

© Springer Nature Switzerland AG 2019
F. Daniel et al. (Eds.): BPM 2018 Workshops, LNBIP 342, pp. 281–293, 2019.
https://doi.org/10.1007/978-3-030-11641-5_22

Several approaches for modeling uncertainty exist, notably probabilistic and fuzzy modeling [12]. Probabilistic models rely on frequency-based quantitative analysis while fuzzy models deal with uncertainties in a qualitative way, modeling the imprecise way of human reasoning [12,15]. We propose to model the uncertainties in decision-intensive processes using fuzzy rather than probabilistic models since there is a lack of rich data to generate probabilistic models and a need to support the use of knowledge from human experts.

Fuzzy models describe the universe of discourse using *linguistic* terms with corresponding linguistic labels instead of specifying precise numerical values (including 0–1 binary representations) [12]. Each linguistic label has a membership function defined in the universe of discourse. These linguistic labels may be "large", "high", or "most", depending on the context. Another important point of use of fuzzy modeling is to model the decision-making of knowledge workers which leads to the extraction of their domain expertise. Modeling expert knowledge properly is clearly beneficial to enable automated decisions of high quality.

For instance, suppose a loan is classified as large if it exceeds 100\$, and not large otherwise. In reality, it can be uncertain what exactly comprises the class of large loans. A loan of $99K\$$ is not large according to the classification, but obviously "more" large than a loan of $10K\$$. Knowledge workers may have their own understanding of when a loan is large or small, which fuzzy modeling can capture [15].

The goal of this paper is to explore the use of fuzzy modeling for declarative artifact-centric process models. We define how to "fuzzify" such artifact models by using fuzzy rules. As host notation for modeling declarative artifact-centric schemas, we use Guard-Stage-Milestone (GSM) schemas [11]. The language GSM schemas has inspired CMMN, the OMG standard on Case Management [4], so the results can provide a stepping stone for incorporating fuzzy reasoning in CMMN schemas. As we explain in Sect. 4, existing GSM engines can support fuzzy GSM schemas by invoking fuzzy rule evaluators.

The remainder of this paper is organized as follows. Section 2 introduces a running example that is revisited in the remainder of this paper. Section 3 defines GSM schemas and introduces fuzzy modeling. Section 4 explores how fuzzy modeling can be applied to GSM schemas, using the running example as illustration. Section 5 evaluates the use of fuzzy GSM schemas by modeling a fragment of a complex regulated process for handling mortgages. Section 6 discusses related work. Section 7 ends the paper with conclusions and an outlook for future work.

2   Running Example

We use a running example based on a fragment of a real-world process from an international technology company, in which business criteria for partner contracts are assessed [8]. The process fragment has the following behavior. First, data is gathered needed to perform the assessment. Next, three activities are

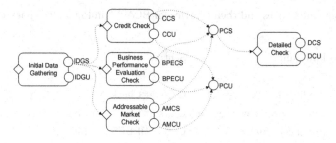

Fig. 1. Business Criteria Assessment process (BCA^{base}) [8]

Table 1. Stages and guards for BCA^{base} in Fig. 1

Stage	Guard
Initial Data Gathering	E:AssessmentRequest
Credit Check	IDGS
Business Performance Evaluation Check	IDGS $\land$ employee_count ≥ 300
Addressable Market Check	IGDS $\land$ annual_revenue $\geq \$500K$
Detailed Check	PCS

performed in parallel as a pre-check. First, the credit is checked to ensure that the credit limit of the partner is still valid. Second, the past performance of the partner is evaluated and checked, but only if the partner has more than 300 employees. Third, the market addressed by the partner is assessed if the annual revenue exceeds $500K. If the three parallel checks are successful, the pre-check succeeds and a detailed check is performed, which may either succeed or fail.

Figure 1 shows the lifecycle part of the GSM schema for this Business Criteria Assessment process BCA [8]. Rounded rectangles denote stages, in which work is performed. Circles denote milestones, which are business objectives achieved by completing the work in a stage to which a milestone is attached or by another external event. Diamonds denote guards, which specify conditions called sentries under which a stage opens. Sentries are also used for milestones to specify the conditions when they are achieved. Table 1 lists the sentries (guards) of the defined stages and Table 2 the sentries of those defined milestones that are revisited in the sequel of this paper; the full list for all milestones can be found elsewhere [8]. Dashed arrows in Fig. 1 denote dependencies between stages and milestones caused by sentries: for instance, the sentry (guard) of stage Detailed Check states that the stage is opened if milestone PCS has been achieved, so stage Detailed Check depends on milestone PCS.

The GSM schema in this paper is modeled in a crisp way: the sentries specify exact conditions for each stage to be opened and each milestone to be achieved. In reality, the knowledge worker may use more fluid conditions. In Sect. 4 we explore how fuzzy logic can model such conditions.

Table 2. Some milestones and their sentries for BCA^{base} in Fig. 1. ';' separates different sentries

Milestone	Full name	Sentry
...	...	...
CCS	Credit Check Successful	C:Credit Check ∧ rating ≥ 8
CCU	Credit Check Unsuccessful	C:Credit Check ∧ rating < 8
PCS	Pre-checks Successful	CCS ∧ BPECS ∧ AMCS ;
		CCS ∧ employee_count < 300 ∧ AMCS ;
		CCS ∧ BPECS ∧ annual_revenue < $500K$;
		CCS ∧ employee_count < 300 ∧ annual_revenue < $500K$
PCU	Pre-checks Unsuccessful	CCU ∨ BPECU ∨ AMCU
..	...	...

3 Preliminaries

We introduce GSM schemas in more detail, as well as fuzzy modeling.

3.1 GSM Schemas

This section presents formal definitions for the variant of GSM schemas used in this paper. Given the focus on the integration of sentries with fuzzy modeling, we have chosen a lightweight GSM variant [8] in which there is no hierarchy and executions are monotonic, i.e., each stage and each milestone changes value only once. However, the classic GSM schemas [5] can be extended in a similar way. Given the restricted space, the presentation is concise, explaining those GSM details that are important to understand the remainder of this paper. For an extended introduction to classic GSM schemas, we refer the interested reader to other papers [5,7,11].

A GSM schema Γ consists of attributes, subdivided into data attributes, stage attributes and milestone attributes. The latter two represent the status of stages and milestones. If a stage attribute is true (false), the stage is open (closed). If a milestone attribute is true (false), the milestone is achieved (invalid). Each data attribute a has a type $type(a)$ which is scalar, e.g., string, character, integer, float, etc. Status attributes have type Boolean.

We assume a propositional condition language $\mathcal{C}$ that includes fixed predicates over scalars (e.g., '≤' over integers or floats), and Boolean connectives. The condition formulas may involve stage, milestone, and data attributes. A status attribute will take the value *True* if one of its sentries goes true.

A *sentry* ψ defines a condition for a status attribute, i.e., a stage or milestone, to become true. A *sentry* ψ has one of the three forms: "φ", "C:S", or "C:$S \wedge \varphi$", where φ is a condition formula ranging over the attributes of Γ. Here "C:S" is called the *completion event* for stage S. Also, C:S (if present) is the *completion event* for ψ and φ (if present) is the *formula* for ψ.

Definition 3.1. A GSM *schema* is a tuple $\Gamma = (\mathcal{A} = \mathcal{A}_D \cup \mathcal{A}_S \cup \mathcal{A}_m, \mathcal{E}, sen)$ where:

- $\mathcal{A}$ is a finite set of attributes.
- $\mathcal{A}_D$ is a finite set of data attributes.
- $\mathcal{A}_S$ is a finite set of stage attributes.
- $\mathcal{A}_m$ is a finite set of milestone attributes.
- $\mathcal{E} = \{ \text{C:}S \mid S \in \mathcal{A}_S \}$ is the set of stage completion events.
- The *sentry assignment sen* is a function from $\mathcal{A}_S \cup \mathcal{A}_m$ to sets of sentries with formulas in the condition language $\mathcal{C}$ ranging over $\mathcal{A}$. Each element of set $sen(v)$ for $v \in \mathcal{A}_S \cup \mathcal{A}_m$ is called a *sentry* of v.

Sentries define dependencies between stage and milestone attributes of Γ. For $a_1, a_2 \in \mathcal{A}_S \cup \mathcal{A}_m$, a dependency (a_1, a_2) signifies that there is a sentry of a_2 that references a_1. The dependencies are visualized as dashed arrows in Fig. 1.

Definition 3.2. For a GSM schema $\Gamma = (\mathcal{A}, \mathcal{E}, sen)$ a *snapshot* is a mapping σ from $\mathcal{A}$ into values of appropriate type. For stage and milestone attributes, the only permitted values are *False* and *True*. Initially, all stage and milestone attributes are *False*.

Let $\Gamma = (\mathcal{A}, \mathcal{E}, sen)$ be a GSM schema. Let $\psi = \text{C:}S \wedge \varphi$ be a sentry for a status attribute $v \in \mathcal{A}_S \cup \mathcal{A}_m$, so $\psi \in sen(v)$. Given a snapshot σ of Γ and a stage completion event $\text{C:}S'$ that occurs, $\sigma \models \psi$ denotes that $\text{C:}S = \text{C:}S'$ and $\sigma \models \varphi$, so σ satisfies the formula φ of sentry ψ.

If in a snapshot a stage completion event occurs of an open stage, then several sentries may become satisfied. If the sentry of a status attribute becomes satisfied, the status attribute becomes true. Then sentries referencing the status attribute may also become true, which in turn means that the status attributes that own the sentries become true. For instance, stage Credit Check is open and milestones BPECS and AMCS have been achieved, then the completion event C:Credit Check with payload $\langle$rating, 9$\rangle$ results in achieving milestone CCS, which in turn leads to achieving milestone PCS, which in turn opens stage Detailed Check. In the resulting snapshot, no sentries are satisfied and the evaluation of sentries stops and the system waits for the next stage completion event of C:Detailed Check. Thus, the evaluation of sentries has a cascading effect, which is called a Business-step, of B-step for short [5].

To properly compute B-steps, dependency graphs are used [5]. For each completion event C:S of stage S, a dependency graph $DG(\text{C:}S)$ is created whose nodes are $V = \mathcal{A}_S \cup \mathcal{A}_m$ and an edge (v_1, v_2) exists if there is a sentry of v_2 that references v_1. If stage completion event C:S occurs, then the sentries of v_1 must be evaluated before those of v_2, in order to ensure that each change in status attribute is justified when examining the starting and ending snapshots of the B-step [5].

3.2 Fuzzy Modeling

In real-world decision making, obtaining precise parameters for decision models may be hardly possible or precise values for some threshold may not be meaningful. A classical example is assessing whether the weather is hot, cold or medium based on the temperature. The same value of temperature can result in different assessments for different countries. For instance a temperature of 30° is considered hot in the Netherlands, but medium in Turkey.

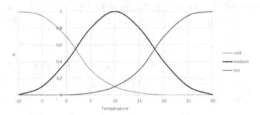

Fig. 2. Membership functions of linguistic terms of *Temperature*

Fuzzy logic has been proposed to capture this uncertainty in precision [15]. Fuzzy logic uses linguistic variables, i.e., variables whose values are different *linguistic terms* that are defined in terms of a base variable. For a linguistic variable x, each linguistic term t is defined in terms of a membership function $\mu_t : dom(X) \rightarrow [0, 1]$, which maps each value of X to the degree in which term t holds, zero meaning t is not applicable at, one that t is completely applicable. Figure 2 shows the membership functions for the linguistic terms "hot", "medium", "cold" for the linguistic variable *Temperature*. These definitions apply for countries like the Netherlands; for a country like Turkey, the membership functions would be different. For instance, $\mu_{hot,TR}(30) = 0.75 < \mu_{hot,NL}(30) = 1$, where TR represents Turkey and NL the Netherlands.

In fuzzy logic, logical statements contain fuzzy operands, indicating that uncertain or poor information is available. The two states of the satisfaction of a logical statement, namely *True* (1) and *False* (0), are extended to take value in the interval $[0, 1]$. This requires defining *fuzzy aggregation operators*, that are the fuzzy counterparts to logical operands like AND and OR.

Fuzzy Aggregation: The logical aggregation of AND and OR operators are generalized in fuzzy logic by using t-norm (conjunctive) and t-conorm (disjunctive) operators, respectively [12,15]. These are most commonly used, although there are other aggregation operators, like compensative, non-compensative, and weighted operators. By definition, any conjunctive (disjunctive) operator should come up with one value that is not greater (smaller) than any individual value in the aggregation. So the highest (smallest) value is delivered in a conjunctive (disjunctive) aggregation, by *minimum* (*maximum*) operator. Besides these basic optimistic operators for an aggregation, other operators are proposed in the literature [12,15]. In this paper, we use *product* operator for the conjunctive aggregation, and the influence of the aggregation operators on the decisions is not contained in the scope of our work.

4   Fuzzifying GSM Schemas

In this section we show how the GSM schema in Sect. 2 can be modeled using fuzzy logic, i.e., how the sentries of both stages and milestones can be fuzzified.

4.1   Linguistic Modeling of Variables

Artifact models are abstractions of real-world procedures that include uncertainty due to several reasons like insufficient sampling size, involvement of human perception and so on. Thus, in classic GSM schemas values of crisp variables can actually carry considerable uncertainty. These variables can therefore be better modeled as linguistic variables with various linguistic labels such as "low", "medium", "high", and so on. The range of values for each linguistic label may vary across different organizational units, hence it is usually not easy, perhaps even impossible, to accurately capture these different states in a quantitative manner.

First, let us consider the relation between number of employees and company size. In the crisp GSM schema in Sect. 2, the sentry "employee_count $\geq$ 300" is used to *classify a company as big*, if the result of this check is true. Now, 300 is a somewhat arbitrary number and specified by a human expert as the threshold value. However, a razor sharp classification of a company to be small or big may not be a wise thing to do. For the sake of extracting more knowledge from an expert, the following question can be asked: *"For what employee numbers is a company considered to be big to what extent?"*.

Then the answer will lead us to define the membership function in part (b) of Fig. 3. Consequently, the binary check "employee_count $\geq$ 300" is replaced with the fuzzy statement *"the company is big"* with a given degree of correctness μ_{big}(employee_count) $\in [0, 1]$, a so-called membership value that can be read from part (b) of Fig. 3.

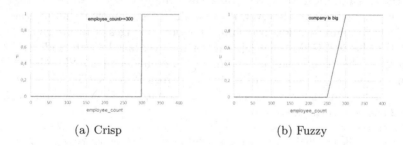

(a) Crisp   (b) Fuzzy

Fig. 3. Binary and fuzzy logic for employee count check

We define a procedure to fuzzify the crisp variables of a given GSM schema:

1: Pick a crisp variable.
2: Find different numerical checks involving that crisp variable.
3: Interpret every check linguistically.
4: Define membership functions of the linguistic labels based on either expert knowledge or a limited sample, if no expert knowledge is available.

Let us apply the above procedure for another crisp variable in our GSM example of Fig. 1.

- Crisp variable: annual_revenue
- Numerical checks: annual_revenue $\geq$ \$500K
- Linguistic interpretation: "company's financial size is *big*"
- Membership function: $\mu_{big}(ar) = \begin{cases} 0, \text{ if } ar \leq 400 \\ \frac{ar-400}{100}, \text{ if } 400 \leq ar < 500 \\ 1, \text{ if } 500 \leq ar, \end{cases}$

  where $ar = $ annual_revenue.

4.2 Condition Formula of Sentries with Fuzzy Aggregation

We explain how to compute the truth value of a sentry with linguistic variables.

First, we show how to write the completion event of a stage in the fuzzy notation. Practically, we keep the completion event crisp, but the notation is adapted to have a single consistent notation for fuzzy sentries. The completion event of stage S has the corresponding fuzzy linguistic statement "Completion event of Stage S has *occurred*". Then we define the discrete fuzzy membership function for the linguistic label "*occurred*" as

$$\mu_{occurred}(\mathsf{C}{:}S) = \{(True, 1), (False, 0)\}.$$

The reason why we interpret the completion event in a binary fashion is that there is no uncertainty regarding the completion of a stage: a stage has either completed or not. However, it can make sense to allow stages to gradually complete. We plan to study gradual completion of stages in future research.

In the classical GSM schemas, sentries may contain logical "and" and "or" operators in their condition formula. In our fuzzy modeling of GSM schema, condition formulas are computed using fuzzy aggregations, introduced in Sect. 3.2. For example, the sentry "C:Credit Check $\wedge$ rating $\geq$ 8" of milestone CCS in Table 2 can be transformed into

"C:Credit Check has *occurred*" and "rating is *high*"

where rating is modeled as linguistic variable with a linguistic label "high" that can be interpreted as "being approximately greater than or equal to 8".

Computation of a fuzzy aggregation is performed by using conjunctive and disjunctive operators, as mentioned in Sect. 3.2. So the condition formula is computed as

$$\mu_{CCS} = \mu_{occurred}(\mathsf{C}{:}\mathsf{CreditCheck}) \times \mu_{high}(r)$$

where r denotes the crisp value of rating, and the "product" t-norm operator is used in the above conjunctive fuzzy aggregation.

Note that condition formulas may also contain the satisfaction value of other condition formulas as a component in the aggregation. For example, the fuzzification of the sentry "CCS $\wedge$ BPECS" of the milestone PCS is done as

$$\mu_{PCS} = \mu_{CCS} \times \mu_{BPECS}$$

where again the "product" t-norm operator is used.

Given a stage or milestone $a \in \mathcal{A}_S \cup \mathcal{A}_m$, evaluating a fuzzy sentry $\varphi \in sen(a)$ results in a fuzzy value between 0 and 1. Defuzzification is needed to derive a crisp Boolean value that denotes whether a becomes true or value. Assuming a general threshold value $\alpha \in [0..1]$ for the entire GSM schema, we define a defuzzification function that assigns a status attribute true if its fuzzy value $\mu(a)$ exceeds the threshold α:

$$defuzzify(a) = \begin{cases} 1, & \text{if } \mu(a) \geq \alpha \\ 0, & \text{otherwise.} \end{cases}$$

4.3   Incorporating Fuzzy Sentries in the GSM Execution

So far, we have presented how sentries can fuzzified. We now sketch a potential solution how the evaluation of fuzzy sentries can be done at run time (cf. Fig. 4). The only difference with the regular GSM semantics, in which sentries are evaluated to perform a B-step (see Sect. 3.1) is that sentries are now evaluated using fuzzy reasoning, consisting of three steps that are standard in evaluating fuzzy rules [15]: fuzzify the crisp input to determine the values of the membership functions of the fuzzy variables referenced in the sentry, then evaluate the sentry, and next defuzzify the input by applying the change for which the sentry is a condition, if the sentry exceeds the threshold value. The functions used in these steps have been defined above.

For classical GSM schemas, the only change is that sentries are evaluated using classical logic [5], so (de)fuzzification does not apply. Thus, the proposed extension is conservative: GSM engines only need to invoke a fuzzy evaluation tool that implements the (de)fuzzification and aggregation functions defined above.

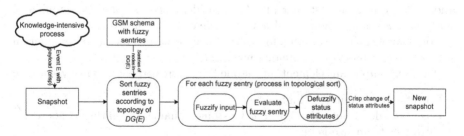

Fig. 4. Evaluation of fuzzy sentries in GSM execution

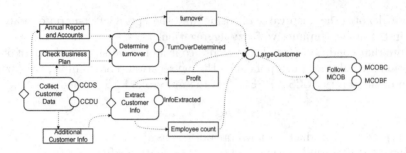

Fig. 5. Part of GSM schema for processing regulated mortgage contracts

5   Evaluation

As a first evaluation step, we apply the approach to some of the rules defined by the Financial Conduct Authority in the UK as part of the Mortgage Code of Business (MCOB) for financial firms that offer regulated mortgage contracts: www.handbook.fca.org.uk/handbook/MCOB. Below we show a few MCOB regulations for regulated mortgage contracts that financial firms use to decide whether they have to adhere to the code of business for the customer. We show how they are formalized with the fuzzy modeling approach advocated in this paper.

1.2.3. In relation to a regulated mortgage contract for a business purpose (1) MCOB applies if the customer is not a large business customer; and (2) if MCOB applies, a firm must comply with MCOB in full, taking into account tailored exceptions and provisions.

1.2.6. In determining whether a customer is a large business customer, a firm will need to have regard to the figure given for the customer's annual turnover in the customer's annual report and accounts or business plan. In addition, a firm may rely on information provided by the customer about the annual turnover, unless, taking a common-sense view of this information, it has reason to doubt it.

Figure 5 shows part of a fuzzy GSM schema that operationalizes these regulations. As additional customer info, we consider the profit and the number of employees an organization has. First, we define the linguistic terms and corresponding labels in Table 3. Then for every linguistic label a membership function is defined based on expert/domain knowledge.

The most important element, milestone LargeCustomer, has two fuzzy sentries:

– Business sales volume is high
– Business sales volume is high *and* Customer financial situation is good *and* Customer team size is big

Table 3. Converting crisp variables to linguistic terms and labels

Crisp variable	Linguistic term	Label	Membership function
turnover	Business sales volume	high	μ_{high}(turnover)
profit	Customer financial situation	good	μ_{good}(profit)
employee count	Customer team size	big	μ_{big}(employeecount)

The second sentry requires a three-component fuzzy aggregation:

$$\mu_{\text{LargeCustomer}} = \mu_{high}(\text{turnover}) \times \mu_{good}(\text{profit}) \times \mu_{big}(\text{employeecount})$$

where the product operator is used for the t-norm operation. Next, defuzzification as defined below results in an binary output of the milestone LargeCustomer which directly determines if the stage Follow MCOB, which has sentry "¬LargeCustomer" turns to *True*.

$$defuzzify(\text{LargeCustomer}) = \begin{cases} 1, & \text{if } \mu_{\text{LargeCustomer}} \geq \alpha \\ 0, & \text{otherwise,} \end{cases}$$

where α is the threshold introduced in Sect. 4.

In a classical GSM schema, the decision of when a customer is large is naturally modeled by having an atomic stage, that has the relevant data attributes as input, and upon completion achieves either milestone LargeCustomer or Non-LargeCustomer. The actual decision logic is then hidden in the atomic stage. Using fuzzy logic, the reasoning behind the decision can be represented in sentries, so the fuzzy GSM model forces to make the reasoning rules explicit.

This first evaluation step shows that fuzzy GSM schemas can be used to represent uncertainty that is inherent in real-world regulations. We plan to evaluate the approach in industrial case studies.

6 Related Work

Fuzzy modeling has been applied to many different application domains, such as control and decision making and optimization [15]. However, only a few papers have applied fuzzy modeling in the context of BPM.

Thomas et al. [1,17] explore the combination of fuzzy modeling and Event-driven Process Chains (EPCs). They advocate that the resulting fuzzy workflows replace the classical, crisp workflows. Ye et al. [20] propose the use of fuzzy logic to handle exceptions in workflow management, using fuzzy Petri nets as underlying model. They focus on extending existing crisp workflow models and crisp workflow systems to incorporate fuzzy reasoning. In contrast to these works [1,17,20], we propose that the fuzzy sentries are only used as front end to decide the truth value of sentries, so a non-fuzzy GSM engine is still used to drive the execution. Furthermore, these works consider procedural process models, whereas we consider declarative, artifact-centric process models.

Fuzzy logic has also been applied to represent uncertainty for simulation [2,19] and analysis [21] of business processes. Landry et al. [13] apply fuzzy logic to distributed workflows to support collaborative crisis management under uncertainty. Slavícek [16] presents a fuzzy ontology-based workflow system. The focus is on integrating fuzzy workflow systems with fuzzy ontologies. None of these papers focuses on declarative, artifact-centric processes.

In the broader area of BPM, fuzzy modeling has been applied as well. For instance, Hakim et al. [10] present a methodology based on fuzzy logic to select processes for reengineering while Bazhenova et al. [3] define how to extra fuzzy decision models from event logs. However, in these papers the processes themselves are not modelled in a fuzzy way.

7   Conclusion

We have proposed fuzzy GSM schemas as technique for modeling uncertainty in declarative artifact-centric process models. The added value of fuzzy GSM schemas over classical GSM schemas is twofold. Firstly, they are capable of capturing expert knowledge via defining rules in linguistic terms that match human understanding of the problem domain better than the rules in classical GSM schemas. Secondly, fuzzy GSM schemas are more useful for the cases where it is hard, even impossible, to collect enough information to make concrete decisions while performing the business process. The resulting uncertainty is better captured using linguistic variables in decision making, as done in fuzzy models, rather than relying on crisp variables as used in classical GSM schemas. We also discussed how existing GSM engines can support fuzzy GSM schemas by invoking fuzzy rule evaluators. Thus, the proposed GSM extension is lightweight and does not require any redefinition of existing GSM engines.

For future work, we plan to explore the use of fuzzy stages and milestones for fuzzy GSM schemas. This also will affect the execution of GSM schemas, i.e., the notion of Business step needs to be adjusted accordingly. Moreover, we plan to incorporate fuzzy inference systems into fuzzy GSM schemas involving highly complicated decision making parts. Finally, we plan to apply the approach in several case studies in organizations.

References

1. Adam, O., Thomas, O., Martin, G.: Fuzzy workflows - enhancing workflow management with vagueness. In: Proceedings of the EURO/INFORMS 2003 (2003)
2. Azadeh, A., Haghnevis, M., Khodadadegan, Y., Madadi, M.: Modeling and improvement of the integrated business and production processes by fuzzy simulation. In: Proceedings of the SpringSim 2009. SCS/ACM (2009)
3. Bazhenova, E., Haarmann, S., Ihde, S., Solti, A., Weske, M.: Discovery of fuzzy DMN decision models from event logs. In: Dubois, E., Pohl, K. (eds.) CAiSE 2017. LNCS, vol. 10253, pp. 629–647. Springer, Cham (2017). https://doi.org/10.1007/978-3-319-59536-8_39

4. BizAgi, et al.: Case Management Model and Notation (CMMN), v1.1, OMG Document Number formal/2016-12-01, Object Management Group (2016)
5. Damaggio, E., Hull, R., Vaculín, R.: On the equivalence of incremental and fixpoint semantics for business artifacts with guard-stage-milestone lifecycles. Inf. Syst. **38**, 561–584 (2013)
6. Di Ciccio, C., Marrella, A., Russo, A.: Knowledge-intensive processes: characteristics, requirements and analysis of contemporary approaches. J. Data Semant. **4**(1), 29–57 (2015)
7. Eshuis, R., Hull, R., Sun, Y., Vaculín, R.: Splitting GSM schemas: a framework for outsourcing of declarative artifact systems. Inf. Syst. **46**, 157–187 (2014)
8. Eshuis, R., Hull, R., Yi, M.: Property preservation in adaptive case management. In: Barros, A., Grigori, D., Narendra, N.C., Dam, H.K. (eds.) ICSOC 2015. LNCS, vol. 9435, pp. 285–302. Springer, Heidelberg (2015). https://doi.org/10.1007/978-3-662-48616-0_18
9. Goedertier, S., Vanthienen, J., Caron, F.: Declarative business process modelling: principles and modelling languages. Enterp. IS **9**(2), 161–185 (2015)
10. Hakim, A., Gheitasi, M., Soltani, F.: Fuzzy model on selecting processes in business process reengineering. Bus. Proc. Manag. J. **22**(6), 1118–1138 (2016)
11. Hull, R., et al.: Introducing the guard-stage-milestone approach for specifying business entity lifecycles. Proc. WS-FM **2010**, 1–24 (2010)
12. Klir, G.J.: Uncertainty and Information: Foundations of Generalized Information Theory. Wiley, Hoboken (2006)
13. Landry, J.-F., Ulmer, C., Gomez, L.: Fuzzy distributed workflows for crisis management decision makers. In: Ortiz-Arroyo, D., Larsen, H.L., Zeng, D.D., Hicks, D., Wagner, G. (eds.) EuroISI 2008. LNCS, vol. 5376, pp. 226–236. Springer, Heidelberg (2008). https://doi.org/10.1007/978-3-540-89900-6_23
14. Nigam, A., Caswell, N.S.: Business artifacts: an approach to operational specification. IBM Syst. J. **42**(3), 428–445 (2003)
15. Ross, T.J.: Fuzzy Logic with Engineering Applications. Wiley, Chichester (2010)
16. Slavíček, V.: An ontology-driven fuzzy workflow system. In: van Emde Boas, P., Groen, F.C.A., Italiano, G.F., Nawrocki, J., Sack, H. (eds.) SOFSEM 2013. LNCS, vol. 7741, pp. 515–527. Springer, Heidelberg (2013). https://doi.org/10.1007/978-3-642-35843-2_44
17. Thomas, O., Dollmann, T., Loos, P.: Rules integration in business process models - a fuzzy oriented approach. Enterp. Model. Inf. Syst. Arch. **3**(2), 18–30 (2008)
18. Vaculín, R., Hull, R., Heath, T., Cochran, C., Nigam, A., Sukaviriya, P.: Declarative business artifact centric modeling of decision and knowledge intensive business processes. Proc. EDOC **2011**, 151–160 (2011)
19. Völkner, P., Werners, B.: A simulation-based decision support system for business process planning. Fuzzy Sets Syst. **125**(3), 275–287 (2002)
20. Ye, Y., Jiang, Z., Diao, X., Du, G.: Extended event-condition-action rules and fuzzy Petri nets based exception handling for workflow management. Expert Syst. Appl. **38**(9), 10847–10861 (2011)
21. Zakarian, A.: Analysis of process models: a fuzzy logic approach. Int. J. Adv. Manuf. Technol. **17**(6), 444–452 (2001)

Extracting Workflows from Natural Language Documents: A First Step

Leslie Shing[1]([✉])(iD), Allan Wollaber[1]([✉])(iD), Satish Chikkagoudar[2]([✉]),
Joseph Yuen[3,4]([✉]), Paul Alvino[4]([✉])(iD), Alexander Chambers[4]([✉])(iD),
and Tony Allard[4]([✉])(iD)

[1] MIT Lincoln Laboratory, Lexington, MA, USA
{leslie.shing,allan.wollaber}@ll.mit.edu
[2] Naval Research Laboratory, Washington, D.C., USA
satish.chikkagoudar@nrl.navy.mil
[3] Commonwealth Bank of Australia, Sydney, Australia
joseph.yuen@cba.com.au
[4] Defence Science and Technology Group, Edinburgh, SA, Australia
{paul.alvino,alexander.chambers,tony.allard}@dst.defence.gov.au

Abstract. Business process models are used to identify control-flow relationships of tasks extracted from information system event logs. These event logs may fail to capture critical tasks executed outside of regular logging environments, but such latent tasks may be inferred from unstructured natural language texts. This paper highlights two workflow discovery pipeline components which use NLP and sequence mining techniques to extract workflow candidates from such texts. We present our Event Labeling and Sequence Analysis (ELSA) prototype which implements these components, associated approach methodologies, and performance results of our algorithm against ground truth data from the Apache Software Foundation Public Email Archive.

Keywords: Workflow discovery · Natural language · Sequence mining

1 Introduction

Visibility into organizational workflows via business process models enables the completion of critical tasks in a predictable and measurable way, and is especially relevant for organizations which must manage risk and prioritize their tasks in a contested environment. However, it is difficult to properly reconstruct workflows whose tasks are executed outside of regular logging environments. We hypothesize that latent tasks can be learned, at least in part, from unstructured natural language documents (NLDs) (e.g. emails, chats and blogs) using a technique that discovers topics and recurrent event sequences in an automated fashion. Approaches exist for workflow extraction from semi-structured NLDs [6,14]. Our work parallels MailOfMine [4], which aims to build workflow models from unstructured NLDs (i.e. email). The objective of this paper is to highlight approaches for two of the components in the workflow discovery

© Springer Nature Switzerland AG 2019
F. Daniel et al. (Eds.): BPM 2018 Workshops, LNBIP 342, pp. 294–300, 2019.
https://doi.org/10.1007/978-3-030-11641-5_23

pipeline, namely: (1) keyword extraction and topic clustering of events using semantic analysis of the NLDs, and (2) sequence rule mining to identify partial workflow candidates given these keywords and topic clusters.

1.1 Related Work

Email, one type of unstructured NLD (i.e. user-generated written content that lacks a predefined data model), is often a means by which informal tasks are carried out [4] and is the subject of our initial evaluation due to its availability and accessibility. Many approaches have been developed to infer activities from emails and other NLDs [5,6,14]. As our use case requires automated, unsupervised information extraction from unstructured text, we seek topic extraction and clustering approaches similar to [4].

Work in [6,14] use the verbiage inherent in procedural texts to sequence tasks. Di Ciccio and Mecella describe a declarative approach for mining control-flow constraints between tasks [4]. In addition to these methods, process mining is a viable alternative approach for task sequencing. Process mining, such as Sequential Pattern Mining [9], is used to extract exemplar *cases* from information system event logs. It is widely used by organizations to identify patterns and trends in business workflows and gain other insights into workflow processes.

This paper is structured as follows. Section 2 introduces our approach for components of workflow discovery and its implementation. Section 3 provides an overview of a few experiments we conducted and their performance measures on a benchmarking dataset. Section 4 details the results attained from our experiments. Section 5 discusses the current performance of our prototype, as well as the opportunities, challenges, and lessons learned for this first iteration of our prototype, and Sect. 6 concludes the paper.

2 Methods and Approach

We have created a prototype system, Event Labeling and Sequence Analysis (ELSA), that integrates techniques from both natural language processing (NLP) and sequence process mining to automatically generate partial workflow candidates for events inferred from NLDs.

The NLP text processor component clusters NLDs based on topics inferred from related documents. We use latent semantic indexing (LSI), an unsupervised technique widely used in NLP research, and density-based spatial clustering of applications with noise (DBSCAN) [7], an algorithm that determines the number of clusters dynamically. The email subject lines and text bodies are preprocessed, transformed into a TF-IDF weighted document-term matrix for LSI, and input to DBSCAN as a k-dimensionally reduced matrix of document vectors. DBSCAN parameters are initialized with $\varepsilon = 0.2$ and $minpts = 5$ based on empirical evaluation. Each email is either labeled with a cluster ID or discarded as noise by DBSCAN, removing emails that are not representative of tasks in the overarching workflows. The top N keywords of each cluster are extracted to identify the main topics of each cluster.

The sequence process mining component uses sequence rules to extract partial workflow candidates from document clusters. Sequence rule mining identifies temporal rules $A \rightarrow B$ (A followed by B, where A and B are itemsets) from input sequences, support (i.e. the fraction of the sequences that contain the rule), and confidence (i.e. the fraction of sequences that contain the rule out of the sequences that contain the 'trigger' of the rule). The Top-K Non-Redundant Sequential rules (TNS) algorithm [9] is a sequence mining technique that prunes the search space to avoid redundant generation of sequential rules and determines the minimum support for a rule dynamically. This algorithm was selected due to its success in several logistical domains relevant to our research. TNS was configured to keep the top 30 rules with a minimum confidence threshold of 0.5 and a Δ value of 2, based on [9]. Input sequences for TNS are temporally ordered emails grouped by cluster ID as identified by DBSCAN. The output of the TNS algorithm is a set of the top 30 sender sequence rules and their associated support and confidence values.

ELSA's modular design consists of the following key modules: data ingestor, NLP text processor, sequence database transformer, and sequence process miner. Each component uses generalized APIs for communication to allow for the maturation and development of modules within minimal constraints. ELSA is written primarily in Python and uses a Python wrapper for the Java implementation of the TNS algorithm [8]. All data, including the output from pipeline components are stored in a SQLite database.

3   Experimental Design

To verify ELSA's performance, we compare output at each step of the analysis pipeline against a known ground truth, curated from open source data from the Apache Camel project [1]. The process for this is described in [2]. The ground truth consists of a total of 250 manually-evaluated emails, each tagged with five keywords and assigned to one of 65 email traces.

For each email, ELSA produces a vector of terms and associated weights, whereas the ground truth identifies five keywords. For standard set-based metrics like the Jaccard similarity (J) and Sørensen-Dice coefficient/F_1 score, we take only the top five ELSA keywords. To employ vector-based metrics such as the generalized Jaccard (GJ), cosine (C) and soft cosine ($\tilde{C}$) similarities, we assume equal weighting of the ground truth keywords. Since we only require that the keywords be semantically similar for clustering, we use 'soft' versions of the Jaccard and Sørensen-Dice similarities (denoted with a tilde) analogous to the soft cosine,

$$\tilde{J}(A,B) \equiv \frac{\sum_{a \in A, b \in B} S(a,b)}{\sum_{a,a' \in A} S(a,a') + \sum_{b,b' \in B} S(b,b') - \sum_{a \in A, b \in B} S(a,b)}, \quad (1)$$

$$\tilde{F}_1(A,B) \equiv \frac{2 \sum_{a \in A, b \in B} S(a,b)}{\sum_{a,a' \in A} S(a,a') + \sum_{b,b' \in B} S(b,b')}, \quad (2)$$

where S is a similarity measure between pairs of keywords, chosen to be the Wu-Palmer (WP) path-similarity [15] calculated through WordNet [12]. We choose WordNet senses for the keywords to maximize the WP path-similarity between each keyword pair.

Whereas the ground truth identifies event traces, ELSA produces sender rules. To compare these, we use the concepts of support and confidence introduced in Sect. 2, additionally defining 'soft' versions given by

$$\text{soft support}(X \to Y) \equiv \frac{\tilde{N}(X \to Y)}{|S|}, \tag{3}$$

$$\text{soft confidence}(X \to Y) \equiv \frac{\tilde{N}(X \to Y)}{N(X)}. \tag{4}$$

Here $|S|$ is the total number of traces, $N(X)$ is the number of traces where the 'trigger' X of the rule is seen, and $\tilde{N}$ is a count of partially-observed rules,

$$\tilde{N}(X \to Y) \equiv \sum_S \frac{|\Delta Y|}{|Y|} \quad \text{for the largest } \Delta Y \subset Y \text{ after } X \text{ in } S. \tag{5}$$

4   Results

The left table in Fig. 1 shows the proportion of emails with at least n matching keywords between the ground truth and ELSA's top five. This matching is either direct (keyword-to-keyword) or soft (through the soft Jaccard index). ELSA's performance in this area is promising, with 75% of the emails having at least one keyword directly matching, and 60% having at least three keywords softly matching. A comparison between the direct and soft matchings shows that the semantic meaning is often similar in many cases where keywords do not match exactly. The right plot in Fig. 1 shows the fraction of emails with similarity scores of at least s between the ELSA and ground truth keywords, for the metrics discussed in Sect. 3.

The average support and soft support for ELSA's sender rules are relatively low, 0.8% and 1.2% respectively, indicating that the rules are very specific. The table and plot in Fig. 2 show the proportion of sender rules found to have a

n_matches	Direct	Soft
≥ 1	76.4%	94.8%
≥ 2	56.0%	85.6%
≥ 3	22.8%	59.6%
≥ 4	8.8%	16.4%
≥ 5	0.8%	1.2%

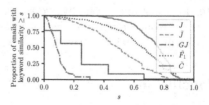

Fig. 1. Proportion of emails with (left) at least n keywords matching (directly or softly) and (right) similarity scores of at least s between ELSA and the ground truth.

Confidence	Direct	Soft
≥ 0.2	45.5%	54.5%
≥ 0.4	27.3%	40.9%
≥ 0.6	13.6%	13.6%
≥ 0.8	13.6%	13.6%
≥ 1.0	13.6%	13.6%

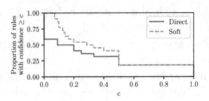

Fig. 2. Proportion of rules with a (direct or soft) confidence of at least c when measured against the ground truth traces.

(direct or soft) confidence of at least c, where we note that 40% of the rules have a soft confidence greater than 0.4. That is, we are seeing the rules (which we are considering to be candidate trace fragments) represented to some extent in the ground truth traces, indicating an important step towards trace construction.

5   Discussion

ELSA experiences several shortcomings that we will address in future iterations. Firstly, LSI uses a bag-of-words method, which does not consider word ordering or associations, and hence polysemy and synonymy are not captured. This can lead to inaccuracies in topic clustering. Additionally, LSI does not perform well on short documents. To address these issues, we aim to add Word Sense Disambiguation techniques, such as [13] which use external knowledge bases like Wordnet [12] to determine the intended word senses and enrich the documents with contextual data. Additionally, we would like to remove Subject Matter Expert (SME) input for initializing algorithm parameters.

Some components of ELSA's pipeline are as yet unimplemented. ELSA has no additional layer of abstraction (metalabels) from the keywords, which provide additional context for the types of actions completed. We will apply techniques such as Explicit Semantic Analysis [10] and lexical graph similarity metrics to extract these metalabels. ELSA also lacks a trace assignment step between the NLP processing and sequence mining components, with input to the TNS algorithm simply being temporal sequences sorted by topic. We will use a combination of Allen's logic [3] and straightforward 'group-by-conversation' rules to generate trace instances. Finally, we require a method to better extract workflow instances from sequences of events. One possibility is to use recurrent neural networks, owing to their effectiveness in learning rules from sequential data [11].

6   Conclusion

In this work, we have highlighted approaches for two components in the workflow discovery pipeline. We have used quantitative metrics to assess the extent to which our software prototype (ELSA) was able to successfully cluster emails, extract keywords, and discover repeated patterns in a ground truth dataset. Although ELSA performed fairly well at keyword labeling and was acceptable at discovering sender-sequences in traces, it is only the first step towards discovering workflows, as discussed in Sect. 5. This approach will now drive the development for the second version of ELSA, which will use further abstractions beyond traces to discover workflows and be tested against the same ground truth dataset, as well as a system with more mission context.

Acknowledgment. This material is based upon work supported under Air Force Contract No. FA8721-05-C-0002 and/or FA8702-15-D-0001. Any opinions, findings, conclusions or recommendations expressed in this material are those of the authors and do not necessarily reflect the views of the U. S. Air Force.

References

1. Apache camel. https://camel.apache.org/. last accessed 23 Jan 2018
2. Allard, T., Alvino, P., Shing, L., Wollaber, A., Yuen, J.: A novel dataset to facilitate automated workflow analysis. PLOS ONE (2018) (submitted)
3. Allen, J.F., Ferguson, G.: Actions and events in interval temporal logic. J. Log. Comput. **4**(5), 531–579 (1994)
4. Di Ciccio, C., Mecella, M., Scannapieco, M., Zardetto, D., Catarci, T.: MailOfMine – analyzing mail messages for mining artful collaborative processes. In: Aberer, K., Damiani, E., Dillon, T. (eds.) SIMPDA 2011. LNBIP, vol. 116, pp. 55–81. Springer, Heidelberg (2012). https://doi.org/10.1007/978-3-642-34044-4_4
5. Dredze, M., Lau, T., Kushmerick, N.: Automatically classifying emails into activities. In: Proceedings of the 11th International Conference on Intelligent User Interfaces, pp. 70–77. ACM (2006)
6. Dufour-Lussier, V., Le Ber, F., Lieber, J., Nauer, E.: Automatic case acquisition from texts for process-oriented case-based reasoning. Inf. Syst. **40**, 153–167 (2014)
7. Ester, M., et al.: A density-based algorithm for discovering clusters in large spatial databases with noise. Kdd **96**, 226–231 (1996)
8. Fournier-Viger, P., Gomariz, A., Gueniche, T., Soltani, A., Wu, C.W., Tseng, V.S.: SPMF: a Java open-source pattern mining library. J. Mach. Learn. Res. **15**(1), 3389–3393 (2014)
9. Fournier-Viger, P., Tseng, V.S.: TNS: mining top-k non-redundant sequential rules. In: Proceedings of the 28th Annual ACM Symposium on Applied Computing, pp. 164–166. ACM (2013)
10. Gabrilovich, E., Markovitch, S.: Computing semantic relatedness using wikipedia-based explicit semantic analysis. In: Proceedings of the 20th International Joint Conference on Artificial Intelligence, pp. 1606–1611 (2007)
11. Lipton, Z.C., Berkowitz, J., Elkan, C.: A critical review of recurrent neural networks for sequence learning. arXiv preprint arXiv:1506.00019 (2015)
12. Miller, G.A.: Wordnet: a lexical database for English. Commun. ACM **38**(11), 39–41 (1995)
13. Navigli, R., Lapata, M.: An experimental study of graph connectivity for unsupervised word sense disambiguation. IEEE Trans. Pattern Anal. Mach. Intell. **32**(4), 678–692 (2010)

14. Schumacher, P., Minor, M., Walter, K., Bergmann, R.: Extraction of procedural knowledge from the web: a comparison of two workflow extraction approaches. In: Proceedings of the 21st International Conference on World Wide Web, pp. 739–747. ACM (2012)
15. Wu, Z., Palmer, M.: Verbs semantics and lexical selection. In: Proceedings of the 32nd Annual Meeting on Association for Computational Linguistics, pp. 133–138 (1994). https://doi.org/10.3115/981732.981751

DCR Event-Reachability via Genetic Algorithms

Tróndur Høgnason and Søren Debois[⊠]

IT University of Copenhagen,
Rued Langgaards Vej 7, 2300 Copenhagen S, Denmark
{thgn,debois}@itu.dk

Abstract. In declarative process models, a process is described as a set of rules as opposed to a set of permitted flows. Oftentimes, such rule-based notations are more concise than their flow-based cousins; however, that conciseness comes at a cost: It requires computation to work out which flows are in fact allowed by the rules of the process. In this paper, we present an algorithm to solve the Reachability problem for the declarative Condition Response (DCR) graphs notation: the problem, given a DCR graph and an activity, say "Payout reimbursement", is to find a flow allowed by the graph that ends with the execution of that task. Existing brute-force solutions to this problem are generally unhelpful already at medium-sized graphs. Here we present a *genetic algorithm* solving Reachability. We evaluate this algorithm on a selection of DCR graphs, both artificial and from industry, and find that the genetic algorithm with one exception outperforms the best known brute-force solution on both whether a path is found and how quickly it is found.

Keywords: Genetic algorithm · Reachability · Declarative model · DCR

1 Introduction

In the field of business process modelling, there are currently two major modelling paradigms. In declarative notations such as such as DECLARE [2,23], DCR graphs [8,15], GSM [17] and CMMN [21] a model comprises the rules governing process execution. In imperative notations such as Petri- and Workflow nets [1,3] and BPMN [22], a model comprises a more explicit representation of the set of possible executions.

The choice between the two paradigms is in essence a trade-off of conciseness for computational requirements. A declarative model may, via its rules, concisely represent a very large number of possible process executions: A DCR graph or a DECLARE model may represent a space of process executions exponentially larger than the graph itself. Many interesting problems of DCR graphs are consequently computationally hard [10].

© Springer Nature Switzerland AG 2019
F. Daniel et al. (Eds.): BPM 2018 Workshops, LNBIP 342, pp. 301–312, 2019.
https://doi.org/10.1007/978-3-030-11641-5_24

In this paper, we explore the problem of determining whether given a model M and an activity a there exists a sequence of activity executions admitted by M which ends with a. We study this problem in the setting of DCR graphs, where activities are typically conflated with events. Here, the problem is called "Event reachability" [7,10].

Solving this problem is practically important for model development. E.g., suppose we are constructing a DCR model of an insurance process involving an event "Payout reimbursement", and we are interested in knowing whether it is possible to reach the "Payout reimbursement" event from a given state of the model.

Definition 1. *Given a DCR graph G and an event e of G, we say that e is reachable from G if there exists a path $e_1, \ldots, e_n$ of event executions admitted by G such that $e_n = e$. We call the path $e_1, \ldots, e_n$ a* witness *for reachability of e.*

Existing solutions to this problem are either brute-force approaches and impractical on all but the smallest models [8,9,13] or approximations applicable only in special cases of graph structure [7,10]. In this paper we investigate an alternative approximation using *genetic algorithms* [5] to heuristically find such paths, or give up if a time-bound expires. Genetic algorithms are viable because we can quantify how far off a path not reaching G is as a function of the number of constraints relating to G that have been satisfied at the end of the path. Presumably, fewer inhibiting constraints means we are closer to a solution. While genetic algorithms have previously been applied in other BPM subdomains, e.g., to process mining [4] and prediction [14,19], we are unaware of other applications to reachability problems within the BPM domain.

We make in this paper the following contributions.

1. We provide a Genetic Algorithm for approximating DCR Event Reachability, including an implementation;
2. We report on a comparison with the existing brute-force solver reported in [8,9] on both artificial and real-world models.

While the genetic algorithm is inferior to the brute-force one on examples constructed specifically to be difficult for it, it is remarkably effective on the real-world examples. It finds paths where the brute-force algorithm times out, and when both succeed, the genetic algorithm does so orders of magnitude faster than the brute-force one. We conclude that the genetic algorithm appears to be a practical approximation to Event Reachability in DCR graphs.

2   Dynamic Condition Response Graphs

DCR Graphs is a declarative notation for modelling processes superficially similar to DECLARE [23,24] or temporal logics such as LTL [25]. The present paper makes two restrictions on DCR graphs: (1) we consider only finite executions, and (2) we assume each event is labelled by a unique activity. None of these restrictions are controversial, the latter is pervasive in the literature on DCR graphs. Because of the latter assumption we do not distinguish between (DCR) "events" and "activities" in the sequel, speaking only of "events".

We give an example DCR graph in Fig. 1; formal definitions follow below. This example has five events, A, B, C, D, and *Goal*. Because there is a condition from A to D, A must be executed before we can execute D, and D must be executed before we can execute *Goal*. Executing B excludes A, making A not executable and voiding the condition from A to D.

The event *Goal* is not initially enabled for execution: it is prevented by the condition from D. D is similarly not enabled, prevented by the condition from A. One way to reach *Goal* is thus the sequence A, D, *Goal*. Alternatively, one may exploit the exclusion from C to void the condition from D, arriving at the sequence C, *Goal*.

Since an event in a DCR graph can be executed multiple times, there are infinitely many ways to reach the event *Goal*; e.g., the set of strings characterised by the regular expressions $A^+D(A|D)^+ Goal$ or $(B|A|C)^* C Goal$.

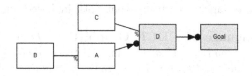

Fig. 1. Example DCR graph

Previous approaches to event reachability in DCR graphs have simply explored the state space of a DCR model [8,9]. Since the state of a DCR event is fully characterised by its three boolean attributes (executed, included, pending), an event can be in at most 2^3 different states; a DCR graph with n events therefore maximally has 2^{3n} distinct states. Not all of these will necessarily be reachable, however, experience suggests that for real-life graphs, the state space is large enough to prohibit brute force approaches [9,13].

Definition 2 (DCR Graph [15]). *A DCR graph is a tuple* (E, R, M) *where*

- E *is a finite set of (activity labelled) events, the nodes of the graph.*
- R *is the edges of the graph. Edges are partitioned into five kinds, named and drawn as follows: The* condition $(\rightarrow\bullet)$, response $(\bullet\rightarrow)$, inclusion $(\rightarrow+)$, exclusion $(\rightarrow\%)$ *and* milestone $(\rightarrow\diamond)$.
- M *is the* marking *of the graph. This is a triple* (Ex, Re, In) *of sets of events, respectively the previously executed* (Ex), *the currently pending* (Re), *and the currently included* (In) *events.*

The marking of a DCR graph is its run-time state: it records which events have been executed (Ex), which are currently included (In) in the graph, and which are required to be executed again in order for the graph to be accepting (Re). We first define when an event is enabled.

Notation. When G is a DCR graph, we write, e.g., $E(G)$ for the set of events of G, $Ex(G)$ for the executed events in the marking of G, etc. In particular, we write $M(e)$ for the triple of boolean values $(e \in Ex, e \in Re, e \in In)$. We write $(\rightarrow\bullet e)$ for the set $\{e' \in E \mid e' \rightarrow\bullet e\}$, write $(e\bullet\rightarrow)$ for the set $\{e' \in E \mid e \bullet\rightarrow e'\}$ and similarly for $(e\rightarrow+)$, $(e\rightarrow\%)$ and $(\rightarrow\diamond e)$.

Definition 3 (Enabled events [15]). *Let $G = (E, R, M)$ be a DCR graph, with marking $M = (Ex, Re, In)$. An event $e \in E$ is* enabled, *written $e \in$ enabled(G), iff (a) $e \in In$ and (b) $In \cap (\rightarrow\bullet e) \subseteq Ex$ and (c) $(Re \cap In) \cap (\rightarrow\diamond e) = \emptyset$.*

That is, enabled events (a) are included, (b) their included conditions have been executed, and (c) they have no included milestones with an unfulfilled response.

Executing an enabled event e of a DCR Graph results in a new marking where (a) e is added to the set of executed events, (b) e is removed from the set of pending response events, (c) the responses of e are added to the pending responses, (d) the events excluded by e are removed from included events, and (e) the events included by e are added to the included events.

From this we can define the language of a DCR Graph as all finite sequences of events ending in a marking with no event both included and pending.

Definition 4 (Language of a DCR Graph [15]). *Let $G_0 = (E, R, M)$ be a DCR graph with marking $M_0 = (Ex_0, Re_0, In_0)$. A* trace *of G_0 is a finite sequence of events $e_0, \ldots, e_n$ such that for $0 \leq i \leq n$, (a) e_i is enabled in the marking $M_i = (Ex_i, Re_i, In_i)$ of G_i, and (b) G_{i+1} is a DCR Graph with the same events and relations as G_i but with marking $(Ex_{i+1}, Re_{i+1}, In_{i+1}) = (Ex_i \cup \{e_i\}, (Re_i\backslash\{e_i\}) \cup (e_i\bullet\rightarrow)), (In_i\backslash(e_i\rightarrow\%)) \cup (e_i\rightarrow+)).$*

We call such a trace accepting *if for all $0 \leq i \leq n$ we have $Re_{i+1} \cap In_{i+1} = \emptyset$. The* language lang$(G_0)$ *of G_0 is then the set of all such accepting traces.*

3 Genetic Algorithms

Genetic algorithms (GA) imitate natural selection. Solutions to a problem are structured as individuals with one or more chromosomes, each consisting of genes. For the present problem, one chromosome per individual will suffice. Initially, in generation 1, the individuals comprise a set of randomly generated chromosomes. Individuals are ranked by their fitness, based on how good the solution encoded in their chromosome is. We then select two or more individuals (parents) and create a child whose chromosome is the crossover of the parents chromosomes. We create children until we reach a specified number of individuals, commonly double the initial population. At this point we proceed to kill off the worst individuals until we reach our initial population size. The remaining individuals constitute generation 2 of individuals.

This process is repeated until a stop condition is met. Common stop conditions are: a solution is found, a timeout is reached or n number of generation have been generated. When it is unknown whether or not a solution is optimal, we can also set an optimisation threshold oT and run the algorithm until the best individual improves less than oT in some amount of generations.

To implement a GA, we will need to implement the main building blocks of any genetic algorithm: a fitness function, a selection function, a crossover function and a mutation function. However, first we have to decide how to encode the problem as genes in a chromosome. If our problem is to construct a bit string of length 6 with the maximum number of 1's[1], our genes will be either a 1 or a 0, and our chromosome will be a list or 1's and 0's of length 6 as in Fig. 2.

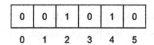

Fig. 2. A possible instantiation of a chromosome

Fitness Function. The fitness function of a GA takes an individual as input and returns a value based on how good the solution in its chromosome is. A good fitness function for a GA that constructs a bit string with the maximum number of 1's simply returns the number of ones in the string. Such a fitness function would return a score of 2 for the chromosome in Fig. 2.

Selection Function. We select individuals at random, with the probability of selecting a particular individual weighed by its fitness. We have a greater chance of choosing individuals with better fitness score, but occasionally we get a worse individual, which is beneficial for escaping local maximums in optimisation algorithms [5]. Some selection functions always retain the top 5–10% of individuals, the elite. This ensures that the top individuals never get worse.

Crossover Function. A crossover function in most cases takes two individuals (it may take an arbitrary number) as input and returns a new individual that is the child of the two parents. The crossover function depends greatly on which problem we are solving, and is sometimes trial and error. Crossovers are usually performed by splitting both parents' chromosome at one or more point and gluing the parts from both parents together.

Mutation Function. The last of the main functions used in a GA is the mutation function. This function is called from inside the crossover function before the newly created child is returned. The mutation function changes one or more genes with a certain mutation probability.

4   GA to Solve Reachability in DCR Graphs

To solve reachability in DCR graphs, we must find a sequence of events to execute before the goal event is in an executable state, so a gene will be the id of an event to execute. A chromosome will be a list of these ids.

---

[1] While this problem may seem trivial, because we already know the answer, one can think of the 1's as the binary representation of yes to the answer of a more complex question. Our problem therefore expands into: What is the maximum number of questions I can get a yes from.

Fitness Function. Suppose we are trying to execute an event G, the "Goal" event. Looking at Definition 3, we know that an event G can be executed if:

1. G is included.
2. All events that have a condition to G have been executed or excluded.
3. All events that have a milestone to G are excluded or are not pending.

Therefore, we score the execution of an event X when it has:

1. an include to G. This score is only considered if G is excluded.
2. a condition to G. This score is only considered if X has not been executed previously. An event Y that has a condition to X must also be given a score and so on.
3. a milestone to G. This score is only considered if the X is pending. An event Y that has a milestone to X must also be given a score, in the same way as the condition above.
4. an exclude relation to events that would have given some score according to 2 or 3 if executed.

(1) is easy to check. Before we consider (2), (3) and (4) it is helpful to recall the example graph in Fig. 1. Here, it is more beneficial to execute C than B: Executing C instantly enables execution of *Goal*, while executing B also requires executing D to enable execution of *Goal*. The fitness function must therefore reflect the distinction that C is better to execute than B.

We achieve (2) and (3) by performing a Breadth-First Search (BFS) in the graph of events and relations from G following only condition and milestone relations and store a score on the events we encounter and return that score in the fitness function when the event is executed. We decrease the score the further we are from G. The score stored on events that should be scored according to (2) or (3) is $\frac{1}{\text{depth of BFS}}$. This ensures that the execution of an events that is closer to G returns a higher fitness score than an event further away.

We achieve (4) by giving events that exclude other events the score of the event that they exclude, if the excluded event has a score according to (2) or (3). To be eligible for a score the event that becomes excluded must fulfil the requirements e.g. for condition score that the event has never been executed and for milestone score that the event is pending.

The events in the graph in Fig. 1 would be scored the following way. D has a condition to *Goal* and will get a score of 1. Executing C will exclude D, so C will also get a score of 1. A however, is two conditions away from *Goal*, and will only get a score of $\frac{1}{2}$. B excludes A and will therefore also get a score of $\frac{1}{2}$.

The *selection function* simply distributes probabilities according to fitness, then selects a parent randomly.

The *crossover function* is implemented by choosing alternating genes from the two selected parents. If we reach the end of one parents chromosome, we just append the rest of the genes of the second parent to the child's chromosome. This is equivalent to having a crossover point after every gene[2].

---

[2] We attempted other types of crossover, none of which improved performance.

The *mutation function* has three equiprobable mutations:

1. Change a gene (execution) to another random execution value.
2. Append a gene with a random value to the end of the chromosome.
3. Remove a gene at a random position in the chromosome.

These mutations allow chromosomes to contain genes that represent executions of events that are not executable.

The last parameters of the genetic algorithm are the mutation probability, which is set to 10% (we tried 5% with slightly worse results), the initial population size which is set to 50 and the starting length the individuals' chromosomes which is set to 2. The rather high mutation probability is necessary, as difficult to find paths require more than 2 preceding executions, so we must force the chromosomes to grow rather often.

The low initial population size allows us to generate new generations quicker. Higher initial population sizes were tried, but generally performed worse. The starting length the individuals' chromosomes is set low, as the algorithm provides better solutions faster for events where few preceding executions are necessary.

Post-processing. The mutation function above allows paths that actually cannot execute under DCR semantics. We remove such events before returning a solution, and they do not contribute to the fitness score; however, they are important because they may become legal after crossover or mutation.

5   Experiments

We evaluated the genetic algorithm on six real-world models and three artificial ones, comparing results with those of the brute-force state exploration tool dcri [8,9].

We list the 9 models in Table 1. The real-world models were kindly provided by Exformatics A/S and DCR Solutions A/S, except for BigBelt, which was the result of a case study [11]. We note that the BigBelt model is the largest publicly DCR model available, and that the Dreyer model is known to be unfeasible for the dcri tool [13]. The artificial models "hard" and "harder" were constructed by dcri authors and pre-date the present study. Finally, we are grateful to Tijs Slaats for suggesting the model "milestones" as a model that has minimal paths exponentially larger than the model itself.

The tests were run on a machine equipped with an Intel Core i7-6700 processor. The genetic algorithm used the same amount of memory around 18–20 MB on all models whereas the dcri tools memory footprint depends on the state-space of a model (~5 GB for the model "harder"). Because the GA is non-deterministic, reported timings are the average over 100 queries; because dcri is deterministic, reported timings are the result of a single run.

Table 1. Overview of graphs used in experiments

Graph	Events[a]	Relations[b]	Comment
Real-World Models			
Dreyer	31(4)	205	Executable model running Dreyer systems [12,13,27]
BigBelt	65(3)	293	Danish Railways emergency response processes [11]
CreateCase	12	24	Used in SPIN-based model-checking case-study [20]
Court	44(1)	135	Evaluation form for a Danish Arbitration Court [26]
Complaint	49	275	Complaints process from property evaluation
Pension	25(3)	134	Application process of a major Danish pension fund
Artificial Models			
hard	25	42	Hard for state-space exploration
harder	29	49	Even harder for state-space exploration
milestones	8	56	Has paths exponentially larger than the model itself

[a] Number of events with no self-conditions; number with self-condition in parenthesis.
[b] Number of relations after flattening out nestings [16].
Real-world models (except BigBelt) kindly provided by Danish vendors of process technology, Exformatics A/S and DCR Solutions A/S.

Real-world model timing results are reported in Table 2. For both algorithms, on each model, we ran the algorithm on each event of the model[3], with a timeout of 10 s per event. For those graphs where one or more event was not reached, we reran all searches with a timeout of 60 s per event.

Each row in the table reports the aggregate time consumption of consecutively searching for each event in the model. The columns in the table are:

1. "Total", the total time spent searching for paths for each event in the graph.
2. "Max", the maximum time spent on a single event (i.e., on the most difficult event); or DNF (Did Not Finish) in case of timeouts.
3. "Avg.", the average time spent per event.
4. "Misses", the percentage of queries failing to find a path.

When the 10-s timeout search succeeded on all events, we did not repeat the search with a 60-s timeout, hence the cells containing "–".

Note that whereas dcri deterministically either always finds a path or never does, the GA randomly finds or fails to find certain paths. Also note that for GA, "Total" is the average over all runs, whereas "Max" is maximum, so it is possible for "Max" to be larger than "Total". Finally, we note that the GA does not consistently miss: Every event was reached in the majority of runs.

Artificial model timing results are reported in Table 3. We ran both tools with timeouts of 50 min, searching only for a designated goal event.

---

[3] A common trick in DCR models is to have events prevented from ever executing by having a condition to themselves. We did not search for paths to such events.

Table 2. Comparison of results, Real-world models

Graph	10 s timeout				60 s timeout			
	Total (s)	Max (s)	Avg. (s)	Misses (%)	Total (s)	Max (s)	Avg. (s)	Misses (%)
Genetic algorithm								
Dreyer	36.18	DNF	1.17	6.06	49.82	DNF	1.61	0.06
Union	0.28	1.24	0.02	0	–	–	–	–
BigBelt	6.45	2.78	0.10	0	–	–	–	–
Court	0.70	0.19	0.02	0	–	–	–	–
Complaint	61.46	8.24	1.25	0	–	–	–	–
Pension	4.47	DNF	0.15	0.20	6.18	10.25	0.23	0
Brute-force (dcri)								
Dreyer	154.00	DNF	4.97	35.48	595.98	DNF	19.23	25.80
Union	0.08	0.05	0.01	0	–	–	–	–
BigBelt	386.57	DNF	5.94	47.69	1405.40	DNF	21.62	26.15
Court	6.00	1.23	0.14	0	–	–	–	–
Complaint	247.00	DNF	5.06	40.81	966.64	55.12	19.73	0
Pension	45.01	DNF	1.80	16.00	155.94	32.91	6.00	0

Table 3. Comparison of results, Artificial models (50 m timeout)

Graph	Goal	GA (s)	dcri (s)
hard	GOAL	0.15	976.00
harder	GOAL	0.32	DNF
milestones	A_8	257.39	0.03

6 Discussion

Real-World Models. For *all* of these, the GA finds solutions more often than dcri.

1. With the 10-s timeout, of the 6 models, the GA fails on 2, missing 6% and 0.2%; whereas dcri fails 4, missing 16%, 35%, 41%, and 48%.
2. With the 60-s timeout, GA fails on 1 model, missing only 0.06% on Dreyer; whereas dcri fails on 2 models, missing ca. 25% of events on Dreyer and BigBelt.

On all but the "Union" model, the GA is also faster:

1. Total time for the GA is between 4 and 12 times smaller than dcri at the 10-s timeout. It is always an order magnitude smaller than dcri.
2. Average time for the GA is between 4 and 594 times smaller than dcri at the 10-s timeout.

Timing results differ on Union likely because the model is small enough to be in the "sweet spot" for a brute force solution. We see in Table 1 that Union is indeed the smallest model, both in terms of number of events and number of relations. In particular, it has an order of magnitude fewer relations than other models, and so likely has a much smaller state-space.

In summary, our data indicates that for real-world models, the GA is universally better on all but very small models.

Artificial Models. As expected, dcri performed very poorly on the models "hard" and "harder" designed to have large state spaces, solving "hard" after ca. 16 min and failing to terminate on "harder". Perhaps surprisingly, the GA performs excellently on both, solving either in substantially less than a second.

The "milestones" model consists of 8 events $\{A_1 \ldots A_8\}$. Each event has a milestone to all events with a higher number and a response to all events with a lower number. This results in an execution pattern where the executions necessary to execute A_{n-1} must be repeated to execute A_n; it easy to prove by induction that in a graph of this shape, the shortest path to executing A_k has length 2^{k-1}, so executing A_8 requires a path of length 128.

However, the state-space of this model is tiny: Each event in the graph can be in only 3 distinct states, namely "not executed and pending", "executed and not pending" or "executed and pending"; with 8 events, it follows that the state-space contains at most a tiny $3^8 = 6461$ possible states, making it easy to solve for dcri. The GA struggles with this model, only finding a path in ca. 4.2 min. To see why, note that the longer the shortest path, the more often the GA has to randomly find a good execution. Moreover, the infinite execution pattern $(A_1 A_2)^+$ will return a higher fitness score the more times it is run.

The paths discovered by the GA in the real-world graphs is generally in the single digits for BigBelt and around 20 for Dreyer. However, these graphs contain many more events and relations—cf. Table 1—and therefore presumably exponentially more states to explore, making them more challenging for dcri.

GAs are not an exact science: a small change to the fitness, mutation or crossover function might substantially change results. We have found suitable parameters for these functions by trial and error, and it is likely that a more efficient implementation can be found. Furthermore smaller optimisations can possibly make the performance of the algorithm better. One example is to implement the events as bit vectors as done in other DCR engines [18].

Shortest Paths. We note that our GA algorithm does not necessarily find the shortest path; not even typically. For example, refer to the graph in Fig. 1: The current fitness function would give executions $\langle B, D \rangle$ the combined condition scores of A and D while executing C only would give the condition score of D. Therefore in the eyes of the GA, the path $\langle B, D, Goal \rangle$ is better than $\langle C, Goal \rangle$.

7   Conclusion

We have implemented event reachability for DCR graphs using a Genetic Algorithm, and evaluated the resulting algorithm against an existing brute-force solutions. On medium-sized real-world models and up, the Genetic Algorithm both finds a solution more often, and does so more quickly.

Future Work. The event reachability problem can be captured directly in DECLARE by the constraint Existence$(1, G)$, however, it is unclear that a computationally feasible means of checking consistency of a model with that constraint exists. As such, it seems promising to apply the ideas of the present paper also to DECLARE.

For practical applications, an even stronger algorithm may be obtained by *combining* a brute-force algorithm with the genetic approach, in an effort to find shortest paths, and to discover the major variations of possible paths.

Finally, there is room to improve existing brute-force solutions by employing traditional model checking techniques [6], e.g., symbolic verification techniques.

Acknowledgements. We gratefully acknowledge helpful comments from anonymous reviewers, and insightful discussions with Tijs Slaats.

References

1. van der Aalst, W.M.P.: The application of Petri nets to workflow management. J. Circuits Syst. Comput. **8**, 21–66 (1998)
2. van der Aalst, W.M.P., Pesic, M.: DecSerFlow: towards a truly declarative service flow language. In: Bravetti, M., Núñez, M., Zavattaro, G. (eds.) WS-FM 2006. LNCS, vol. 4184, pp. 1–23. Springer, Heidelberg (2006). https://doi.org/10.1007/11841197_1
3. van der Aalst, W.M.P.: Verification of workflow nets. In: Azéma, P., Balbo, G. (eds.) ICATPN 1997. LNCS, vol. 1248, pp. 407–426. Springer, Heidelberg (1997). https://doi.org/10.1007/3-540-63139-9_48
4. van der Aalst, W.M.P., de Medeiros, A.K.A., Weijters, A.J.M.M.: Genetic process mining. In: Ciardo, G., Darondeau, P. (eds.) ICATPN 2005. LNCS, vol. 3536, pp. 48–69. Springer, Heidelberg (2005). https://doi.org/10.1007/11494744_5
5. Anderson-Cook, C.M.: Practical genetic algorithms (2005)
6. Baier, C., Katoen, J.: Principles of Model Checking. MIT Press, Cambridge (2008). https://books.google.dk/books?id=nDQiAQAAIAAJ
7. Basin, D.A., Debois, S., Hildebrandt, T.T.: In the nick of time: proactive prevention of obligation violations. In: IEEE 29th Computer Security Foundations Symposium, CSF 2016, pp. 120–134. IEEE Computer Society (2016)
8. Debois, S., Hildebrandt, T.: The DCR workbench: declarative choreographies for collaborative processes. In: Behavioural Types: from Theory to Tools. Automation, Control and Robotics, pp. 99–124. River Publishers, June 2017
9. Debois, S., Hildebrandt, T., Marquard, M., Slaats, T.: Hybrid process technologies in the financial sector: the case of BRFkredit. In: vom Brocke, J., Mendling, J. (eds.) Business Process Management Cases. MP, pp. 397–412. Springer, Cham (2018). https://doi.org/10.1007/978-3-319-58307-5_21

10. Debois, S., Hildebrandt, T., Slaats, T.: Replication, refinement & reachability: complexity in dynamic condition-response graphs. Acta Inform., 1–32 (2017)
11. Debois, S., Hildebrandt, T.T., Sandberg, L.: Experience report: constraint-based modelling and simulation of railway emergency response plans. In: ANT 2016/SEIT 2016: Affiliated Workshops, pp. 1295–1300 (2016)
12. Debois, S., Hildebrandt, T.T., Slaats, T., Marquard, M.: A case for declarative process modelling: agile development of a grant application system. In: EDOC Workshops 2014, pp. 126–133. IEEE Computer Society (2014)
13. Debois, S., Slaats, T.: The analysis of a real life declarative process. In: IEEE Symposium Series on Computational Intelligence, SSCI 2015, Cape Town, South Africa, 7–10 December 2015, pp. 1374–1382. IEEE (2015)
14. Di Francescomarino, C., et al.: Genetic algorithms for hyperparameter optimization in predictive business process monitoring. Inf. Syst. **74**, 67–83 (2018)
15. Hildebrandt, T., Mukkamala, R.R.: Declarative event-based workflow as distributed dynamic condition response graphs. In: Post-proceedings of PLACES 2010. EPTCS, vol. 69, pp. 59–73 (2010)
16. Hildebrandt, T., Mukkamala, R.R., Slaats, T.: Nested dynamic condition response graphs. In: Arbab, F., Sirjani, M. (eds.) FSEN 2011. LNCS, vol. 7141, pp. 343–350. Springer, Heidelberg (2012). https://doi.org/10.1007/978-3-642-29320-7_23
17. Hull, R., et al.: A formal introduction to business artifacts with guard-stage-milestone lifecycles (2011)
18. Madsen, M.F., Gaub, M., Høgnason, T., Kirkbro, M.E., Slaats, T., Debois, S.: Collaboration among adversaries: distributed workflow execution on a blockchain. In: 2018 Symposium on Foundations and Applications of Blockchain (2018)
19. Márquez-Chamorro, A.E., Resinas, M., Ruiz-Cortés, A., Toro, M.: Run-time prediction of business process indicators using evolutionary decision rules. Expert. Syst. Appl. **87**, 1–14 (2017)
20. Mukkamala, R.R., Hildebrandt, T., Slaats, T.: Towards trustworthy adaptive case management with dynamic condition response graphs. In: Proceedings of the 17th IEEE International EDOC Conference, EDOC 2013, pp. 127–136 (2013)
21. Object Management Group: Case Management Model and Notation. Technical report formal/2014-05-05, Object Management Group, version 1.0, May 2014
22. Object Management Group BPMN Technical Committee: Business Process Model and Notation, Version 2.0 (2013)
23. Pesic, M., Schonenberg, H., van der Aalst, W.M.P.: DECLARE: full support for loosely-structured processes. In: 11th IEEE International Enterprise Distributed Object Computing Conference (EDOC 2007), p. 287, October 2007
24. Pesic, M., van der Aalst, W.M.P.: A declarative approach for flexible business processes management. In: Eder, J., Dustdar, S. (eds.) BPM 2006. LNCS, vol. 4103, pp. 169–180. Springer, Heidelberg (2006). https://doi.org/10.1007/11837862_18
25. Pnueli, A.: The temporal logic of programs. In: 18th Annual Symposium on Foundations of Computer Science (FOCS), pp. 46–57 (1977)
26. Strømsted, R., Lopez, H.A., Debois, S., Marquard, M.: Dynamic evaluation forms using declarative modeling. In: BPM 2018 (Industry track) (2018, submitted for publication)
27. Slaats, T., Mukkamala, R.R., Hildebrandt, T., Marquard, M.: Exformatics declarative case management workflows as DCR graphs. In: Daniel, F., Wang, J., Weber, B. (eds.) BPM 2013. LNCS, vol. 8094, pp. 339–354. Springer, Heidelberg (2013). https://doi.org/10.1007/978-3-642-40176-3_28

Classifying Process Instances
Using Recurrent Neural Networks

Markku Hinkka[1,2(✉)], Teemu Lehto[1,2], Keijo Heljanko[1,3], and Alexander Jung[1]

[1] School of Science, Department of Computer Science,
Aalto University, Espoo, Finland
{markku.hinkka,keijo.heljanko,alex.jung}@aalto.fi
[2] QPR Software Plc, Helsinki, Finland
teemu.lehto@qpr.com
[3] HIIT Helsinki Institute for Information Technology, Espoo, Finland

Abstract. Process Mining consists of techniques where logs created by operative systems are transformed into process models. In process mining tools it is often desired to be able to classify ongoing process instances, e.g., to predict how long the process will still require to complete, or to classify process instances to different classes based only on the activities that have occurred in the process instance thus far. Recurrent neural networks and its subclasses, such as Gated Recurrent Unit (GRU) and Long Short-Term Memory (LSTM), have been demonstrated to be able to learn relevant temporal features for subsequent classification tasks. In this paper we apply recurrent neural networks to classifying process instances. The proposed model is trained in a supervised fashion using labeled process instances extracted from event log traces. This is the first time we know of GRU having been used in classifying business process instances. Our main experimental results shows that GRU outperforms LSTM remarkably in training time while giving almost identical accuracies to LSTM models. Additional contributions of our paper are improving the classification model training time by filtering infrequent activities, which is a technique commonly used, e.g., in Natural Language Processing (NLP).

Keywords: Process mining · Prediction · Classification ·
Machine learning · Deep learning · Recurrent neural networks ·
Long Short-Term Memory · Gated Recurrent Unit ·
Natural Language Processing

1 Introduction

Unstructured *event logs* generated by systems in business processes are used in Process Mining to automatically build real-life process definitions and as-is models behind those event logs. There are growing number of applications for predicting the properties of newly added event log cases, or process instances, based on case data imported earlier into the system [3,4,12,14]. The more the

© Springer Nature Switzerland AG 2019
F. Daniel et al. (Eds.): BPM 2018 Workshops, LNBIP 342, pp. 313–324, 2019.
https://doi.org/10.1007/978-3-030-11641-5_25

users start to understand their own processes, the more they want to optimize them. This optimization can be facilitated by performing predictions. In order to be able to predict properties of the new and ongoing cases, as much information as possible should be collected that is related to the event log traces and relevant to the properties to be predicted. Based on this information, a model of the system creating the event logs can be created. In our approach, the model creation is performed using supervised machine learning techniques.

In paper [7] we explored the possibility to use machine learning techniques for performing classification and root cause analysis for a process mining related classification task. In the paper, we tested the efficiency of several feature selection techniques and sets of features based on process mining models in the context of a classification task. One of the biggest problems with that approach is that, due to the simplicity of the features that are just numeric values, the user still needs to select and generate the set of features from which to select the final subset of features used for classification. For this purpose, we use Natural Language Processing techniques together with recurrent neural network techniques such as Long Short-Term Memory (LSTM) and Gated Recurrent Unit (GRU), the latter of which has not been used in process mining context before. These techniques can also learn more complicated causal relationships between the features in *activity sequences* in event logs. We have tested several different approaches and parameters for the recurrent neural network techniques and have compared the results with the results we collected in our earlier paper. As in our previous paper, we focus on classification tasks yielding boolean results which can be seen as responding to a query: Does this trace have the selected properties or not? This approach provides a very flexible basis for implementing additional functionalities such as, e.g., predicting the eventual duration, category, or resource usage required for the trace to complete.

The primary motivation for this paper is the need to perform prediction and classification based on *activity sequences* in event logs as accurately as possible and while simultaneously maximizing the throughput. This motivation comes from the need to build a system that can perform classification and prediction activities accurately on user configurable phenomena based on huge event logs collected and analyzed, e.g., using *Big Data processing frameworks* and methods such as those discussed in our earlier paper [6]. Again, we focus on classification response times by targeting web browser based interactive process mining tool where user wants to perform classifications and expects classification results to be shown within a couple of seconds. Due to this requirement, we also performed some additional experiments for a couple of techniques in order to speed up the classification process: Filtering out infrequent activities and truncating repeated infrequent activities.

Based on a the number of released papers on the subject of predictions and process mining, the interest in combining these subjects has been rapidly increasing. However, only in very recent years, deep learning techniques have been used to perform the actual process mining prediction tasks. [3] and [14] describe techniques for prediction cycle times and next activities of ongoing traces using

LSTM. In [12], the authors further improve the LSTM based prediction technique by also incorporating a mechanism for including attributes associated to events. In contrast, our experimental system is designed to be used as a foundation to solve any classification problem based on *activity sequences*, including the prediction of the next activity or cycle time using either LSTM or GRU. In [4], the authors present a framework for predicting outcomes of user specified predicates for running cases using clustering based on control flow similarities and then performing classification using attributes associated to events. This two phased process and the usage of event attributes are their biggest difference to our one phase process using only activity sequences.

The rest of this paper is structured as follows: Sect. 2 introduces main concepts related to this paper as well as our goals. Section 3 presents the test system, framework, and the data sets used in implementing our experiments. The results of these experiments will be presented in Sect. 4. Finally Sect. 5 draws the conclusions from the test results.

2   Problem Setup

The concepts and terminology used throughout this paper mostly follow those commonly used in process mining and machine learning communities. However, the following subsections will provide short introduction to the most important concepts related to this paper. For more detailed examples and discussion about *event logs*, *activity sequences* and other related terminology, see, e.g., the book by van der Aalst [15].

2.1   Classification and Prediction

A common *supervised analysis task* solved by *machine learning* techniques is to *predict* or *classify data points* based on their properties. The properties of data points are often called also as *predictors* or *features*. Predictors as well as *outcomes* of the prediction can be either continuous or they may be of enumerated types which often are also known as *categorical values* or *labels*. When the outcome of a prediction is this kind of a categorical value, such as a binary value, the performed analysis task is often called *classification*.

Usually classification in machine learning consists of two phases: *training* a *model* and performing the actual predictions using the trained model. In the model training phase, a supervised machine learning algorithm is used to create a model which produces a predicted outcome for a data point given in a form of predictors. This model building is performed by repeatedly feeding the algorithm with training data points consisting of predictors of an actual data point and actual *outcomes* that the modeled system produced for that data point. Eventually the model *learns* to simulate the system it is modeling by becoming better and better in predicting the outcomes for the training set. As a good *training data set* is a representative sample of the actual *test data* to be used on the model, the trained model will also be able to predict also the outcomes of the actual test data. If the accuracy of predictions for a training data is much better than the accuracy achieved for the test data, the model is said to be *overfitting*: The model has been trained with the biases in the training data and it is not able to *generalize* its predictions for test data.

2.2 Recurrent Neural Networks

Artificial Neural Networks are computing systems inspired by biological neural networks constituting animal brains. They consist of simple interconnected units, also known as *neurons*. The whole network can be trained to provide desired outputs for desired inputs. *Recurrent Neural Networks* (RNN) are a kind of *deep neural networks* that are connected in a way that provides the neural network a capability to remember earlier inputs fed into the network or when producing text, the network is capable of remembering what it has produced before. For example, recurrent neural networks can be used to train to produce text sentences. In this case it is essential to know what words have been produced before. RNNs have been used for large variety of problems, such as speech recognition, machine translation and automatic image captioning. Traditional RNNs have an inherent problem called *vanishing gradient problem* that makes it very hard for them to learn long distance dependencies [2].

Long Short-Term Memory and Gated Recurrent Unit. To overcome vanishing gradient problem, more complicated cell types have been developed, such as Long Short-Term Memory (LSTM) [9] and Gated Recurrent Units (GRU) [1].

Both GRU and LSTM solve the problem using a *gating mechanism* that has multiple layers of gates, which are actually a *layers of neurons*, that optionally let information through. In LSTM, the purposes of the gates are: Forget gate determines what information to throw away from the current hidden state, input gate layer decides which values to update and output gate decides which values the cell should output. In GRU, there are only two gates: Update gate determines how much of the previous hidden state needs to be passed along to the future, whereas reset gate determines how much of the previous hidden state to forget.

All of these layers are trained as any other neural network which usually involves defining a cost function and using some method of gradient descent to find out the optimal parametrization. The size of the hidden state defines how long vector of numeric values is used to store the internal state of the unit. The larger the hidden state size is, the more the model has potential for learning while also taking more time to train. When using too large hidden states compared to the actual modeled phenomenon, there is also a risk of overfitting the training data.

According to the empirical evaluations [2,10], there is no clear winner on whether GRU or LSTM is the preferred choice. Both the architectures yield models with similar performance characteristics. However, due to GRU having fewer parameters to train, it has the reputation of being somewhat faster to train. In this paper we want to see if these observations carry over to process mining.

2.3   Natural Language Processing

Natural Language Processing is a field of study that focuses on studying inter-actions between human languages and computers. It is used to tackle problems involving speech recognition, understanding natural language and generating natural language. In the context of this paper, we use similar approach often used in tasks requiring understanding of a natural language. In a way, we produce a new language that consists of sentences consisting of words that represent activities within traces. Using these artificial sentences and labels, representing the desired labeling of the trace attached to each sentence, we train a deep neural network model to predict the eventual labels for these sentences, even for activity sequences that represent process instances that have not yet finished.

2.4   Process Instance Classification

The goal of this paper was to produce a classification label using a trained RNN for any given activity sequence based only on the activity identifiers contained in the sequence itself. The actual labels for activity sequences used in this paper were of boolean-type, but the used algorithms should work equally well also with more than two possible labels. We did not set any limitations for the actual property being labeled. However, in this paper we concentrated especially in trace throughput time related properties, but it can as easily be related to, e.g., used resources, trace value or its type. We also experimented with a couple of RNN-based approaches in predicting the eventual classification for unfinished traces.

3   Experimental Setup

Tests were performed using five publicly available data sets. Table 1 shows the details of each tested dataset including the number of traces, number of positive classifications, the maximum activity sequence lengths of traces and the number of unique activities. For all the other datasets except BPIC14, we used all the available rows. For BPIC14 we used 40000 first cases of all the available 466616 traces in order for the results to be comparable with our earlier work in [7], which had this limitation. For every data set, we selected at least one property that somehow split the model into two segments with roughly 20%–40% of all the traces in the positive segment and the rest in the negative. For BPIC14 model we used two boolean labellings: Is the total duration of the case longer than 7 days, and does the case represent a "request for information" or something else. Case duration-based labeling relies only on the contents of the events in the event log, whereas the categorization uses a separate case attribute. For all the other data sets we decided to test only case durations in order for the results to be comparable with the tests performed in [7] and its extended version [8]. In BPIC12 and BPIC13, the duration threshold was set to 2 weeks. In BPIC17, this threshold was set to 4 weeks and in Hospital data set to 20 weeks.

Table 1. Used event logs and their relevant statistics

Event log	# Traces	# Positive	% Positive	Seq. length	# Activities
BPIC14-40k [18]	40000	8108/7473	20%/19%	179	39
BPIC12 [17]	13087	3330	25%	176	36
BPIC13, incidents [13]	7554	1579	21%	124	12
BPIC17 [19]	31509	11584	37%	181	26
Hospital [16]	1143	372	33%	1201	624

The input given to the test framework was a CSV file that was formatted in such a way that every row in the file had one column for the labeling and another column for the activity sequence of a single trace in the source data set. These CSV files were created using QPR ProcessAnalyzer Excel Client-process mining tool[1]. The used CSV files are available in support materials [5].

After reading these CSV files into memory, we used standard Natural Language Processing techniques. I.e., every activity sequence is treated as a sentence and every activity identifier as a word in a sentence. These sentences are then converted by assigning a unique integer identifier for each unique activity identifier and also for each classification label. Finally, when sending the activity sequences into the RNN, both in the training and in the actual validation phase, these integers representing activities were "one-hot" encoded. The actual "one-hot" encoded classification label for the trace was used as the expected classification label in the training phase.

In order to enhance the training time performance, we experimented with limiting the number of activity identifiers by only accepting N most common activity identifiers in the training set and using a special *unknown* activity identifier to represent all the rest of the activity identifiers. We also ran an experiment applying an additional truncation step where all continuous sequences of these *unknown* activity identifiers were replaced with just one occurrence of the said activity identifier.

Testing was performed on a single system having Windows 10 operating system. The used hardware consisted of 3.5 GHz Intel Core i5-6600K CPU with 32 GB of main memory and NVIDIA GeForce GTX 960 GPU having 4 GB of memory. The testing framework was built on the test system using Python programming language. The actual recurrent neural networks were built using Lasagne[2] library that works on top of Theano[3] which is an efficient mathematical expression evaluation library that can transparently perform computations in GPU and can also perform symbolic differentiation efficiently. Theano was configured to use GPU via CUDA for expression evaluation. The framework allowed testing several different hyperparameter combinations. The source code of the testing framework is available in support materials [5].

---

[1] https://www.qpr.com/products/qpr-processanalyzer.
[2] https://lasagne.readthedocs.io/.
[3] http://deeplearning.net/software/theano/.

The model was trained using Adam-gradient descent optimizer that has been found performing well with various types of neural networks [11]. We also used fixed learning rate through all the test runs referred to in this paper. Cross-entropy between the predicted and true labeling is used as the model training cost function. Gradient clipping was also used to avoid exploding gradients problem. All the training and prediction was performed in batches of configurable size by creating a batch of sentences and then sending these batches as the training or test data for the RNN to process. Batching is used to improve the efficiency since it enables Theano to distribute calculations in bigger chunks to GPU for parallel processing. All RNN unit gates, nonlinearities and weight matrices were initialized with the default initialization values built-in to Lasagne library.

In most of the test runs, model was trained for 50 test iterations, each consisting of total of 100000 traces, which translated roughly to minimum of 166 and maximum of 5834 *epochs*, depending on the used dataset. After every iteration, prediction accuracies, Area Under the Receiver Operating Characteristic curves (AUROC) and confusion matrices were calculated for the whole validation data set. The accuracy prediction was performed separately for traces of length 25%, 50%, 75% and 100% from the original length so that a continuous subsequence starting from the first activity is used. Every test iteration consisted of one hundred thousand training runs to train the model. One training run consisted of one activity sequence and its associated outcome.

4   Experimental Results

Figure 1 shows the maximum validation set classification accuracy results for all the tested datasets separately on both the RNN types having 32 as the size of the hidden state and the number of layers set to 1. Similarly Fig. 2 shows the AUROC values in the same test runs. From these results, it can be seen that LSTM and GRU both manage to get almost the same classification accuracies in both the measurements. AUROC values indicate that both RNN types have been created in a way that the model is able to classify data quite accurately and also that the model is not a trivial one such as always predicting a certain classification.

Next we measured the time usage when training and testing models. Figure 3 shows how long it took to train the model for total of 100000 traces. Similarly Fig. 4 illustrates the average time usage for one test iteration in the test involving running the predictions four times for all the validation data set traces with all the tested activity sequence subsets. From these figures it can clearly be seen that GRU is faster to train and perform classifications with, than LSTM with similar hidden state sizes and the default initializations. Based on these results, we decided to use only GRU in our further tests.

The next step was to figure out whether it is of any use to use more than one GRU layer for our classification task. Based on our tests, it was found out that having two layers brings only very minimal value in our test case. In some of the tested datasets, it takes longer for the two layer model to even start getting any

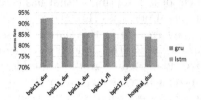

Fig. 1. Maximum classification accuracy for data sets for both the experimented RNN types

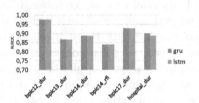

Fig. 2. Maximum AUROC values for data sets for both the experimented RNN types

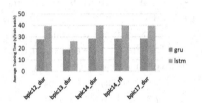

Fig. 3. Training time usage by RNN type

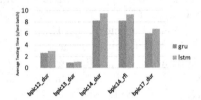

Fig. 4. Testing time usage by RNN type

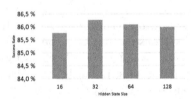

Fig. 5. The effect of hidden state size to test accuracy

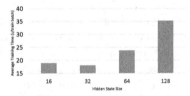

Fig. 6. The effect of hidden state size to training time usage

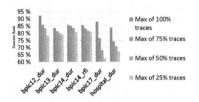

Fig. 7. Prediction accuracy for incomplete traces

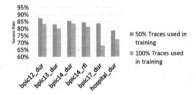

Fig. 8. Prediction accuracy for incomplete traces using model trained using incomplete traces

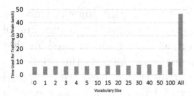 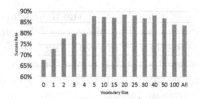

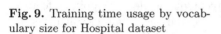

Fig. 9. Training time usage by vocabulary size for Hospital dataset

Fig. 10. Maximum classification accuracy by vocabulary size for Hospital dataset

real advantage over the always predicting the most common outcome, whereas the one layer model learns clearly faster. In addition to this, it was seen that training two layers required double the amount of time. While the maximum accuracy was in some cases slightly better for the two layer model, we chose to continue our tests only with one layer model. It is also characteristic of the test runs that the accuracy first rises from the trivial classification accuracy to its maximum, after which it starts to slowly degrade. This indicates that after certain point, the model starts to over-fit the data and does not generalize that well any more.

Next, we wanted to test the effect of the hidden state size into the accuracy of the classification. Figure 5 illustrates this by showing the average achieved accuracy for all the tested datasets, except Hospital. It can be seen that using hidden state size of 32 yields the most accurate results in the experimented cases. Since also Fig. 6 shows that the average time usage for each training iteration is the smallest when using 32 as the hidden state size, we chose 32 as the hidden state size for all the remaining experiments.

Next task we wanted to experiment with was the prediction: What is the accuracy of trace classification when the trace is still ongoing? Figure 7 shows the maximum prediction accuracies for all the experimented datasets when the model was trained only with full length traces but the validation was performed with continuous subsequences of 25%, 50%, 75% and 100% of all the activities within traces. Based on this figure, one can draw a conclusion that the prediction accuracy clearly depends on the data set and the classification task being performed. E.g., in BPIC12, BPIC13 and BPIC14 data set, it is possible to achieve over 80% accuracy when the classification is performed based on the duration of the cases even when the trace being predicted has only 50% of the activities of a full trace. However, BPIC17 and Hospital perform much worse both providing over 80% accuracy only when given full traces.

The next Fig. 8 compares the classification prediction accuracies of models built using two different methods. In the first method, the model is trained with full traces as in Fig. 7 whereas in the second method the model is trained with first 50% of the activities in the traces. In both the cases, 50% traces are used as test data set. Based on these results, a conclusion can be drawn that it is best to train the model using traces having as similar characteristics as possible to the traces used in the testing. Thus, predicting the labeling of an ongoing trace, it is recommended that the model has also been trained with ongoing traces in similar phase. The phase could be measured, e.g., by measuring a life-time of the trace thus far or even using a separate neural network model trained for that purpose. It should, however, be noted also that for some datasets the prediction works nearly as well with full traces as with partial traces. Next we compared the maximum prediction accuracies achieved with GRUs to those achieved using *Gradient Boosting Machine* (GBM)-based technique while also applying feature selection as discussed in paper [7]. The differences in the accuracies achieved in these experiments are quite consistently in favor of GRU technique. Especially in the Hospital data set the accuracy improvement was exceptionally good.

We also compared the time required for GBM to reach its maximum accuracy for each dataset, in the experiments made for [8], with the time required for GRU to train a model that has at least similar accuracy as the GBM. In this test, GBM had better response times in BPIC12 and BPIC17. Hospital training time performance was also clearly worse in GRU. Partially the reason for that was the fact that we had to use four times smaller batch size in training since the GPU in the test system did not have enough memory to use larger batch sizes used in other datasets. Another issue to be noted especially in Hospital dataset is that by using feature selection the amount of features can be brought down to a very small number, for which GBM can be performed very efficiently. However, the GRU still has to work with full activity sequences and full vocabularies.

For these purposes, we also experimented with vocabularies that were limited to a selected number of the most common activity identifiers. The results of these tests for Hospital data set are shown in Figs. 9 and 10, which illustrate that the best training times as well as accuracies were achieved using a limited vocabulary. The time required to build the model with the vocabulary size of 20, which achieved the most accurate classification results, is only about 16% of the time required to train with full vocabulary. Using this same model, the time required to reach GRU's best performance was 119 s, which still was slower than GBM's 29 s. Thus it seems that the length of the sequence is also a bottleneck. Finally, we made a test runs where vocabulary size was set to 20 and we also truncated all the sequences so that successive infrequent words that were not in vocabulary were replaced with only one occurrence of the word representing an infrequent word. In this case, the model training took about 20% less time with very small effect to the accuracy. This way, we managed to reach the GBM's best performance in 46 s. Finally using similar approach for a model built using vocabulary of size 5, the result was achieved faster than GBM. In this case, the maximum sequence length was almost halved due to the truncation down to 610 unique activity identifiers.

Thus, in the end, GRU managed to outperform more traditional GBM in all the measured metrics based on the classification accuracy and training time. GRU, which has not been earlier used in process mining context, also clearly outperformed LSTM in the required training time while still achieving similar accuracy. GRU based solutions offer also various other simple means to improve the accuracy and required training time, such as using different gradient descent optimization algorithms, modifying the learning rate and also using bigger batch sizes if there is enough memory available in the GPU. Also, in order to avoid overfitting, a regularization method, such as dropout, could be applied.

5   Conclusions

Employing Recurrent Neural Network based classification for process mining traces, processed by Natural Language Processing techniques into sequences of *words*, can achieve at least similar level of performance as feature selection and GBM based classification. One big advantage for RNN based solution is that the amount of input data required is very small; just the list of traces with their activity sequences and the classification information for the training data. For more traditional classification solutions, user needs to provide the full set of features based on which the training and classifications are made. Extracting these features out of activity sequences can be a very expensive process in itself. This feature selection was the core of our earlier paper [7]. Our experiments in this paper clearly showed that GRU based models yield more accurate classification results faster than more traditional GBM based classification using any of the structural feature combinations experimented in our previous work. All the experiments were performed on a framework that offloaded most of the calculations to GPU for improved performance and scalability.

One of our main results is the suggestion to use GRU models for predicting process instance outcomes as GRU, that has not previously been used in process mining context, is usually better choice for RNN type than LSTM mostly due to it being faster to train and its ability to achieve almost identical classification and prediction accuracy. We experimented also two approaches for the prediction of eventual classification label for still ongoing traces. From this test we found out that it is always clearly better to train a model with traces at as similar phases of their lifetime as possible to the traces being tested.

We also investigated how to improve the required training time, especially when using data sets having long activity sequences and a lot of activities compared to the number of training data activity sequences. We found out that the number of activities can be decreased by treating all the infrequent activities as one activity without it having a big effect to the classification accuracy, while still having a noticeable effect in throughput time and GPU memory requirements. We also found out that replacing long sequences of infrequent activities with just one activity representing all the infrequent activities can further improve the throughput time without it affecting dramatically into classification accuracy.

All the raw test results gathered from the performed experiments, some of which was not discussed nor explored in this paper in detail, together with the developed python source code for the test framework, can be found in the support materials [5].

Acknowledgements. We want to thank QPR Software Plc for funding our research. Financial support of Academy of Finland projects 139402 and 277522 is acknowledged.

References

1. Cho, K., van Merrienboer, B., Bahdanau, D., Bengio, Y.: On the properties of neural machine translation: encoder-decoder approaches. In: Wu, D., Carpuat, M., Carreras, X., Vecchi, E.M. (eds.) Proceedings of SSST@EMNLP 2014, Eighth Workshop on Syntax, Semantics and Structure in Statistical Translation, Doha, Qatar, 25 October 2014, pp. 103–111. Association for Computational Linguistics (2014)
2. Chung, J., Gülçehre, Ç., Cho, K., Bengio, Y.: Empirical evaluation of gated recurrent neural networks on sequence modeling. CoRR, abs/1412.3555 (2014)
3. Evermann, J., Rehse, J.-R., Fettke, P.: Predicting process behaviour using deep learning. Decis. Support Syst. **100**, 129–140 (2017)
4. Francescomarino, C.D., Dumas, M., Maggi, F.M., Teinemaa, I.: Clustering-based predictive process monitoring. CoRR, abs/1506.01428 (2015)
5. Hinkka, M.: Support materials for articles (2018). https://github.com/mhinkka/articles. Accessed 11 Mar 2018
6. Hinkka, M., Lehto, T., Heljanko, K.: Assessing big data SQL frameworks for analyzing event logs. In: 24th Euromicro International Conference on Parallel, Distributed, and Network-Based Processing, PDP 2016, Heraklion, Crete, Greece, 17–19 February 2016, pp. 101–108. IEEE Computer Society (2016)
7. Hinkka, M., Lehto, T., Heljanko, K., Jung, A.: Structural feature selection for event logs. In: Teniente, E., Weidlich, M. (eds.) BPM 2017. LNBIP, vol. 308, pp. 20–35. Springer, Cham (2018). https://doi.org/10.1007/978-3-319-74030-0_2
8. Hinkka, M., Lehto, T., Heljanko, K., Jung, A.: Structural feature selection for event logs. CoRR, abs/1710.02823 (2017)
9. Hochreiter, S., Schmidhuber, J.: Long short-term memory. Neural Comput. **9**(8), 1735–1780 (1997)
10. Józefowicz, R., Zaremba, W., Sutskever, I.: An empirical exploration of recurrent network architectures. In: Bach, F.R., Blei, D.M. (eds.) Proceedings of the 32nd International Conference on Machine Learning, ICML 2015, Lille, France, 6–11 July 2015. JMLR Workshop and Conference Proceedings, vol. 37, pp. 2342–2350. JMLR.org (2015)
11. Kingma, D.P., Ba, J.: Adam: A method for stochastic optimization. CoRR, abs/1412.6980 (2014)
12. Navarin, N., Vincenzi, B., Polato, M., Sperduti, A.: LSTM networks for data-aware remaining time prediction of business process instances. In: 2017 IEEE Symposium Series on Computational Intelligence, SSCI 2017, Honolulu, HI, USA, 27 November–1 December 2017, pp. 1–7. IEEE (2017)
13. Steeman, W.: BPI challenge 2013, incidents (2013)
14. Tax, N., Verenich, I., La Rosa, M., Dumas, M.: Predictive business process monitoring with LSTM neural networks. In: Dubois, E., Pohl, K. (eds.) CAiSE 2017. LNCS, vol. 10253, pp. 477–492. Springer, Cham (2017). https://doi.org/10.1007/978-3-319-59536-8_30
15. van der Aalst, W.M.P.: Process Mining - Discovery, Conformance and Enhancement of Business Processes. Springer, Heidelberg (2011). https://doi.org/10.1007/978-3-642-19345-3
16. Van Dongen, B.: Real-life event logs - hospital log (2011)
17. Van Dongen, B.: BPI challenge 2012 (2012)
18. Van Dongen, B.: BPI challenge 2014 (2014)
19. Van Dongen, B.: BPI challenge 2017 (2017)

Leveraging Regression Algorithms for Predicting Process Performance Using Goal Alignments

Karthikeyan Ponnalagu[1(✉)], Aditya Ghose[2], and Hoa Khanh Dam[2]

[1] Robert Bosch, Bangalore, India
karthikeyan.ponnalagu@in.bosch.com
[2] University of Wollongong, Wollongong, Australia
aditya.ghose@gmail.com, hoa@uow.edu.au

Abstract. Industry-scale context-aware processes typically manifest a large number of variants during their execution. Being able to predict the performance of a partially executed process instance (in terms of cost, time or customer satisfaction) can be particularly useful. Such predictions can help in permitting interventions to improve matters for instances that appear likely to perform poorly. This paper proposes an approach for leveraging the process context, process state, and process goals to obtain such predictions.

Keywords: Variability · Contextual factor analysis ·
Business process mining

1 Introduction

Execution of complex business processes that are specifically knowledge driven, generally leads to significant amounts of event records corresponding to the execution of activities in the processes. Most of the current literature assumes that the performance of a process instance is entirely determined by what happens over the course of the execution of the process instance. We see limitations in such assumptions [7], when applied in knowledge intense process models, where the specific instance executions are dictated by other factors that are not part of process executions. In this paper, we propose a novel approach that inter-operates with cloud based cognitive systems towards predicting process performance. Towards this, we leverage contextual factors and goal alignments associated with the actual execution of processes.

We assume the following inputs to our proposed approach: (a) a goal model hyper graph with goals and sub-goals (AND, OR) represented as a collection of boolean conditions in conjunctive normal form (CNF), (b) an event log containing multiple process instance execution data and (c) a process design annotated with normative end effects. In this paper, we consider an `Incident management` process design as our running example. A process log[1] containing 1400 executed

---

[1] https://www.scribd.com/document/333254045/IncidentLog.

© Springer Nature Switzerland AG 2019
F. Daniel et al. (Eds.): BPM 2018 Workshops, LNBIP 342, pp. 325–331, 2019.
https://doi.org/10.1007/978-3-030-11641-5_26

instances of this process design is considered for the evaluation of our proposed approach. A total of over 25000 task execution records is available as part of this process log. Each process instance in this log indicates how after receiving a complaint from a customer, an incident ticket is created, resolved and closed. We leverage annotated goal models with end effects. Such a goal model can be constructed through a goal refinement machinery as discussed in [2].

A variety of outcome predicting process monitoring techniques have been proposed in the literature [6]. In [4], the authors clearly establish the need for a general framework for mining and correlating business process characteristics from event logs. In [1], the authors discuss construction of a configurable process model as a family of process variants discovered from a collection of event logs. The existing works in the area of contextual correlation of business processes have addressed different challenges related to collaboration, contract conformance, process flexibility [5]. In comparison our work uses contextual factors and semantic effect traces on both partial and completed executions to correlate and predict execution deviation based on goal alignments. Works such as [3] focuses on generating performance predictions leveraging process simulation data. Works such as [9] focus on generating hybrid process model creation by leveraging event log clusters. In comparison, we focus on an orthogonal approach of discovering multiple process designs that are goal aligned variants of the original process design.

2   Identifying Process Context, Goals and Process State

Contextual information can be traced from process instances to a range of time-stamped information sources, such as statements being made on enterprise social media, financial market data, weather data and so on. Process log time-stamps can be correlated with time-stamps in these repositories of information to derive a wealth of information about the context within which a process instance was executed. In our proposed approach, we leverage this specific category of contextual information.

The performance indicators associated with process effect assertions are typically influenced with the entailment to specific OR-refinement sub goals (Email confirmation or Telephonic confirmation with customer) in the goal model. Given a state S and a set of effect assertions e obtained from events accruing from the execution of a task, the resulting partial state is given by $S \oplus e$, where $\oplus$ is a state update operator [8]. Similarly, given a normative state S_N and a set of effect assertions e_N obtained from events accruing from the execution of a task in a process, the resulting partial state is given by $S_N \oplus e_N$ where $\oplus$ is a state update operator. We also use a knowledge-base KB of domain constraints. If $S \cup e \cup KB$ is consistent, then $S \oplus e = S \cup e$. Otherwise, $S \oplus e = e \cup \{s \mid s \subseteq S, s \cup e \cup KB$ is consistent, and there does not exist any s' where $s \subset s' \subseteq S$ such that $s' \cup e \cup KB$ is consistent$\}$. We start with an initial partial state description (which may potentially be empty) and incrementally update it (using $\oplus$) until we reach the partial state immediately following the final task in the process instance. Towards

achieving this, the proposed machinery leverage the OR-refinement goal correlations associated with each state transition from the process event log. For
generating goal correlations based on the end effects (at the process or task levels), we have leveraged the Process Instance Goal Alignment Model (PIGA)
discussed in our previous work [8]. Therefore, given a *goal-realizing effect group*
S, finding correlation with a goal G in formal terms is simply finding the truth
assignments in the CNF expression of G using the cumulative end effects of S.
Towards generating PIGA, the list of state transitions and the goal decomposition model as input are considered. Then, for each event group in the process
log, the truth assignments of all goals in the goal model are validated. This is
repeated for all event groups in the process log to identify the "valid process
instances". The representation of each process instance as a list of *maximally
refined correlated goals* constitutes the completion of generating Process Instance
Goal Alignment (PIGA).

Table 1. Context Correlated Goal Models (CCGM)

No. of instances	Observed state effects	OR-refined goal entitlement	Context name (Value)
62	T4: (Resolution_Suggested)	(Link to Existing Problem, Close Problem)	CM1 = Connection('Remote', 'NotAvailable', 'BehindFirewall'), CustomerExpertise('High'), CustomerPriority('Low')
155	T3: (Resolution_Known)	(Link to Existing Problem, Close Problem)	CM2 = Solution('Known', 'AutoFix', 'BroadCast'), CustomerAffected('Group')
11	T5: (Resolution_Cancelled)	(Close Problem)	CM3 = Agent('New'), ProblemOrigin('3rd Party', 'NotUnderContract')
51	T5: (Ticket_NotEnriched)	(Escalate Problem)	CM4 = CustomerProvided ('NoEventTrace', 'NotReproduced')
10	T1: (Problem_NotCategorized), T9: (Problem_DetailIncomplete)	(Escalate Problem, Link to Existing Problem)	CM5 = Agent('New'), ProblemAutoCategory('Failed')
5	T2: (Problem_SeverityWarning), T3: (Set_TicketPriorityHigh)	(Escalate Problem, Enrich Problem)	CM6 = Agent('Expert') , ProblemAutoCategory('Complex')
31	T4: (Customer_NotNotified)	(Escalate Problem, Enrich Problem)	CM7 = CustomerSupport('Rare'), CustomerProvided('NoEventTrace', 'NotReproduced')

The CCGM generated for our running example is illustrated in Table 1. For example as observed in row 3, 11 process instances are partially executed without a resolution to a reported incident due to a collection of contextual factors (CM3). To support predictions both at the process and individual task levels, we have leveraged two categories of effect log data sets: **Process Data Set(PD)**, where record in this data set is a tuple { Process Instance Identifier, a semantic trace, process execution time, context, aligned OR-refinement sub-goals } and **Task Data Set(TD)**: Each record in this data set is a tuple { Process Instance Identifier, Task Identifier, semantic trace from the execution of task, task execution time, total process execution time, context, task aligned goals, process aligned goals}.

For our evaluation in this paper, we used Watson Analytics Engine's Deep QA pipeline, to generate insights for some very interesting questions. The training data set belongs to two categories of process log data sets **PD** and **TD**. The questions that were asked using both these data sets are listed in Table 2.

3   Empirical Evaluation

Our evaluation is conducted in two phases: **Phase 1:** This is basically a pre-processing step that enables generation of effect logs, which are provided as input data to the Watson Analytics Engine (discussed in Phase 2). The `VAGAI tool` [8] annotates semantic traces from process logs with goal alignments to generate process effect logs (PD) and task effect logs (TD) respectively[2]. **Phase 2:** `Watson Analytics Engine` for generating performance and goal alignment predictions using the PD and TD data sets respectively as depicted in Table 2. For individual task level executions, the alignment predictions are at OR-refinement sub goal levels (providing alternate realization of its parent goal) for a given goal model. This is based on the accumulated effects at the completion of corresponding task execution.

The consolidated view of predictive insights as a visualization is depicted in Fig. 1. Here the performance prediction in terms of total process execution time is depicted for each observed effect at completion of a task. We started with questions of type **Q01, Q02** to generate the predictions of process performance time (in minutes) for each of the six contextual factors `DataIssues + AgentExplow`, `DataIssues + Highseverity`, `RemoteResolution + CustomerNew`, `RemoteResolution + AlertsComplete`, `SoftwareUpgrade`, `PasswordReset + AgentExplow`, `PasswordReset + Severity High` at specific semantic traces in the execution of process instances. This consolidated representation generated using the *Watson Analytics Engine* helps in predicting performance at different partial states of an instance execution. This demonstrates the impact of contexts on the execution of otherwise similar process

---

[2] https://www.scribd.com/document/333254045/IncidentLog.

Table 2. Questions to Watson Analytic Engine

Question ID	Question text	Used data set	Question type
Q01	Given a performance limit – what are the most commonly occurring semantic effect traces?	TD	Exploratory
Q02	What are the context sets associated with processes taking high performance time?	TD	Exploratory
Q03	Given the effect sequence E1E2E3, what is the probability of the process being aligned for a given goal G?	PD	Predictive
Q04	Given the current effect sequence taking performance time N, what is the projected completion time of the process	PD	Predictive
Q05	Given the current context, and the current effect sequence, what is the remainder of the effect sequence for a successful (goal-aligned) execution	TD	Predictive
Q06	Given the current context, what will be the number of instances that are aligned with Goal G1?	PD	Predictive
Q07	Given the current context, what is the probability of this instance to conclude with a specific effect sequence?	PD	Predictive
Q08	Given the tickets with current effect sequence, what is the average total performance time of completion of these tickets?	TD	Predictive
Q09	Given the current context, how many executed instances will be valid?	TD	Predictive
Q10	Given the current effect sequence, which process designs the completed instances will be aligned with?	TD	Predictive

execution instances. Similarly using this prediction model represented in Fig. 1, we can make predictions of performances at multiple states of process execution. This eventually can lead the organization to evaluate their resource deployment strategies, shifting to a different process design variant.

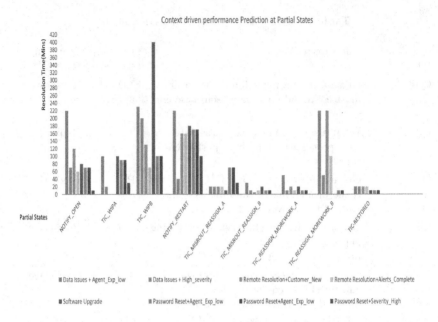

Fig. 1. Performance predictions at partial states

4   Conclusion

Organizations increasingly tend to analyze the performance drifts in day to day execution of customer and context sensitive business processes. In our proposed approach, we leverage goal correlated process variations and contextual factors mined from process log and goal correlated state transitions mined from effect logs. In our future work, we will focus on correlating dynamic run-time variations in contextual factors with shifts in goal alignment.

References

1. Buijs, J.C.A.M., van Dongen, B.F., van der Aalst, W.M.P.: Mining configurable process models from collections of event logs. In: Daniel, F., Wang, J., Weber, B. (eds.) BPM 2013. LNCS, vol. 8094, pp. 33–48. Springer, Heidelberg (2013). https:// doi.org/10.1007/978-3-642-40176-3_5
2. Ghose, A.K., Narendra, N.C., Ponnalagu, K., Panda, A., Gohad, A.: Goal-driven business process derivation. In: Kappel, G., Maamar, Z., Motahari-Nezhad, H.R. (eds.) ICSOC 2011. LNCS, vol. 7084, pp. 467–476. Springer, Heidelberg (2011). https://doi.org/10.1007/978-3-642-25535-9_31
3. Heinrich, R., Merkle, P., Henss, J., Paech, B.: Integrating business process simulation and information system simulation for performance prediction. Softw. Syst. Model. **1**, 1–21 (2015)

4. de Leoni, M., van der Aalst, W.M.P., Dees, M.: A general framework for correlating business process characteristics. In: Sadiq, S., Soffer, P., Völzer, H. (eds.) BPM 2014. LNCS, vol. 8659, pp. 250–266. Springer, Cham (2014). https://doi.org/10. 1007/978-3-319-10172-9_16
5. Magdaleno, A.M., de Oliveira Barros, M., Werner, C.M.L., de Araujo, R.M., Batista, C.F.A.: Collaboration optimization in software process composition. J. Syst. Softw. **103**, 452–466 (2015)
6. Maggi, F.M., Di Francescomarino, C., Dumas, M., Ghidini, C.: Predictive monitoring of business processes. CAiSE 2014. LNCS, vol. 8484, pp. 457–472. Springer, Cham (2014). https://doi.org/10.1007/978-3-319-07881-6_31
7. Mrquez-Chamorro, A.E., Resinas, M., Ruiz-Corts, A.: Predictive monitoring of business processes: a survey. IEEE Trans. Serv. Comput. (2017)
8. Ponnalagu, K., Ghose, A., Narendra, N.C., Dam, H.K.: Goal-aligned categorization of instance variants in knowledge-intensive processes. In: Motahari-Nezhad, H.R., Recker, J., Weidlich, M. (eds.) BPM 2015. LNCS, vol. 9253, pp. 350–364. Springer, Cham (2015). https://doi.org/10.1007/978-3-319-23063-4_24
9. Yu, Y., et al.: Case analytics workbench: platform for hybrid process model creation and evolution. In: Motahari-Nezhad, H.R., Recker, J., Weidlich, M. (eds.) BPM 2015. LNCS, vol. 9253, pp. 226–241. Springer, Cham (2015). https://doi.org/10. 1007/978-3-319-23063-4_16

First International Workshop on Emerging Computing Paradigms and Context in Business Process Management (CCBPM)

First International Workshop on Emerging Computing Paradigms and Context in Business Process Management (CCBPM)

BPM has been referred to as a "holistic management" approach to align an organization's business processes with the needs of users. It promotes business effectiveness and efficiency while striving for innovation, flexibility, and integration with technology. However, the challenge for a large-scale use of business process is the failure in addressing both the dynamic execution environment (e.g., cloud and fog) and the elastic requirement of users (i.e., logic of use). Two streams of research emerge to address this problem. In the upstream, researchers try to make explicit the contextualization process in designing flexible and elastic business process for process optimization and reuse. In the downstream, emerging computing paradigms such as mobile-cloud and fog computing could bring a promising orchestration of business process but involve a revision of BPM architecture.

The goal of CCBPM 2018 was to promote the role of emerging computing paradigms such as mobile-cloud computing, edge computing, especially fog computing, and context in business process management (BPM) by discussing (1) what the distributed computing and context community can bring to the BPM community, including business and scientific workflow management; and (2) what are the challenges of BPM and workflow system that the BPM community think these emerging computing paradigms and context (e.g., context-aware fog computing-based workflow system) may solve. After the previous CCBPM events including CCBPM 2013 (October 28, 2013, Annecy, Haute-Savoie, France), CCBPM 2014 (August 27, 2014, Beijing, China), and CCBPM 2015 (November 4, 2015, Shanghai, China), CCBPM 2018 extended its scope to accommodate the latest progress in computing paradigms especially in fog computing.

This year, the workshop received a good number of international submissions. Each paper was reviewed by at least three members of the Program Committee. Finally, the top five were accepted as full papers for presentation at the workshop. These papers provide a good coverage of hot topics in BPM. Rongbin Xu, Yeguo Wang, Yongliang Cheng, et al. propose a workflow scheduling algorithm based on improved particle swarm optimization for seeking the best trade-off between makespan and cost in a cloud-fog environment. Junhua Zhang, Dong Yuan, Lizhen Cui, and Bing Bing Zhou investigate the trade-off problem of resource utilization and propose a provenance candidates elimination algorithm that can efficiently find the minimum cost strategy for data storage, transfer, and regeneration BPM in multiple clouds. Christian Sturm, Jonas Szalanczi, Stefan Schonig, and Stefan Jablonski present a novel lean architecture of a blockchain-based process execution system with smart contracts to dispense with a trusted third party in the context of inter-organizational collaborations. Yuanchun Jiang, Cuicui Ji, Yang Qian, and Yezheng Liu present a two-phase approach to discover related cloud services by jointly leveraging the descriptive texts of services and their associated tags to achieve better recommendation results than traditional service clustering and recommendation methods. Finally, Tianhong Xiong, Yang Yu, Maolin Pan, and Jing Yang propose a workflow modeling approach based on state machines to design a crowdsourcing model that can support complex crowdsourcing tasks, especially creative tasks. This year, the workshop also featured a keynote speech delivered by Professor Shiping Chen on "Blockchain—Yesterday, Today, and Tomorrow."

We hope that the audience will find this year's papers interesting and useful to keep track some of the latest topics in the area of computing paradigms and context technologies for BPM, and we look forward to bringing you the future editions of the CCBPM workshop.

November 2018

Xiao Liu
Xi Zheng
Michael Sheng
Shiping Chen

Organization

Program Committee

Xiaoliang Fan	Xiamen University, China
Marc Frincu	West University of Timisoara, Romania
Christian Janiesch	Karlsruhe Institute of Technology, Germany
Xuejun Li	Anhui University, China
Xiao Liu	Deakin University, Australia
Xi Zheng	Macquarie University, Australia
Shiping Chen	CSIRO DATA61, Australia
Haoyu Luo	South China Normal University, China
Massimo Mecella	Sapienza University, Italy
Flavia Santoro	Federal University of the State of Rio de Janeiro, Brazil
Fei Teng	Southwest Jiaotong University, China
Lijie Wen	Tsinghua University, China
Dong Yuan	University of Sydney, Australia
Marielba Zacarias	University of Algarve, Portugal
Gaofeng Zhang	Hefei University of Technology, China
Xuyun Zhang	Auckland University, New Zealand
Wei Zheng	Xiamen University, China

Improved Particle Swarm Optimization Based Workflow Scheduling in Cloud-Fog Environment

Rongbin Xu[1,2], Yeguo Wang[1], Yongliang Cheng[1], Yuanwei Zhu[1], Ying Xie[1,2(✉)], Abubakar Sadiq Sani[3], and Dong Yuan[3]

[1] School of Computer Science and Technology, Anhui University, Hefei 230601, China
{xurb_910,xieying}@ahu.edu.cn,
e1620106l@stu.ahu.edu.cn, 1399603339@qq.com,
575072641@qq.com

[2] Co-Innovation Center for Information Supply and Assurance Technology, Anhui University, Hefei 230601, China

[3] School of Electrical and Information Engineering, University of Sydney, Sydney, Australia
{sadiq.sani,dong.yuan}@sydney.edu.au

Abstract. Mobile edge devices with high requirements typically need to obtain faster response on local network services. Fog computing is an emerging computing paradigm motivated by this need, which currently is viewed as an extension of cloud computing. This computing paradigm is presented to provide low commutation latency service for workflow applications. However, how to schedule workflow applications for seeking the tradeoff between makespan and cost in cloud-fog environment is facing huge challenge. To address this issue, in current paper, we propose a workflow scheduling algorithm based on improved particle swarm optimization (IPSO), where a nonlinear decreasing function of inertia weight in PSO is designed for promoting PSO to gain the optimal solution. Finally, comprehensive simulation experiment results show that our proposed scheduling algorithm is more cost-effective and can obtain better performance than baseline approach.

Keywords: Cloud computing · Fog computing · Workflow scheduling · PSO

1 Introduction

The explosive growth of information and data in the Internet age has brought more attention to cloud computing. In recent years, with the continuous popularity of Internet of Things (IoT) technology, traditional network architecture of cloud computing framework is facing great challenge. Currently, a large number of smart IoT devices located at the edge of network require data processing with low latency, location-aware and mobility requirements. To cope with huge number of end-user IoT devices and big data volumes for real-time low-latency applications, Bonomi et al. [1] first proposed the

© Springer Nature Switzerland AG 2019
F. Daniel et al. (Eds.): BPM 2018 Workshops, LNBIP 342, pp. 337–347, 2019.
https://doi.org/10.1007/978-3-030-11641-5_27

concept of "fog computing" aiming to extend the application scenarios of computing. Fog computing is a new computing paradigm that maintains the advantages of cloud computing, which can be introduced as an "edge network cloud". Fog resources can provide local computing, storage and network for end-user applications. The location of fog devices is close to end-user applications so that it can process tasks with low-latency in high response. On the other hand, the flexibility and scalability of cloud computing can help fog computing to cope with the growing demands for large-scale computation-intensive business applications when the processing capacity of fog computing are insufficient. It is obvious that fog computing can compensate the shortage of cloud computing [2].

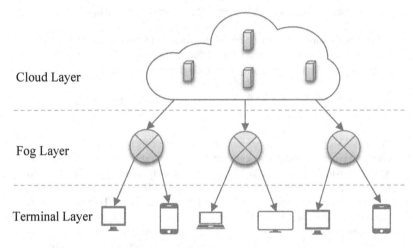

Fig. 1. Architecture of fog computing

In general, the architecture of fog computing is composed of three layers of network structure. As shown in Fig. 1, the bottom layer is regarded as terminal layer, which mainly consists of IoT devices such as smartphone, smart wearable devices, smart home appliances and so on. These terminal devices send service requests to upper layer. The middle layer is viewed as fog layer, which contains fog devices with the capabilities of computing and storage such as routers, switches and gateways. These fog devices can be employed by near end-user devices so as to quickly process those time-sensitive tasks. Meanwhile, fog devices also need to connect to the cloud layer for offloading those latency-tolerant and computation-intensive tasks on the demand of users. The upper layer is called as cloud layer, which hosts a large number of heterogeneous virtual machines that provide abundant resources to process the task requests dispatched from fog layer.

Although fog computing has many advantages, it also faces enormous challenges. One of challenges is the allocation of different resources for the scheduling of business tasks [3], especially workflow applications scheduling in fog environment where any workflow application contains many tasks with communication constraint or temporal dependency [4]. In recent years, with fast development of big data, cloud, fog and edge computing, many business workflow applications have been emerged and efficient scheduling of workflow applications has become a hot spot for scholars at home and abroad. For example, the video intelligent surveillance application typically is a low latency workflow application which is composed of five modules [5], including motion detector, object detector, object tracker, user interface and pan, tilting, and zoom control (PTZ). In such application, the motion detector module and PTZ control module are normally executed in the fog nodes, and the user interface is always in the remote cloud. The remaining two modules are executed in the fog nodes or cloud nodes according to the decision-making policies. Therefore, how to design an efficient workflow scheduling algorithm in a cloud and fog environment is crucial, which can highly improve the quality of service (QoS) for providing better user experience.

In this paper, we mainly focus on workflow scheduling in cloud-fog environment and present our scheduling method based on improved particle swarm optimization (IPSO) for workflow applications. The update way of the inertia weight in original PSO is changed by a novel nonlinear decreasing function, which can balance and adjust the search capability of particles in search process. Then a scheduling algorithm is designed by considering the actual problem of workflow applications in cloud-fog environment.

The remainder of this paper is organized as follows. Section 2 presents the related work. Section 3 describes the scheduling problem in cloud-fog environment and proposes our solution based on IPSO. Section 4 elaborates the detail experimental settings and results. Section 5 concludes the paper and puts forward some future work.

2 Related Work

There are many existing works regarding the task or workflow scheduling in distributed environment, especially task scheduling under cloud computing platform [6–8]. The authors in [6] present a market-oriented hierarchical scheduling algorithm in cloud workflow. The scheduling algorithm contains two levels, include the service-level scheduling which deals with the Task-to-Service assignment and the task-level scheduling which deals with the optimization of the Task-to-VM assignment. The authors in [7] propose a near-optimal dynamic priority scheduling (DPS) strategy for instance-intensive business workflows that have a larger number of parallel workflow instances. The authors in [8] propose a hybrid PSO algorithm based on non-dominance sort for handling the workflow scheduling problem with multiple objective in the cloud.

Meanwhile, there is a small amount of studies about task scheduling in the cloud and fog environment [9–12]. Hoang et al. [9] first design a fog-based region architecture for providing nearby computing resources. They investigate efficient scheduling algorithms to allocate tasks among regions and remote clouds. Pham et al. [10] propose a cost-makespan aware workflow scheduling for achieving the balance between performance of application execution and monetary cost for using cloud resources. Besides, an efficient task reassignment strategy based on critical path is also presented to satisfy the user-defined QoS constraints. Tang et al. [11] present a mobile cloud-based scheduling method for the industrial internet of things, which views energy consumption as the main optimization objective of the task scheduling problem while taking into account task dependency, data transmission and other constraint conditions. Zeng et al. [12] consider fog computing as support software-defined embedded system. The authors mainly address three issues: (1) how to balance the task workload and computation servers; (2) how to place task images on storage servers; (3) how to balance the I/O interrupt requests from storage servers.

All the above researches mainly focus on task scheduling problem which have not considered task scheduling with temporal dependency and workflow task scheduling based on traditional method in hybrid cloud and fog environment. As we know, task or workflow scheduling in distributed computing environment is viewed as an NP-hard problem. PSO, as one of intelligent meta-heuristic algorithms, has been applied in addressing the scheduling problem in many fields [13]. However, there are only very few researches about workflow application scheduling problem based on PSO in cloud-fog environment. Thus, this paper will consider PSO-based scheduling algorithm as our basic model.

3   Problem Formalization and Solution

Workflow application scheduling in cloud-fog environment is defined as the formulated problem of assigning computing resources with different processing abilities to the tasks of workflow application, which can minimize the makespan and cost of workflow application scheduling. To formulate this issue, we apply directed acyclic graph (DAG) to represent a workflow application. In addition, a PSO based scheduling algorithm is proposed for solving the mapping process between tasks and computing resources. Then we also present the solution for the proposed issue in this section.

3.1   Problem Formalization

In general, a workflow application is composed of a set of tasks, which has similar structure to DAG. So we employ the workflow application by DAG as shown in Fig. 2. We denote a DAG as $G = (T, E)$, where T indicates a set of tasks and E represents the set of temporal dependency or communication constraint between pairs of tasks. Specifically, each task $t_s \in T(T = \{t_1, t_2, \cdots, t_n\})$ has its own computation workload cw_s. Accordingly, each directed edge $e_{ij} = <t_i, t_j> \in E$ means that t_j cannot be executed until t_i is completed and e_{ij} has its nonnegative weight value cv_{ij} which represents the communication data transferring t_i to t_j. The task t_i without direct predecessors is denoted as T_{start} and the task t_j without direct successors is denoted as T_{end}. Here, we assume that a task cannot be implemented until all the direct precedent tasks have been completed.

A. Makespan

In addition, computing resources in cloud-fog environment are divided into two types, i.e., fog servers and cloud servers. For a workflow task, it is either processed by fog servers or cloud servers. If the task t_s is submitted to a server, the execution time of t_s can be calculated as follows:

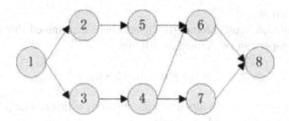

Fig. 2. A simple DAG example of workflow application

$$T_{t_s}^l = \frac{cw_s}{\delta_l} \tag{1}$$

where δ_l represents the processing rate of the computing server l. Let $ct\left(e_{ij}^l\right)$ be the commutation time for transferring data from t_i to t_j, which is defined by

$$ct\left(e_{ij}^l\right) = \frac{cv_{ij}}{B} \tag{2}$$

where B in Eq. (4) is the network bandwidth between two servers. Duo to task t_s will be allocated to one server, this task has its start time ST_{t_s} and finish time FT_{t_s}, which are represented by Eqs. (3) and (4), respectively.

$$ST_{t_s} = \max\{FT_{t_p} + ct\left(e_{ps}^l\right), t_p \in pre(t_s)\} \tag{3}$$

$$FT_{t_s} = ST_{t_s} + T_{t_s}^l \tag{4}$$

where $pre(t_s)$ denotes the set of direct predecessors of task t_s.

The time period from the start of the first task to the completion of the last task is called the makespan of the entire workflow, which is calculated by Eq. (5).

$$T_{total} = \max\{FT_{t_s}, t_s \in \mathrm{T}\} \tag{5}$$

B. Economic cost

The corresponding computation cost of server l is

$$C_{t_s}^l = p_l * (FT_{t_s} - ST_{t_s}) \tag{6}$$

where p_l is the unit price of server l. In this paper, we only consider the computing cost of renting servers. So the total cost can be denoted as C_{total}, which is the computing cost of servers. That is

$$C_{total} = \sum_{l=1}^m \sum_{s=1}^n C_{t_s}^l \tag{7}$$

where m and n indicate the number of servers and the number of tasks, respectively.

C. Objective function

Based on the total makespan and cost which have been determined above, the objective function of this paper can be defined as follows:

$$f = 0.5 * T_{total} + 0.5 * C_{total} \tag{8}$$

Equation (8) takes two main factors into consideration, which can help to maintain the balance between makespan and economic cost.

3.2   Solution

In this subsection, we further discuss IPSO and IPSO-based workflow scheduling algorithm in cloud-fog environment.

A. Improved particle swarm optimization algorithm

The original particle swarm optimization (PSO) algorithm is derived by the study of predation behaviors of flock birds, which is an intelligence evolutionary computing technology. The basic idea of PSO is to search the optimal solution through the cooperation and information sharing among individuals in a group. Suppose that one population has N particles and the searching space is D dimensional. For a particle $P_i(i = 1, 2, \cdots, N)$, it has two typical parameters, i.e., the position $X_i = (x_{i1}, x_{i2}, \cdots, x_{iD})$ and the velocity $V_i = (v_{i1}, v_{i2}, \cdots, v_{iD})$. The optimal position of particle individual is represented as *pbest* and the global optimal position of the whole population is denoted as *gbest*. In k^{th} iteration of PSO, the velocity and position of particle will be updated by the following two equations:

$$V_i^k = w^{(k)} \cdot V_i^{k-1} + c_1 \cdot r_1 \cdot \left(pbest_i - X_i^{k-1}\right) + c_2 \cdot r_2 \cdot \left(gbest - X_i^{k-1}\right) \tag{9}$$

$$X_i^k = X_i^{k-1} + V_i^k \tag{10}$$

where c_1 and c_2 denote to learning factor. r_1 and r_2 are random numbers from the range of [0,1]. $w^{(k)}$ is called inertia weight that influences search capability of particles. The inertia weight w in the original PSO is a liner decreasing function, which is not conducive to balance the global and local search capabilities of particles. Thus, a novel update method in this paper is designed as follows:

$$w^{(k)} = w_{end} + (w_{ini} - w_{end}) \cdot \sin(\frac{\pi}{2} \sqrt{(1 - \frac{k}{Tmax})^3}) \tag{11}$$

where w_{ini} and w_{end} is the initial value and ending value of inertia weight w, respectively. *Tmax* indicates the maximum iteration times in PSO. We can know that w is a nonlinear decreasing function from Eq. (11). In the early period of IPSO algorithm search, the value of w is large, which facilitates to enhance the global search capability of particles. In the later period of the search, the value of w is small, which is beneficial to the improvement of the local search ability of particles.

B. IPSO-based workflow scheduling algorithm

Here, the solution space of a particle in PSO is employed to represent a task scheduling plan. Each task node in DAG corresponding to a dimension of one particle. In other words, the dimension of a particle encoding is equal to the number of workflow tasks. The solution for each dimension of one particle indicates a service mapping between task and server.

As shown in Fig. 2, the DAG of a workflow application contains 8 tasks, thus the corresponding dimension of a particle is also 8. As shown in Table 1, one of the particles is constructed according to Fig. 2. The value of this particle in each dimension indicates the ID of server (cloud or fog server) which each task will be assigned to. Thus, the encoding of position for a particle can be seen in Table 2.

Table 1. The mapping between tasks and servers

Task	1	2	3	4	5	6	7	8
Server	4	2	3	1	2	1	4	2

Table 2. The position of each particle encoding

4	2	3	1	2	1	4	2

Table 3. IPSO-based workflow scheduling algorithm

Input: number of tasks TN, maximum numbers of iteration $Tmax$, size of particle swarm K, a set of cloud and fog servers.

Output: the optimal scheduling solution S_{best}

---

1.   for i=1 to K do
2.   Randomize the initialization of scheduling plan S_i and search velocity V_i;
3.   end for
4.   for i=1 to K do
5.   Compute the fitness of the scheduling plan S_i;
6.   end for
7.   Select the global optimal scheduling plan of minimum fitness from K scheduling plan;
8.   for i=1 to $Tmax$ do
9.   Update the velocity of particle;
10.   Update the scheduling plan according to the velocity;
11.   Redistribute servers for tasks in the scheduling plan;
12.   for i=1 to K do
13.   Compute the fitness value of the scheduling plan S_i;
14.   end for
15.   Select the global optimal scheduling plan of minimum fitness from K scheduling plan;
16.   end for
17.   return S_{best}.

Furthermore, Eq. (8) is regarded as the fitness function in PSO. Our object is to minimize the fitness value. Based on the above particle encoding and the fitness function, we design an IPSO-based scheduling algorithm for workflow application as shown in Table 3. This algorithm first initializes randomly position (i.e., the scheduling plan) and velocity of all particles, and some other necessary parameters (see lines 1–7). Next, update scheduling plan according to the velocity of current generation and allocate tasks to servers again (see lines 8–11). After that, compute the fitness value for each scheduling plan and update the global optimal scheduling plan (see lines 12–16). Finally, the algorithm returns the optimal scheduling plan (see line 17).

4   Evaluation

In this section, we first describe the experimental environment and settings. Then, the experimental results are discussed and elaborated.

4.1   Experimental Environment and Settings

In this experiment, first, we test our proposed scheduling method using Matlab 2016a software based on the running environment of Intel core i5 3.0 GHz CPU and 8 GB RAM.

Next, some details about experimental settings are described as shown in Table 3. We specify the performance parameters regarding cloud servers and fog servers where various servers have different processing capabilities. The number of cloud servers and fog servers are 6 and 4, respectively. For tasks in workflow, the computation workload of each task is ranging from 500 to 15000 MI and the I/O data of a task has a size from 100 to 500 MB. In addition, the population size of IPSO is set as 30. The dimension of particle is equal to the number of workflow tasks in a DAG. The learning factor $c_1 = c_2 = 2$. The initial value and ending value of inertia weight $w_{ini} = 0.9$ and $w_{end} = 0.4$, respectively. The value of maximum iteration times $Tmax$ in PSO is set as 100. In order to reduce the impact of experimental uncertainties, finally, each round of experiment running is set to 30 times repeatedly to get average results (Table 4).

Table 4. Parameter list of cloud and fog servers

Parameter	Processing rate (MIPS)	Processing cost per unit time	Bandwidth (Mbps)	Number of servers
Cloud servers	6000	0.8	1000	2
	4500	0.6	1000	2
	3500	0.4	1000	2
Fog servers	2500	0.3	2000	2
	2500	0.2	2000	2

4.2 Discussion on Experimental Results

To test the influence of different number of tasks on experimental results, workflow is randomly generated by DAG generator, where the number of tasks in a DAG varies from 50 to 250. The results for the total makespan of workflow and economic cost of server rental under different number of tasks are shown in Table 5. We can see that the total economic costs of the two methods are close with each other in the same scale. With regards to the total makespan of workflow, the difference between results of the two methods is obvious. For example, when the number of tasks is 200, the total economic costs of PSO-based algorithm and IPSO-based algorithm are 162.6287 and 160.7947, respectively, while the total makespan of workflow about the two methods are 71.2534 and 63.1079, respectively. However, as a whole, the value of the result of IPSO-based algorithm is always smaller than the value of the result of PSO-based algorithm, regardless of the total economic cost or makespan.

To compare the experimental results regarding the two methods more vividly, relative percentage error is defined as follows:

$$RPE(var) = \left| \frac{RV_1 - RV_2}{RV_2} \right| \times 100\% \tag{12}$$

where RV_1 and RV_2 indicate the values of the total makespan or cost of IPSO-based algorithm and PSO-based algorithm, respectively. We can see the differences between results of the two methods from Fig. 3 clearly. For one thing, the bar chart in Fig. 3(b) shows that the values of $RPE(cost)$ are always small among all the different number of tasks. This phenomenon indicates that our method can guarantee that the cost does not increase compared with the baseline method. Although it cannot significantly reduce the economic cost of renting servers. For another thing, the differences between results of the two methods are easy to be seen in Fig. 3(a). For instance, the value of $RPE(makespan)$ is 10.51% when the number of tasks is 100, which shows that our method reduces the 10.51% of the total makespan of workflow application than PSO-based method.

Table 5. The total makespan and economic cost for different number of tasks

Number of tasks	Makespan		Cost	
	PSO-based	IPSO-based	PSO-based	IPSO-based
50	25.2029	20.7818	34.7404	33.7818
100	44.4373	39.7671	75.6929	73.2540
150	56.1236	51.2107	120.6615	117.9896
200	71.2534	63.1079	162.6287	160.7947
250	89.4044	82.0063	205.1188	201.2285

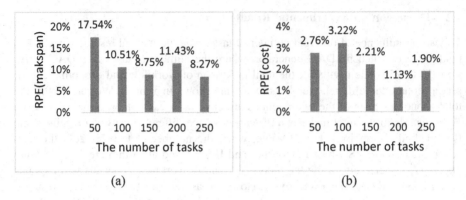

Fig. 3. Relative percentage error of the results of two algorithms

From what has been discussed above, we can make a clear conclusion that our proposed method can effectively reduce the total makespan of workflow applications with low latency in cloud-fog environment while the economic cost of renting cloud or fog servers is guaranteed under certain cost constraint.

5   Conclusion and Future Work

The total completion time of the workflow application and economic cost of cloud or fog server rental are two key issues of workflow scheduling in emerging and hybrid cloud-fog environment. In current study, we propose a scheduling algorithm for workflow application based on the improved PSO aiming to maintain the tradeoff between two objectives. In our IPSO, first, the novel update method of inertia weight is designed as a nonlinear decreasing function, which facilitates to balance and adjust the global and local abilities of particles. Next, each particle in PSO is encoded as a scheduling plan according to the number of workflow tasks. Finally, the experimental results show that our proposed scheduling algorithm reduces the overall completion time of workflow compared with the original PSO-based scheduling method while the economic costs of the two methods are close with each other.

In the future, we will further consider workflow scheduling algorithm based on multiple objectives optimization in hybrid cloud-fog environment.

References

1. Bonomi, F., Milito, R., Zhu, J., et al.: Fog computing and its role in the internet of things. In: Proceedings of the first edition of the MCC workshop on Mobile cloud computing, pp. 13–16. ACM, Helsinki (2012)
2. Lin, Y., Shen, H.: CloudFog: leveraging fog to extend cloud gaming for thin-client MMOG with high quality of service. IEEE Trans. Parallel Distrib. Syst. **28**(2), 431–445 (2017)

3. Puliafito C., Mingozzi E., Anastasi G.: Fog computing for the internet of mobile things: issues and challenges. In: IEEE International Conference on Smart Computing (SMART-COMP), pp. 1–6, IEEE, Hong Kong (2017)
4. Xu, R., Wang, Y., Luo, H., et al.: A sufficient and necessary temporal violation handling point selection strategy in cloud workflow. Futur. Gener. Comput. Syst. **86**, 464–479 (2018). https://doi.org/10.1016/j.future.2018.03.056
5. Bittencourt, L.F., Diaz-Montes, J., Buyya, R., et al.: Mobility-aware application scheduling in fog computing. IEEE Cloud Comput. **4**(2), 26–35 (2017)
6. Wu, Z., Liu, X., Ni, Z., et al.: A market-oriented hierarchical scheduling strategy in cloud workflow systems. J. Supercomput. **63**(1), 256–293 (2013)
7. Xu, R., Wang, Y., Huang, W., et al.: Near-optimal dynamic priority scheduling strategy for instance-intensive business workflows in cloud computing. Concurr. Comput. Pract. Exp. **29** (18), 1–12 (2017)
8. Verma, A., Kaushal, S.: A hybrid multi-objective particle swarm optimization for scientific workflow scheduling. Parallel Comput. **62**, 1–19 (2017)
9. Hoang, D., Dang, T.D.: FBRC: Optimization of task scheduling in fog-based region and cloud. In: IEEE Trustcom/BigDataSE/ICESS, pp. 1109–1114. IEEE, Sydney (2017)
10. Pham, X.Q., Man, N.D., Tri, N.D.T., et al.: A cost-and performance-effective approach for task scheduling based on collaboration between cloud and fog computing. Int. J. Distrib. Sens. Netw. **13**(11), 1550147717742073 (2017)
11. Tang, C., Wei, X., Xiao, S., et al.: A mobile cloud based scheduling strategy for industrial internet of things. IEEE Access **6**, 7262–7275 (2018)
12. Zeng, D., Gu, L., Guo, S., et al.: Joint optimization of task scheduling and image placement in fog computing supported software-defined embedded system. IEEE Trans. Comput. **65** (12), 3702–3712 (2016)
13. Pandey, S., Wu, L., Guru, S.M., et al.: A particle swarm optimization-based heuristic for scheduling workflow applications in cloud computing environments. In: 24th IEEE International Conference on Advanced Information Networking and Applications (AINA), pp. 400–407. IEEE, Perth (2010)

An Efficient Algorithm for Runtime Minimum Cost Data Storage and Regeneration for Business Process Management in Multiple Clouds

Junhua Zhang[1], Dong Yuan[2], Lizhen Cui[1(✉)], and Bing Bing Zhou[2]

[1] School of Computer Science and Technology, Shandong University,
Jinan, China
z.jh@mail.sdu.edu.cn, clz@sdu.edu.cn
[2] School of Information Technology, The University of Sydney,
Sydney, Australia
{dong.yuan,bing.zhou}@sydney.edu.au

Abstract. The proliferation of cloud computing provides flexible ways for users to utilize cloud resources to cope with data complex applications, such as Business Process Management (BPM) System. In the BPM system, users may have various usage manner of the system, such as upload, generate, process, transfer, store, share or access variety kinds of data, and these data may be complex and very large in size. Due to the pas-as-you-go pricing model of cloud computing, improper usage of cloud resources will incur high cost for users. Hence, for a typical BPM system usage, data could be regenerated, transferred and stored with multiple clouds, a data storage, transfer and regeneration strategy is needed to reduce the cost on resource usage. The current state-of-art algorithm can find a strategy that achieves minimum data storage, transfer and computation cost, however, this approach has very high computation complexity and is neither efficient nor practical to be applied at runtime. In this paper, by thoroughly investigating the trade-off problem of resources utilization, we propose a Provenance Candidates Elimination algorithm, which can efficiently find the minimum cost strategy for data storage, transfer and regeneration. Through comprehensive experimental evaluation, we demonstrate that our approach can calculate the minimum cost strategy in milliseconds, which outperforms the exiting algorithm by 2 to 4 magnitudes.

Keywords: Cloud computing · Business Process Management ·
Datasets storage and regeneration

1 Introduction

In recent years, the emergence and proliferation of cloud computing provides users on demand, redundant, inexpensive and scalable resources [1]. However, along with the convenience brought by using on-demand cloud services, users have to pay for the resources used according to the pay-as-you-go model, which can be substantial for complex applications and data intensive applications [2], such as BPM Systems [3],

F. Daniel et al. (Eds.): BPM 2018 Workshops, LNBIP 342, pp. 348–360, 2019.
https://doi.org/10.1007/978-3-030-11641-5_28

which aim to be a "holistic management" approach to satisfy the needs of users in organization's business process and can generate variety of datasets of large amount. These generated data contain important intermediate or final results of computation, which may need to be stored for reuse and sharing [4]. The fast-growing cloud computing market along with more and more cloud service providers enable BPM system to have flexible ways to utilize multiple cloud services with different prices of computation, storage and bandwidth resources [5]. An efficient storage strategy which can cut the cost of multi-cloud-based data management in a pay-as-you-go fashion is in need for deploying applications in multi-cloud computing environment.

Furthermore, due to the dynamic property of usage of data, some data in the application could be more popular to the users at a certain time, some other data could be less popular, the usage frequency of data could vary from time to time, such as the data in BPM system [3], the efficiency of the data storage strategy that was efficient in a previous time could also degrades. To this end, an efficient algorithm that can generate the minimum cost storage strategy at runtime to keep low resource cost is very important for online data intensive applications in multi-cloud environment.

Finding the trade-off among computation, storage and bandwidth costs to achieve minimum total cost in multi-clouds is a complicated problem [6]. Different cloud service providers have different prices on their resources and datasets have different resource usage and generation dependencies. Even worse, the dynamic data usage frequencies demand that the storage strategy should be updated in time to avoid performance degradation. For this problem, our previous work [6] has proposed GT-CSB which can find the optimal storage strategy that has the minimum overall cost, however, this approach is impractical for runtime storage strategy due to high computation complexity. Therefore, it is necessary to design a highly efficient runtime algorithm that can find optimal storage strategy at runtime to adjust the data storage status in real time.

In this paper, by studying the intrinsic property of the minimum cost storage problem, we propose a dynamic programming algorithm which can reduce the searching space and find the optimal storage strategy in nearly linear time. We also propose optimizing strategies, which can help us calculate the (1) minimum regeneration cost in $O(m^2)$ and (2) the sum overall cost rate of dataset in $O(m)$ (m is the number of Cloud Service Providers). By conducting extensive experimental studies, we find that our algorithm has a very good performance and is scalable with large number of datasets and Cloud Service Providers.

The remainder of this paper is organized as follows: Sect. 2 discusses the related work. Section 3 analyses the problem and presents some preliminaries. Section 4 introduces the detail of PCE algorithm. Section 5 evaluate PCE algorithm. Section 6 concludes this paper.

2 Related Work

The resource management in clouds becomes a very important research topic, much work has been done about resource negotiation [7], replica placement [8] and multi-tenancy in clouds. Foster et al. [9] propose the concept of virtual data in the Chimera system, which enables the automatic regeneration of data when needed. Recently, research on data provenance in cloud computing systems has also appeared [10].

Plenty of research has been done with regard to the tradeoff between computation and storage. The Nectar system [11] is designed for automatic management of data and computation in data centers, where obsolete data are deleted and regenerated whenever reused in order to improve resource utilization. In [12], authors firstly propose a cost-effective strategy based on the trade-off of computation and storage cost. In [13], the authors propose a dynamic on-the-fly minimum cost benchmarking approach by pre-storing calculated results with a specially designed data structure.

As the trade-off among different costs is an important issue in the cloud, some research has already embarked on this issue to a certain extent. In [14], Joe-Wong et al. investigate computation, storage and bandwidth resources allocation in order to achieve a trade-off between fairness and efficiency. In our prior work [15], we propose the T-CSB algorithm which can find a trade-off among Computation, Storage and Bandwidth costs (T-CSB). In our another prior work [6], we propose the GT-CSB algorithm, which can find a Generic best Trade-off among Computation, Storage and Bandwidth in clouds.

In this paper, to address above problem, we propose the PCE algorithm, which can efficiently find a Generic best Trade-off among Computation, Storage and Bandwidth in multiple clouds with a computation complexity of $O(n*|cand|*(m^2+log(|cand|)))$.

3 Preliminaries

In this Section, we first introduce some preliminaries and then the GT-CSB algorithm.

3.1 Preliminaries

In general, there are two types of data stored in clouds, *original data* and *generated data*, in this paper, we only consider *generated data*.

In this paper, we use *DDG* [16] (Data Dependency Graph) to represent datasets generation relationships. *DDG* [16] is a *DAG* which is based on data provenance in applications. Figure 1 depicts a simple DDG, where a node in the graph denotes a dataset. Edge denotes the generation relationship between datasets, i.e., d_4 and d_6 are needed for generation of d_7. If there exists a path from d_i to d_j in the DDG, we say d_i and d_j have a generation relationship, and $d_i(d_j)$ is the predecessor (successor) of $d_j(d_i)$, we denote it as $d_i{\rightarrow}d_j$, e.g., $d_1{\rightarrow}d_4$, $d_5{\rightarrow}d_7$.

In a commercial cloud computing environment, there are generally three basic types of resource cost in the cloud: computation cost, storage cost and bandwidth cost:

Total Resource Cost = Computation Cost + Storage Cost + Bandwidth Cost.

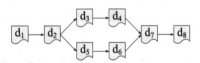

Fig. 1. A simple Data Dependency Graph (*DDG*)

Assumptions: We assume that the application be deployed with m Cloud Service Providers, denoted as $CSP = \{c_1, c_2 \ldots c_m\}$. Furthermore, we assume there are n datasets in the DDG, denoted as $DDG = \{d_1, d_2, \ldots d_n\}$. For every dataset $d_i \in DDG$, it can be either stored with one of the cloud service providers or be deleted.

Denotations: We use X, Y, Z to denote the computation cost, storage cost and bandwidth cost of datasets respectively. Specifically, for a dataset $d_i \in DDG$:

$X_{d_i}^{c_j}$ denotes the cost of computing d_i from its direct predecessors with cloud c_j;

$Y_{d_i}^{c_j}$ denotes the storage cost per time unit for storing dataset d_i with cloud c_j;

$Z_{d_i}^{c_k,c_j}$ denotes the cost of transferring dataset d_i from cloud service provider c_k to c_j.

v_{d_i} denote the usage frequency of d_i, which means how often d_i is accessed.

Definition 1: In a multi-cloud computing environment, in order to regenerate a deleted dataset, we need first to find its stored provenance dataset(s), then to choose a cloud service provider to regenerate it. We denote the minimum regeneration cost of dataset d_i as $minGenCost(d_i)$.

Definition 2: *Cost Rate* of a dataset is the average cost spent on this dataset per time unit in clouds. For $d_i \in DDG$, we denote its *Cost Rate* as $CostR(d_i)$, which is:

$$CostR(d_i) = \begin{cases} minGenCost(d_i) \times v_{d_i}, // \ d_i \text{ is deleted} \\ Y_{d_i}^{c_j}, // \ d_i \text{ is stored in } c_j \end{cases}.$$

The *Total Cost Rate* of a *DDG* is the sum *Cost Rate* of all the datasets: $TCR = \sum_{d_i \in DDG} CostR(d_i)$.

Definition 3: Storage strategy of a *DDG* is the storage status of all datasets in the *DDG*, i.e. whether dataset is stored, and which cloud the dataset is stored.

Definition 4: Minimum cost of a *DDG* is the minimum *Total Cost Rate* for storing and regenerating datasets in the *DDG*, which is denoted as $TCR_{min} = min\left(\sum_{d_i \in DDG} CostR(d_i)\right)$.

3.2 GT-CSB Algorithm

The GT-CSB algorithm proposed in our prior work [6] can find the best trade-off among computation, storage and bandwidth costs in multi-clouds. The core idea of GT-CSB is to convert a minimum cost storage problem to a shortest path problem over a Cost Transitive Graph (CTG) graph. In the CTG graph, for each dataset in *DDG*, there are m nodes each representing that the dataset is stored in the corresponding cloud, and two virtual vertexes, start vertex and end vertex, are used to represent the start point and end point of the shortest path problem. For any two vertexes belonging to different datasets, there is an edge between them. An edge signifies that the datasets between the edge are deleted while the end datasets of the edge are stored in the corresponding cloud. Each path from the start vertex to the end vertex in the CTG corresponds to a storage strategy of the datasets in the clouds. By sophistically setting the edge weight, which represents the sum *Cost Rate* of those datasets between the end nodes of the edge, we can get the minimum cost storage strategy by solving shortest path problem over the graph, the length of the shortest path corresponding to the minimum *Total Cost Rate* of datasets in *DDG*.

4 PCE Algorithm

In this section, we first detailed introduce our PCE algorithm and some optimizing strategies in Section; then we analyze the complexity of PCE algorithm.

4.1 Provenance Candidates Elimination (PCE) Algorithm

In this section, we will first elaborate the minimum cost dataset regeneration in Multiple Clouds Environment and baseline approach for optimal data storage strategy, and then introduce the detail of PCE Algorithm and optimizations.

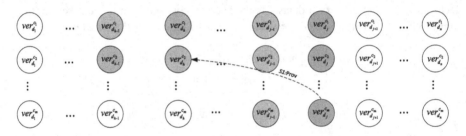

Fig. 2. *DDG* with multiple clouds

Dataset Regeneration with Multiple Clouds. We use *Prov*(*d*) to denote the provenance of dataset *d*, the provenance of *d* is the nearest stored predecessor(s) of *d* and is used to generate *d* when *d* is reused. The Minimum cost to regenerate a dataset is the minimum cost of generating the dataset from its provenance with multiple clouds, which includes the bandwidth cost for transferring datasets among the clouds and the computation cost for regenerating datasets from its predecessors.

Definition 5: We use $ver^{c_s}_{(d_j,c_k)d_i}$ to denote the minimum cost of generating d_i on cloud c_s from its provenance d_j which is stored in c_k, or simplify it as $ver^{c_s}_{d_i}$ in the context without ambiguity.

Based on the definition, if a provenance d_i is stored in cloud c_s, the minimum generation cost of dataset on cloud can be iteratively computed as:

$$\begin{cases} ver^{c_k}_{d_{i+1}} = Z^{c_s,c_k}_{d_i} + X^{c_k}_{d_{i+1}} \\ ver^{c_k}_{d_j} = \min_{h=1}^{m}\left\{ ver^{C_h}_{d_{j-1}} + Z^{c_h,c_k}_{d_{j-1}} \right\} + X^{c_k}_{d_j} \end{cases} \tag{1}$$

where $d_j \in DDG \wedge d_{i+1} \to d_j \wedge \text{Prov}(d_j) = d_i, c_k \in \{c_1, c_2,...c_m\}$.

Based on Definition 5, the minimum regeneration cost of d_j with provenance d_i is:

$$minGenCost(d_j) = \min_{h=1}^{m}\left\{ver_{d_j}^{c_h}\right\}\tag{2}$$

Baseline Algorithm

Lemma 1. *In a linear DDG, if dataset $d_i \in DDG$ is stored in cloud, then the sum Cost Rate of d_i's successors (predecessors) is independent of the storage status of d_i's predecessors (successors).*

According to the definition and the iterative calculation of the minimum regeneration cost of a dataset in Eqs. (1) and (2), a deleted dataset is computed from its provenance, since d_i is stored in cloud, any of d_i's predecessor cannot be a provenance of d_i's successor, so the overall cost of d_i's successors is independent of the storage status of d_i's predecessors. The regeneration cost or storage cost of d_i's predecessors is also independent of d_i's successor. Hence, if a dataset, e.g. d_i, is stored in cloud, we can compute its predecessors' storage strategy and its successors' storage strategy independently.

Assume a dataset d_i is stored in cloud c_k, we use $d_i.preCost$ to represent the minimum total cost of d_i's predecessors, and a tuple (d_i, c_k) to represent that dataset d_i is stored in cloud c_k, and the storage strategy S of a DDG in multi-clouds is represented by a set of tuples $S = \{(d_i, c_k)|d_i \in DDG \wedge c_k \in CSP \wedge d_i \text{ is stored in } c_k\}$. The provenance, e.g. d_j, and the provenance stored place, e.g. c_k, of d_i is represented by a tuple $d_i.Prov = (d_j, c_k)$.

Algorithm 1. Baseline Algorithm
Input: a set of *CSPs* $\{cs_1...cs_m\}$ and a linear *DDG* $\{d_1...d_n\}$;
Output: a set of stored datasets and their storage clouds
1.   Construct vitual dataset $d_0(d_{n+1})$ to be direct predecessor(successor) of $d_1(d_n)$;
2.   Set $d_0.size$, $d_0.computcost$, $d_0.preCost$ and $d_1.preCost$ to 0;
3.   **for each** dataset d_i in *DDG* from d_1 to d_{n+1} **do**
4.   Set $d_i.preCost$ to Infinite;
5.   **for each** dataset d_j in *DDG* from d_0 to d_{i-1} **do**
6.   **for** each Cloud Service Provider c_k in *CSP* **do**
7.   Let d_j stored in c_k;
8.   **if**($d_j.preCost + \sum_{h=j+1}^{i-1} minGenCost(d_h) \times v_{d_h} + Y_{d_j}^{c_k}$)$< d_i.preCost$ **then**
9. $d_i.preCost = \sum_{h=j+1}^{i-1} minGenCost(d_h) \times v_{d_h} + d_j.preCost + Y_{d_j}^{c_k}$;
10. $d_i.Prov = (d_j, c_k)$;
11.   Delete d_j from c_k;
12.   // Collect the stored datasets and their stored clouds by backward traverse
13. $S = \emptyset$;
14. $d_s = d_{n+1}$;
15.   **while** $d_s.Prov.dataset \neq d_0$ **do**
16. $S = S \cup \{d_s.Prov\}$;
17. $d_s = d_{n+1}.Prov.dataset$;

Baseline Algorithm starts by creating two virtual nodes d_0 and d_{n+1} as starts dataset and end dataset respectively (line 1), the two datasets have 0 size and 0 computation cost, they are created only for ease of illustration. For each dataset in DDG and d_{n+1}, e.g. d_i, Baseline-Algorithm computes its minimum $preCost$ and $Prov$ (line 5–11). After the iteration process on all datasets, $d_{n+1}.preCost$ is the minimum total cost of all d_{n+1}'s predecessors and is also the minimum total cost of DDG, then the optimal storage strategy can be collected by a reverse traverse from d_{n+1} with $Prov$ (line 13–17). When computing $preCost$ and $Prov$ of a dataset, e.g. d_i, $preCost$ is first initialed as infinite, then Baseline-Algorithm iterates on all d_i's predecessors and all CSPs to determine d_i's provenance and the stored cloud service, e.g. $d_i.Prov = (d_j, c_k)$, that can make $d_i.preCost$ minimum (line 4–11).

In Baseline Algorithm, let n be the number of dataset and m be the number of CSPs, $minGenCost$ can be compute in $O(m^2n)$. When deciding $Prov$ of a dataset, Baseline-Algorithm have to iterate all its predecessors and all Cloud Service Providers, this procedure can be done in $O(m^3n^3)$, and there are n datasets, so the final time complexity of Baseline-Algorithm is $O(m^3n^4)$.

Provenance Candidates Elimination Strategy. Based on the definition of minimum regeneration cost in multiple clouds, we find that the more distant the $Prov(d_j)$ is from d_j, the higher the minimum regeneration cost of d_j will be.

Theorem 1: *In multi-cloud scenarios, without loss of generality, if exists an optimal storage strategy $S1^*$ for datasets $\{d_1, d_2...d_j\}$, i.e., $\sum_{i=1}^{j} CostR(d_i)$, is minimum with $S1^*$, assuming the last stored dataset of $S1^*$ is d_h and is stored in cloud c_r, then the last stored dataset and its stored cloud of optimal storage strategy $S2^*$ for datasets $\{d_1, d_2...d_j, d_{j+1}\}$ cannot be (d_k, c_i) with $ver^{c_s}_{(c_r,d_h)d_{j+1}} < ver^{c_s}_{(c_i,d_k)d_{j+1}}$ for all $c_s \in CSP$.*

Proof: Assuming the last stored dataset of $S2^*$ is (d_k, c_i) with $ver^{c_s}_{(c_r,d_h)d_{j+1}} < ver^{c_s}_{(c_i,d_k)d_{j+1}}$ for all $c_s \in CSP$. We can construct a strategy $S3$ for $\{d_1, d_2, ..., d_{j+1}\}$ with same storage strategy of $S1^*$ for $\{d_1, d_2, ..., d_j\}$ and d_{j+1} is deleted with lower sum $Cost\ Rate$ than $S2^*$. Since $\sum_{i=1}^{j} CostR_{S1^*}(d_i) < \sum_{i=1}^{j} CostR_{S2^*}(d_i)$ and $ver^{c_s}_{(c_2,d_h)d_{j+1}} < ver^{c_s}_{(c_i,d_k)d_{j+1}}$ for all $c_s \in CSP$, $CostR_{S3}(d_{j+1}) = v_{d_{j+1}} \times \min_{c_s \in CSP} ver^{c_s}_{(c_2,d_h)d_{j+1}} < CostR_{S2}(d_{j+1}) + v_{d_{j+1}} \times \min_{c_s \in CSP} ver^{c_s}_{(c_2,d_k)d_{j+1}}$, hence $\sum_{i=1}^{j+1} CostR_{S3}(d_i) = \sum_{i=1}^{j} CostR_{S1}(d_i) + CostR_{S3}(d_{j+1}) < \sum_{i=1}^{j+1} CostR_{S2}(d_i) = \sum_{i=1}^{j} CostR_{S2}(d_i) + CostR_{S2}(d_{j+1})$, which contradicts the premise. Theorem 1 holds.

According Theorem 1, we propose following Provenance Candidates Elimination Rules (*PCERs*).

Consider the Baseline-Algorithm, assume the provenance of a dataset d_i is d_i. $Prov = (d_j, c_k)$, for d_i's successors, i.e., d_k, the initial provenance candidates set of d_k is $d_k.cand = \{(d_h, c_l) | d_h \rightarrow d_k \wedge d_h \in DDG \wedge c_l \in CSP\}$, we can use the following rules to pruning the candidates set:

1. For $(d_h, c_l) \in d_k.cand$, where $d_h \rightarrow d_i$, if $ver^{c_s}_{(d_h,c_l)d_i} > ver^{c_s}_{(d_k,c_j)d_i}$ for all $c_s \in CSP$, then (d_h, c_l) can be eliminated from $d_k.cand$.

2. For (d_h, c_l) $d_k.cand$, where $d_h \to d_i$, if exists $(d_{h'}, c_{l'})$ $d_k.cand$, $d_{h'} \to d_i$, that

$$\left(d_h.preCost + \sum_{p=h+1}^{i} \left(\min_{c_s \in CSP} \left(ver^{c_s}_{(d_h,c_l)d_p} \right) \times v_{d_p} \right) + Y^{c_l}_{d_h} \right) > (d_{h'}.preCost +$$

$$\sum_{p=h+1}^{i} \left(\min_{c_s \in CSP} \left(ver^{c_s}_{(d_{h'},c_{l'})d_p} \right) \times v_{d_p} \right) + Y^{c_{l'}}_{d_{h'}}) \text{ and } ver^{c_s}_{(d_h,c_l)d_i} > ver^{c_s}_{(d_{h'},c_{l'})d_i}, \text{ then } (d_h,$$

$c_l)$ can be eliminated from $d_k.cand$.

To better illustrate the PCE Algorithm, we first introduce some new data structures:

- *cand* is the candidates set to record the possible provenances of the datasets. In the algorithm, maintaining one *cand* is sufficient for all datasets, because, for example, the reduction on a dataset's provenance candidates $d_i.cand$ also applies on d_i's successors.
- $(d_j, c_k).MGC$ is an array where $(d_j, c_k).MGC[c_s]$ is the value of $ver^{c_s}_{(d_j,c_k)d_{i-1}}$ when d_j is stored in cloud c_k and d_{i-1} is generated on cloud c_s.
- $(d_j, c_k).sucCost$ is similar to $d_j.preCost$, it is the sum $CostR$ of datasets from d_{j+1} to d_{i-1} : $(d_j, c_k).sucCost = \sum_{h=j+1}^{i-1} minGenCost(d_h) \times v_{d_h}$.

Algorithm 2. PCE Algorithm
Input: a set of Cloud Service Providers CSP $\{cs_1, cs_2 ... cs_m\}$ a linear DDG $\{d_1, d_2... d_n\}$;
Output: a set of stored datasets and their storage clouds
01.   Construct vitual dataset $d_0(d_{n+1})$ to be direct predecessor(successor) of $d_1(d_n)$;
02.   Set $d_0.size, d_0.computcost, d_0.preCost, (d_0,c_0).sucCost, d_1.preCost$ to be zero;
03. $(d_0, c_0).MGC[c_s]$=Infinit for all c_s in CSP except 0 when $c_0 == c_s$;
04.   cand$=\{(d_0, c_0)\}$;
05.   **for each** dataset d_i in DDG from d_1 to d_{n+1} **do**
06. $d_i.preCost$=Infinite;
07.   **for each** (d_j, c_k) in cand **do**
08.   **if** $(d_j.preCost+(d_j, c_k).sucCost+Y^{c_k}_{d_j})<d_i.preCost$ **then**
09. $d_i.preCost= d_j.preCost+(d_j, c_k).sucCost+Y^{c_k}_{d_j}$;
10. $d_i.Prov=(d_j, c_k)$;
11.   //Incremental Update
12.   **for each** (d_j, c_k) in cand **do**
13.   Swap $(d_j, c_k).MGC$ with $(d_j, c_k).MGCO$;
14.   **for each** c_s in $CSPs$ **do**
15. $(d_j, c_k).MGC[c_s]= \min_{h=1}^{m}\{(d_j,c_k).MGCO[c_h]+Z^{c_h,c_s}_{d_{i-1}}\}+X^{c_s}_{d_i}$;
16. $(d_j, c_k).sucCost=(d_j, c_k).sucCost+\min_{h=1}^{m}\{(d_j,c_k).MGC[c_h]\}\times v_{d_i}$;
17.   Performing Provenance Candidates Elimination Rule 1 and 2 on *cand*;
18.   //Adding new candidates for d_{i+1}
19.   **for each** c_k in CSP **do**
20. $(d_i, c_k). MGC[c_s]$=Infinit for all c_s in CSP except 0 hen $c_k == c_s$;
21. $(d_i, c_k).sucCost$=0;
22.   cand$=$cand $\cup \{(d_i, c_k)\}$;
23.   Collect the stored datasets and their stored clouds by backward traverse

In PCE algorithm, the *cand* is first initialized as $\{(d_0, c_0)\}$ (line 4). For each d_i in *DDG*, d_i.Prov and d_i.preCost computed in line 6–10, after updating *MGC* and sucCost of all the candidates (line 12–16), the *PCERs* are performed on *cand* (line 17). At last in line 19–22, the new candidates, i.e., (d_i, c_k) for all c_k in *CSP*, are initialized and added to *cand*.

For example, in Fig. 2, the provenance of d_j is (d_h, c_2), the *cand* now is $\{(d_{h-1}, c_1)$, (d_{h-1}, c_2), (d_h, c_1), (d_h, c_2), (d_{j-1}, c_1), (d_{j-1}, c_2), $(d_{h-1}, c_m)...\}$ marked with grey and green circles. After performing the elimination rules, (d_{h-1}, c_2) and (d_h, c_1) marked with grey circles are deleted from *cand*. Then before searching Prov of d_{j+1}, (d_j, c_1), (d_j, c_2) ... (d_j, c_m) marked with blue circles are added to *cand*.

Incremental Minimum Regeneration Cost and Sum Successors' Cost. For the computation of $\sum_{h=j+1}^{i-1} minGenCost(d_h) \times v_{d_h}$, we propose incremental computation for it, it contains two parts: the incremental computation of $minGenCost(d_h)$ and $\sum_{h=j+1}^{i-1} minGenCost(d_h) \times v_{d_h}$, as was illustrated in line 12–16 of PCE algorithm.

First, for the computation of $minGenCost(d_h)$, we use a data structure *MGC*, introduced before, to store the minimum regeneration cost of successors of datasets, e.g. (d_j, c_k).*MGC* stores the minimum regeneration cost of successors of d_j. In the each round, *MGC* is updated accordingly (line 15).

Second, for the computation of $\sum_{h=j+1}^{i-1} minGenCost(d_h) \times v_{d_h}$, similar to the incremental computation of $minGenCost(d_h)$, we use *sucCost*, introduced before, to store the sum cost rate of successors of a datasets, e.g., d_j. In each round, *sucCost* is updated accordingly (line 16).

4.2   Analyses

In PCE Algorithm, let n be the number of datasets, m be the number of Cloud Service Providers and $|cand|$ be the average size of *cand*, searching of *Prov* (line 6–10) can be done in $O(|cand|)$, incremental update(line 12–16) can be done in $O(|cand|*m^2)$, elimination rules (line17) in $O(|cand|*m+|cand|*log(|cand|))$, adding new candidates (line 26–29) can be done in $O(m^2)$, so the overall time complexity of the Algorithm is $O(n*|cand|*(m^2+log(|cand|)))$. For the size of *cand*, it mainly depends on the computation cost rate and storage cost rate of datasets and is independent of the number of datasets n. Our experimental results in Sect. 5.2 (Fig. 4(b)) also demonstrate the independence of the size of *cand* and the number of dataset n.

5   Experiments

Our experiment is conducted on Desktop PC with Intel(R) Core(TM) i5-4200M CPU, RAM 8 GB. The algorithm is implemented in the Java and is run on Windows.

In real world applications, generated datasets may vary dramatically in terms of size, generation time, usage frequency and the structure of *DDG*. Hence, we randomly generate *DDGs* with different number of datasets, each with a random size from 1 GB to 100 GB. The computation time of dataset is also random, from 10 h to 100 h.

The usage frequency is again random, from once per month to once per year. This setting is based on the scenarios of applications of scientific workflow [16] and BPM system [3].

Table 1. The pricing models of 10 cloud services providers

Cloud Service ID	0	1	2	3	4	5	6	7	8	9
Compute cost rate ($/hour)	0.11	0.12	0.15	0.09	0.13	0.15	0.12	0.13	0.12	0.16
Storage cost rate ($/GB*month)	0.1	0.06	0.05	0.08	0.07	0.07	0.06	0.09	0.05	0.04
Transfer cost rate for outbound ($/GB)	0.01	0.03	0.15	0.05	0.06	0.03	0.07	0.02	0.06	0.08

In addition, we randomly generate 10 cloud service providers with different compute, storage and out-bandwidth price (see Table 1)[1].

Our prior work [6] has thoroughly investigated the minimum cost strategy, the algorithm in this paper calculates the same minimum cost strategy as GT-CSB, the effectiveness of PCE algorithm will not be evaluated here.

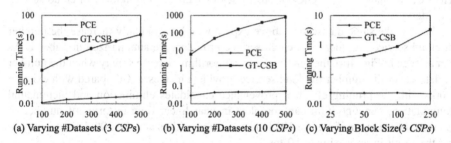

(a) Varying #Datasets (3 *CSPs*)　　(b) Varying #Datasets (10 *CSPs*)　(c) Varying Block Size(3 *CSPs*)

Fig. 3. Comparison of performance with varying settings

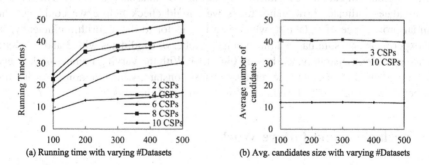

(a) Running time with varying #Datasets　　(b) Avg. candidates size with varying #Datasets

Fig. 4. Evaluation with varying settings

---

[1] The prices are set based on popular cloud service provider's pricing model, e.g., Amazon Web Services' prices are: $0.10 per instance-hour for the computation resources, $0.10 per GB-month for the storage resource and $0.09 per GB bandwidth resources for data downloaded from Amazon via the Internet. https://aws.amazon.com 2018.

5.1 Comparison with Existing Algorithms

We first compare the performance of our strategy with GT-CSB. In this experiment, we use 5 randomly generated DDGs with 100 to 500 datasets and 3 cloud service providers with the pricing models listed in Table 1.

The experiment result shown in Fig. 3(a) and (b) demonstrates our strategy can always finished within 1 s, while the running time of GT-CSB increases fast with the increase of number of datasets.

In the next experiment, based on the philosophy of our prior work [17], we devise a method which can derive localized minimum cost instead of a global one. The method is dividing the DDG into several blocks of the same size, and using the algorithm to find local optimal storage strategy for each block. We use a DDG with 500 datasets and divide it into blocks with different block size. Figure 3(c) demonstrate the speed up of GT-CSB algorithm with small block size, however, it is still not as efficient as PCE algorithm.

5.2 Evaluation of PCE with Varying Settings

Then we evaluate the efficiency of our strategies with varying number of cloud service providers.

We use the same datasets as above experiment, but gradually increase the number of cloud service providers. All cloud service providers are summarized in Table 1. As can be seen in Fig. 4(a), the run time of our algorithm increase slowly when the number of datasets or the number of cloud service providers increases. Compared with existing work, with the pruning effect of provenance candidates elimination and incremental computation, our algorithm can complete in near linear time in terms of number of datasets, hence, even if we use 10 cloud service providers and the 500 datasets, we can get the result in approximate 50 ms.

We demonstrate the effect of provenance elimination strategy by studying the average number of candidates with varying number of datasets (100–500). The number of candidates indicates how many times we should check before we could get the optimal provenance of a dataset, which is a key factor to the algorithm efficiency. In this experiment, we summarized the average number of candidates with 3 and 10 cloud service providers separately, as show in Fig. 4(b). With the varying number of datasets, the average number of candidate remains almost constant, which demonstrate that the number of candidates is independent of the number of datasets.

6 Conclusions and Future Work

In this paper, we proposed a provenance elimination strategy which can identify a small set of possible optimal provenance and reduce the search space. Besides, we propose incremental computations which speed up the algorithm a lot. The experimental results show that the running time of our algorithm is significantly reduced compared to that of the GT-CSB algorithm and our algorithm also scales well even the number of dataset is very large.

In our current work, we only consider the datasets with linear *DDG*. However, in the real world, dependencies between datasets can be very complex; they may contain blocks, sub-blocks and crossed-blocks, the data storage strategy can be very tough to obtain. Furthermore, extra cost might be caused by the "vender lock-in" issue among different cloud service providers, large number of requests from input/output (I/O) intensive applications, etc. In the future, we will consider complex *DDG* and incorporate more complex pricing models in our datasets storage and regeneration cost model.

Acknowledgment. The research work was supported by the National Key R&D Program (2017YFB1400102, 2016YFB1000602), NSFC (61572295), SDNSFC (No. ZR2017ZB0420), and Shandong Major scientific and technological innovation projects (2018YFJH0506).

References

1. Zhang, Q., Zhani, M.F., Boutaba, R., Hellerstein, J.L.: Dynamic heterogeneity-aware resource provisioning in the cloud. IEEE Trans. Cloud Comput. **2**(1), 14–28 (2014)
2. Szalay, A., Gray, J.: 2020 computing: science in an exponential world. Nature **440**(7083), 413–414 (2006)
3. Weske, M.: Business process management architectures. Business Process Management, pp. 333–371. Springer, Heidelberg (2012). https://doi.org/10.1007/978-3-642-28616-2_7
4. Burton, A., Treloar, A.: Publish my data: a composition of services from ANDS and ARCS. In: Fifth IEEE International Conference on e-Science, pp. 164–170. IEEE (2009)
5. Agarwala, S., Jadav, D., Bathen, L.A.: iCostale: adaptive cost optimization for storage clouds. In: 4th International Conference on Cloud Computing, pp. 436–443. IEEE (2011)
6. Yuan, D., Cui, L., Li, W., Liu, X., Yang, Y.: An algorithm for finding the minimum cost of storing and regenerating datasets in multiple clouds. IEEE Trans. Cloud Comput. **6**, 519–531 (2015)
7. Deng, K., Song, J., Ren, K., Yuan, D., Chen, J.: Graph-cut based coscheduling strategy towards efficient execution of scientific workflows in collaborative cloud environments. In: Proceedings of the 2011 IEEE/ACM 12th International Conference on Grid Computing, pp. 34–41. IEEE Computer Society (2011)
8. Li, W., Yang, Y., Chen, J., Yuan, D.: A cost-effective mechanism for cloud data reliability management based on proactive replica checking. In: Proceedings of the 2012 12th IEEE/ACM International Symposium on Cluster, Cloud and Grid Computing (ccgrid 2012), pp. 564–571. IEEE Computer Society (2012)
9. Foster, I., Vockler, J., Wilde, M., Zhao, Y.: Chimera: a virtual data system for representing, querying, and automating data derivation. In: Proceedings of 14th International Conference on Scientific and Statistical Database Management, pp. 37–46. IEEE (2002)
10. Muniswamy-Reddy, K.-K., Macko, P., Seltzer, M.I.: Provenance for the cloud, pp. 14–15 (2010)
11. Gunda, P.K., Ravindranath, L., Thekkath, C.A., Yu, Y., Zhuang, L.: Nectar: automatic management of data and computation in datacenters. In: OSDI, pp. 1–8 (2010)
12. Yuan, D., Yang, Y., Liu, X., Chen, J.: A cost-effective strategy for intermediate data storage in scientific cloud workflow systems. In: Parallel & Distributed Processing (IPDPS), pp. 1–12. IEEE (2010)
13. Yuan, D., Liu, X., Yang, Y.: Dynamic on-the-fly minimum cost benchmarking for storing generated scientific datasets in the cloud. IEEE Trans. Comput. **64**(10), 2781–2795 (2015)

14. Joe-Wong, C., Sen, S., Lan, T., Chiang, M.: Multiresource allocation: fairness-efficiency tradeoffs in a unifying framework. IEEE/ACM Trans. Netw. (TON) **21**(6), 1785–1798 (2013)
15. Yuan, D., et al.: An algorithm for cost-effectively storing scientific datasets with multiple service providers in the cloud. In: 2013 IEEE 9th International Conference on eScience (eScience), pp. 285–292 (2013)
16. Yuan, D., Yang, Y., Liu, X., Chen, J.: On-demand minimum cost benchmarking for intermediate dataset storage in scientific cloud workflow systems. J. Parallel Distrib. Comput. **71**(2), 316–332 (2011)
17. Yuan, D., et al.: A highly practical approach toward achieving minimum data sets storage cost in the cloud. IEEE Trans. Parallel Distrib. Syst. **24**(6), 1234–1244 (2013)

A Lean Architecture for Blockchain Based Decentralized Process Execution

Christian Sturm[⊠], Jonas Szalanczi, Stefan Schönig, and Stefan Jablonski

Universität Bayreuth, Bayreuth, Germany
{christian.sturm,jonas.szalanczi,stefan.schonig,
stefan.jablonski}@uni-bayreuth.de

Abstract. Interorganizational process management bears an enormous potential for improving the collaboration among associated business partners. A major restriction is the need for a trusted third party implementing the process across the participating actors. Blockchain technology can dissolve this lack of trust due to consensus mechanisms. After the rise of cryptocurrencies, the launch of *Smart Contracts* enables the *Ethereum Blockchain* to act beyond monetary transactions due to the execution of these small programs. We propose a novel lean architecture of a Blockchain based process execution system with Smart Contracts to dispense with a trusted third party in the context of interorganizational collaborations.

Keywords: Business Process Management · Blockchain · Collaborative process management · Choreography processes · Process execution

1 Introduction

Blockchain technology currently triggers a revolution in the way we store our data: data move from centralized (cloud) storage towards decentralized data management and information systems. The technology has proven to bear a great potential for disruptive change in many domains. Our aim is to exploit and leverage on this technology when organizational processes in context of Business Process Management has to be enacted.

Blockchain technology became famous as backbone behind the cryptocurrency Bitcoin [11], followed by a vast number of alternative cryptocurrencies. While these 1st-generation Blockchains were originally focused on monetary transactions, next-generation Blockchains like *Ethereum* were established further on [3]. The latter provides the turing-complete programming language *Solidity* on top for executing small programs, called *Smart Contracts* directly on the Blockchain. The idea of Smart Contracts was first described 1997 in [15] and was resurged with the Ethereum Blockchain where Smart Contracts are account holding objects following several principals: They provide code functions, they

© Springer Nature Switzerland AG 2019
F. Daniel et al. (Eds.): BPM 2018 Workshops, LNBIP 342, pp. 361–373, 2019.
https://doi.org/10.1007/978-3-030-11641-5_29

can interact with other contracts or users and they are able to make decisions based on stored data.

The key fundamental concept of using a Blockchain instead of legacy systems is the tamper-proof character of the underlying database (distributed ledger) without the need of having a trusted third party included. Different applications had proven the feasibility of Smart Contracts ranging from government regulations regarding manufacturing of pharmaceuticals [12] to organizing rescue packages in crisis areas [14].

With respect to Business Process Management, especially the collaboration between companies through choreography processes seems a suitable and profitable application domain. Unlike orchestration processes in inter-organizational process management, where the participants are also owner of the processes, we assume an adversarial setting where participants may blame each other as pointed out in [9]. The consensus mechanism in Blockchain technology is said to annul the need for a trusted third party as an arbitrator, because the nodes in the network will reach accordance by their selves and thus are impervious to potential attackers.

Our contribution composes of a novel architecture using Blockchain technology as a backbone for inter-organizational process execution. The lean architecture enables a lightweight full-featured on-chain implementation of a decentralized process execution system. We exploited the latest advanced concepts of the Solidity programming language for our Smart Contract. Compared to existing approaches, the omission of additional software artefacts in our architecture leads to an entire on-chain execution. For large process models comprising a big number of tasks, we believe that our approach scales best. With our solution, we state that most of the execution steps should be secured on the Blockchain. We further critically analyse our approach, compare it with existing alternatives, and discuss advantages and disadvantages.

The remainder of the paper is structured as follows. In Sect. 2 we provide a comprehensive overview of the principles, how Blockchain technology works internally. Then we summarize Related Work of using Blockchain technology in Business Process Management in Sect. 3. Our approach is the main constituent of Sect. 4 and after a qualitative Evaluation in Sect. 6, we conclude the paper in Sect. 7.

2   Background

This section provides an overview on the most important concepts of Blockchain technology. We describe, how transactions are validated and how consensus can be reached and point out differences between a public Blockchain and a Consortium Blockchain. The main reason that different branches in industry investigate Blockchain technology is the non-necessity of a trusted third party. In contrast, traditional execution of financial transactions between two parties always requires the bank of the sender as well as the bank of the receiver as intermediary players. We refer to [1] for more detailed information.

The backbone of the Blockchain, a linked list of blocks comprising transactions, is a Peer-to-Peer network of participants, so-called *(full) nodes*. Each of them holds the entire Blockchain locally including all transactions ever processed. A new transaction is propagated within the whole network and each node includes it into its pool of unconfirmed transactions. Always, nodes try to *mine* a new block by solving a cryptographic puzzle. All unconfirmed transactions in the pool as well as additional information (ID of the latest block, the timestamp, an adjustable *nonce*) serves as input for a hashing algorithm. The nonce is generated at random, until the output of the hashing algorithm falls below a certain limit. If such a nonce is found, the node wins this laborious mining game and informs the network. The unconfirmed transactions that were affecting the hash output are confirmed, and all nodes remove them from their pools. As a reward, the miner receives the fees associated with the transactions. The confirmation of transactions relies on the principle, that every node can doublecheck a new propagated block against a set of specific rules. Hence, if invalid transactions, e.g. fraudulent double spends of monetary units, are included in the new block, the network rejects it and miners would waste a vast amount of computations for finding an expedient nonce and thus waste electricity and thus money. Additionally, also the mining fees are lost.

The term *Blockchain* is often referred to the public Blockchain, which is open and free to access to anyone willing to participate in the P2P network. This may cause trouble regarding privacy especially when storing data payload on-chain. As an alternative, private or Consortium Blockchains are isolated from the global network and thus read/write accesses can be permissioned and adjusted. Consortium Blockchains are often confused with private Blockchains, but the latter denies the participation of external actors whereas the former just restricts the permission to mine and verify. In private Blockchains, mining and so consensus power is much more concentrated than in public Blockchains. Therefore, one has to pay attention that a participant of the consortium does not gain too much power. Janne Hansen, IT-architect and developer from Microsoft, states that at least four parties should form a Consortium Blockchain for reasons of trust[1]. Some more differences to public Blockchains and the reasons why we rely rather on private or Consortium Blockchains in our context are discussed further in Sect. 6. However, the proposed architecture is not limited in the type of the underlying Blockchain.

3   Related Work

In [10], the authors emphasize the chances of using Blockchain Technology in the context of Business Process Management. They provide a broad overview of open research challenges w.r.t. the process life cycle. They also discuss critical issues of applying this technology in the area of Business Process Management, for instance network performance, security and usability. They further stated that inter-organizational processes are affected by a lack of mutual trust, which can be

---

[1] https://jannehansen.com/where-consortium-blockchains-fit/.

solved by providing a trustworthy environment through Blockchain technology. They believe that a system can support global process monitoring where the encryption mechanism is used to handle privacy. However, their work is rather conceptually and they do not focus on or provide any implementation aspects.

In [16] the feasibility of Blockchain technology for process management was proven for the first time. This work includes a **Translator** for an automated translation of BPMN process models to Smart Contracts as well as two different artefacts for the execution: a choreography monitor (**C-Monitor**) for passive monitoring and an active **mediator**. A **trigger** establishes a connection between the artefacts on the Blockchain and the external workflow management systems of the involved participants. The Smart Contract stores two lists, which are used for encoding the process status. Their system enables collaborative process execution over untrusted nodes where all transactions are stored immutable and only conforming cases are executable. They completely rely on executions on a public Blockchain and thus their architecture is affected by some restrictions. For instance, they propose that not all aspects of collaborative process management should be transferred onto the Blockchain due to costs for data storage, transactions and computation. On the other hand, they also had to implement triggers to connect the Blockchain technology world with the enterprise internal process engine, because Smart Contracts could not call external APIs. However, Solidity was developed further and supports function callbacks to the user-interface and error-handling in the meantime. Further on, they have to create a specific Smart Contract for each process model, whereas our Smart Contract is more generic. The work of [16] was revived in [5] and again highlights the need for resource usage optimization driven by the execution on a public Blockchain. This is done by detouring the translation of BPMN models to Smart Contracts via Petri Nets which are minimized. They also took care of the number of deployed contracts as well as throughput rates by optimizing runtime components.

The authors of [7] adopted the approach of artefact-driven business modelling and make use of Blockchain technology for a shared ledger Business Collaboration Language. They discussed the advantages of domain specific languages on a rather conceptual level without any implementation aspects regarding Smart Contracts. [8] introduces Caterpillar, a Business Process Management System that runs on top of the Ethereum Blockchain and makes use of the work in [16], for instance the BPMN-to-Solidity translator. The work does not give enough insights how the internal structures are build and work together or how the system can be extended. Madsen et al. demonstrated in [9] a declarative workflow execution based on DCR-Graphs [6] in an adversarial setting in which Smart Contracts on Ethereum solves the lack of a trusted third party. They also had taken cost issues into concern due to the execution on the public Blockchain. In contrast to them, we see our approach not solely in adverse settings, but also where documentation and traceability plays a key role, for instance regarding quality management in the supply chain. The problem of the immaturity of the Ethereum Blockchain is discussed in [13] where the decision was made in favour of the Bitcoin Blockchain to address the shortcomings regarding major stability issues. However, this 1st-generation Blockchain restricts the opportunities by far. Hence, the authors focus rather on runtime verification of processes than on a sophisticated execution engine.

4   Lean Architecture with a Generic Smart Contract

We present a source-code optimised solution for executing business processes on the Blockchain. At the time of writing, our implementation comprises just around 100 lines of code (independent from the underlying process model) and yet, the deployed Smart Contract is able to run a fully - albeit rudimentary - system for collaborative process execution on the Ethereum Blockchain. Additionally, due to brevity, the contract is easy to maintain, to extend and to integrate via its *Application Binary Interface*. Contrary to the work of [16], we use a generic Smart Contract, i.e. our Smart Contract serves as scaffolding for each process instance and process model. Once deployed, the logic of the implemented process is poured into the Smart Contract by sending transactions (see below). Using this approach, also the process logic itself is on-chain and thus transparent and trustworthy, whereas in the architecture of [16] a superior authority creates and deploys a process-specific Smart Contract. We refer to this issue in Sect. 6.

Infrastructure of the Architecture. As a prerequisite, every participant who wants to join the collaboration, i.e. execute a task, has to establish a connection to the Ethereum Peer-to-Peer-network. This is done with the help of a so-called *wallet*, which assigns the user an unique identifier (160-bit hash value). The user can interact through this *account* with the Ethereum Blockchain, i.e. create transactions and interact with Smart Contracts. Consider Fig. 1. A *Supervisor* S deploys as a first step the Smart Contract out of his wallet (1a) with a transaction (1b). After then, he can add users by registering the wallet addresses via addCollaborators (2) and create tasks in a similar way (3). Step (2) is optional but enables an advanced permission system (see below). The implementation of the core architecture is described in detail in Sect. 5. Our approach is easy to connect to organisation-internal structures. The Ethereum Virtual Machine allows callbacks and error-handling which can be used in a JavaScript interface on the client side. The role of the *Supervisor* can be assigned to any wallet and is responsible for the initial setup (e.g. addTasks) but the full transparency of the Blockchain does not limit the level of trust and tamper protection.

Process Logic Through Requirements. For the implementation of the process logic, the generic Smart Contract provides a struct Task. The struct contains a list requirements holding a reference to every task which must be set on completed before the next task can be completed.

In the process model depicted in Fig. 3 the task B have the requirements [A]. Referring to Fig. 4, the Tasks C and D have [A] as a requirement and B has the requirements [C, D]. Figure 2 describes the situation after the execution of the activity A. C and D are now ready for execution. Based on the gateway, it is sufficient that at least one task is completed (*Exclusive Gateway (XOR)*) or both tasks in the requirements must be set on completed (*Parallel Gateway*). The Smart Contract implements the gateway logic as described in Sect. 5.

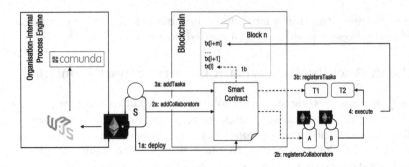

Fig. 1. Architecture of the Blockchain based process execution

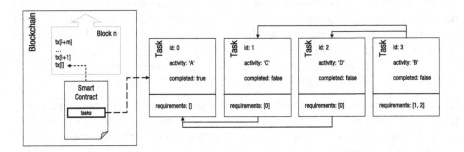

Fig. 2. Requirements

Execution. For each process instance, one Smart Contract must be deployed on the Blockchain. After than, all collaborators (their wallet addresses) can be registered in the Smart Contract for access control purposes. To generalize the execution, the wallet addresses could be mapped organization-internal to a number of employees or IoT-devices etc. The next step is to build the process logic by registering the tasks, with the specific requirements. This is controlled by further transactions, i.e. function calls to the Smart Contract. This procedure can also be automated (cf. Sect. 5). To finish a task, another method call to the Smart Contract is required. To achieve conformance, this method checks if this task can be executed, i.e. if the requirements are fulfilled and if the user is allowed to complete the task.

Fig. 3. Simple sequence flow

Fig. 4. Process model with BPMN-gates

5 Implementation

The architecture of the implementation is depicted in Fig. 1. We only focus on the essential elements of BPMN to proof the feasibility of our approach. However, at spots where we omitted advanced concepts, we show up possible extensions we plan to implement in future.

From BPMN to the Smart Contract. We start from a BPMN process model which encodes the collaboration. The process model is parsed to find the requirements and generate the transactions. Contrary to [16], we do not generate a Smart Contract in this step. As stated, our generic Smart Contract is the same for every process (instance) at time of deployment, and is filled with logic by transactions. Thus, instead of creating Solidity programming code, the translator module outputs the transactions containing the requirements for instance.

The Smart Contract Scaffolding. One Smart Contract represents exactly one process model (or instance) and its structural elements are depicted in Listing 1 exemplary.

Listing 1. Structural concerns of the Collaboration Manager

```
1  contract ContractCollaborationManager {
2      address supervisor;
3      enum Tasktype {TASK, AND, OR};
4      struct Collaborator { address resource; string organisation; }
5      struct Task { string activity; address taskresource; bool completed; Tasktype
       tasktype; uint[] requirements; }
6      mapping(uint=>Task) tasks; uint[] public tasksArray;
7      mapping(uint=>Collaborator) collaborators; uint[] public collaboratorArray; }
```

The struct Collaborator defines the template for any organizational resource or actor, which is included in the process execution and potentially wants to execute a task. Currently, such a collaborator is described by address resource, the wallet address of the performer, and string organization that associates the resource with the employing organization. This can be used in future to restrict the execution not only to a specific person, but also to an organization in general instead. The execution permission mechanism (cf. Permission System) is described below.

The struct Task defines the template for any task in the process model. A task is identified by int id which is used in combination with the address taskresource field for checking the permission to execute the task or to store the actor which has actually executed the task. string activity gives a textual representation of the task and uint[] requirements empowers the conformance of the process execution (cf. Conformance System).

Note the behavioural concerns of the Smart Contract in Listing 2. Mainly two functions are responsible for the initialization or deployment of the Smart

Contract. Firstly, function addCollaborator to register every wallet address who wants to execute at least one task and secondly function addTask, to make the required tasks of the process model available. Note the third parameter of the function createTask. This enum can assume the values TASK, AND and OR and describes the represented BPMN element. This is an essential feature that is used in the Conformance System, described in Listing 3.

Listing 2. Behavioural concerns of the Collaboration Manager

```
1  contract ContractCollaborationManager {
2      function addCollaborator(address _collab, string _org) public { /*...*/ }
3      function createTask(string _activity, address _taskresource, Tasktype
       _tasktype, uint[] _requirements) public {
4          require(msg.sender == supervisor);
5          Task storage task = tasks[taskcount++];
6          task.taskresource = _taskresource;
7          task.requirements = _requirements;
8          /*...*/
```

Permission System. Our Permission System follows currently the guidelines described next, but is easy to adapt. The decision, who is permissioned to execute a task, is made at design time. When a new task is registered at the Smart Contract, the address _taskresource parameter (cf. function createTask in Listing 2, Line 6) specifies the sole wallet address which is permissioned to place the transaction of completing the task (with function setTaskOnCompleted, cf. Line 2 in Listing 3). This is essential when using the public Blockchain, because every wallet could interact with the Smart Contract. The relevant code can be refactored to address less restrictive policies for instance just a requirement that a registered Collaborator belongs to a specific organisation. Note that the organization is encoded in the struct Collaborator. A big advantage of using Blockchain technology is the inbuilt account system with wallet addresses to trace the transactions and interactions with their corresponding performer.

A different conceivable approach is to register a bunch of wallet addresses within the Smart Contract, so that the specific resource allocation can be done at runtime. The resource which wants to execute a task can log in with his credentials to a client-side user-interface which provides an assigned wallet address.

Conformance System. The Conformance System ensures, that the Smart Contract allows only valid process executions according to the initial BPMN model. Line 7 in Listing 2 assigns the ids of tasks which have to be completed, before a Collaborator wants to execute the next task with setTaskOnCompleted. The Conformance System is built upon the _requirements which are extracted from the BPMN model. For instance, in a trivial case the starting event with _id = 0 has no requirement, i.e. it can be executed. The AND-gate is encoded in Listing 3 from Line 8 on. The task is only set on completed (Line 12), when all tasks

Listing 3. Execution concerns of the Collaboration Manager

```
1  function setTaskOnCompleted(uint _id) public returns(bool success) {
2      require(tasks[_id].taskresource == msg.sender);
3      uint[] temprequire = tasks[_id].requirements;
4      /*TASKS */ if(tasks[_id].tasktype == Tasktype.TASK) {
5          if(isTaskCompletedById(temprequire[0]) == true) {
6              tasks[_id].completed = true; return true;
7          else { return false; }}
8      /*AND-GATE */ if(tasks[_id].tasktype == Tasktype.AND) {
9          for(uint i = 0; i ¡ temprequire.length; i++) {
10             if(isTaskCompletedById(temprequire[i])==true) {tempcount++; } }
11         if(tempcount == temprequire.length) {
12             tasks[_id].completed = true; return true; }
13         else { return false; } /*...*/
```

in the requirements are completed. The contract uses a similar solution for the OR-gate, where only one of the requirements has to be fulfilled.

Advanced Implementation Concepts. Contrary to Related Work that we evaluated, we were able to use language constructs which were not introduced back than to design a more sophisticated architecture.

One of the key concepts our proposed architecture is built on, is the require keyword which allows us to introduce a kind of *error-handling* and to have an interaction with the client side. Error-handling is one of the reasons, we could get rid of intermediary structures like triggers [16] to communicate with Smart Contract-external structures and therefore develop a much cleaner concept. We used require at several points, for instance to check if the current sender of a transaction is allowed to execute a task (cf. Listing 3, Line 2). When trying to complete an unassigned task, we receive an error message from the Ethereum Virtual Machine which we can handle in our external application. Also in this context, Solidity provides so-called events for enabling callbacks to a user interface. The concept of mapping helps us to set up a key-value data structure for all Tasks within the Smart Contract. In combination with this, the use of structs and enums gives us a customizable data structure and the final execution on the Smart Contract is according to that much more scalable (cf. Sect. 6). Due to lack of space, we provide the full Smart Contract at our GitHub-Repository[2].

6   Evaluation

In this section, we evaluate how our approach can solve the problem of the need for a trusted third party within collaborative process execution. We also enlarge on a performance and cost evaluation as well as on a comparison among related work.

---

[2] https://github.com/Jonasmpi/PExSCo.

Solving the Lack of Trust. Without Blockchain technology, an automated process-based collaboration would be far more complex and restricted. Each participant may run his own workflow system on a local centralized data storage. After a collaboration is established, a malicious competitor is able to corrupt his local data storage and blame the other actors, if no trusted third party is included. The decentralized data storage in conjunction with the consensus mechanism on Blockchains prevents such a scenario. Albeit the advantages, our solution is not suitable for all choreography processes. Participating business partners have to agree on a globally accepted process model which is difficult, if sensitive data is included in the process or participants do not want to reveal organisation internal process flows. To the best of our knowledge, our approach is the only one, where the process logic is also on-chain, i.e. every participant can comprehend the steps and how the process is defined. In contrast, in [16] the process logic is defined off-chain directly within Solidity code for instance, which requires a kind of a trusted third party for initializing the process run and misses the point of Blockchain a bit.

Execution Costs. Transactions on the (public) Ethereum Blockchain are not free of charge. Miners are awarded for using their computational power to solve the proof-of-work. The amount of GAS refers to the complexity of the computations. The GAS multiplied with the user-set GAS price (a higher price leads to a faster validation) then indicates the price to pay. We propose a very different approach compared to former work, wherefore we have to investigate the execution costs for our implementation. The result of our benchmarks is depicted in Table 1. The (time) measurements bear on our local network solely, as we suggest the usage of a private/Consortium Blockchain anyhow for several reasons (see below). As Vitalik Buterin stated, the costs on a Consortium Blockchain can be significantly lower as just a few nodes must verify the transactions, instead of a world-wide network [4]. Then again, a Consortium Blockchain weakens the tamper-proofness a little bit, as the voting power in the system is concentrated on selected nodes. A sophisticated discussion on public vs. private/Consortium Blockchains related to BPM is still missing in research. Table 1 shows that even for oversized process executions with 8000 tasks, the architecture scales very well and the average duration for adding and completing a task respectively remains constant. The overall GAS consumption rises on a linear basis with the amount of tasks and is 127,000 for the transaction of adding one task to the Smart Contract and 28,006 for the function call of completing one task. The initial deployment of the Smart Contract requires $1,265,261$ GAS. In [16], the execution of an Incident Management process containing 9 tasks costed on average 0.0347 Ether (ETH). The execution costs of a process with 9 tasks using our solution are the following: $1,265,261 + 9 \cdot 127,000 + 9 \cdot 28,009 = 2,660,342$. Based on the GAS price set, i.e. how fast the transactions are validated, the overall costs range from 0.00532 ETH and 0.0532 ETH which is on a similar level compared to the values of [16]. However, as they generate one function for each task and gate, and further on each function holds three different arrays for *PreviousElements*, *NextElements* and *NextJoins*, we believe that for more complex process models our approach with key-value stores scales much better.

Table 1. Time and GAS consumption for different number of tasks on our test network

Number of tasks	100	1000	5000	8000
Duration (function addTask)	14.36	148.98	742.5	1237.6
Average	0.1463	0.1490	0.1485	0.1547
GAS (10^6)	12.7	1270	6350	10160
Duration (function setTaskOnCompleted)	11.38	114	482.6	802.5
Average	0.1138	0.1140	0.0965	0.1003
GAS (10^6)	2.8	280	1400	2240

Privacy and Security. Distributing all process related data affects the privacy of the data. Other publications address this issue by keeping the sensible data off-chain, or encrypt the data and use hash values to establish data integrity at least [16]. However, our architecture is designed to store as much data as possible on-chain (e.g. just extend the struct Task). This will further raise traceability, transparency and to make Blockchain technology attractive to advanced disciplines of Business Process Management like Process Mining etc. The usage of a Consortium Blockchain helps to address also this issue. All participants are known in this isolated network, thus the data sensibility must only be evaluated against the known business collaboration partners. On the other hand, security is also enhanced. Referring to the *DAO attack*[3] or the theoretical *51% attack*[4] in the Bitcoin network, Consortium Blockchains are not affected from these due to the isolation from the public network.

7 Conclusion

The usage of Blockchain technology empowers business partners to facilitate the collaboration within choreography processes. Having a Consortium Blockchain in action reduces the costs as the proof-of-work mechanism to reach consensus is spread over a preselected number of nodes. Hence, our solution is not forced to limit the number of transactions or the amount of data stored on the Blockchain. Thus, our proposal can be used in future research to include more concepts of BPMN for once and second, to include data attributes to support data-aware process execution. Further research has to keep up with latest innovations in the Blockchain universe. The technology is still immature and new ideas are discussed permanent. For instance, the consensus mechanism on the Ethereum

---

[3] https://www.coindesk.com/understanding-dao-hack-journalists/.

[4] https://www.coindesk.com/bitcoin-miners-ditch-ghash-io-pool-51-attack/.

Blockchain is planned to be changed from proof-of-work to proof-of-stake [2]. For the future we plan to support the mentioned data-aware processes and evaluate the integration of IoT devices in our process system on the Blockchain. We also want to develop further tool support like process-aware user-interfaces for wallets and the integration of our implementation into workflow management systems.

References

1. Antonopoulos, A.M.: Mastering Bitcoin: Programming the Open Blockchain. O'Reilly Media, Inc., Sebastopol (2017)
2. Bentov, I., Gabizon, A., Mizrahi, A.: Cryptocurrencies without proof of work. CoRR (2014). http://arxiv.org/abs/1406.5694
3. Buterin, V.: A next-generation smart contract and decentralized application platform. Technical report (2014). https://github.com/ethereum/wiki/wiki/White-Paper
4. Buterin, V.: On public and private blockchains (2015). https://blog.ethereum.org/2015/08/07/on-public-and-private-blockchains/
5. García-Bañuelos, L., Ponomarev, A., Dumas, M., Weber, I.: Optimized execution of business processes on blockchain. In: Carmona, J., Engels, G., Kumar, A. (eds.) BPM 2017. LNCS, vol. 10445, pp. 130–146. Springer, Cham (2017). https://doi.org/10.1007/978-3-319-65000-5_8
6. Hildebrandt, T.T., Mukkamala, R.R.: Declarative event-based workflow as distributed dynamic condition response graphs (2010). https://doi.org/10.4204/EPTCS.69.5
7. Hull, R., Batra, V.S., Chen, Y.-M., Deutsch, A., Heath III, F.F.T., Vianu, V.: Towards a shared ledger business collaboration language based on data-aware processes. In: Sheng, Q.Z., Stroulia, E., Tata, S., Bhiri, S. (eds.) ICSOC 2016. LNCS, vol. 9936, pp. 18–36. Springer, Cham (2016). https://doi.org/10.1007/978-3-319-46295-0_2
8. López-Pintado, O., García-Bañuelos, L., Dumas, M., Weber, I.: Caterpillar: a blockchain-based business process management system (2017)
9. Madsen, M.F., Gaub, M., Hgnason, T., Kirkbro, M.E., Slaats, T., Debois, S.: Collaboration among adversaries: distributed workflow execution on a blockchain. In: Symposium on Foundations and Applications of Blockchain (2018)
10. Mendling, J., et al.: Blockchains for business process management - challenges and opportunities. CoRR (2017). http://arxiv.org/abs/1704.03610
11. Nakamoto, S.: Bitcoin: a peer-to-peer electronic cash system (2008). http://bitcoin.org/bitcoin.pdf
12. Neubauer, D.M., Goebel, A.: Blockchain for off-chain smart contracts in an SAP environment (2018)
13. Prybila, C., Schulte, S., Hochreiner, C., Weber, I.: Runtime verification for business processes utilizing the bitcoin blockchain. CoRR (2017). http://arxiv.org/abs/1706.04404
14. Rohr, J.: Blockchain for disaster relief: creating trust where it matters most (2017). https://news.sap.com/blockchain-disaster-relief/
15. Szabo, N.: Formalizing and securing relationships on public networks. First Monday (1997). http://firstmonday.org/htbin/cgiwrap/bin/ojs/index.php/fm/article/view/548

16. Weber, I., Xu, X., Riveret, R., Governatori, G., Ponomarev, A., Mendling, J.: Untrusted business process monitoring and execution using blockchain. In: La Rosa, M., Loos, P., Pastor, O. (eds.) BPM 2016. LNCS, vol. 9850, pp. 329–347. Springer, Cham (2016). https://doi.org/10.1007/978-3-319-45348-4_19

Mining Product Relationships for Recommendation Based on Cloud Service Data

Yuanchun Jiang[✉], Cuicui Ji, Yang Qian, and Yezheng Liu

School of Management, Hefei University of Technology,
Hefei 230009, Anhui, People's Republic of China
ycjiang@hfut.edu.cn

Abstract. With the rapid growth of cloud services, it is more and more difficult for users to select appropriate service. Hence, an effective service recommendation method is need to offer suggestions and selections. In this paper, we propose a two- phase approach to discover related cloud services for recommendation by jointly leveraging services' descriptive texts and their associated tags. In Phase 1, we use a non-parametric Bayesian method, DPMM to classify a large number of cloud services into an optimal number of clusters. In Phase 2, we recommend a personalized PageRank algorithm to obtain more related services for recommendation among the massive cloud service products in the same cluster. Empirical experiments on a real data set show that the proposed two-phase approach is more successful than other candidate methods for service clustering and recommendation.

Keywords: Cloud service · Cluster · DPMM · Personalized PageRank

1 Introduction

The emerging cloud computing technology offers a new computing environment which enables us to access computing resources, storage and network infrastructure through the Internet without up-front infrastructure costs [1, 2]. With the rapid development of cloud computing technology, many information resources are wrapped and released as cloud services on public servers [3] and companies such as Google, IBM, Microsoft and Amazon opt to provide cloud service products through the public servers [4]. Because a public server usually has massive cloud service products, cloud service recommendation is necessary to provide right services to right users.

Many methods have been proposed to construct selection and ranking models for service products. Among them, QoS (quality of services)-based service selection model [7–9], AHP-based cloud service ranking model [10], trust-aware service selection model [11] and selection method based on collaborative filtering mechanism [12] are popular models. In these models, quantitative criteria are employed to evaluate service quality and the textual information (e.g. service descriptions) is rarely considered.

© Springer Nature Switzerland AG 2019
F. Daniel et al. (Eds.): BPM 2018 Workshops, LNBIP 342, pp. 374–386, 2019.
https://doi.org/10.1007/978-3-030-11641-5_30

This paper proposes an approach to recommend cloud services with the textual description information and tags. We first propose a non-parametric Bayesian model to cluster cloud services. The model is constructed based on Dirichlet process mixture model (DPMM), which can infer the number of clusters automatically without specifying the number of clusters in advance and work well with large-scale datasets [6]. Then, we proposed a personalized PageRank algorithm to generate cloud service rankings based on service tags and clusters we obtained.

The major contributions of this paper are summarized as follows:

(1) This paper employs textual information to recommend cloud services. Compared with service title and click records, the textual information implies rich service features which can help us understand the service functions and make accurate recommendations. To the best of our knowledge, this is the first research to recommend cloud services based on textual description information.

(2) We propose a nonparametric DPMM to classify cloud services into an optimal number of clusters while the number of clusters is identified endogenously. To cluster cloud services, managers usually do not have knowledge on how many clusters exist and which cloud services belong to which cluster. The nonparametric model is particularly suitable for cloud service clustering because it requires no predefined number of clusters, instead it optimizes the number automatically based on data.

(3) We propose a personalized PageRank algorithm to rank the cloud services in each cluster obtained by the proposed DPMM method. The personalized PageRank algorithm can rank cloud services by tags and textual descriptions, and recommend services to meet users' personalized requirements.

(4) We conduct a set of experiments based on a real-world dataset from Programmable Web. Our experiment shows, compared with the baseline methods, the proposed model achieves a significant improvement.

The remainder of this paper is organized as follows: Sect. 2 reviews the related works in literature. Section 3 introduces the proposed approach. Then, in Sect. 4, carries out experiments on some real-world data sets to validate the performance of our approach. Finally, we conclude our work by presenting summary and future directions in Sect. 5.

2   Related Work

2.1   Cloud Service Recommendation

Since Weiss [13] first proposed the concept of cloud computing, research on cloud computing is becoming more and more popular. Formerly, most of the researches on service selection and recommendation were based on the QoS values. However, sometimes it is difficult for us to get the exact QoS values, so scholars began to focus on evaluating and predicting the missing QoS values [14]. In [7], they presented an evaluation approach of QoCS (Quality of Cloud Service) in service-oriented cloud computing which combines the cloud users' preferences evaluation of cloud service providers employing fuzzy synthetic decision with uncertainty calculation of cloud services based on monitored QOCS data for cloud users. Han [8] proposed a recommendation system which creates ranks of different cloud services based on the network QoS and Virtual Machine (VM) platform factors of different cloud providers. Considering that collaborative filtering technology (CF) is the most mature and widely used technology in the recommend system, CF is also widely used in service recommendation based on QoS [12, 15]. In reality, collaborative filtering is vulnerable to the sparse data and is extremely time-consuming with the enlargement of data.

In [16], the author introduced the cloud broker who is responsible for the service selection and developed impactful service selection algorithms to rank potential service providers and aggregate them. Yu [17] put forward a new train of thought that integrates Matrix Factorization (MF) with decision tree learning to bootstrap service recommendation systems. Ding [18] proposed a ranking-oriented prediction method and the method consists of two parts: ranking similarity estimation and cloud service ranking prediction that takes the customer's attitude and expectations for service quality into account.

2.2 Text Clustering Based on Topic Model

Clustering is a widely researched data mining problem in text domain and the popular method in probabilistic description clustering is topic modeling [19]. Topic model is a probabilistic generation model for finding abstract topics in a series of descriptions and it has been widely applied in information retrieval, natural language processing and machine learning.

Topic models, such as Probabilistic Latent Semantic Analysis (PLSA), has been applied to service discovery [20]. Zhang [22] applied the LDA model to cluster the services and extracted service goals from the textual descriptions of services so that they can help users improve their initial queries by recommending similar service goals. The above service clustering models need to specify the number of clusters in advance. Given the limitations of managers' expertise, time and energy, they may not be flexible enough.

Existing cloud service selection approaches rarely consider some important data sources, such as tags, which have been proved to be very powerful in many domains and have been widely used in search engines, social medias, such as Facebook [23].

For cloud service recommendation, we develop a novel model consisting of two phases: cloud services clustering based on Dirichlet Process Multinomial Mixture model (DPMM) and cloud service ranking based on service tags and clusters we obtained. Details of our model are discussed next.

3   The Proposed Model

Our cloud service recommendation system recommends a set of related cloud service products for users by jointly leveraging the textual description information and tag data. Our approach consists of two main phases. In Phase 1, we propose a nonparametric DPMM model to cluster cloud services based on the textual information.

In Phase 2, we propose the Personalized PageRank algorithm to rank the cloud services in each cluster obtained by the proposed DPMM method. The approach framework is illustrated in Fig. 1.

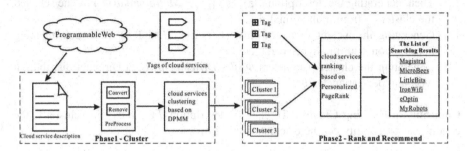

Fig. 1. The framework of the cloud services recommendation.

3.1   Phase1-The Topic Modeling of Web Cloud Service Using DPMM

The DPMM Model. The DPMM is a powerful non-parametric Bayesian method [24] which means that the method can cluster according to the actual situation without specifying the number of clusters in advance. The probabilistic graph of DPMM is shown in Fig. 2 Here, d represents each cloud service description. z represents the cluster label of cloud service description. Multinomial Φ is distributed according to Dirichlet prior β. Multinomial Θ is distributed according to stick-breaking prior α (Table 1).

Table 1. Notations

D	Number of the whole cloud service descriptions set
V	Size of the vocabulary
d	Descriptions in the cloud service descriptions set
z	Cluster labels of each description
m_z	Number of descriptions in cluster z
N_d	Number of words in description d
N_d^ω	Number of occurrences of word ω in description d
n_z	Number of words in cluster z
n_z^ω	Number of occurrences of word ω in cluster z

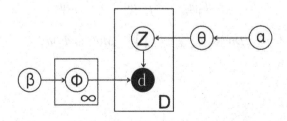

Fig. 2. The probabilistic graph of DPMM.

The generative process of our DPMM is described as follows:

(1) When generating description, the DPMM first selects the cluster $z_d | \Theta \sim$ *Multinomial*(Θ) for description d and z_d is distributed according to multinomial Θ.

(2) Then, generating the description $d | z_d, \{\Phi_k\}_{k=1}^{\infty} \sim$ *Multinomial*(Φ_z) by the selected the cluster z_d from multinomial Φ_{z_d}.

(3) Generating the weight vector of clusters, $\Theta | \alpha \sim GEM(1, \alpha)$ by a stick-breaking construction with the hyper-parameter α.

(4) Generating the cluster parameters $\Phi_z | \beta \sim Dirichlet(\beta)$ by a Dirichlet distribution with a hyper-parameter β.

Choosing an Existing Cluster. To classify description d to an existing cluster z, the conditional probability can be calculated as follows:

$$p(z_d = z | z_{\neg d}, d, \alpha, \beta)$$

$$\propto p(z_d = z | z_{\neg d}, d_{\neg d}, \alpha, \beta) p(d | z_d = z, z_{\neg d}, d_{\neg d}, \alpha, \beta)$$

$$\propto p(z_d = z | z_{\neg d}, \alpha) p(d | z_d = z, d_{z, \neg d}, \beta) \tag{1}$$

Here, we apply the Bayes Rule in Eq. (1) and use the properties of D-Separation [24] in Eq. (1) where $\neg d$ means the description d does not include and $d_{z, \neg d}$ represents other descriptions allocated to cluster z.

The first expression in Eq. (1) means the probability of description d choosing cluster z given the cluster assignments of other descriptions. It can be derived as follows:

$$p(z_d = z | z_{\neg d}, \alpha)$$

$$= \int p(\Theta | z_{\neg d}, \alpha) p(z_d = z | \Theta) d\Theta$$

$$= \int Dir(\Theta | m_{\neg d}) Mult(z_d = z | \Theta) d\Theta$$

$$= \frac{m_{z, \neg d}}{D - 1 + \alpha} \tag{2}$$

The second expression in Eq. (1) indicates a predictive probability of description d given $d_{z, \neg d}$. We can derive the second expression as follows:

$$p(d | z_d = z, d_{z, \neg d}, \beta)$$

$$= \int p(\Phi_z | d_{z, \neg d}, \beta) p(d | \Phi_z, z_d = z) d\Phi_z$$

$$= \int Dir(\Phi_z | n_{z, \neg d} + \beta) \prod_{\omega \in d} Mult(\omega | \Phi_z) d\Phi_z$$

$$= \frac{\prod_{\omega \in d} \prod_{j=1}^{N_d^{\omega}} \left(n_{z,\neg d}^{\omega} + \beta + j - 1 \right)}{\prod_{i=1}^{N_d} \left(n_{z,\neg d} + V\beta + i - 1 \right)} \tag{3}$$

Now we can get the probability of description d choosing an existing cluster z when we know the information of other descriptions and their cluster assignments as follows:

$$p(z_d = z | z_{\neg d}, \boldsymbol{d}, \alpha, \beta) \propto \frac{m_{z,\neg d}}{D - 1 + \alpha} * \frac{\prod_{\omega \in d} \prod_{j=1}^{N_d^{\omega}} \left(n_{z,\neg d}^{\omega} + \beta + j - 1 \right)}{\prod_{i=1}^{N_d} \left(n_{z,\neg d} + V\beta + i - 1 \right)} \tag{4}$$

Choosing a New Cluster. We denote a new cluster as $K + 1$, the conditional probability description d belonging to a new cluster z can be calculated as follows:

$$p(z_d = K + 1 | z_{\neg d}, \boldsymbol{d}, \alpha, \beta)$$

$$\propto p(z_d = K + 1 | z_{\neg d}, \boldsymbol{d}_{\neg d}, \alpha, \beta) p(d | z_d = K + 1, z_{\neg d}, \boldsymbol{d}_{\neg d}, \alpha, \beta)$$

$$\propto p(z_d = K + 1 | z_{\neg d}, \alpha) p(d | z_d = K + 1, \boldsymbol{d}_{z,\neg d}, \beta) \tag{5}$$

We can derive the first expression in Eq. (5) as follows:

$$p(z_d = K + 1 | z_{\neg d}, \alpha) = 1 - \sum_{k=1}^{K} p(z_d = k | z_{\neg d}, \alpha) = \frac{\alpha}{D - 1 + \alpha} \tag{6}$$

Then, the second expression in Eq. (5) can be derived as follows:

$$p(d | z_d = K + 1, \boldsymbol{d}_{z,\neg d}, \beta)$$

$$= \int Dir(\Phi_{K+1} | \beta) \prod_{\omega \in d} Mult(\omega | \Phi_{K+1}) d\Phi_{K+1}$$

$$= \frac{\prod_{\omega \in d} \prod_{j=1}^{N_d^{\omega}} (\beta + j - 1)}{\prod_{i=1}^{N_d} (V\beta + i - 1)} \tag{7}$$

Finally, we can get the probability of description d choosing a new cluster:

$$p(z_d = K + 1 | z_{\neg d}, \boldsymbol{d}, \alpha, \beta) \propto \frac{\alpha}{D - 1 + \alpha} * \frac{\prod_{\omega \in d} \prod_{j=1}^{N_d^{\omega}} (\beta + j - 1)}{\prod_{i=1}^{N_d} (V\beta + i - 1)} \tag{8}$$

After Gibbs Sampling, we can get the representation of clusters by Φ. For each cluster z, we can derive the posterior of Φ_z as follows:

$$p(\Phi_z | d, z, \alpha, \beta) = \frac{1}{\Delta(n_z + \beta)} \prod_{\omega=1}^{V} \Phi_{z,\omega}^{n_z^{\omega} + \beta - 1} = Dir(\Phi_z | n_z + \beta) \qquad (9)$$

where $n_z = \left\{ n_z^{\omega} \right\}_{\omega=1}^{V}$.

Using the expectation of the Dirichlet distribution, we can infer $\Phi_{z,\omega}$ as follows:

$$\Phi_{z,\omega} = \frac{n_z^{\omega} + \beta}{n_z + V\beta} \qquad (10)$$

3.2 Phase2-Cloud Service Ranking Using Personalized PageRank Algorithm

In Phase1, cloud service products are classified into different clusters based on the proposed DPMM algorithm. However, it is still difficult to recommend the appropriate services to users among the massive cloud service products in same cluster. Here we propose the Personalized PageRank algorithm [25] to rank the cloud service products in same cluster.

The proposed Personalized PageRank algorithm employs random walk to rank nodes of a graph consisting of cloud services and tags as nodes and it is a variation of PageRank [26]. PageRank model random-walk process on the web graph composed of numerous pages as nodes and during the process a random surfer will stay the current page i as the next step with probability 1-ε and access to other pages with probability ε. Once the surfer decides to access to other pages, he will uniformly choose a hyperlink contained in the current page. Thus, the random access probability of each page can be calculated as:

$$PR(i) = \frac{(1 - \varepsilon)}{N} + \varepsilon \sum_{j \in in(i)} \frac{PR(j)}{|out(j)|} \qquad (11)$$

where $PR(i)$ represents the probability of a node to be selected. N is the number of all nodes. $in(i)$ represents the node set pointing to node i and $out(j)$ represents the node set pointed by node j. The first part of Eq. (11) means the probability of the surfer staying on the current page i when it is the starting pointing and the second part means the probability of the surfer jumping back to the current page i by clicking on other pages.

For calculating the access probability of a cloud service node in Personalized PageRank, we substitute $\frac{(1-\varepsilon)}{N}$ to $(1 - \varepsilon)\gamma_i$ where γ_i is 1 if the node is our target service and others γ_i is 0. In this way, we can get the relevance of all services relative to the target cloud service.

The Personalized PageRank algorithm will quickly converge to a stable state by recursively calculating and updating the probability of each node. As a result, we can use the value $PR(i)$ of each node as the rank score and recommend Top-k cloud services by selecting cloud service nodes in the node set for the target cloud service.

4 Experiments and Results

4.1 Data Sets and Preprocessing

Experimental data is obtained from Programmable Web, which provides detailed profile information of massive cloud services. The information of cloud services contains services' name, descriptive text and tags. Our data set consists of 799 cloud services and 790 distinct tags. Many tags exist in multiple services, totally 2,745 tags are included in these services. In addition, the average length (i.e., number of words) of each text description is 71.

Because the raw data of the descriptive texts are very noisy, we conduct the following preprocessing: (1) Convert letters into lowercase; (2) Remove meaningless words such as stop words, low frequency words, high frequency words and characters not in Latin.

4.2 Baseline Methods

In the experimental study, we compare DPMM with two typical service clustering methods for service texts nowadays. The details of them are shown below.

K-Means: K-means [27] is probably the most widely used method for clustering. Before being able to utilizing k-means on a set of text descriptions, the texts must be represented as mutually comparable vectors. To achieve this task, each text description can be represented using the TF-IDF score [28].

LDA: We consider the topics found by LDA [29] as clusters and assign each cloud service to the cluster with the highest value in its topic proportion vector.

Some automatic evaluation metrics are proposed in the past few years to measure the quality of the clusters discovered. The typical metric is the coherence score [30], which indicates that a cluster (or topic) is more coherent if the most probable words in it co-occurring more frequently in the corpus. We can calculate the coherence value of a cluster k as follows:

$$C_k = \sum_{m=2}^{M} \sum_{l=1}^{m-1} log \frac{D\left(v_m^{(k)}, v_l^{(k)}\right)}{D\left(v_l^{(k)}\right)} \tag{12}$$

where $v_m^{(k)}$ is one of the most M probable words in cluster k; $D\left(v_l^{(k)}\right)$ represent the description frequency of word l; and $D\left(v_m^{(k)}, v_l^{(k)}\right)$ is the co-description frequency of words.

4.3   Parameter Setting

For DPMM, we set $K = 1$, $\beta = 0.01$. We also assume $Gamma(1, 1)$ priors over the parameters α_0 that can be optimized in Gibbs sampling procedure [31]. In LDA model, we place $\alpha = 50/k$ and $\beta = 0.1$ where K is the number of topics assumed by LDA.

4.4   Results of Service Clustering

Before presenting the final comparisons of baseline methods, we first show the results of cloud services clustering discovered by DPMM. We run Gibbs samplers for 3000 iterations and finally obtain 26 clusters. Figure 3 shows our cluster results with word cloud. Our methods exhibit effectiveness in grouping related cloud services and semantically coherent words together. For instance, Cluster 1 includes cloud-based services designed to handle description, optical character recognition (OCR), and email formats. Cluster 2 offers cloud-based software-as-a-service platforms for enterprise or business. Cluster 3 presents dedicated servers and cloud hosting services for computing. Cluster 4 is about Internet of Thing (IoT) platforms for connections between the clouds and different kinds of devices or appliances. Cluster 5 is about communication technologies that can integrate voice, messaging and email into application.

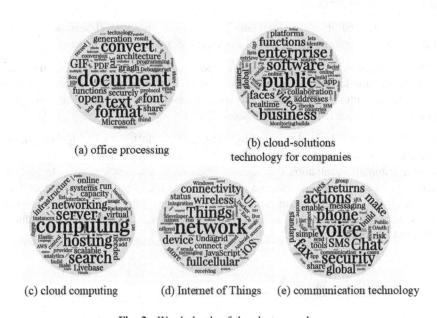

(a) office processing

(b) cloud-solutions technology for companies

(c) cloud computing   (d) Internet of Things   (e) communication technology

Fig. 3. Word clouds of the cluster results.

To evaluate the overall quality of a cluster set, we analyze the average coherence score, namely $\frac{1}{K}\sum_{k=1}^{K} C_k$, for each method. The result is listed in Table 2, where the number of top words ranges from 5 to 25. As shown in Table 2, we find that DPMM obtains the highest coherence score in all the settings. It demonstrates that the DPMM is able to achieve better performance for cluster quality compared with K-means and LDA.

Table 2. Comparison of coherence scores among different methods. A larger score indicates better performance for cluster quality.

Method	Kmeans (K = 10)	Kmeans (K = 20)	Kmeans (K = 30)	LDA (K = 10)	LDA (K = 20)	LDA (K = 30)	DPMM
Top5	−9.16	−7.03	−8.21	−8.04	−8.06	−10.19	**−5.18**
Top10	−64.73	−53.79	−54.66	−55.48	−60.26	−65.30	**−43.39**
Top15	−179.18	−152.96	−155.78	−145.70	−176.94	−178.71	**−142.01**
Top20	−338.39	−314.32	−323.83	−299.303	−356.74	−360.40	**−308.05**
Top25	−562.25	−530.30	−540.17	−521.32	−599.56	−602.32	**−522.03**

4.5   Results of Recommendation

In this section, we show the results of cloud services recommendation. Using personalized PageRank algorithm for each cluster discovered by DPMM, we obtain a ranking list for each cloud service based on the relevance score. For assessing the performance of our results, we adopt Jaccard coefficient, which is an alternative approach to measuring the correlation between products [32, 33]. The Jaccard coefficient is defined as:

$$Jaccard(A, B) = \frac{|d_A \cap d_B|}{|d_A \cup d_B|} \tag{13}$$

Where A is the given product and B the recommended product; d_A and d_B are the textual descriptions of product A and B respectively. $d_A \cap d_B$ is the intersection between two sets d_A and d_B. Thus $d_A \cap d_B$ reveals all words which are in both sets. $d_A \cup d_B$ is the union between two sets d_A and d_B, which represents all words in two sets.

In our tasks, we calculate the averaged Jaccard coefficient of different recommendation lists which are obtained by three methods (Cosine similarity with TF-IDF on textual descriptions, Personalized PageRank on tags, our two-phase approach by jointly leveraging textual descriptions and tags). Each recommendation list contains L highest recommended cloud service resulting. For a given L, the result with a higher averaged Jaccard coefficient is better, and vice versa. The averaged Jaccard coefficient for some typical lengths of recommendation list are shown in Fig. 4, as shown in the Figure, our recommendation results achieve better performance than other two methods, which strongly guarantee the validity of our two-phase approach.

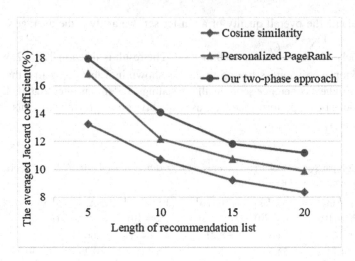

Fig. 4. Comparison of the averaged Jaccard coefficient of different methods.

5   Conclusion

In this paper, we have presented a novel two-phase method by utilizing service text descriptions and tags, to extract latent relations among different cloud services, to generate relevant cloud service recommendation results for aiding users in discovering the available combination of cloud services. Our method is designed to successfully address the cloud service clustering and recommendation. With experiments on a real-world dataset consisting of 799 cloud services and 790 distinct tags obtained from Programmable Web, we demonstrate the effectiveness of this method.

References

1. Michael, A., Fox, A., et al.: Above the clouds: a Berkeley view of cloud computing. Electr. Eng. Comput. Sci. **53**(4), 50–58 (2009). EECS Department University of California Berkeley
2. Katzan, H.: Cloud software service: concepts, technology, economics. Serv. Sci. **1**(4), 256–269 (2013)
3. Chan, H., Chieu, T.: Ranking and mapping of applications to cloud computing services by SVD. In: Network Operations and Management Symposium Workshops, pp. 362–369. IEEE (2010)
4. Marston, S., Li, Z., Bandyopadhyay, S., et al.: Cloud computing - the business perspective. In: Hawaii International Conference on System Sciences, pp. 1–11. IEEE Computer Society (2011)
5. https://en.wikipedia.org/wiki/Mashup_(web_application_hybrid)
6. Yin, J., Wang, J.: A model-based approach for text clustering with outlier detection. In: IEEE, International Conference on Data Engineering, pp. 625–636. IEEE (2016)
7. Wang, S., Liu, Z., Sun, Q., et al.: Towards an accurate evaluation of quality of cloud service in service-oriented cloud computing. J. Intell. Manuf. **25**(2), 283–291 (2014)

8. Han, H., Mehedi, M., et al.: Efficient service recommendation system for cloud computing market. Commun. Comput. Inf. Sci. **63**, 839–845 (2009)

9. Newton, P.C., Arockiam, L.: A Novel Prediction Technique to Improve Quality of Service (QoS) for Heterogeneous Data Traffic. Springer New York, Inc., New York (2011)

10. Garg, S.K., Versteeg, S., Buyya, R.: A framework for ranking of cloud computing services. Future Gener. Comput. Syst. **29**(4), 1012–1023 (2013)

11. Kong, D., Zhai, Y.: Trust based recommendation system in service-oriented cloud computing. In: International Conference on Cloud and Service Computing, pp. 176–179. IEEE Computer Society (2012)

12. Adomavicius, G., Tuzhilin, A.: Toward the next generation of recommender systems: a survey of the state-of-the-art and possible extensions. In: Multimedia Services in Intelligent Environments, pp. 734–749. Springer International Publishing (2013)

13. Weiss, A.: Computing in the clouds. Networker **11**(4), 16–25 (2007)

14. Ding, S., Xia, C., Wang, C., et al.: Multi-objective optimization based ranking prediction for cloud service recommendation. Decis. Support Syst. **101**, 106–114 (2017)

15. Li, J., Zeng, X., Xia, J., et al.: Recent advances in approaches of Web service selection based on QoS. Appl. Res. Comput. (2015)

16. Sundareswaran, S., Squicciarini, A., Lin, D.: A brokerage-based approach for cloud service selection. In: IEEE International Conference on Cloud Computing, pp. 558–565. IEEE (2012)

17. Yu, Q.: Decision tree learning from incomplete QoS to bootstrap service recommendation. In: IEEE International Conference on Web Services, pp. 194–201. IEEE (2012)

18. Ding, S., Wang, Z., Wu, D., et al.: Utilizing customer satisfaction in ranking prediction for personalized cloud service selection. Elsevier Science Publishers B. V. (2017)

19. Aggarwal, C.C., Zhai, C.X.: A Survey of Text Clustering Algorithms. Mining Text Data, pp. 77–128. Springer, US (2012)

20. Ma, J., He, J., He, J.: Efficiently finding web services using a clustering semantic approach. In: International Workshop on Context Enabled Source and Service Selection, Integration and Adaptation: Organized with the, International World Wide Web Conference, p. 5. ACM (2008)

21. Chen, L., Wang, Y., Yu, Q., Zheng, Z., Wu, J.: WT-LDA: user tagging augmented LDA for web service clustering. In: Basu, S., Pautasso, C., Zhang, L., Fu, X. (eds.) ICSOC 2013. LNCS, vol. 8274, pp. 162–176. Springer, Heidelberg (2013). https://doi.org/10.1007/978-3-642-45005-1_12

22. Zhang, N., Wang, J., He, K., et al.: An approach of service discovery based on service goal clustering. In: IEEE International Conference on Services Computing, pp. 114–121. IEEE (2016)

23. Lin, M., Cheung, D.W.: An automatic approach for tagging Web services using machine learning techniques (2016)

24. Bishop, C.M.: Pattern Recognition and Machine Learning. Information Science and Statistics. Springer New York, Inc., New York (2006)

25. Kleinberg, J.M.: Authoritative sources in a hyperlinked environment. J. ACM **46**(5), 604–632 (1999)

26. Kamvar, S.D., Haveliwala, T.H., Manning, C.D., et al.: Extrapolation methods for accelerating PageRank computations. In: International Conference on World Wide Web. pp. 261–270. ACM (2003)

27. Hartigan, J.A., Wong, M.A.: Algorithm AS 136: a k-means clustering algorithm. J. Roy. Stat. Soc.. Ser. C (Appl. Stat.) **28**(1), 100–108 (1979)

28. Larson, R.R.: Introduction to information retrieval. J. Am. Soc. Inform. Sci. Technol. **61**(4), 852–853 (2010)

29. Blei, D.M., Ng, A.Y., Jordan, M.I.: Latent dirichlet allocation. J. Mach. Learn. Res. **3**, 993–1022 (2003)
30. Mimno, D., Wallach, H.M., Talley, E., et al.: Optimizing semantic coherence in topic models. In: Proceedings of the Conference on Empirical Methods in Natural Language Processing, pp. 262–272. Association for Computational Linguistics (2011)
31. Escobar, M.D., West, M.: Bayesian density estimation and inference using mixtures. J. Am. Stat. Assoc. **90**(430), 577–588 (1995)
32. Netzer, O., Feldman, R., Goldenberg, J., et al.: Mine your own business: Market-structure surveillance through text mining. Mark. Sci. **31**(3), 521–543 (2012)
33. Humphreys, A., Jen-Hui Wang, R.: Automated text analysis for consumer research. J. Consum. Res. (2017)

SmartCrowd: A Workflow Framework for Complex Crowdsourcing Tasks

Tianhong Xiong, Yang Yu[✉], Maolin Pan, and Jing Yang

Sun Yat-sen University, Guangzhou 510006, China
tianhongxiong@qq.com, {yuy,panml}@mail.sysu.edu.cn,
yangj357@mail2.sysu.edu.cn

Abstract. Over the past decade, a number of frameworks have been introduced to support different crowdsourcing tasks. However, complex creative tasks have remained out of reach for workflow modeling. Unlike typical tasks, creative tasks are often interdependent, requiring human cognitive ability and team collaboration. The crowd workers are required not only to perform typical tasks, but also to participate in the analysis and manipulation of complex tasks, hence the number and execution order of tasks are unknown until runtime. Thus, it is difficult to model this kind of complex tasks by using existing workflow approaches. Therefore, we propose a workflow modeling approach based on state machine to design crowdsourcing model that can be translated into SCXML code and executed by an open source engine. This approach and engine are embodied in SmartCrowd. Through two evaluations, we found that Smart-Crowd can provide support for complex crowdsourcing tasks, especially on creative tasks. Moreover, we introduce a set of basic design patterns, and by employing them to compose complex patterns, our framework can support more crowdsourcing research.

Keywords: Crowdsourcing · State machine · Workflow · Complex tasks · Creative tasks · Design patterns

1 Introduction

Crowdsourcing can be defined as an emerging computing paradigm that uses advanced internet technologies to harness the efforts of a virtual crowd to perform specific organizational tasks [1]. The requesters (or employers) publish tasks through crowdsourcing marketplaces (such as Amazon Mechanical Turk [2]), which are completed by crowd workers (or employees). The typical tasks are self-contained, simple and repetitive, crowd workers can directly complete work without worrying about how their contributions affect others [3], such as identifying objects in a photo or video, deduplicating data, transcribing audio recordings, or researching data details. Conversely, the complex tasks, especially about creative tasks [4, 5], are often interdependent, requiring human cognitive ability and team collaboration [6]. Such tasks can not be solved directly and need to be decomposed into subtasks. And the workers are required not only to perform typical tasks, but also to participate in the analysis and manipulation of complex tasks, including judgment and decomposition of tasks. In this situation, the

© Springer Nature Switzerland AG 2019
F. Daniel et al. (Eds.): BPM 2018 Workshops, LNBIP 342, pp. 387–398, 2019.
https://doi.org/10.1007/978-3-030-11641-5_31

number and execution order of tasks are unknown until runtime. Consider for example the task of writing a short article, this is a creative task that involves many subtasks and crowd workers, such as deciding structure of the article, deciding how other parts of the article need to write, deciding which sections need to be decomposed further, taking pictures and laying out the document, and so forth. Furthermore, changing one part of the article may trigger changes to the overall plot and vice versa, hence each subtask also needs to coordinate in order to avoid redundant work and to make the final version of the article coherent. However, this kind of complex creative tasks has remained out of reach for workflow modeling, and it will be resisted by the features of these tasks.

Therefore, we propose a workflow modeling approach to support complex crowdsourcing tasks. More specifically, our approach extends statecharts [7] based on state machine to model crowdsourcing tasks, and the model can be translated into State Chart XML (SCXML) which is proposed by World Wide Web Consortium (W3C) to combine statecharts semantics with XML syntax [8]. And we integrate the modeling approach and an open source Apache engine called Commons SCXML [9] to build a framework called SmartCrowd.

In summary, we make the following contributions:

(1) We present a visual modeling approach, which combines graphical symbols of state machine with SCXML. Moreover, by introducing the concept of task instance tree, we can monitor the running process of crowdsourcing.
(2) We provide the SmartCrowd to support model design and implementation of crowdsourcing tasks.
(3) We introduce a group of basic design patterns based on existing crowdsourcing literatures. We can employ them to compose complex design patterns that can be supported by SmartCrowd.

The remainder of the paper is organized as follows. In Sect. 2 we compare our work with related work. Section 3 introduces the core conception for our approach, and describes our modeling approach in detail. The structure of SmartCrowd will be shown in Sect. 4. And Sect. 5 reports two evaluations about our approach. Section 6 concludes with a brief description of future work.

2   Related Work

Over the past decade, crowdsourcing has received a lot of attention, and the approaches and frameworks that support crowdsourcing have attracted the interest of many researchers.

[10] describes a new toolkit Turkit that executes JavaScript files with APIs, the programmer need to write JavaScript code for deploying iterative tasks to Amazon Mechanical Turk (MTurk). [11] introduces AutoMan that is a programming system based on the Scala programming language. By automatically managing quality control and budgeting, it can drive the tasks to continue to do the computation until a desired confidence level is achieved. However, similar to Turkit, complex task must be manually decomposed into smaller tasks by programmers. [12] proposes CrowdDB that provides a declarative approach to solve problems by using an extension of SQL

called CrowdSQL. It is good at dealing with tasks that are of pure data processing nature, such as subjective comparisons and ordering of datasets. However, it is limited to pure data processing problems.

[13] presents Crowdforge that is inspired by the MapReduce distributed computing approach, and it decompose the complex task into three types of subtasks, including partition, mapping and reducing. Yet it cannot support recursion of whole task [15]. [14] presents Jabberwocky, a crowd computing framework that contains three components: the Dormouse is a human and machine resource management system, the ManReduce is a parallel programming framework, and the Dog is a high-level programming language. The programmer can handle tasks with a support of cross-platform programming languages, routing tasks from one platform to another one. However, it is assumed that task requesters (not workers) will determine how tasks are broken down in all cases, hence it is not suitable for creative tasks. [15] introduces a tool called Turkomatic, which emphasizes the decomposition, resolution and merging of solutions. It enables the requesters and workers to collaborate on execution of tasks. Specifically, the workers execute tasks following the instructions of requesters. However, this approach forces the workers must follow the instructions of requesters, and it lacks support for creative tasks that are determined by workers.

These approaches mentioned above are dependent on different programming languages, different types of tasks require different programs to match, involving a lot of hard code. Recently, some workflow-based approaches have been introduced for crowdsourcing research. [17] describes CrowdLang, a programming framework that defines a set of operators, including Reduce, Aggregate, Multiply, and so on, and it can employs these operators to compose complex patterns. [18] presents CrowdSearcher, a search paradigm that defines a query language as a bridge between input and output, which can improve the quality of search results in complex seeking tasks. And the most recent version of CrowdSearcher [19] introduces the modeling concept, and defines a set of task types (e.g. labeling, liking, sorting, classifying, grouping) to solve different problems. Moreover, it can support some specific crowdsourcing patterns. However, it restricts crowdsourcing tasks to simple operations, and the workers are limited to simple and repetitive tasks. [20] provides Crowd Computer that adopt a business process modeling approach based on BPMN [21] to support crowdsourcing workflow. Concretely, it introduces BPMN4Crowd, an extension of BPMN that can model crowdsourcing process. And it extends the standard workflow to relax the constraints imposed on the task assignment, so that it can assign tasks to workers through different tactics (e.g., marketplace, contest, auction, mailing list). However, these approaches mentioned above are rigid that all of tasks still must be decomposed by requesters in advance and executed in a given order, thus, they are also not suitable for creative tasks described previously.

3 Modeling Approach

3.1 Statechart

As we know, the traditional finite state machine cannot model large and complex problems because of three main reasons. Firstly, it has no hierarchy or module concepts. Secondly, there may be a "state explosion" problem. Thirdly, it cannot describe concurrency. Therefore, David Harel presented statecharts, an extension of finite state machine that provides a broadcast mechanism for communication between concurrent components [7]. It conquers the limitations of traditional finite state machine while inherits its main strengths.

In general, a statechart (state machine) contains State, Event, Condition, Action and Transition. The state is a description of the status of a system that can execute action. Transition is a kind of relationship between two states, when a condition is fulfilled or an event is received, the action of transition will be executed, and the source state will transfer to the target state. A transition is often expressed like this: event [condition]/ action. Figure 1 shows an example of statecharts. And we invite the reader who wants to get more information about statechart to read the literatures [7, 8].

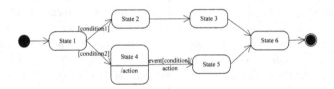

Fig. 1. An example of statecharts.

3.2 Approach Features

In essence, a complex crowdsourcing task often involves one or more subtasks, some of which may continue to be decomposed into subtasks, this may lead to a large number of tasks in the crowdsourcing solution. So how to model these tasks, how to coordinate the interrelated subtasks, these are the key questions for crowdsourcing. In our approach, each task or subtask includes multiple steps from start to end, and each step is treated as a certain state. Hence, all steps of a task can be mapped to different states, which form a complete lifecycle from the start state to the final state. Clearly, lifecycles of different task types are different. Therefore, we consider a crowdsourcing task as a process, which consist of several subprocesses (subtasks). And we employ state machines to model task types as the appropriate building blocks of crowdsourcing model, and then we aggregate them to yield the crowdsourcing result. The approach is described in detail below.

Firstly, we extend action of state machine to meet the requirements of task. Specifically, we add multiple structured elements into action, such as name, attribute, event, data model, and so on, so that we can configure our crowdsourcing solution in these elements, including task decomposition, number of workers, subtasks model, condition of next step, evaluation of results, and so on. In this way, we can not only assign simple tasks to workers, but also arrange workers to participate in the execution of complex tasks. When the task is in different states, the workers are required to participate in different work items, including voting for task decomposition, deciding the next step, processing and evaluating the results of other workers. These structured elements will be embedded into the executable content (EC) of state machine in SCXML. The Fig. 2 indicates the EC can be executed by state and transition of state machine. When the state machine takes a transition, it executes the EC of exit action in the states it is leaving, followed by the EC in the transition, followed by the EC of entry action in the states it is entering.

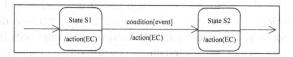

Fig. 2. The EC of state machine.

Secondly, the events of state machine are usually used for internal communication and triggering state transition. Since the statecharts and SCXML provide a flexible mechanism for communication between state machines, the events are used not only for a single state machine, but also for coordination between multiple state machines, this means that different tasks can communicate with each other by sending and receiving events. Therefore, our approach can support interdependent tasks, not just independent tasks.

Lastly, we introduce the task instance tree to monitor the running of crowdsourcing in our framework. As we mentioned previously, a crowdsourcing solution often involves one or more tasks. When it is running, some tasks (called parent tasks) can be broken down into several subtasks (called child tasks), and all tasks will be instantiated in chronological order. A parent task will trigger its child tasks, and the child tasks will continue to trigger their child tasks, and go on. Finally all task instances will form a tree structure called task instance tree. The Fig. 3 shows some examples of task instance trees, Fig. 3(a) shows that a task is complex, while it is non-decomposable, such as Macro Tasks [16]. Thus, only one node in task instance tree. Figure 3(b) shows a complex task that can be decomposed into subtasks. When a task node is finished, it sends an event to its parent node as the signal to finish. It is worth noting that our approach also supports a crowdsourcing model similar to tournament form, Fig. 3(c) shows it as below. Note that the task instance tree of a crowdsourcing solution may not be fixed, when workers are asked to participate in the task decomposition, different decomposition schemes will lead to different instance trees.

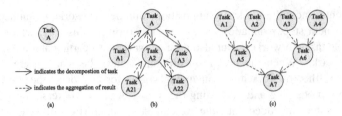

Fig. 3. Some examples of task instance trees.

4 Framework

We integrate our modeling approach and Commons SCXML to build SmartCrowd. The Commons SCXML is an implementation aimed at creating and maintaining a Java SCXML engine that is capable of executing a state machine defined using a SCXML document, while abstracting out the environment interfaces [9]. The Smart-Crowd mainly includes design, compilation, execution and management & maintenance. Figure 4 indicates the architecture of SmartCrowd.

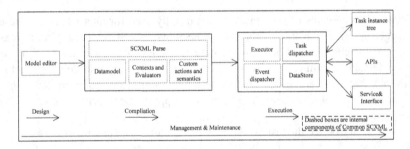

Fig. 4. The high level architecture of SmartCrowd.

Design. We provide a model editor for users to design their crowdsourcing solution. As we mentioned previously, we add several structured elements to support crowd-sourcing solution, such as name, attribute, event, data model, and so on. And the model editor can verify whether the model conforms to the syntax rules of the XML speci-fication, and export the SCXML code document to compilation. Figure 5 shows one part of crowdsourcing design process and corresponding SCXML code.

Compilation. In this stage, the SCXML code document that defines the state machines will be parsed into Commons SCXML Java object model. More specifically, the ele-ments of data and temporary variables will be defined in the Data model. The Contexts and Evaluators can support expression evaluation and context explanation. In addition, the Custom actions and semantics provide an extension of the Commons SCXML for specialized uses, such as supporting custom elements and custom processing logics.

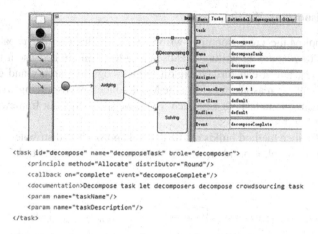

```
<task id="decompose" name="decomposeTask" brole="decomposer">
    <principle method="Allocate" distributor="Round"/>
    <callback on="complete" event="decomposeComplete"/>
    <documentation>Decompose task let decomposers decompose crowdsourcing task
    <param name="taskName"/>
    <param name="taskDescription"/>
</task>
```

Fig. 5. One part of crowdsourcing design process and corresponding SCXML code.

Execution. The Executor, a SCXML engine that can drive the crowdsourcing model to run. And the Task dispatcher can support the task assignment. The Triggering events (event dispatcher) will deliver events for state machines. The DataStore is a data repository that provides the required data information for other parts of the framework.

Management and Maintenance. When a task is in a certain state that needs workers to participate in a work item, such as voting, the work item will be instantiated as a micro task that can be published to traditional crowdsourcing markets (such as Amazon Mechanical Turk) by APIs and task templates. The task instance tree provides a visual UI to show the running process of crowdsourcing. And the Services&Interfaces will be used for custom page, functional interface and web services, etc. Note that the two-way arrows indicate these parts can interaction with Execution.

5   Evaluation

The SmartCrowd places an emphasis on how to support different crowdsourcing tasks. On the one hand, it has a powerful flexibility, allowing the users to focus on the design and optimization of solution, and do not need to develop an additional, dedicated prototype, especially without hard coding. On the other hand, it has enough adaptability to support complex crowdsourcing tasks which are composed by several basic design patterns. In this section, we report two evaluations on exploring whether or not this framework has achieved the desired goals.

More specifically, the first evaluation describes the design and implementation of complex creative tasks. It demonstrates the capability of SmartCrowd for crowdsourcing. In the second evaluation, we show a set of basic design patterns, which can be used to compose more complex design patterns to accommodate different types of crowdsourcing problems, and our approach can also support these complex patterns. This evaluation suggests that SmartCrowd has a great adaptability.

5.1 Crowdsourcing Writing

Here, we employ the crowdsourcing writing task to illustrate the capability of our framework. Crowdsourcing writing is chosen as a test domain because it is a complex creative task [3], and it can further exploit human cognitive ability and collaborative innovation. For these reasons, we chose the MapReduce crowdsourcing writing process from [13] as example to test our approach. Here, we show how our framework supports it without hard coding.

The MapReduce method builds on the general approach to distributed computing, it brokes down a complex problem into a sequence of simpler subtasks, as shown in Fig. 6.

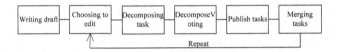

Fig. 6. The MapReduce write processes.

For a better readability, the design model only show the key actions and conditions, and "[]" means the condition, "/" represents the executable content of action in state and transition.

The MapReduce model will judge (vote) for drafts or outline (section headings) in state "Judging", as shown in Fig. 7(a), if the outline is complex, and can not be done by workers directly, it enters the state "Decomposing". According to the decomposition strategy, several workers independently decompose the task into subtasks, and then there will be multiple alternative decomposition schemes. Thus in state "Decompose Voting", the workers vote for the best decomposition scheme, and the framework assigns the number of corresponding subtasks to variable "DecTaskCount". Different decomposition schemes have different number of subtasks, so the number of subtasks here is not known until runtime. And then the SmartCrowd instantiates new state machines for subtasks by executing the "NewSubtasks(n)" in action, the parameter n represents the number of subtasks. Note that since all tasks, including subtasks, must first be judged whether they need to be broken down in this scenario, we employ the same model for these tasks, namely the state machines of subtasks are the same as their parent state machine, it makes our design more succinct and effective. After that, the parent task enters the state "Waiting" to wait for results of its subtasks, when all subtasks are finished, the condition "DecTaskCount == SubtaskFinishCount" is true, it enters the state "Merging" and aggregates the results to produce a final result, and sends an event as a finish signal to parent task. When the task is determined as simple, it do not need to decompose, and is directly allocated to multiple workers to solve in state "Solving". After all workers finish their jobs, the condition "TCount == WorkerFinishCount" is true, a group of workers vote for the best answer in state "SolveVoting", and finally send an event to inform the parent task. The Fig. 7(b) shows one of the task instance trees of MapReduce model, states of state machine are color-coded to indicate their status: in progress (red, underlined), finished (green). Different outlines of article yield different task instance trees, and even each execution might generate a different task instance tree since the task decomposition is uncertain until runtime. For example, the task A is decomposed into 3 subtasks, while its subtask A2 is decomposed into 2 subtasks (Fig. 7(b)), the different decomposition schemes voted by workers result in this result.

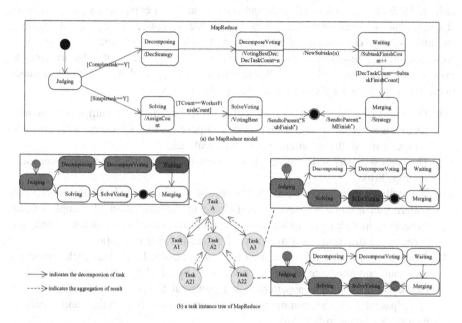

Fig. 7. The MapReduce model and one of its task instance trees (Color figure online)

Through this case, SmartCrowd allows workers to participate in the task decomposition, supporting the collaboration between requesters and workers. Through the task instance tree, we can easily monitor the running process of crowdsourcing, including tasks types, the status of tasks, the number of task instances.

It is worth noting that, when workers are required to determine the task decomposition with no supervision, the task was likely to be constantly decomposed and iterated, resulting in excessive task instances. It may cause tasks to get out of control, thus SmartCrowd provides three ways to avoid this kind of situation, and the first one is to allow the requester to end task directly, which is the highest priority, the second allows the workers to decide when to end the task, within a set period of time. And the last one is to stop task after a predefined number of task decomposition rounds, namely preset depth of the task instance tree. And limited to the pages of paper, detailed SCXML code exported by our framework for this crowdsourcing writing case is displayed on the web site [22].

5.2 Adaptability Study

The Amazon Mechanical Turk team has indicated that the investigations on crowd-sourcing should be designed, conducted, and published in a manner such that the research experiments can be repeated, potentially yielding standard design patterns and methods to achieve high quality, consistent results for a variety of human computation tasks [23]. So far, crowdsourcing design pattern has been explored by previous work [19, 24], but still has remained out of reach. We reviewed the existing literatures, including many empirical studies and reports [6, 25], and introduce four basic crowdsourcing design patterns based on them. Furthermore, we employ SmartCrowd to express these patterns, and for a better readability, we report them using task instance trees. The Fig. 8 shows these basic design patterns.

(1) The collection pattern, as shown in Fig. 8(a), in which the number of subtasks decomposed is fixed, and each task is independent of each other, with no inter-action. It meets the requirements of a typical task.
(2) The contest pattern will choose a best answer for task, similar to tournament form, see Fig. 8(b).
(3) The collaboration pattern usually appears in creative crowdsourcing work. This pattern can arrange workers to participate in the execution of complex tasks. Especially in task decomposition, the number of subtasks is unknown until run-time, hence different tasks may have different number of subtasks.
(4) The interaction pattern is often used for creative tasks. In our approach, the parent task can communicate with its child tasks by sending and receiving events. However, the interaction pattern has a powerful feature that subtasks with the same parent task can communicate with each other by sending and receiving events, as shown in Fig. 8(d).

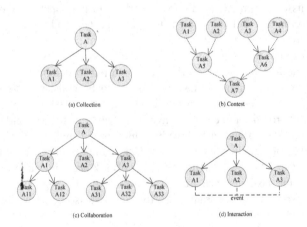

Fig. 8. The basic design patterns

In general, the complex design model is composed by different basic design patterns, such as crowdsourcing writing model is shown in Sect. 5.1, which deployed some basic patterns, including collaboration and contest (voting for the best scheme). This evaluation shows that our framework can support complex patterns that are composed by basic design patterns. Note that we do not think these basic design patterns to be complete. In addition, we have tested a variety of crowdsourcing cases, such as purchasing decisions, evaluating a design and so on, and more detailed information can be seen in [22].

6 Conclusion

In this paper, we present SmartCrowd based on state machine and Common SCXML to support crowdsourcing research, especially in complex creative tasks. It allows researchers to put more energy on the solution formulation rather than on the development and maintenance of system. Due to the use of the state machine and Common SCXML, the users of our approach also need to understand the state machine and the specification of SCXML.

There are a number of directions we are exploring for future work. We will continue to exploit new design patterns of crowdsourcing, and the performance analysis of crowdsourcing model is also a part of the follow-up work.

Acknowledgments. This work was supported by the National Key Research and Development Program of China under Grant No. 2017YFB0202200; the National Natural Science Foundation of China under Grant No. 61572539; the Research Foundation of Science and Technology Major Project in Guangdong Province under Grant Nos. 2015B010106007, 2016B010110003; the Research Foundation of Science and Technology Plan Project in Guangdong Province under Grant No. 2016B050502006.

References

1. Saxton, G.D., Oh, O., Kishore, R.: Rules of crowdsourcing: models, issues, and systems of control. Inf. Syst. Manag. **30**(1), 2–20 (2013)
2. Amazon Mechanical Turk. https://www.mturk.com/. Accessed 10 May 2018
3. Kim, J., Sterman, S., Cohen, A.A.B., Bernstein, M.S.: Mechanical novel: crowdsourcing complex work through reflection and revision. In: Proceedings of the 2017 ACM Conference on Computer Supported Cooperative Work and Social Computing, pp. 233–245. ACM (2017)
4. Orlowska, M.E.: Challenges for workflows technology to support crowdsourcing activities. In: International Conference on Business Intelligence and Technology, pp. 88–92. Bustech (2015)
5. Kim, J., Cheng, J., Bernstein, M.S.: Ensemble: exploring complementary strengths of leaders and crowds in creative collaboration. In: Proceedings of CSCW, pp. 745–755. ACM, New York (2014)
6. Malone, T.W., Laubacher, R., Dellarocas, C.: The collective intelligence genome. MIT Sloan Manag. Rev. **51**(3), 21 (2010)

7. Harel, D.: Statecharts: a visual formalism for complex systems. Sci. Comput. Program. **8**(3), 231–274 (1987)

8. SCXML. https://www.w3.org/TR/scxml/. Accessed 10 June 2018

9. Apache Commons SCXML. http://commons.apache.org/proper/commons-scxml. Accessed 15 June 2018

10. Little, G., Chilton, L.B., Goldman, M., Miller, R.C.: Turkit: tools for iterative tasks on mechanical turk. In: Proceedings of the ACM SIGKDD Workshop on Human Computation, pp. 29–30. ACM (2009)

11. Barowy, D.W., Curtsinger, C., Berger, E.D., McGregor, A.: Automan: a platform for integrating human-based and digital computation. ACM Sigplan Not. **47**(10), 639–654 (2012)

12. Franklin, M.J., Kossmann, D., Kraska, T., Ramesh, S., Xin, R.: CrowdDB: answering queries with crowdsourcing. In: Proceedings of the 2011 ACM SIGMOD International Conference on Management of data, pp. 61–72. ACM (2011)

13. Kittur, A., Smus, B., Khamkar, S., Kraut, R.E.: Crowdforge: crowdsourcing complex work. In: Proceedings of the 24th Annual ACM Symposium on User Interface Software and Technology, pp. 43–52. ACM (2011)

14. Ahmad, S., Battle, A., Malkani, Z., Kamvar, S.: The Jabberwocky programming environment for structured social computing. In: Proceedings of the 24th Annual ACM Symposium on User Interface Software and Technology, pp. 53–64. ACM (2011)

15. Kulkarni, A., Can, M., Hartmann, B.: Collaboratively crowdsourcing workflows with turkomatic. In: Proceedings of CSCW, CSCW 2012, pp. 1003–1012. ACM, New York (2012)

16. Chittilappilly, A.I., Chen, L., Amer-Yahia, S.: A survey of general-purpose crowdsourcing techniques. IEEE Trans. Knowl. Data Eng. **28**(9), 2246–2266 (2016)

17. Minder, P., Bernstein, A.: *CrowdLang*: a programming language for the systematic exploration of human computation systems. In: Aberer, K., Flache, A., Jager, W., Liu, L., Tang, J., Guéret, C. (eds.) SocInfo 2012. LNCS, vol. 7710, pp. 124–137. Springer, Heidelberg (2012). https://doi.org/10.1007/978-3-642-35386-4_10

18. Bozzon, A., Brambilla, M., Ceri, S.: Answering search queries with CrowdSearcher. In: Proceedings of the 21st International Conference on World Wide Web, pp. 1009–1018. ACM (2012)

19. Bozzon, A., Brambilla, M., Ceri, S., Mauri, A., Volonterio, R.: Designing complex crowdsourcing applications covering multiple platforms and tasks. J. Web Eng. **14**(5–6), 443–473 (2015)

20. Tranquillini, S., Daniel, F., Kucherbaev, P., Casati, F.: Modeling, enacting, and integrating custom crowdsourcing processes. ACM Trans. Web (TWEB) **9**(2), 7 (2015)

21. Object Management Group: Business Process Model and Notation (BPMN) Version 2.0 (2011). http://www.omg.org. Accessed 15 Mar 2018

22. SmartCrowdCode. https://github.com/rinkako/RenWFMS/tree/master/exampleProcess/Crowdsourcing. Accessed 10 June 2018

23. Chen, J.J., Menezes, N.J., Bradley, A.D., North, T.: Opportunities for CrowdSourcing research on amazon mechanical turk. Interfaces **5**(3), (2011)

24. Lofi, C., El Maarry, K.: Design patterns for hybrid algorithmic-crowdsourcing workflows. In: 2014 IEEE 16th Conference on Business Informatics (CBI), vol. 1, pp. 1–8. IEEE (2014)

25. Quinn, A.J., Bederson, B.B.: Human computation: a survey and taxonomy of a growing field. In: Proceedings of the SIGCHI Conference on Human Factors in Computing Systems, pp. 1403–1412. DBLP (2011)

Third International Workshop on Process
Querying (PQ 2018)

Third International Workshop
on Process Querying (PQ 2018)

Third International Workshop on Process Querying (PQ 2018)

Process-related information grows exponentially in organizations via workflows, guided procedures, business transactions, Internet applications, real-time device interactions, and other coordinative applications underpinning commercial operations. Event logs, application databases, process models, and business process repositories capture a wide range of process data, e.g., activity sequences, document exchanges, interactions with customers, resource collaborations, and records on product routing and service delivery. Process querying studies automated methods for the inquiry, manipulation, and update of models and data of the real-world and designed processes, as well as relations between the processes, with the ultimate goal of converting process-related information into decision-making capabilities. Process querying research spans a range of topics from theoretical studies of algorithms and the limits of computability of process querying techniques to practical issues of implementing process querying technologies in software. Examples of practical problems tackled using process querying include process compliance, standardization, reuse, variance management, comparison, and monitoring.

The Third International Workshop on Process Querying (PQ 2018) aimed to provide a high-quality forum for researchers and practitioners to exchange research findings and ideas on technologies and practices in the area of process querying. Two full papers were presented at the workshop. In his paper entitled "Checking Business Process Models for Compliance Comparing Graph Matching and Temporal Logic," Riehle presents his research endeavor in the area of business process compliance management (BPCM). In particular, he examines the two main approaches in the field, namely, graph-based pattern matching and pattern matching based on temporal logic. Furthermore, he compares the approaches by implementing four compliance patterns taken from the literature. In "From Complexity to Insight: Querying Large Business Process Models to Improve Quality," Madsen reports on an industrial case study in the context of automotive manufacturing. More specifically, he presents the application of his approach which, by querying, manipulating, and transforming the process models in the enterprise collection, allows the stakeholders to gain insight into improving their current quality of models.

November 2018

Artem Polyvyanyy
Arthur H. M. ter Hofstede
Claudio Di Ciccio

Organization

Program Committee

Agnes Koschmider	Karlsruhe Institute of Technology, Germany
Ahmed Awad	Cairo University, Egypt
Artem Polyvyanyy	The University of Melbourne, Australia
Arthur H. M. ter Hofstede	Queensland University of Technology, Australia
Boudewijn van Dongen	Eindhoven University of Technology, The Netherlands
Chun Ouyang	Queensland University of Technology, Australia
Claudio Di Ciccio	Vienna University of Economics and Business, Austria
David Knuplesch	Ulm University, Germany
Dirk Fahland	Eindhoven University of Technology, The Netherlands
Fabrizio Maria Maggi	University of Tartu, Estonia
Gero Decker	Signavio, Germany
Hagen Völzer	IBM Research – Zurich, Switzerland
Hajo A. Reijers	VU University Amsterdam, The Netherlands
Henrik Leopold	VU University Amsterdam, The Netherlands
Hyerim Bae	Pusan National University, South Korea
Jochen De Weerdt	KU Leuven, Belgium
Jorge Munoz-Gama	Pontificia Universidad Católica de Chile, Chile
Luciano García-Bañuelos	University of Tartu, Estonia
Manfred Reichert	Ulm University, Germany
Marcello La Rosa	The University of Melbourne, Australia
Marco Montali	Free University of Bozen-Bolzano, Italy
Massimiliano de Leoni	Eindhoven University of Technology, The Netherlands
Matthias Weidlich	Humboldt University of Berlin, Germany
Minseok Song	Pohang University of Science and Technology, South Korea
Remco Dijkman	Eindhoven University of Technology, The Netherlands
Seppe Vanden Broucke	KU Leuven, Belgium
Stefan Schönig	University of Bayreuth, Germany
Stefanie Rinderle-Ma	University of Vienna, Austria
Wil M. P. van der Aalst	RWTH Aachen University, Germany

Checking Business Process Models for Compliance – Comparing Graph Matching and Temporal Logic

Dennis M. Riehle[✉]

European Research Center for Information Systems, University of Münster,
Münster, Germany
dennis.riehle@ercis.uni-muenster.de

Abstract. Business Process Compliance Management (BPCM) is an integral part of Business Process Management (BPM). A key objective of BPCM is to ensure and maintain compliance of business processes models with certain regulations, e.g. governmental laws. As legislation may change fast and unexpectedly, automated techniques for compliance checking are of great interest among researchers and practitioners. Two dominant concepts in this area are graph-based pattern matching and pattern matching based on temporal logic. This paper compares these two approaches by implementing four compliance patterns from literature with both approaches. It discusses what requirements both approaches have towards business process models and shows how to meet them. The results show that temporal logic is not able to fully capture all four patterns.

Keywords: Process querying · Pattern matching · Compliance patterns · Graph matching · GMQL · Temporal logic · CTL

1 Compliance Checking of Business Process Models

With the rise of Business Process Management (BPM) over the last decades, organizations have created large amounts of conceptual models [1], making conceptual models an integral and important part of modern organizations. One important aspect of BPM is Business Process Compliance Management (BPCM). BPCM is about checking if business processes adhere to agreed-upon objectives, to gain and to maintain such compliance [2]. Objectives which define a compliant process may come from several sources, as internal and external requirements. Though organizations may specify internal rules and objectives which business processes need to comply with, many compliance requirements come from legislation (e.g. [3, 4]). As governmental regulations may change frequently and at unexpected times, BPCM must react to such external changes in a fast, dynamic and efficient way. These requirements towards BPCM call for tools, techniques and methodologies, to support BPCM, by finding compliance violations in an automatic and simple manner [5].

One approach towards automatic process analysis is Business Process Querying [6], where a set of (structural) patterns is queried against a repository of process models

© Springer Nature Switzerland AG 2019
F. Daniel et al. (Eds.): BPM 2018 Workshops, LNBIP 342, pp. 403–415, 2019.
https://doi.org/10.1007/978-3-030-11641-5_32

(e.g. [4]). This requires a query language, where a structural pattern is used as an input and queried against an arbitrary amount of models (the repository), to check whether these models contain the searched pattern, and is seen as an enabler for business intelligence [7]. In literature, there are plenty of query languages that can be used to identify compliance violations in process models (e.g. [8–11]). Query languages can be divided into structural and behavioural query languages [12], where behavioural query languages tend to operate directly on process logs, while structural query languages tend to operate on a process model. This work focuses on compliance checking of process models and not on the analysis of event logs. Two well-known but methodically different concepts are graph matching for structural querying and temporal logic for both structural and behavioural querying. While authors of either approaches describe the advantages of their approach (see [8–11] for details), the research question of this paper is, if graph matching and temporal logic can both be used to express the same compliance patterns. For this, four patterns from literature are specified using both approaches – as far as possible – and the results are compared.

This paper is structured as follows. In Sect. 2 shortly discusses modelling query languages in general, while in Sect. 3, the transformation of business models to structures suitable for analysis is discussed. Section 4 presents the specification of different queries using the techniques of graph matching and temporal logic. Section 5 compares graph matching and temporal logic and Sect. 6 concludes with an outlook.

2   Background and Related Work

Business Process Management (BPM) is about the "concepts, methods, and techniques to support the design, administration, configuration, enactment, and analysis of business processes" ([13], p. 5). One key part of BPM is the representation of business processes through business process models, for which a variety of modelling languages exists (see, e.g., [14] or [15] for an overview).

One core concern in the analysis of business processes is the identification of structural and behavioural patterns in process models. In literature, there is a variety of model query languages, which enable the specification of such patterns and provide a technique to identify occurrences of them in process models (e.g., [16]). One common concept is graph matching, which tries to match a pattern (a small graph) within a model (a large graph). When found, this small graph would be called a match and represent part of the whole model (a subgraph that is either equal or similar to the pattern). Consequently, if graph matching does not return any results, the pattern was not found in the model. Contrastingly, temporal logic formulates rules and assertions, which are evaluated against a formalized version of a process model. These rules and assertions are then processed and result in a binary decision, true or false. The contextual interpretation of this result of course depends on the way the query was formulated.

2.1   Compliance Checking with Graph Matching

One exemplary model querying language based on graph matching is the Generic Model Query Language (GMQL), which provides pre-defined sets of elements and a batch of functions to apply to these elements [11]. Functions in GMQL can be nested to any level, to provide a maximum of flexibility. Functions to identify elements include *ElementsOfType (EOT)* and *ElementsWithAttributeOfValue (EWAOV)*, which do what their name suggests and otherwise are defined in [11]. Functions to identify elements including their related neighbours include *ElementsWith{NumberOf}{Pred|Succ} Relations{Of-Type} (EW{N}{P|S}R{OT}*, *ElementsDirectlyRelated (EDR)* and *AdjacentSuccessors (AS)*. Additionally, there are functions to identify paths and loops within the model, including *{Directed}Paths{[Not]ContainingElements} ({D}P{{N} CE})* and *{Directed}Loops{[Not]Containing-Elements} ({D}L{{N}CE})*. Further operators exist, which can combine two result sets into one set. These operators include *{Self}Union, Join, {Inner|Self}Intersection* and *{Inner}Complement*. All of these functions are de-fined in the relevant literature.

GMQL has already been applied to the scenario of compliance checking, where compliance patterns have been derived from legislation and where these patterns have been searched in real-world process models [3].

2.2 Compliance Checking with Temporal Logic

Another way of model querying is the Computational Tree Logic (CTL). CTL originates from the specification and verification of software systems. An introduction to model checking with CTL can, for instance, be found in [17]. CTL includes several statements, which describe states. The operator X can be used to describe the next state; the operator F describes any state in the future. To describe all states, G can be used (global) and U can be used describe all states until a certain condition appears to be true. Additionally, the operators X, F and G can be preceded by an A, which requires all paths to fulfil the criteria, or with E, which requires that there is at least one path matching the criteria. The exclamation mark can be used to negate conditions.

Linear Temporal Logic (LTL) and CTL have already been used for compliance checking of process models. For example, [4] and [18] have specified compliance patterns using temporal logic. Additionally, [4] specifies a Compliance Request Language (CRL) as a set of abbreviations for commonly used CTL/LTL expressions. This makes the specification of compliance patterns shorter and easier to understand.

3 Transforming Process Models to State-Machines

Graph matching and temporal logic pose different demands on the business process models they shall be applied to. While the pattern is specified with GMQL on the one hand and CTL, LTL or CRL on the other hand, there are different requirements towards the model as well. Graph matching requires the model to be in a graph structure, that is, a structure consisting of vertices, edges, and attributes, and temporal logic requires the process models to be in a formal representation like a computation tree that can be processed with CTL. Conceptual models like business process models consist of elements and relations between these elements, hence they fulfil the general criteria of a graph structure. However, depending on the Business Process Modelling Language (BPML) used, process models may not be computation trees and, hence, require further pre-processing and transformations, before temporal logic can be applied. For demonstration purposes, business processes modelled with a domain-specific BPML (icebricks, see [19] for an introduction) are used. Since domain-specific BPMLs usually follow similar concepts of frameworks (e.g., [20]), other BPMLs are likely to face similar challenges during the transformation. There are two important characteristics of icebricks: First, main processes and detail processes can consist of multiple main process variants or detail process variants. These variants are used to reflect different or partly different activities performed in different instances of the same business process. Such a variant might reflect the same business process, like opening a bank account, for either a private customer or a business customer. Since one process instance only reflects one variant, different variants can be compared to XOR splits in other BPMLs. icebricks models do not have directed edges, as models are – by convention – read from top to bottom.

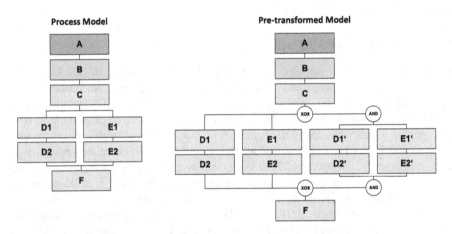

Fig. 1. Original and pre-transformed process model.

Second, several process elements can succeed one process element and one process element can be preceded by multiple elements (see process model in Fig. 1). For this modelling scenario, icebricks does not define which of the process branches needs to be executed, as long as at least one branch appears in a process instance. Consequently, they are to be considered as an inclusive OR. While one may argue that this unclear execution semantic is a design flaw, the originators of icebricks argue that this simplification enables the creation of process models which are easy to understand, even for non-technicians. Additionally, other popular process modelling languages like Event-driven Process Chains [21] or its variants [22] do include a similar OR connector and whose semantics have been discussed thoroughly (e.g., [23]). As a model check with CTL requires a formally specified process model as input, the execution semantics of the icebricks models needs to be clarified first. This can be easily done by replacing OR constructs through a combination of XOR and AND.

The example given in Fig. 1 shows a simple process. It consists of one main process (A), one detail process (B) and six process activities. The process activities D1 and E1 follow the process activity C, which means that after executing C, either only D1 and D2 can be executed, or only E1 and E2 can be executed, or all of them (D1, D2, E1 and E2) can be executed. To keep the computation tree simple, it is assumed that if multiple process branches are executed, each process branch is fully executed before another process branch starts (this means that there is no real parallel execution). With this assumption in mind, the tree shown in Fig. 2 can be created, showing all possible executions paths throughout the process model from Fig. 1.

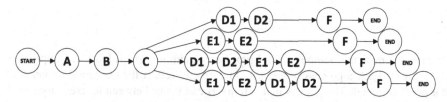

Fig. 2. Process model as computation tree.

While the example above with the two process activities D1 and E1 following C is rather simple, the pre-transformation becomes drastically more complex if there are more process branches. In particular, n process branches will result in a total amount of $(2^n)-1$ combinations when still disregarding the order of execution. If we would further regard the order in which process branches are executed, the term k! would be added as a multiplication inside the sum, which explains why the amount of execution paths regarding the execution order of activities explodes with rising numbers of process branches. This issue is also known as the state explosion problem [24].

For a short demonstration, a process database from a financial service provider has been transformed from icebricks to computation trees using the logic described above. This database contained 31 main processes and 216 detail processes, with 2,894 process activities altogether (Table 1 only shows four sample processes). For the process model before and after expansion, the number of activities (Act.) and gateways (Gate.) in the model is depicted. Obviously, the number of states in a computation tree would be even lager in the end, as many activities are duplicated.

Table 1. Complexity of process models after transformations.

Process model	Process model		After expansion	
	Act.	Gate.	Act.	Gate.
Current account	39	4	3,042	1.030
Savings account	19	4	603	252
Construction loan	56	2	393	24
Consumer loan	13	2	29	10

Besides the unclear execution semantics introduced by an inclusive OR operator, BPMLs can also contain other structures which need special attention when process models are to be transformed in computation trees. Probably the most prominent structure to be found in many BPMLs are loops. For demonstration purposes, the commonly known language Business Process Model and Notation (BPMN) [25] is used.

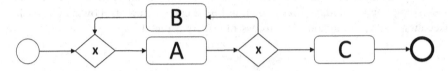

Fig. 3. Process model containing a loop.

Figure 3 depicts a sample BPMN process containing a small loop. A computation tree built for such a process model should correctly reflect the execution semantic of that process model. This means that activity B can occur between to executions of A and activity A can occur between two executions of B. Of course, A can also occur without any execution of B. These assumptions also holds true if there were more than just one activity on the loop. In addition, the loop can be executed unlimited times, which makes an adequate representation in a computation tree impossible. As branches in trees only split but never merge, a tree cannot reflect a loop.

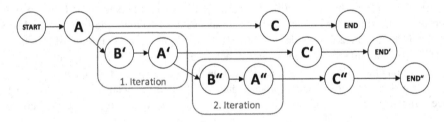

Fig. 4. Process model with loop as computation tree.

As a work-around to convert process models containing loops to computation trees, only a limited amount of loop executions is simulated. More precisely, in this paper exactly two loop executions are simulated, as depicted in Fig. 4. This is just enough to fulfil the previously mentioned criteria: A can occur for itself, A can occur between two executions of B and B occurs between two executions of A. Again, the problem of state explosion is still present, as all elements following the loop need to be duplicated two times. Therefore, loops still create computational overhead.

4 Exemplary Compliance Patterns

GMQL and CTL provide different features for compliance checking. To enable a comprehensive comparison of both languages, four different compliance patterns are presented and, where possible, specified for both languages. While the first two patterns directly originate from literature, i.e., they were already specified with either GMQL or CTL, the last two patterns are motivated from requirements towards compliance checking approaches or textual descriptions. For these patterns, both the GQML query and the CTL statement have been newly developed.

The first compliance pattern originates from [3] where it was named infringement pattern #1 and was originally specified in GMQL. This compliance pattern derives from the German securities trading act [26], which states that when banks consult customers in order to sell them new products, they need to hand out all necessary product information to the customer, before the product is sold, meaning before any transactions are made. In this example, such process activities are named "Consult customer", "Talk to customer", "Hand out contract", "Hand out documents", "Hand out preliminary contract", "Make account transaction" and "Perform account transaction."

Table 2. Example Pattern "Infringement pattern".

Name: Infringement pattern	**Reference:** [3]
The pattern searches for a process element, where an advice is given as well as an account transaction made. If no contract is handed out in between the rule is violated. (in accordance to WpHG § 31 (4a))	

GMQL	DPNCE (
	    UNION (EWAOV (O,'Caption','Consult customer'), EWAOV (O,'Caption','Talk to customer')),
	    UNION (EWAOV (O,'Caption','Make account transaction'),
	        EWAOV (O,'Caption','Perform account transaction')),
	    UNION (UNION (
	    EWAOV (O, 'Caption', 'Hand out contract'),
	    EWAOV (O, 'Caption', 'Hand out documents')),
	    EWAOV (O, 'Caption', 'Hand out preliminary contract'))))

CTL	M, s0 |= AG (("Consult customer" ∨ "Talk to customer" →
	    AG (
	        EG (¬"Make account transaction" ∧ ¬ "Perform account transaction") ∨
	        A [¬"Make account transaction" ∧ ¬ "Perform account transaction") U
	            ("Hand out contract" ∨ "Hand out documents" ∨ "Hand out preliminary contract")))

The GMQL query shown in Table 2 returns compliance violations. This is done by searching for all paths that start with a process activity named either "Consult customer" or "Talk to customer" and end in an activity named "Make account transaction" or "Perform account transaction", where the directed path does not contain any element named either "Hand out contract", "Hand out documents" or "Hand out preliminary contract." An empty result set does consequently imply that no violations are found.

A corresponding CTL statement is also shown in Table 2. The statement checks if there is an activity called "Consult customer" or "Talk to customer" somewhere in the process model. If so, for all paths reachable from there must hold that it is not allowed to reach an activity called "Make account transaction" or "Perform account transaction" before there was not another activity called "Hand out contract" or "Hand out documents" or "Hand out preliminary contract". If an element "Make account transaction" or "Perform account transaction" was found earlier, the statement evaluates to FALSE and indicates a compliance violation.

Table 3. Example Pattern "Too Many Routing Paths per Element".

	Name: Too Many Routing Paths per Element	**Reference:** [27]
	The pattern refers to situations where a process contains too many different variants	
	starting at the same object. (Three or more variants in this example)	

GMQL	COMPLEMENT (O,
	  UNION (
	    UNION (EWNSR (O, 0), EWNSR (O, 1)),
	    EWNSR (O, 2)))	

CTL	Not possible	

The second example originates from [27] and addresses the issue that models may be too complex for the participants to efficiently work with them. The pattern addresses this problem in the context of process models which consist of multiple variants. Such models may become overly complicated due to the occurrence of too many variants at a certain position in the model. In this pattern, the threshold of variants starting at any given element is set to three.

In the corresponding GMQL query (see Table 3) this is done by building the complement of all available elements in a model and the elements of the model which have zero, one or two succeeding relations. As it can be seen for each number of outgoing relations below the threshold two additional statements (union and EWNSR) have to be used. Increasing the threshold therefore leads to a more complex query. In contrast to the GMQL this pattern cannot be built with CTL, as the language does not support the identification of the number of succeeding or preceding elements.

Table 4. Example Pattern "Check documents repetitively".

	Name: Check documents repetitively	**Reference:** -
	The weakness pattern addresses situations where a document is checked more than once in a loop of a process	

GMQL	DLCE (O,
	  UNION (
	    EWAOV (O,'Caption','Check document'),
	    EWAOV (O,'Caption','Analyse document')))	

CTL	M, s0 $\models$ AG (("Check document" $\lor$ "Analyse document") $\rightarrow$ AG ($\neg$"Check document" $\land$ $\neg$"Analyse document"))	

A mandatory requirement of a compliance query language in practice is the identification of complex graph structures as stated in [3], e.g., loops. In the following, two patterns are used to showcase the limitations of the languages regarding such structures. Therefore, the third pattern is used to identify processes in which an activity is conducted multiple times in a process due to a loop although the activity is performed redundantly after the first execution. In this example a document check named "Check document" or "Analyse document" should only be checked once. In the GMQL query, depicted in Table 4, all loops are selected which do contain an element with the name "Check document" or "Analyse document". In case one or more loops are found, the model violates the pattern.

The CTL statement also shown in Table 4 works differently. It does not explicitly select loops. Instead it is searched if an element called "Check document" or "Analyse document" exists. In case it does, all following paths are not allowed to have any states labelled "Check document" or "Analyse document". In comparison to the GMQL query the usage of the CTL statement has two weaknesses. Firstly, during the transformation of the pre-transformed process model into a computation tree, the loop has to be passed through at least two times. If the loop is only traversed once, the loop's states are only included one time in the computation tree and can thus not be detected. Secondly, the CTL statement does not only detect the elements within loops, but also within the whole process. This may lead to false negative results, in which multiple occurrences of the states not being in a loop are detected. The result of the expression would in turn be FALSE, although the pattern is not violated.

Table 5. Example Pattern "Observe iteration efficiency".

Name: Observe iteration efficiency	**Reference:** -
The pattern identifies if the efficiency of a loop is observed each iteration	

GMQL	DLNCE (O,
	    UNION (
	        EWAOV (O, 'Caption', 'Check if exception handling necessary'),
	        EWAOV (O, 'Caption', 'Check for exception handling')))

CTL	Not possible

The fourth example pattern is again concerned with complex structures. It states that a loop should always contain an activity "Check if exception handling necessary" or "Check for exception handling", which has the aim to determine if the loop's iteration should be cancelled and an exception handling process should be started. The purpose of the pattern is the identification of possible inefficiencies and delays in a process, caused by loops not being observed closely and interrupted, if to many iterations occur.

As depicted in Table 5 the GMQL query looks for loops which do not contain any of the two activities "Check if exception handling necessary" or "Check for exception handling". The pattern is violated, if the result set is not empty.

Using CTL this pattern cannot be expressed. In contrast to the previous example it is necessary to be able to identify loops, as it is the basic prerequisite implying the other conditions. The creation of a partially correct CTL statement is therefore not possible.

5   Comparison of Graph Matching and Temporal Logic

Graph matching and temporal logic are two different methodologies, which both can be applied in the context of compliance management and which both already have been applied in practice. Because of their different underlying theoretical concepts, the two methodologies provide a different set of functions and operations (cf. Sects. 2.1 and 2.2). As displayed in Sect. 4, the result of this difference is that two of the example patterns are not expressible in CTL and one does not entirely reflect the pattern. In particular, CTL is missing the features to count elements and to detect loops. The latter can in some cases be circumvented by using the transformation procedure described in Sect. 3 and additionally formulating the CTL statement as a path constraint as shown with the pattern "Check documents repetitively" in Sect. 4. Nonetheless, the resulting statement produces only partially correct results and its applicability is practice is hence questionable. These weaknesses of the CTL approach seem to arise from the necessity to use a computation tree. During the transformation process information gets lost, i.e., loops since they have to be unwound, or blurred, i.e., preceding or succeeding elements cannot be simply detected anymore.

Another disadvantage arising from the transform process is the state explosion problem. As shown in Sect. 3, this transformation can lead to large state machines, which grow exponentially with the amount of branching elements (e.g., OR, XOR) in a process model. In return, GMQL can operate directly on the process model and is thus not affected by this issue. The author further argues that compliance checking need to be applicable in process models with multiple concurrent executed process flows, as process modelling languages such as icebricks and EPC contain branching elements.

Lastly, regarding the visualization of compliance violations, there is a significant difference comparing graph matching with temporal logic. As graph matching is able to directly operate on the process model, the pattern is directly mapped to elements of the process model using algorithms of graph matching. If the pattern is found in the model, there is a relation of elements in the pattern to the matched elements in the model, which allows the GMQL implementation provided by [3] to highlight the matched parts directly in the model. This is possible, because [3] provides an implementation in form of a plugin as part of a modelling tool. The visualization of matched results is a very convenient feature for users, as it makes it easy to find the part of a process which has caused the compliance pattern to match – even for very large process models. Contrastingly, to the best of the author's knowledge, a comparable visualization of results is not possible, as CTL, LTL and CRL only return a Boolean value whether a process contains a certain compliance pattern or not. Furthermore, it is questionable if such a visualization feature could be implemented using CTL, as CTL only operates on the computation trees and therefore, there is no direct relation of identifiers used in the compliance pattern and elements in the process model.

6 Conclusion and Outlook

This paper has shown and discussed different approaches for checking business process models for compliance violations. CTL as a representative of temporal logic was used and compared it to the GMQL, representing graph matching. By specifying compliance patterns for both approaches, it was possible to point out a gap in the capabilities of both languages in regards to compliance checking. The examples presented revealed several disadvantages of CTL in comparison to GMQL.

Firstly, two patterns could not expressed using CTL, which shows that CTL is not fully capable to capture the functionality of GMQL. As only four patterns were specified, the author suggests that further research should focus on these edge cases, by comparing the functionalities of GMQL and CTL further. This may reveal situations in which GMQL is not applicable, whereas CTL is. In general, further research in this area would allow to state more accurately the limitations of GMQL and CTL compared to each other, to improve both approaches in the future. Secondly, the importance of the visualisation of compliance violations stressed. As CTL only returns TRUE or FALSE for a given statement, this is not supported by CTL. The author therefore asks researchers to focus on this issue, as an easy-to-use tool with a comprehensible visualization could greatly improve the usage of pattern-based compliance management in practice.

Lastly, there are several compliance patterns mentioned in literature. However, these patterns are widely spread over different papers and usually are only specified for one approach, if not only specified textually. Therefore, no integrated catalogue exists providing an overview of all compliance patterns researchers have described up till today. Future research should develop such an integrated catalogue of compliance patterns and to invent techniques for automatically transferring compliance patterns, e.g., from CTL to GMQL and vice versa. Additionally, practitioners could benefit from the presence of such a catalogue and researchers could conduct empirical studies on the effectiveness of BPCM in practice with regard to compliance patterns.

Acknowledgements. The research leading to these results received funding from the European Union's Horizon 2020 research and innovation programme under the Marie Skłodowska-Curie grant agreement No 645751 (RISE_BPM).

References

1. Dijkman, R., Rosa, M.L., Reijers, H.A.: Managing large collections of business process models - current techniques and challenges. Comput. Ind. **63**, 91–97 (2012)
2. Sadiq, S., Governatori, G., Namiri, K.: Modeling control objectives for business process compliance. In: Alonso, G., Dadam, P., Rosemann, M. (eds.) BPM 2007. LNCS, vol. 4714, pp. 149–164. Springer, Heidelberg (2007). https://doi.org/10.1007/978-3-540-75183-0_12
3. Becker, J., Delfmann, P., Dietrich, H.-A., Steinhorst, M., Eggert, M.: Business process compliance checking – applying and evaluating a generic pattern matching approach for conceptual models in the financial sector. Inf. Syst. Front., 1–47 (2014)

4. Elgammal, A., Turetken, O., van den Heuvel, W.-J., Papazoglou, M.: Formalizing and appling compliance patterns for business process compliance. Softw. Syst. Model., 1–28 (2014)
5. Höhenberger, S., Riehle, D.M., Delfmann, P.: From legislation to potential compliance violations in business processes-simplicity matters. In: European Conference on Information Systems, Istanbul, Turkey (2016)
6. Polyvyanyy, A.: Business process querying. In: Sakr, S., Zomaya, A. (eds.) Encyclopedia of Big Data Technologies. Springer, Cham (2018)
7. Polyvyanyy, A., Ouyang, C., Barros, A., van der Aalst, W.M.P.: Process querying: enabling business intelligence through query-based process analytics. Decis. Support Syst. **100**, 41–56 (2017)
8. Awad, A.: BPMN-Q: a language to query business processes. In: Proceedings of the 2nd International Workshop on Enterprise Modelling and Information Systems Architectures, St. Goar, Germany, pp. 115–128 (2007)
9. Störrle, H.: VMQL: a visual language for ad-hoc model querying. J. Vis. Lang. Comput. **22**, 3–29 (2011)
10. ter Hofstede, A.H.M., Ouyang, C., La Rosa, M., Song, L., Wang, J., Polyvyanyy, A.: APQL: a process-model query language. In: Song, M., Wynn, M.T., Liu, J. (eds.) AP-BPM 2013. LNBIP, vol. 159, pp. 23–38. Springer, Cham (2013). https://doi.org/10.1007/978-3-319-02922-1_2
11. Delfmann, P., Steinhorst, M., Dietrich, H.-A., Becker, J.: The generic model query language GMQL – conceptual specification, implementation, and runtime evaluation. Inf. Syst. **47**, 129–177 (2015)
12. Deutch, D., Milo, T.: A structural/temporal query language for Business Processes. J. Comput. Syst. Sci. **78**, 583–609 (2012)
13. Weske, M.: Business Process Management: Concepts, Methods. Technology. Springer, Berlin (2007)
14. List, B., Korherr, B.: An evaluation of conceptual business process modelling languages. In: Proceedings of the 2006 ACM Symposium on Applied Computing, Dijon, France, pp. 1532–1539 (2006)
15. Lu, R., Sadiq, S.: A survey of comparative business process modeling approaches. In: Abramowicz, W. (ed.) BIS 2007. LNCS, vol. 4439, pp. 82–94. Springer, Heidelberg (2007). https://doi.org/10.1007/978-3-540-72035-5_7
16. Becker, J., Delfmann, P., Eggert, M., Schwittay, S.: Generalizability and applicability of model-based business process compliance-checking approaches - a state-of-the-art analysis and research roadmap. BuR - Bus. Res. **5**, 221–247 (2012)
17. Clarke, E.M., Grumberg, O., Peled, D.: Model Checking. MIT Press, Cambridge (2000)
18. Awad, A., Decker, G., Weske, M.: Efficient compliance checking using BPMN-Q and temporal logic. In: Dumas, M., Reichert, M., Shan, M.-C. (eds.) BPM 2008. LNCS, vol. 5240, pp. 326–341. Springer, Heidelberg (2008). https://doi.org/10.1007/978-3-540-85758-7_24
19. Becker, J., Clever, N., Holler, J., Shitkova, M.: Icebricks - business process modeling on the basis of semantic standardization. In: International Conference on Design Science Research in Information Systems, Helsinki, Finland, pp. 394–399 (2013)
20. Jannaber, S., Riehle, D.M., Delfmann, P., Thomas, O., Becker, J.: Designing a framework for the development of domain-specific process modelling languages. In: Maedche, A., vom Brocke, J., Hevner, A. (eds.) DESRIST 2017. LNCS, vol. 10243, pp. 39–54. Springer, Cham (2017). https://doi.org/10.1007/978-3-319-59144-5_3
21. Keller, G., Nüttgens, M., Scheer, A.-W.: Semantische Prozeßmodellierung auf der Grundlage "Ereignisgesteuerter Prozeßketten (EPK)," (1992)

22. Riehle, D.M., Jannaber, S., Karhof, A., Thomas, O., Delfmann, P., Becker, J.: On the de-facto standard of event-driven process chains: how EPC is defined in Literature. In: Modellierung 2016, 2–4 März 2016, Karlsruhe, pp. 61–76. Köllen Druck + Verlag, Bonn (2016)
23. Cuntz, N., Kindler, E.: On the semantics of EPCs: efficient calculation and simulation. In: van der Aalst, W.M.P., Benatallah, B., Casati, F., Curbera, F. (eds.) BPM 2005. LNCS, vol. 3649, pp. 398–403. Springer, Heidelberg (2005). https://doi.org/10.1007/11538394_30
24. Valmari, A.: The state explosion problem. In: Reisig, W., Rozenberg, G. (eds.) ACPN 1996. LNCS, vol. 1491, pp. 429–528. Springer, Heidelberg (1998). https://doi.org/10.1007/3-540-65306-6_21
25. OMG: Business Process Model and Notation (BPMN) Version 2.0. (2011)
26. BMJV: Wertpapierhandelsgesetz (WpHG) (1994)
27. Delfmann, P., Höhenberger, S.: Supporting business process improvement through business process weakness pattern collections. In: Proceedings of the 12. Internationale Tagung Wirtschaftsinformatik, Osnabrück, Germany, pp. 378–392 (2015)

From Complexity to Insight:

Querying Large Business Process Models to Improve Quality

Kurt E. Madsen[(⊠)]

MetaTech, Inc., Tampa, FL, USA
kmadsen@metatech.us

Abstract. This industry study presents work in querying manufacturing processes using a portal to navigate query results in the broader context of enterprise architecture. The approach addresses the problem of helping stakeholders (e.g., management, marketing, engineering, operations, and finance) understand complex BPM models. Stakeholders approach models from different viewpoints and seek different views. Gleaning insight into improving model quality is challenging when models are large and complex. The focus of this work was on process models only, not process execution, because many legacy organizations have done initial BPM modeling but do not have BPM systems in production or have yet to realize the benefits of coupling log mining with incremental model refinement. The use cases presented address the complexity of multi-year, process models used by $\sim 10,000$ workers globally to develop new products. These models were queried to find quality issues, isolate stakeholder data flows, and migrate BPMN [1] activities to cloud-based, micro-services. The approach presented creates filtered process views, which serve as starting points for stakeholders to navigate interconnected models within TOGAF [2] enterprise architecture models. The process querying lifecycle—query, manipulate, and transform—was also applied to other enterprise models. Lessons learned from introducing BPM into a legacy organization, model refinement, limitations of research, and open problems are summarized.

Keywords: Process querying · Business process management ·
Business intelligence · Process compliance · Process standardization ·
Architecture portal

1 Introduction

This paper examines the question of how to query, filter (i.e., manipulate), and transform large, complex process models to gain insight into improving model quality. The focus is on process models only; process execution and log mining are out of scope, as many of the world's largest organizations have fragmented process management efforts, in various stages of maturity, scattered across disconnected departments. There is value in applying process querying to business process models without considering downstream process execution.

© Springer Nature Switzerland AG 2019
F. Daniel et al. (Eds.): BPM 2018 Workshops, LNBIP 342, pp. 416–427, 2019.
https://doi.org/10.1007/978-3-030-11641-5_33

The query method presented was tested on enterprise architecture models in automotive manufacturing. The development lifecycle for new vehicles – from initial-marketing-concepts through to ready-for-mass-production – spans years. At any given time, there are dozens of concurrent vehicle programs in progress globally. The life-cycle was modeled as one process with variants by program scale and vehicle model. The process models shared 700+ different activity types (both sub-processes and tasks), instantiated as 4,000+ activity instances, interconnected via 7,000+ workflows, per-formed by 10,000+ process workers, assuming two dozen roles.

Modeling was performed in OpenText ProVision [3], a commercial, enterprise architecture modeling tool. ProVision integrates with Excel, providing rapid model creation and manipulation using tabular data. The process querying lifecycle involves three steps, model inquiry, manipulation, and update. Model inquiry involves searching process XML data to focus on key process areas, generally to improve model quality. Model manipulation alters the process model XML using XQuery, XSL, and Excel macros. Model update applies the changes so that ProVision renders revised process model data to highlight the query results.

Prior to establishing a BPM practice with ProVision, the product development lifecycle was planned as a Gantt-style program schedule in Microsoft Project with occasional efforts to model processes in Visio. Neither Project nor Visio adequately met process modeling needs, as model size and complexity were overwhelming. Even after the introduction of ProVision models, some stakeholders resisted enterprise modeling. To secure BPM buy-in, stakeholders needed a way to move from process complexity to insight so they could incrementally refine model quality. The solution came in the form of process queries that filter out irrelevant details to focus attention on problems and opportunities within process models. Once value was established, demand for BPM and process querying increased.

This paper is structured as follows. Section 2 states the problem being addressed, namely the challenges of understanding, analyzing, and improving large, complex manufacturing processes when the subject matter experts responsible for authoring the process definitions approach them from different perspectives. Section 3 describes our method for querying these process models (i.e., model inquiry, manipulation, and update) to generate filtered process views from multiple perspectives that focus attention on opportunities for improvement. Section 4 presents results and extensions of work into related areas. Section 5 concludes with limitations, open problems, and lessons learned.

2   The Problem: Large, Complex Process Models

The automotive product development pipeline (from marketing concept to ready-for-mass-manufacturing) takes years. In this case, the entire process spanned 50 + process maps. Each map was printed on a one by two meter poster and displayed at least 70 activities. These posters – taped across the walls of a large room – presented a degree of complexity such that even with a magnifying glass, workflows were nearly impossible to follow. These maps were unwieldy, difficult to read, with too much unfiltered, detailed process information (see Fig. 1).

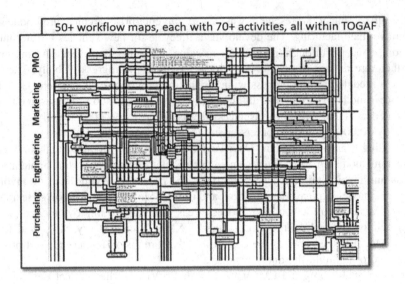

Fig. 1. Complex process maps, before query and filtering.

2.1 Differing Stakeholder Perspectives

Stakeholders, particularly the authors responsible for process design, approached models from different viewpoints (e.g., management, engineering, purchasing). From these viewpoints, they searched for different views (e.g., value streams, program schedules, workflow simulations, functional, business information, applications & services). The combination of viewpoints and views established search perspectives.

2.2 Quality Issues

BPM models were created by aggregating program management data from different teams using Microsoft Project, Visio, and custom, in-house applications. Once collected, data was loaded into Excel and split across object tables (e.g., activities) and link tables (e.g., workflows to interconnect activities). The resulting Excel files were imported into ProVision's inventory of modeling objects and links between objects. This approach to creating BPM models was faster than creating them by hand, but there were quality issues with the input data that led to model inconsistencies such as:

- Graph completeness problems [4]: missing inputs or outputs.
- Temporal problems [5]: inputs available after activity start, outputs produced too late, missing activity duration.
- Attribute quality problems: missing author/resources, typing errors in description.

3 Querying Business Processes

Process queries helped stakeholders to navigate interconnected models and to discover model improvement opportunities. For example,

- Given an author, find all of his/her activities within a workflow model.
- Given an artifact, find all activity usages (i.e., instantiations of an activity).
- Given a milestone, find the distinct list of artifacts that cross swim lane boundaries.
- Given a milestone, find the distinct set of artifacts that are associated with workflows that cross swim lane boundaries. (e.g., bill-of-material information handed off from one team to another).
- Given a role, highlight all activities performed by that role (e.g., marketing).
- Given a parameter, find model objects a matching attribute (e.g., find activities).
- Quality query: given a process model, verify that there are no workflows with one end detached (i.e., no dangling workflows).
- Quality query: given a process model, assert (number of activities where author is NOT missing = total number of activities).
- Compliance query: given a process model, compare it to the APQC reference model for automotive manufacturing to assess standards compliance noting drift [6].
- Navigation queries. Use query results as input to the next query. Repeat to navigate within and across models. For example, given an author, find all of his/her activities. Then, for each activity, trace all input workflows to find only those upstream activities owned by a different author. In this way, two authors can discover their dependencies other and coordinate their teams' planned process details.

3.1 Model Inquiry

Process data conformed to ProVision's common interchange format (CIF.xsd), an XML schema supporting model portability across vendor platforms (Fig. 2).

```
01  <activity id="157896" name="CAD-849">
02    <descr>Build prototype car parts</descr>
03    <parent refID="435524"/>
04    <workTime></workTime>
05    <performer refID="467908"/>
06    <customProperties>
07      <property name="Author">
08        <value>John Doe</value>
09      </property>
10    </customProperties>
11  </activity>
```

Fig. 2. An XML fragment of a process activity

On lines 7–9 of this XML fragment, the stakeholder responsible for this activity is stored in the author custom property, an important query search key. On line 1, the activity id "157896" is referenced throughout the process model to refer back to this activity (e.g., to connect it to workflows). These reference IDs chained together to enable navigation queries and nested searches and were invoked repeatedly as users traversed workflows. Missing data on line 4, <workTime>, is an example of poor quality input data that hinders queries such as finding critical paths to reduce time-to-market. Consider the query in Fig. 3: given an author, find all of her activities.

```
01  declare variable $author:="John Doe";
02  for $activity in /process/activities/activity
03  let $activity-id := $activity/@id
04  where $activity/customProperties/property
05    [@name="Author"]/value[matches(., $author)]
06  return <member refID="{$activity-id}" />
```

Fig. 3. XQuery to return a collection of all activities owned by $author

When this query runs, the result is a set of zero or more <member> elements as shown on lines 3–5 in Fig. 4. The set <members> in <modelScenario> includes only those activities owned by $author.

```
01  <modelScenario name="Process layer, author filter">
02    <members>
03      <member refID="157896"/>
04      <member refID="...etc..."/>
05    ...etc...
06    </members>
07  </modelScenario>
```

Fig. 4. XQuery result set with references to all activities owned by $author = "John Doe"

ProVision uses the <modelScenario> element to manage process simulation scenarios, which it renders as graph layers. We discovered that this element can be overloaded to create a collection of model layers, which, when superimposed on each other, filter out irrelevant process details to focus attention on query results.

3.2 Model Manipulation

This stage of the querying life cycle alters XML in a process model. For instance:

- Given query results, insert the results as rows into a new process model layer
- Given task duration data, populate the work time for each activity
- Given inter-activity timing data, populate transit time for each Workflow

This stage was challenging because most of this work was performed manually by editing boilerplate CIF.xml process data similar in format to XPDL and inserting query results. Note that the process definition files were often over 100 GB in size.

3.3 Model Update

Updating a model involved uploading a manipulated model definition into ProVision. In some cases, post-processing was applied to color workflows using JavaScript, which had the effect of highlighting workflows to draw attention to gaps, overlaps, and errors. See Fig. 5.

```
function highlightWorkflowsByStereotype(model) {
  var bpmParts = model.getComponents();
    for (i = 0; i < bpmParts.length; i++) {
      if (bpmParts[i].getType() == "Workflow")  {
        var nextStereotype = bpmParts[i].getStereotype();
        if (nextStereotype == "WfOverlap")
          bpmParts[i].Line.setColor(BLUE);
        if (nextStereotype == "WfGap")
            bpmParts[i].Line.setColor(GREEN);
        if (nextStereotype == "WfError" )
            bpmParts[i].Line.setColor(RED); }}}
```

Fig. 5. JavaScript to highlight process model elements during model update

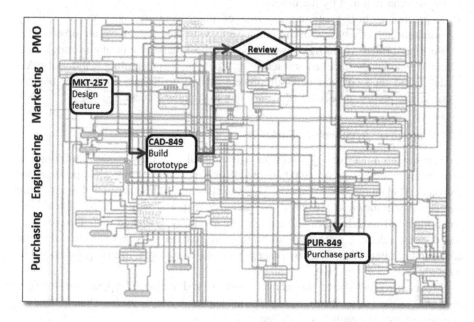

Fig. 6. Filtered workflow map with query results highlighted for visibility. (Color figure online)

When the collection of activities owned by a given author was combined with color highlighting of the workflows by stereotype, the resulting filtered process layer overlaid the ghosted process layer and was rendered as shown in Fig. 6.

4   Results and Extensions of Work

This process querying work helped a multi-disciplinary team realize a significant, undisclosed reduction in time-to-market for a multi-year manufacturing process, while improving overall model quality. Further, BPM gained acceptance among skeptics within the organization. Accordingly, the success of this work expanded beyond the original scope of improving BPM model quality.

4.1   Queries Applied to Other Enterprise Models

TOGAF, as it was used, specified seven types of enterprise models: strategy, organization, capability, process, information, application, and technology. This approach to querying process models was equally useful when applied to querying other enterprise models. Examples follow:

- An executive might start with a process model, and then search for the TOGAF business capability it implemented, and in turn, navigate to the associated TOGAF value stream.
- An enterprise architect might start with a process model, navigate to the information model it required, and then navigate to the application models that produced the information required by the process.
- A process author, assuming the role of purchasing manager, might start by searching an activity within the purchasing swim lane, then navigate upstream to work performed within other swim lanes (such as marketing or engineering). S/he could examine the attached artifacts (e.g., inputs such as marketing features or CAD data), and then create a new sub-process activity to handle bottlenecks (e.g., if substitute parts had become necessary due to supplier issues).

4.2   Enterprise Architecture Portal

Process models were part of a broader collection of enterprise model portfolios [8], collectively containing over 800,000 model objects and covering all aspects of the company. The approached to process querying also applied to other model types within the TOGAF framework. An enterprise architecture portal was built so stakeholders could query all models types and navigate interconnected, filtered model views. This was done by selecting a model portfolio (e.g., vehicle design) and then selecting from configurable filters (e.g., filter by process map stage and organization perspective) as shown in Fig. 7.

Model metadata drove the portal's content and included support for multiple portfolios of models as shown in Fig. 8. This approach can be useful when comparing two sets of models, each from different authoring tools.

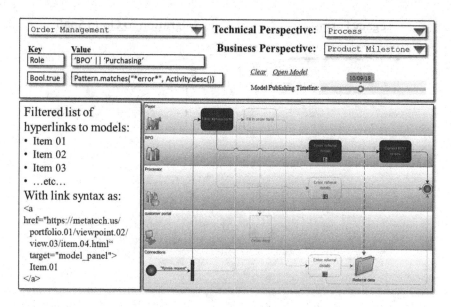

Fig. 7. Portal for exploring TOGAF enterprise models which included process models.

```
01   <Portfolio name="GlobalProductDev">
02     <Notebook name="OrderMgt" vers="" DateTm="">
03     <AuthoringTool product="ProVision" release="9.2"/>
04     <Models>
05       <Model>
06         Fltr01Key="Role" Fltr01Val="BPO || Purchasing"
07         Fltr02Key="RegExp" Fltr02Val="Pattern …etc…"
08         <Title>Order Entry</Title>
09         <Description>This order is…etc…</Description>
10         <URL>model_image.htm</URL>
11       </Model>
12     </Models>
13   </Notebook>
14   </Portfolio>
```

Fig. 8. XML data to populate model entries in enterprise architecture portal

4.3   Querying Workflows and Artifacts to Discover Micro-services

A firm-wide effort existed to replace legacy information systems with cloud-based, micro-services. Part of this work involved identifying workflows where process participants used email to hand off information across swim lane boundaries, a practice which led to document management issues and rework. Combining query results that identified sets of candidate workflows with lists of end-of-life systems provided a short-list of migration-eligible system as shown in Fig. 9.

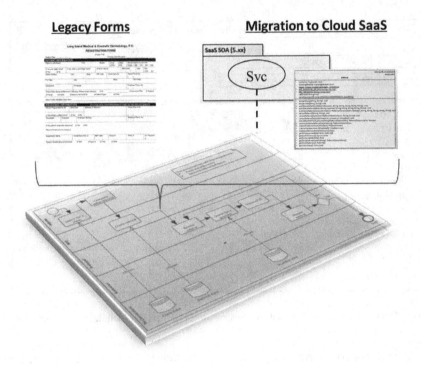

Fig. 9. Querying information flows to map legacy information systems to microservices

An example of a candidate information flow is the activity "PR-849 Purchase Parts" in the purchasing swim lane of Fig. 6. Such activities were identified by exporting process models to Excel and searching the descriptions of system, workflow, artifact, and activity objects via regular expressions to find target data (e.g., parts data, order data, CAD files, etc.). While this worked, a better approach would have been to use a process query language [7] with a search query along the lines of the following SQL-like pseudo code:

```
SELECT id FROM workflows AS w WHERE crossesSwimlaneBound-
ary(w.id) = true AND w.id IN (SELECT id FROM workflows AS
w WHERE w.endLink.refId IN (SELECT id FROM activities AS
a WHERE has_artifact(a.id) = true AND regExp(a.id,
partsDataPattern) = true))
```

Ideally, such a PQL statement would be able to invoke regular expression searches (e.g., "[part|BOM].*data") of artifacts outside the process model being searched in a manner similar to Transact-SQL's xp_cmdshell() [9], but without security issues.

The end goal was an inventory of service-ready activities (SrActivities). In Pro-Vision, when a process modeler drags an SrActivity onto the process designer canvas, the tool validates and instantiates the web services interface to the correct micro-service. A proof-of-concept was produced in ProVision using the Excel approach.

SrActivites were inventoried separately from non-service-ready activities, and web service interfaces were generated. Thus, it was possible to measure progress towards migrating activity inputs from legacy information systems and external MS Office documents (passed by email) to micro-services.

4.4 Filters to Normalize Models for Vendor-Neutrality

The long-term preservation of model assets is an important part of corporate records retention. Enterprise model artifacts must be portable across evolving modeling tools. Unfortunately, model fidelity is sometimes lost when exporting/importing models between tools (e.g., using BPMN or XPDL). There is an inherent problem in relying on import/export features of BPM modeling tools because tools to manage enterprise content (ECM) and models (e.g., BPM) have different objectives than archival tools [13]. Anticipating the long-term need to preserve models across tools, a repository was created of rendered model views in both PDF and HTML formats. To create these static model views, process queries filtered out tool-specific branding (e.g., ProVision, Activiti, Sparks Enterprise Architect) and aggregated models published by all vendors. Thus, stakeholders could focus on unified views of enterprise models independently of the tools used to produce them. In this way, the modeling lifecycle and modeling-tool vendor management lifecycle could evolve independently.

5   Conclusion

This work focused exclusively on process querying as it relates to BPM modeling, not the mining of process logs, because many organizations are not yet ready for log mining. Even with this narrow focus, there was still much value in applying process querying to models.

5.1   What Worked Well; What Did Not

The most useful query was finding timing gaps (i.e., leads), overlaps (i.e., lags), and errors in workflows between activities. A gap exists when an upstream activity finishes one or more weeks before a downstream activity starts. An overlap exists when both activities execute concurrently for one or more weeks. A workflow error exists when the downstream activity starts before the upstream activity starts, or when one end of a workflow is unattached to a process element (activity, start, end, or gateway). Reducing time-to-market involved iteratively refining process models to close gaps, maximize overlaps, and eliminate errors.

Process models were planned backwards so the first activity (Design vehicle concept) had negative start and finish times, and the last activity (Confirm ready-to-manufacture) had a finish time of 0. Thus, given two activities Ai [Si, Fi] and Aj [Sj, Fj], a Gap exists when $Sj < Fi$ an Overlap exists when $Si \geq Sj > Fi$, and an Error exists when $Sj > Si$, where A = Activity, S = Start, F = Finish, and Aj depends on input from Ai.

ProVision version 9.2 has a defect when importing model data from Excel: it does not load the activity.workTime column into the process model's activities, which prevents automated, critical path analysis. To circumvent this problem, workflows between activities where inventoried in Excel with one row per workflow, sorted by activity.workTime to prioritize leads, lags, and errors between adjacent process activities. With this prioritized list in hand, filtered views of process maps were created to highlight timing problems.

Regarding performance, the models exported by ProVision in its common interchange XML format were big, often over 100 MB. Loading them into OxygenXML Designer and ProVision led to non-linear processing delays (and occasional crashes), which seemed to grow exponentially with file size as the model was loaded into memory. In XML processing, streaming has better performance than loading large DOMs. [10] This is a consideration when designing process query languages and PQL processors.

The manipulation stage of process querying involved labor-intensive, batch work for developers, which proved challenging. The turn-around time to produce a filtered process layer could be as much as 20 min. Using shell scripts with regular expressions, experiments with XSL, and manual processing, layers were created. An area of future work would be to automate model manipulation so that stakeholders could execute ad-hoc PQL queries on the fly to explore and navigate models. Process modeling tools would have to support a PQL-compliant API in order to dynamically render ad-hoc queries.

5.2   Limitations, Open Problems, and Lessons Learned

Resources such as code snippets are available on github [11]. Areas for future research include:

- Improving XML processing performance—not enough effort was spent on measuring model size vs. processing time.
- PQL portability across modeling tools—based on this experience using a specific BPM modeling tool (ProVision and its common interchange format), it is clear that PQL portability across BPM tools will be in demand in industry.
- Support for ad-hoc queries and model navigation—the enterprise portal was popular among stakeholders. However, as implemented, it was limited in its ability to render dynamically-generated model views on the fly in response to ad-hoc queries. Such queries support exploration and discovery of interconnected models and are valuable to stakeholders.
- Model design drift and compliance—just as process instances drift during execution, so too does design intent drift when subject matter experts design and maintain large, complex process models over years. There was interest in archiving model changes for corporate history and in measuring the cost of model drift [12]. A big driver is tracking process compliance and alignment to industry standards, particularly the APQC standard for automotive manufacturing processes.

- The biggest lesson learned from stakeholder feedback was that PQL techniques are applicable to all TOGAF enterprise architecture layers, not just the process layer. Thus, an open problem and area for future research is whether and how to apply PQL techniques to performing queries in the broader context of enterprise architecture models.

References

1. Business Process Model and Notation Specification, version 2.0.2, sect. 7.3.1. https://www.omg.org/spec/BPMN/2.0/. Accessed 30 July 2018
2. TOGAF version 9.2, Part IV. http://pubs.opengroup.org/architecture/togaf92-doc/arch/. Accessed 30 July 2018
3. ProVision product data sheet. https://www.opentext.com/file_source/OpenText/en_US/PDF/ProVision%20Enterprise%20and%20Product%20Architecture%20Software%20Product%20Overview%20.pdf. Accessed 30 July 2018
4. Dettmer, H.W.: Goldratt's theory of constraints: a systems approach to continuous improvement. ASQC Quality Press, Milwaukee (1997). Section: The categories of legitimate reservation
5. Kumar, A., Sabbella, S.R., Barton, R.R.: Managing controlled violation of temporal process constraints. In: BPM 2015, pp. 280–296 (2015)
6. APQC Process Classification Framework (PCF) - Automotive (OEM) - Excel Version 7.0.5. https://www.apqc.org/knowledge-base/documents/apqc-process-classification-framework-pcf-automotive-oem-excel-version-705. Accessed 13 July 2018
7. http://processquerying.com/pql-grammar/. Accessed 01 Aug 2018
8. Polyvyanyy, A., Ouyang, C., Barros, A., van der Aalst, W.: Process Querying: Enabling Business Intelligence Through Query-Based Process Analytics, p. 3. Queensland University of Technology, Brisbane. https://doi.org/10.1016/j.dss.2017.04.011
9. Transact-SQL, xp_cmdshell(). docs.microsoft.com/en-us/sql/relational-databases/system-stored-procedures/xp-cmdshell-transact-sql. Accessed 30 July 2018
10. Wei, Z.: Efficient XML stream processing and searching, pp. 1, 18–21. Florida State University (2012). http://ww2.cs.fsu.edu/~wzhang/dissertation.pdf. Accessed 11 July 2018
11. Github site for this project. https://github.com/curiouskurt/pq2018
12. Alizadeh, M., de Leoni, M., Zannone, N.: History-based Construction of Log-Process Alignments for Conformance Checking: Discovering What Really Went Wrong? p. 9, Eindhoven University of Technology, 19–21 November 2014
13. Katuu, S.: Enterprise content management and digital curation applications–maturity model connections. UNESCO Memory of the World in the Digital Age Conference, p. 1029 (2012)

Second International Workshop
on BP-meet-IoT (BP-meet-IoT 2018)

Second International Workshop on BP-meet-IoT (BP-meet-IoT 2018)

Business process management (BPM) is a well-established discipline that deals with the identification, discovery, analysis, (re-)design, implementation, execution, monitoring, and evolution of processes. The Internet of Things (IoT) is a network of interconnected computing devices that are seamlessly embedded in objects, animals, people, which are called the things in the IoT. By embedding these computing devices, we enable things to sense and respond to their surrounding environment and to build a bridge between the digital and the physical worlds (e.g., all effort around cyber-physical systems). While BPM and the IoT are very different domains, they can mutually benefit from each other. However, several challenges need to be tackled. Particularly, it has to be understood:

- How BPM can improve the IoT by (a) taking a process-oriented perspective and considering the process history, (b) bridging the abstraction gap between raw sensor data and higher level knowledge extracted from this event data, and (c) optimizing decision-making in the large
- How to exploit the IoT to improve processes along their lifecycle by (a) considering sensor data for automatically detecting the start and end of activities, (b) using event data for making decisions in a pre-defined process model, and (c) detecting discrepancies between the pre-defined model and actual enactment using event data for online process compliance checking and exception management.

In the second edition of the BP-meet-IoT workshop, the following three full papers were presented: "Retrofitting of Workflow Management Systems with Self-X Capabilities for Internet of Things" by Ronny Seiger, Peter Heisig and Uwe Assmann; "On the Contextualization of Event-Activity Mappings" by Agnes Koschmider, Felix Mannhardt and Tobias Heuser; and "A Classification Framework for IoT Scenarios" by Sankalita Mandal, Marcin Hewelt, Maarten Oestreich and Mathias Weske.

November 2018

Agnes Koschmider
Massimo Mecella
Estefanía Serral
Victoria Torres

Organization

Program Committee

Jan Mendling	Vienna University of Economics and Business, Austria
Manfred Reichert	University of Ulm, Germany
Vicente Pelechano	Universitat Politècnica de València, Spain
Francisco Ruiz	University of Castilla-La Mancha, Spain
Antonio Ruiz-Cortés	University of Seville, Spain
Barbara Weber	Technical University of Denmark, Denmark
Bart Baesens	KU Leuven, Belgium
Ferry Pramudianto	North Carolina State University, USA
Adrian Mos	Naver LABS, Grenoble, France
Matthias Weidlich	Imperial College London, UK
Mathias Weske	Hasso-Plattner-Institut at the University of Potsdam, Germany
Pnina Soffer	University of Haifa, Israel
Gero Decker	Signavio GmbH, Germany
Anne Monceaux	Airbus Group Innovations, France
Sylvain Cherrier	University Marne-la-Vallée, France
Armando Walter Colombo	University of Applied Sciences Emden/Leer, Schneider Electric, Germany
Alaaeddine Yousfi	Hasso-Plattner-Institut at the University of Potsdam, Germany
Selmin Nurcan	Université Paris 1-Pantheon-Sorbonne, France
Andrea Delgado	INCO, Universidad de la República, Uruguay
Christian Janiesch	University of Wurzburg, Germany
Udo Kannengießer	eneon IT-solutions GmbH, Austria
Andreas Oberweis	Karlsruhe Institute of Technology, Germany
Jianwen Su	University of California at Santa Barbara, USA
Liang Zhang	Fudan University, China
Andrea Marrella	Sapienza Università di Roma, Italy

Retrofitting of Workflow Management Systems with Self-X Capabilities for Internet of Things

Ronny Seiger[✉], Peter Heisig, and Uwe Aßmann

Software Technology Group, Technische Universität Dresden, Dresden, Germany
{ronny.seiger,peter.heisig,uwe.assmann}@tu-dresden.de

Abstract. The Internet of Things (IoT) introduces various new challenges for business process technologies and workflow management systems (WfMS's) to be used for managing IoT processes. Especially the interactions with the physical world lead to the emergence of new error sources and unanticipated situations that require a self-adaptive WfMS able to react dynamically to unforeseen situations. Despite a large number of existing WfMS's, only few systems feature self-x capabilities to be used in the dynamic context of IoT. We present a retrofitting process and generic software component based on the MAPE-K feedback loop to add autonomous capabilities to existing WfMS's. Using a smart home example process, we show how to retrofit different WfMS's in an invasive and non-invasive way. Experiments and a brief discussion confirm the feasibility of our retrofitting processes and software component to add self-x capabilities to service-oriented WfMS's in an IoT context.

Keywords: Workflow management systems · Self-management ·
Internet of Things · Retrofitting

1 Introduction

The application of Business Process Management (BPM) technologies in the context of the Internet of Things (IoT) is a new and vibrant research field as it promises easily configurable, flexible and reusable processes to be modelled, executed and analysed among the typical IoT entities including sensors, actuators, smart objects and humans. However, with these novel interactions and application domain, new challenges for both research fields arise that need to be addressed [4,9,13]. Especially the new dimension of interactions with the physical world and associated sensors and actuators introduces additional requirements for Workflow Management Systems (WfMS's) as the IoT devices are mobile, embedded and more constraint regarding their resources. The IoT entities and environment are the source of new errors, imprecisions and unforeseeable situations for the process execution that an *IoT WfMS* has to cope with.

© Springer Nature Switzerland AG 2019
F. Daniel et al. (Eds.): BPM 2018 Workshops, LNBIP 342, pp. 433–444, 2019.
https://doi.org/10.1007/978-3-030-11641-5_34

A common approach for dealing with these new kind of unanticipated situations is to implement feedback loops and adaptation mechanisms to make a system self-aware and self-adaptive [7]. These mechanisms usually rely on goals specifying aspects regarding the expected outcome of particular actions, adaptation strategies for dealing with unexpected behaviour, and interactions with sensors and effectors of the respective target systems to monitor and manipulate the components of the self-managed system [11]. Several approaches investigate the implementation of autonomic capabilities for specific WfMS's–with and without relations to IoT. Despite a very large number of existing WfMS's being actively used in industry and academia, these implementations are tied to specific proprietary WfMS's and not reusable within other systems, though.

In this work, we present a general framework and process for retrofitting existing (*legacy*) WfMS's with self-x capabilities (self-awareness, self-adaptation, self-healing, etc.) based on the *MAPE-K* control loop from engineering self-adaptive software systems and autonomic computing [5, 7]. We discuss two ways of retrofitting service-oriented WfMS's and existing processes using an implementation of the MAPE-K loop by a generic software component (*Feedback Service*) to realize self-x mechanisms with respect to arbitrary quality criteria. We show how to retrofit four different WfMS's with the help of a simple scenario process from the smart home domain as an example of an IoT environment.

2   Smart Home Scenario Process

The smart home is an excellent example of an IoT environment. It consists of various sensors for environmental factors (e.g., light levels, temperature, humidity) and actuators for controlling domestic appliances (e.g., light switches, thermostats, service robots) as well as smart objects equipped with sensing technologies (e.g., RFID and NFC). All these IoT devices interact with each other, with the physical environment and other objects, and most importantly with the residents of the smart home. Due to the nature of these interactions with the physical world and the entities being more imprecise, unreliable and unforeseeable, new sources of errors emerge for smart home control applications. Therefore, a WfMS orchestrating the interactions among all involved smart home entities on the business process level needs to be able to adapt the processes and itself to deal with unanticipated situations and errors.

Figure 1 shows a simple smart home process in BPMN 2.0 notation. The process only switches on a specific light via a service call. In our case, a middleware for IoT (*OpenHAB*[1]) provides a RESTful web service interface to all sensors and actuators, including a HomeMatic dimmer switch for controlling the light. The process shows the typical interaction between the virtual and physical world in IoT environments. The WfMS calls the IoT middleware and with that the control software of the light dimmer to influence the physical environment. After executing this process, the light levels are expected to have reached a certain threshold in the physical world and the room is assumed to be lit up (*assumed state*). However, simple errors like

---

[1] https://www.openhab.org/.

a burnt or worn off light bulb may lead to an incorrect assumption of the lighting state (*light is on*) compared with the *actual state* in the physical world (*light is off*), which may not be detected by the WfMS or the dimmer switch. To detect and remedy this kind of issue, the workflow execution has to be self-aware and self-adaptive to support the self-healing of the process. For our scenario, an additional light sensor is used to detect the broken light bulb and an alternative dimmer switch is triggered to light up the room. We call this detection and repair of inconsistent physical and virtual process execution states *Cyber-physical Synchronization* [20].

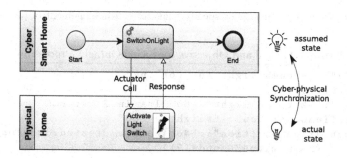

Fig. 1. Scenario process showing the synchronization of cyber and physical worlds.

3 MAPE-K as Framework for Autonomous Capabilities

The MAPE-K control loop is the most common approach to implement self-x capabilities for software systems operating in IoT and Cyber-physical Systems (CPS) [16]. In this section, we introduce the MAPE-K (Monitor-Analyse-Plan-Execute over a shared Knowledge) feedback loop [5] as the basic framework for adding autonomic capabilities to WfMS's.

3.1 The MAPE-K Feedback Loop Applied to Processes

The application of the MAPE-K feedback loop [5] to manage processes and individual process steps is discussed in detail in [20]. In general, data from additional virtual and physical sensors is used to *Monitor* and *Analyse* the process execution with respect to certain success and error criteria defined in *Goals*. In case of unexpected behaviour, a compensation strategy is selected by the *Planner* and performed by the *Executor*. The necessary *Knowledge* regarding sensors, actuators and compensation strategies is contained in a shared knowledge base [20].

We use *Goals* to specify relevant data and success/error criteria for a process used within the MAPE-K loops [20]. Figure 2 shows our proposal of a generic extension of a component-oriented workflow meta-model [22] to specify these goals. The basic concept is the *ProcessStep*, which can be composed of other process steps or an atomic activity. By adding the *ManagedStep* interface to the abstract *ProcessStep* class, a *Goal* can be specified for the process step

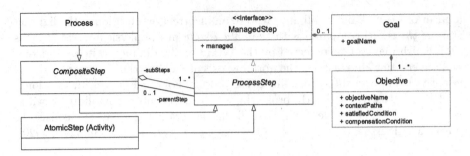

Fig. 2. Meta-model extensions to specify goals for self-management of process steps.

Listing 1.1. Goal and Objective for SwitchOnLight Process Step.

```
1  {"name": "enough light for working",
2  "objectives": [
3  {"name": "kitchen light > 600 lux in 2 seconds",
4  "satisfiedCondition": "#light > 600",
5  "compensationCondition": "#objective.created.isBefore
6        (#now.minusSeconds(2))",
7  "contextPaths": [ "MATCH (thing)-[:type]->
8        (sensor {name: 'LightSensor'})",
9  "MATCH (thing)-[:isIn]->(room {name: 'Kitchen'})",
10 "MATCH (thing)-[:hasState]->(state:LightIntensity)",
11 "MATCH (state)-[:hasStateValue]->(value)",
12 "WHERE toFloat(value.realStateValue) > 0",
13 "RETURN toFloat(value.realStateValue) AS light" ]}]}
```

(i.e., on the level of atomic activities, subprocesses and entire processes). This goal aggregates one or more *Objectives* defining data to be monitored in the *contextPaths*, success criteria for the process execution in the *satisfiedCondition*, and error criteria in the *compensationCondition*. Listing 1.1 presents the exemplary goal "enough light for working" (Line 2) for our smart home process *SwitchOnLight*. The goal contains one objective "kitchen light > 600 lux in 2 seconds" (Line 4). The satisfied condition defines the success criterion for the process step as reaching a light intensity of at least 600 Lux (Line 5). If this light value is not reached within 2 seconds, an error is assumed as defined in the compensation condition (Line 6). The context paths specify that the sensor values of the "LightSensor" in the room "Kitchen" should be monitored and analysed (Lines 7–12).

3.2 Implementation in the Feedback Service

We implemented the MAPE-K loop (*Autonomic Manager*) as a component-based micro-service called *Feedback Service*[2] (FBS) with components for each phase of the MAPE-K loop [20]. During process execution, the "Legacy" WfMS sends the goal and instance information regarding a *managed* process step to the FBS (cf. Fig. 3). Monitoring agents being part of the *Monitor* collect and update sensor data in the *Knowledge Base* continuously. The Knowledge Base is a central graph database storing all information regarding the IoT entities and process execution. Significant changes in relevant (according to the context paths) data (*Symptoms*) are forwarded to the *Analyser* component. The Analyser evaluates the execution based on the criteria defined in the satisfied condition (success) or compensation condition (error). If the satisfied condition is evaluated positively, then the FBS terminates the MAPE-K executions and the WfMS continues with the "regular" process execution. If the compensation condition is evaluated positively, then a *Change Request* is sent to the *Planner* to search for a compensation strategy. The Planner uses the determined *mismatch* contained in the change request and an extensible *Compensation Repository* to find suitable replacement resources and commands to be executed by the respective effectors. The derived *Change Plan* is then enacted by the *Executor* instructing the IoT actuators. In our example, the planner queries the knowledge base for an alternative process resource (dimmer switch) in the same context (kitchen) able to influence the same context factors (light levels) as the original resource and measured by the respective sensors (light sensor) specified in the context paths [20].

4 Retrofitting Process

After introducing the MAPE-K control loop as framework for self-managed software systems and its implementation for process-aware information systems by the Feedback Service (FBS), we show how to add self-x capabilities to existing service-oriented WfMS's. We distinguish between an *invasive* and a *non-invasive* retrofitting process. In general, we exploit the fact that most WfMS's are designed to orchestrate invocations of web services and applications to also call the FBS in parallel to the "original" service calls defined in the business processes.

4.1 Invasive Retrofitting

The invasive retrofitting process requires modifications to the WfMS's underlying workflow meta-model as well as to its execution logics. The meta-model has to be adapted as proposed in Sect. 3 to support the specification of *Goals* and *Objectives* for a process, subprocess and atomic process step. Figure 3 shows the required changes regarding the execution behaviour of the "legacy" WfMS when executing a managed process step. In parallel to executing the basic process

---

[2] https://github.com/IoTUDresden/feedback-service.

activity, the WfMS has to evaluate if this activity should be *managed* and issue a service call containing the respective goal to the FBS. The WfMS then has to wait/listen for a response from the FBS concerning the execution of the MAPE-K loop and the *State* of the goal (success or no success) [20]. The process execution continues w.r.t. this result–either execute the next process steps or initiate the WfMS's error handling mechanisms when the FBS is not able to fix the issue.

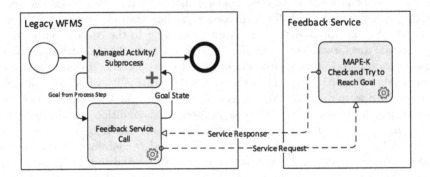

Fig. 3. Retrofitting of existing WfMS's with the feedback service.

4.2   Non-invasive Retrofitting

While the invasive retrofitting process requires modifications to the meta-model and execution logics of the WfMS, the non-invasive retrofitting process relies on modifying the existing process models. The additional task of invoking the FBS depicted in Fig. 3 has to be modelled as an explicit process activity describing an (asynchronous) call to the FBS in parallel to the managed process step. This call contains the respective goal as input parameter. The process execution continues once the "regular" process step and FBS call were executed successfully.

5   Evaluation

In this section, we show the application of both retrofitting processes to existing WfMS's. The chosen WfMS's are open source software projects that rely on different workflow notations and execution engines. The process to be executed is the smart lighting process described in Sect. 2. We "break" the lamp to be switched on by removing its bulb, which cannot be detected by the respective WfMS or HomeMatic switch, i.e., the light is still off. The FBS uses an additional TinkerForge light sensor and a second dimmer switch to check and repair the "broken" process. This process is executed in a controlled lab environment once per WfMS. The WfMS's, FBS and OpenHAB run on a central computer (Intel Core i7, 4 × 3.1 GHz, 8 GB RAM, 32 GB SSD, 2 TB HDD, Ubuntu Linux 14.04). The processes and services used in the experiments can be found on GitHub[3].

---

[3] https://github.com/IoTUDresden/fbs-retrofit.

5.1   Invasive Retrofitting of Existing WfMS's

PROtEUS. The *PROtEUS* WfMS is designed to be used in the context of IoT and CPS [21]. It relies on a component-based architecture [21] and process meta-model [22]. We extended the meta-model and execution behaviour as suggested in Sect. 3. Figure 4 shows the process model of the smart lighting process. The basic process activity is a RESTful service call to the IoT middleware to switch on the light. This process step is augmented with the *goal* attribute from Listing 1.1. The execution of the *LightInvoke* process step–an asynchronous REST service to the middleware–took 193 ms with PROtEUS. The parallel invocation, error detection and compensation of the broken light bulb by the FBS dimming up the second light stepwise from 85 Lux to 657 Lux took approx. 24 s. A more detailed performance evaluation of PROtEUS and the FBS can be found in [20].

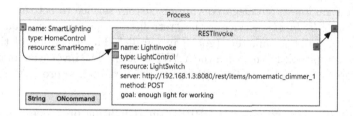

Fig. 4. Retrofitted smart lighting process for PROtEUS.

5.2   Non-invasive Retrofitting of Existing WfMS's

Activiti. The extended BPMN 2.0 smart lighting process executed with Activiti[4] 6.0.0 is depicted in Fig. 5. The *Input* process step is used to provide input parameters (goals). The basic process activity is the *LightInvoke* service task to call a custom service triggering the dimmer switch via OpenHAB. In parallel, the *FBSInvoke* service task calling the Feedback Service with the goal parameters is specified. Activiti relies on intermediate services that are locally deployed for executing external functionality. In our example, these services are custom implementations that delegate the calls to the actual RESTful services provided by the IoT middleware and FBS. The execution of the basic *LightInvoke* step took 459 ms with Activiti. The parallel invocation of the Feedback Service dimming up the second light stepwise from 89 Lux to 766 Lux took approx. 22 s.

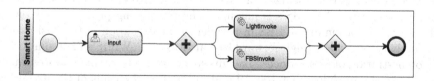

Fig. 5. Retrofitted smart lighting process for Activiti.

---

[4] https://www.activiti.org/.

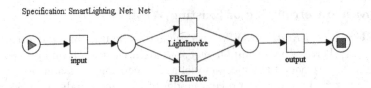

Fig. 6. Retrofitted smart lighting process for YAWL.

YAWL Engine. The extended YAWL smart lighting process executed with the YAWL engine[5] 4.2 is depicted in Fig. 6. The YAWL system also relies on custom services/classes that are deployed locally to invoke external applications or services. The *input* and *output* services print input/output parameters for debugging purposes. The *LightInvoke* step is the basic process activity that calls the IoT middleware delegated via our custom service to activate the light dimmer. The *FBSInvoke* service is the "retrofitting" process step that invokes the FBS with input parameters (goals) in parallel. The execution of the basic *LightInvoke* step took 171 ms with the YAWL system. The invocation of the FBS dimming up the second light from 81 Lux to 685 Lux took approx. 22 s.

Apache ODE. The extended WS-BPEL smart lighting process executed with Apache ODE[6] 1.3.8 is depicted in Fig. 7. The first process step is used to receive input data and assign it to the following requests. The *LightCall* branch contains the assignment of input parameters to the following basic *LightInvoke* step, which calls a custom service to then call OpenHAB to switch on the light. In the parallel *FBSCall* branch, the *FBSInvoke* process step invokes a custom service to execute the FBS with goals provided as input parameters. Apache ODE also requires us to provide intermediate services to delegate the service calls to the actual RESTful web services of the middleware and FBS. The execution of the basic *LightInvoke* step took 98 ms with Apache ODE. The invocation of the FBS dimming up the second light from 93 Lux to 639 Lux took approx. 24 s.

6   Discussion

With the MAPE-K feedback loop as general framework, we are able to provide a flexible way of retrofitting existing WfMS's with autonomic capabilities. Due to the Feedback Service being implemented as a micro-service, a loose coupling with service-oriented WfMS's is possible. The retrofitting processes rely on an additional invocation of the FBS in parallel to the execution of the basic workflow activity, which is straightforward to realize as most WfMS's main purpose is the orchestration of web services. The invasive way of retrofitting requires minor changes to the workflow meta-model and internal mechanisms of the execution

---

[5] http://www.yawlfoundation.org/.
[6] http://ode.apache.org/.

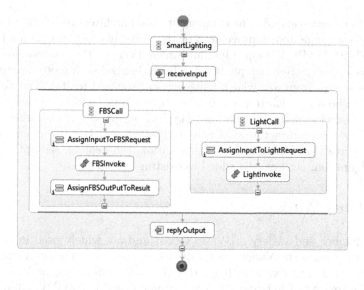

Fig. 7. Retrofitted smart lighting process for Apache ODE.

engine, which is not always possible due to inflexible and proprietary software architectures and the lack of support for service invocations–requiring a tighter integration of the FBS as an additional software component of the respective WfMS. However, compared to the non-invasive way, the modelling of the basic workflows and additional goals can be introduced more naturally into the process landscape and tools. The non-invasive retrofitting process does not require changes to the underlying WfMS but to the existing process models. The modelling of the service call with the goal parameters parallel to the managed process step is less intuitive but maintains compatibility with the basic WfMS. Software or process engineers have to decide about which retrofitting process to choose based on these criteria. The selection of an appropriate "basic" WfMS depends on features and properties the WfMS has to provide and fulfil (e.g., formal verification or scalability) in the respective application domain or enterprise.

The operations executed by the Feedback Service introduce only little overhead to the overall process execution. In our evaluation examples, the major parts of the MAPE-K execution times relate to performing and analyzing actions in the physical world (e.g., increasing light levels by actuators and waiting for the sensors to detect these changes), which are usually much slower than virtual computations. The execution times for the basic service invocations are in similar orders of magnitude for all tested WfMS's. The Feedback Service adds the same execution times to the overall process execution.

Using the Feedback Service on the more abstract level of business processes to manage mostly discrete and asynchronous workflow activities proves to be feasible for our smart home IoT use case. Its suitability for managing more real-time demanding and synchronous processes on layers closer to the hardware

remains to be investigated. The component-based architecture of the FBS facilitates the exchange and improvement of the individual algorithms used in the
phases of the MAPE-K loop. Our implementations of the analysis and planning phases are relatively simple but can be exchanged with more sophisticated
approaches, e.g., from artificial intelligence as proposed in [14]. The Feedback
Service can be used with respect to arbitrary context factors, key performance
indicators, services levels, virtual and physical sensor values and other criteria
to be considered within the MAPE-K feedback loops. Based on that, various
self-x properties (self-awareness, self-adaptation, self-healing, self-optimization,
self-configuration, etc.) can be added to existing WfMS's.

7   Related Work

A large number and variety of WfMS's exist and are widely used in academic
and industrial contexts. Various works discuss the realization of autonomic capabilities for WfMS's in general (e.g., the MABUP system [18] or ViePEP [8])
and in the context of IoT and CPS (e.g., SmartPM [14], SitOPT [23] or Wise-
Ware [15]). These approaches tie their realization of self-x mechanisms to specific self-developed WfMS's, which prevents reuse with other systems. With our
implementation of the MAPE-K control loop in the Feedback Service [20], we are
able to flexibly couple the autonomic service with other "legacy" WfMS's. Various works discuss aspects regarding the implementation of adaptive and flexible
business processes most prominently as part of the *ADEPT* workflow system [6]
and follow-up works [17]. *Worklets* [1] and *Exlets* [2] provide mechanisms for
dynamic flexibility and exception handling in workflows. These approaches do
not relate to IoT and they do not implement feedback loops to achieve self-*
capabilities. However, they could be integrated into the planning and execution phases of the proposed MAPE-K loop for workflows to implement more
sophisticated planning and adaptation strategies.

   A general approach to implement self-managed systems is proposed in [11]
and follow-up works. The retrofitting of a manufacturing system with fault-
tolerance and security based on an external event-coordination layer is proposed
in [24]. In [12] the authors present a way to add adaptivity to an existing scientific workflow engine by new components realizing the MAPE-K loop. Sensors provide data about the status of jobs running on the computation grid;
once resources are available, the workflow engine is instructed to execute new
workflows. The retrofitting of assembly processes with more high-level business
processes based on components and services to facilitate synchronization across
workflows is discussed in [3]. A general methodology for retrofitting autonomic
capabilities onto legacy systems is presented in [19]. Sensors gather information (*Probes*) from the legacy systems that are analysed by *Gauges*, decision and
coordination of adaptations are performed by controllers instrumenting the effectors of the legacy system. These works discuss specific retrofitting processes or

general frameworks for adding autonomous capabilities to legacy systems based on feedback loops. None of the approaches discusses this retrofitting specifically for WfMS's. Our retrofitting processes and software component can be used with arbitrary WfMS's in IoT but also in more traditional business process domains.

8 Conclusion

In this work, we presented an approach for retrofitting existing workflow management systems (WfMS's) with autonomous capabilities to be able to handle unanticipated situations that emerge with new properties of IoT environments. We presented an invasive and a non-invasive way of retrofitting WfMS's using a generic software component that implements the MAPE-K feedback loop from autonomic computing. Our approach proves feasible for adding self-healing mechanisms to existing WfMS's executing smart home processes, but it can also be applied in wider business process contexts with respect to arbitrary self-x capabilities and service-oriented WfMS's. The application of our retrofitting framework to WfMS's operating in production contexts to implement self-adaptive workflows for *Industry 4.0* is subject to future work [10]. This domain usually imposes more real-time and safety-related constraints on the process executions.

Acknowledgements. This research has received funding under the grant number 100268299 by the European Social Fund (ESF) and the German Federal State of Saxony.

References

1. Adams, M., ter Hofstede, A.H.M., Edmond, D., van der Aalst, W.M.P.: Worklets: a service-oriented implementation of dynamic flexibility in workflows. In: Meersman, R., Tari, Z. (eds.) OTM 2006. LNCS, vol. 4275, pp. 291–308. Springer, Heidelberg (2006). https://doi.org/10.1007/11914853_18
2. Adams, M., ter Hofstede, A.H.M., van der Aalst, W.M.P., Edmond, D.: Dynamic, extensible and context-aware exception handling for workflows. In: Meersman, R., Tari, Z. (eds.) OTM 2007. LNCS, vol. 4803, pp. 95–112. Springer, Heidelberg (2007). https://doi.org/10.1007/978-3-540-76848-7_8
3. Barros, A.P., ter Hofstede, A.H., Szyperski, C.: Retrofitting workflows for B2B component assembly. In: 25th Annual International Computer Software and Applications Conference, COMPSAC 2001, pp. 123–128. IEEE (2001)
4. Chang, C., Srirama, S.N., Buyya, R.: Mobile cloud business process management system for the internet of things: a survey. ACM Comput. Surv. (CSUR) **49**(4), 70 (2016)
5. Computing, A., et al.: An architectural blueprint for autonomic computing. IBM White Pap. **31**, 1–6 (2006)
6. Dadam, P., Reichert, M.: The ADEPT project: a decade of research and development for robust and flexible process support. Comput. Sci.-Res. Dev. **23**(2), 81–97 (2009)

7. de Lemos, R., et al.: Software engineering for self-adaptive systems: a second research roadmap. In: de Lemos, R., Giese, H., Müller, H.A., Shaw, M. (eds.) Software Engineering for Self-Adaptive Systems II. LNCS, vol. 7475, pp. 1–32. Springer, Heidelberg (2013). https://doi.org/10.1007/978-3-642-35813-5_1

8. Hoenisch, P., Schulte, S., Dustdar, S., Venugopal, S.: Self-adaptive resource allocation for elastic process execution. In: 2013 IEEE Sixth International Conference on Cloud Computing (CLOUD), pp. 220–227. IEEE (2013)

9. Janiesch, C., et al.: The internet-of-things meets business process management: mutual benefits and challenges. arXiv:1709.03628 (2017)

10. Jazdi, N.: Cyber physical systems in the context of industry 4.0. In: IEEE International Conference on Automation, Quality and Testing, Robotics, pp. 1–4 (2014)

11. Kramer, J., Magee, J.: Self-managed systems: an architectural challenge. In: Future of Software Engineering, 2007, FOSE 2007, pp. 259–268. IEEE (2007)

12. Lee, K., Paton, N.W., Sakellariou, R., Deelman, E., Fernandes, A.A., Mehta, G.: Adaptive workflow processing and execution in pegasus. Concurr. Comput.: Pract. Exp. **21**(16), 1965–1981 (2009)

13. Leotta, F., Mecella, M., Mendling, J.: Applying process mining to smart spaces: perspectives and research challenges. In: Persson, A., Stirna, J. (eds.) CAiSE 2015. LNBIP, vol. 215, pp. 298–304. Springer, Cham (2015). https://doi.org/10.1007/978-3-319-19243-7_28

14. Marrella, A., Mecella, M., Sardina, S.: Intelligent process adaptation in the SmartPM system. ACM Trans. Intell. Syst. Technol. **8**(2), 25:1–25:43 (2016)

15. Mass, J., Chang, C., Srirama, S.N.: WiseWare: a device-to-device-based business process management system for industrial internet of things. In: IEEE International Conference on Internet of Things (iThings), Green Computing and Communications (GreenCom), Cyber, Physical and Social Computing (CPSCom) and Smart Data (SmartData), pp. 269–275 (2016)

16. Muccini, H., Sharaf, M., Weyns, D.: Self-adaptation for cyber-physical systems: a systematic literature review. In: Proceedings of the 11th International Symposium on Software Engineering for Adaptive and Self-managing Systems, SEAMS 2016, pp. 75–81. ACM, New York (2016)

17. Müller, R., Greiner, U., Rahm, E.: AgentWork: a workflow system supporting rule-based workflow adaptation. Data Knowl. Eng. **51**(2), 223–256 (2004)

18. Oliveira, K., Castro, J., España, S., Pastor, O.: Multi-level autonomic business process management. In: Nurcan, S., et al. (eds.) BPMDS/EMMSAD -2013. LNBIP, vol. 147, pp. 184–198. Springer, Heidelberg (2013). https://doi.org/10.1007/978-3-642-38484-4_14

19. Parekh, J., Kaiser, G., Gross, P., Valetto, G.: Retrofitting autonomic capabilities onto legacy systems. Cluster Comput. **9**(2), 141–159 (2006)

20. Seiger, R., Huber, S., Heisig, P., Aßmann, U.: Toward a framework for self-adaptive workflows in cyber-physical systems. Softw. Syst. Model. (2017)

21. Seiger, R., Huber, S., Schlegel, T.: Toward an execution system for self-healing workflows in cyber-physical systems. Softw. Syst. Model., 1–22 (2016)

22. Seiger, R., Keller, C., Niebling, F., Schlegel, T.: Modelling complex and flexible processes for smart cyber-physical environments. J. Comput. Sci. **10**, 137–148 (2015)

23. Wieland, M., Schwarz, H., Breitenbucher, U., Leymann, F.: Towards situation-aware adaptive workflows: SitOPT—a general purpose situation-aware workflow management system. In: PerCom Workshops, pp. 32–37. IEEE (2015)

24. Xiao, K., Ren, S., Kwiat, K.: Retrofitting cyber physical systems for survivability through external coordination. In: Proceedings of the 41st Annual Hawaii International Conference on System Sciences, pp. 465–465. IEEE (2008)

On the Contextualization
of Event-Activity Mappings

Agnes Koschmider[1(✉)], Felix Mannhardt[2], and Tobias Heuser[1]

[1] Institute AIFB, Karlsruhe Institute of Technology, Karlsruhe, Germany
{agnes.koschmider,tobias.heuser}@kit.edu
[2] Department of Economics and Technology Management, SINTEF,
Trondheim, Norway
felix.mannhardt@sintef.no

Abstract. Event log files are used as input to any process mining algorithm. A main assumption of process mining is that each event has been
assigned to a distinct process activity already. However, such mapping
of events to activities is a considerable challenge. The current status-quo
is that approaches indicate only likelihoods of mappings, since there is
often more than one possible solution. To increase the quality of event
to activity mappings this paper derives a contextualization for event-
activity mappings and argues for a stronger consideration of contextual
factors. Based on a literature review, the paper provides a framework for
classifying context factors for event-activity mappings. We aim to apply
this framework to improve the accuracy of event-activity mappings and,
thereby, process mining results in scenarios with low-level events.

1 Introduction

Event log files are used as input to any process mining algorithm. Often, the
aim of these algorithms is to derive an as-is model of the process that created
these logs that can be used to further analyze the actual process execution.
Usually, process mining is applied to historical data and, thus, event log files that
were recorded from IT systems (e.g., ERP systems). Recently, the application of
process mining to real-time data became more common where log files of low-
level sensor data is increasingly being leveraged for process mining and for such
data the event-activity mapping is especially challenging [1].

In process mining, events are commonly defined as the observable and instantaneous occurrences of specific well-defined business activities [2] within the
scope of specific process instances (i.e., cases). In fact, most process mining
methods rely on these two requirements and results have shown to been poor if
any of the requirements is not met [3–6]. Many of the different kind of events,
e.g., a sudden change in sensor data, do not comply with the strict assumptions
of process mining methods. Therefore, we use the term event in a wider meaning
than its typical usage in process mining. Events can be any kind of observations
(e.g., a sensor value changed) with relevance to a certain process and are not
necessarily already linked to a specific (business) process activity.

© Springer Nature Switzerland AG 2019
F. Daniel et al. (Eds.): BPM 2018 Workshops, LNBIP 342, pp. 445–457, 2019.
https://doi.org/10.1007/978-3-030-11641-5_35

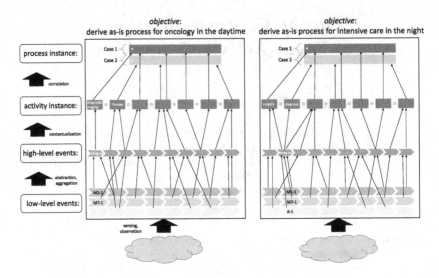

Fig. 1. Two examples for event-activity mappings are shown. The behavior of a process in the physical world is captured through sensing of low-level events from the observable behavior, aggregation and abstraction to high-level events, inference of activity instances from the process context and correlation to a specific process instance. The objective of the derived as-is model influences the event-activity mapping.

Generally, process mining algorithms assume that each event has already been assigned to a distinct meaningful process activity in the context of the process in question. Although several mapping approaches exist in the literature, the status-quo is that approaches can only indicate likelihoods of mappings, since there is often more than one possible solution [7,8]. Figure 1 looks at different levels of events and their mapping to process activities. The first and lowest level of events are **low-level events** generated through sensing or observation of the physical world. Let us assume that in this hospital example the low-level events were generated through sensors. Looking at low-level events in isolation is usually not useful for the analysis of a process since their semantics may be unclear or even ambiguous. Low-level events need to be aggregated or abstracted to **high-level events** through aggregation or abstraction using methods such as CEP, which derives events on a higher level of abstraction from a set of low-level events. High-level events already carry a semantics in the terms of the process under observation since they are often derived through rules based on domain knowledge. However, it is also possible to derive high-level events through unsupervised abstraction techniques [6], in which case their exact semantics may not be clear. Sometimes high-level events may already be correlated with the occurrence of a specific (business) activity that is recognizable in the context of a particular process. Yet, often additional information on the context in which one or several high-level events occurred needs to be taken into account. Through contextualization of events into the realm of a specific process, occurrences of activities, i.e., activity instances can be identified and correlated to a specific process instance. Referring to Fig. 1 the high-level event "image on" requires in the left hand process the aggregation of two events, while on the right hand the event is aggregated by three low-level events due to the stronger light needed for the nightly diagnostic imaging (i.e., in this example *time* is a contextual factor). Also the event "image on" is mapped to more activities of the nightly intensive care process than for the oncology at daytime process. Thus, accuracy of event-activity mappings is difficult to be benchmarked if contextual factors are not fully considered.

Traditionally, event-activity mappings consider the order of events, timestamps and related persons (resource) as sole context attributes. However, a more comprehensive view of the process context in which the event was recorded is necessary in order to increase the quality of event to activity mappings. To understand how events-activities mappings can be contextualized, we studied context taxonomies. From this study we provide a framework for classifying context factors for event-activity mappings and demonstrate the applicability of the framework.

The remainder of this paper is structured as follows. The next section discusses context dimensions. These dimensions are applied in Sect. 3 to our literature search. Section 4 defines the context framework for event-activity mappings and demonstrates its application. The paper ends with a summary and an outlook.

2   Context Dimensions

In [9] context is defined as *"any information that can be used to characterize the situation of an entity. An entity refers to a person, place, or object, which is related to the interaction between user and application."* A *process context* is *"...the set of process context information that characterizes the current execution situation of a process..."* [10]. To understand the process context of event-activity mappings several context taxonomies were studied [11–19]. From this study, we classify context information into four context dimensions as depicted in Fig. 2:

1. **Personal and Social Context:** describes all tasks in which an entity is involved and also mental and physical information about an entity and on her interaction to others [17]. The tasks in which an entity is involved are discriminated by the context property *activity*, the mental and physical information by *ability* and interaction to others is addressed by *relationship* [13]. The personal and social context of entities might be additionally described by properties that are not covered elsewhere (i.e. workload).
2. **Environmental Context:** addresses an entities' surrounding [18] such as tool and device aspects (*equipment*) [17] and the *performance* of the algorithm [14].
3. **Task Context:** is related to the *history* [14], the *goal* or intention behind the process [17], the frequency of tasks or events (*causality*) [19], its *application* [12], and *rules* [17]. The task context might also be described by properties

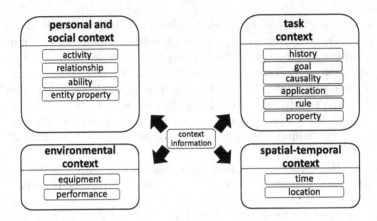

Fig. 2. Context information classified into four context dimensions with properties.

that are not covered elsewhere (such as costs, the cycle time of tasks, security issues) [19,20]. Particularly, historical information on the process are recorded within *history* [19]. *Causality* can be uncovered through the following metrics such as the overall frequency of task, the frequency of task directly preceded, the frequency of task directly succeeded, the frequency of directly or indirectly preceded, but before the next appearance, the frequency of directly or indirectly succeeded, but before the next appearance of tasks [21]. The context property *application* refers to the domain of the task such as health care or mobile banking. Rules refer to business rules that involve tasks (e.g., if more than two persons travel together, the third pays only half price.) and structural constraints and can be specified in a formalized or textual form [22].

4. **Spatial-temporal Context**: is related to the spatial-temporal coordination of the entity and subsumes *location* and *time* [17–19].

Context information results from an aggregation or abstraction of a set of simple context information, which are observed or sensed from raw data and can be classified as:

– **simple context information**, such as location or time, or
– **complex context information** that is aggregated from simple context information, such as that the entity was in a room on the 28rd of February[1].

The next section summarizes the results of a literature review on context in process mining and on event-activity mappings. The context dimensions from Fig. 2 are used to classify the results.

---

[1] This figure is adopted from [10] and extended for our purpose of understanding event-activity mappings.

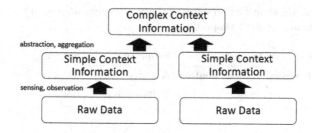

Fig. 3. Hierarchy of context information where simple context information can be captured from raw data and a complex context information is obtained by aggregation.

3 Event to Activity Mapping

To understand context-awareness in event to activity mappings we performed two different literature searches. First, we intended to extract event attributes from log files that are related to each of the four context dimensions (see Fig. 2) in order to give guidelines when developing event-activity mappings. Second, we were interested about the degree to how these event attributes have been tackled for event to activity mapping approaches already. The literature reviews have been conducted between November 2017 and February 2018. We searched the research databases ACM Digital Library, IEEE Xplore, ISI Web of Science, ScienceDirect, Scopus and Springer Link. We additionally used Google Scholar to widen the scope of our search.

3.1 Context Awareness for Process Mining

The literature analysis on context awareness in process mining reveals that several context properties have already been taken into account when mining a process model from an event log. Table 2 summarizes the results.

1. **Personal and Social Context:** The social context of individuals was considering for process mining by work environment [23–26] or organizational structures [20]. For this purpose, events were attached with the attributes PER-FORMER, IDENTITY, ORIGINATOR or some approaches differentiate between ROLE and RESOURCE. These works cover the context properties *activity* and *relationship*. Another contribution uncovered the *abilities* of entities through the analysis of the attributes SERVICE LINE, ENTITY POSITION [27] or CAPA-BILITIES [29]. Interactions of entities can be extracted through the mining of performers and their relationships. Entity properties such as department identifiers (GROUP) have been used to discover sub-processes per department [3] (Table 1).

2. **Task Context:** Objectives behind a task (goal) or a process can be uncovered by the order of activities and activity labels [28]. History and causality are directly determined from a "common" event log as shown in Fig. 1 [2]. The context property application was considered in the attribute-subject

Table 1. Literature identified for event-activity mappings organized according to context dimensions and context properties.

Context dimension	Property	Source
Personal & social context	Activity	[20, 23–25]
	Relationship	[20, 23, 24, 26]
	Ability	[27]
	Entity property	[3]
Task context	History	[2]
	Goal	[28]
	Causality	[2]
	Application	[29]
	Rule	[30, 31]
	Task property	[32]
Environmental context	Equipment	–
	Performance	[33]
Spatial-temporal context	Time	[34]
	Location	[34, 35]

domain [29]. To uncover rules either a RULE attribute was used to point to rules [31] or alignments to LTL-based descriptions or transaction protocols are defined [30].

3. **Environmental Context:** The context property performance (measured by recall or fitness) was determined through meta-data analysis of a log file in case that the user indicated the performance of the mapping algorithm (e.g., exact, approximate) or through the size of a trace [33]. To uncover tool and device aspects, i.e., the *equipment* property, the attribute MEDIUM was attached as event attribute[2].

4. **Spatial-temporal Context:** When mining process models related to spatial-temporal information, the timestamp attribute was refined by START-TIME and ENDTIME [34] (*time* context property) as well as LOCATION AREA, LOCATION LEVEL, and LOCATION CATEGORY (LOCATION context property) [35] were attached to events.

To sum up, mostly the personal and social context of entities, causality, history or rule were considered as context properties when mining a process model. The context properties history and causality are implicitly used by all process mining approaches since they can be determined directly from the minimal event log requirements [36]. Properties such as goal, application or equipment were mostly disregarded, which might be explained due to the difficulty of that challenging task.

---

[2] https://fluxicon.com/blog/2012/02/data-requirements-for-process-mining/.

3.2 Context Awareness for Event-Activity Mappings

This section summarizes the literature review on context awareness for event-activity mappings. Additionally, the results of this review should be used to compare the status-quo of context awareness in process mining (i.e., which context properties have been tackled for event-activity mappings already). The literature results are again crossed with the context dimensions listed in Sect. 2.

1. **Personal and Social Context:** Folino et al. [37] enrich the event logs with the attribute TEAM that indicates the team associated with the first event of a trace and employ a clustering approach to obtain activities. Mannhardt et al. [38] uses information about the department in which events occurred to determine per-department activity patterns for the use of event-activity mapping. Moreover, the event log attribute WORKLOAD as the number of problems open on the time when a trace of an event log started [37] is also related to the property entity property.

2. **Task Context:** Domain knowledge is a key-word often used in the literature which refers to knowledge about the direct task context in the terms of the context property history. Two examples for domain-knowledge based approaches are Baier et al. [39], who uses domain knowledge extracted from process documentation to semi-automatically match events and activities, and Mannhardt et al. [38] who uses activity patterns to capture domain knowledge for event-activity mapping. The behavior defined by the activity patterns is aligned with the observed behavior in the event log, which records historical information about the process. Mounira et al. [14] propose historic related context with regards to patient's immediate members to use for developing a context-aware process mining framework for maximizing business process exibility illustrated in hospital environment. Another context property is causality: Tax et al. [40] propose entropy of an activity in an event log based on its directly follows ratio vector and the directly-precedes ratio vector. Lu et al. describe a semi-supervised approach for log pattern detection. They refine causal dependencies into directly causes and eventually causes. In addition, Diamantini et al. [41] refer to their work to relevant subtraces from an event log by considering process execution patterns. Also, attributes like one-to-many correspondence (an event corresponds to a set of low-level events). The third context property we deal with is application: We identified Folino et al. [37] as related paper dealing with combining the discovery of different execution scenarios with the automatic abstraction of log events. Finally, the last context property we deal with is rule: Goedertier et al. [42] propose the use of first-order logic to define preconditions and time-varying properties to overcome difficulties like the limitation of process mining to a setting of non-supervised learning since negative information is often not available.

3. **Spatial-temporal Context:** Time as a context property is found in several papers [14,37,40,42,43]. Particularly, [37] enrich the time dimensions by attributes such as week-day, month and year. Location of activities can be found via RFID tracking [43].

Table 2. Literature identified for event-activity mappings organized according to context dimensions and context properties.

Context dimension	Property	Source
Personal & social context	Activity	–
	Relationship	[37]
	Ability	–
	Entity property	[3, 37]
Task context	History	[5, 14, 38, 39, 45]
	Goal	–
	Causality	[41–43, 46–49]
	Application	–
	Rule	[5, 38, 39, 49]
	Task property	–
Environmental context	Equipment	–
	Performance	–
Spatial-temporal context	Time	[14, 37, 38, 40, 42, 43]
	Location	[40, 49]

In fact, many context properties have not been covered yet when mapping events to activities, which can be concluded from the comparison to the analysis on context-awareness in process mining. Particularly, the context properties "activity", "ability" and "entity property", "equipment", and "location" are not sufficiently covered by the literature we identified within our review. This might be explained due to the challenging task to retrieve the information. Additionally, more attributes for the personal & social context have been addressed.

Both literature reviews elicit that several contextual factors have been already tackled when mining a process from an event log and developing an event-activity mapping. Properties such as goal, application, privacy or equipment should attract more attention. One solution that allows to identify the application from a log might be the linguistic analysis of activity labels [36]. The development of privacy-aware event-activity mappings might be inspirited by privacy-aware modeling approaches [44] where privacy policies or privacy restrictions are considered. We are convinced that in the further mapping approaches will emerge addressing these context properties. This can be justified by the increase interest in this topic and particularly in the rise of IoT.

4   Framework for Event-Activity Mappings

To improve the accuracy of event-activity mappings we developed a framework based on the literature results found on context awareness in process mining and event-activity mappings. The pillars of the framework are the four context dimensions presented in Sect. 2. The properties of each dimension are those

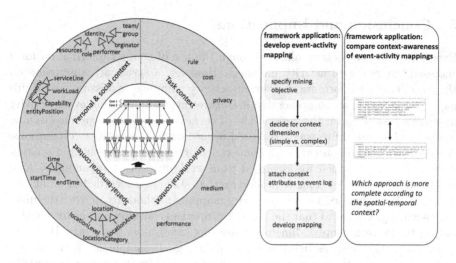

Fig. 4. Context framework for event-activity mappings and the applications of the framework. The benefit of the framework is decision support improving the accuracy of event to activity mapping approaches.

event attributes that are tackled for context awareness in process mining and event-activity mappings. Depending on the objective, these properties might be on different abstraction levels. For instance, in case of event-activity mappings according to personal & social relationships either one might subsume resource and role to performer or consider role and performer as synonym. The sole consideration of performer without any other event attributes is not sufficient in any way.

The benefit of the framework is twofold. For those, who intend to develop an event to activity mapping, it is recommended to first specify the objective of the event-activity mapping. Next, it should be decided whether a simple or complex context information is of relevance. Depending on the extent of context-awareness, event attributes have to be distilled and attached to the trace. Certainly, the decision in favor which attributes to attach to the log file depends on the mining objective or even on the accessibility and availability of information. Definitely, the more complete the log file the more accurate the results obtained from process mining. In this way, this framework is applied as guide towards developing an accurate event-activity mapping.

On the other hand, the application of the framework might be the comparison of event-activity mapping approaches (i.e., find the event-activity mapping that consider more attributes of the spatial-temporal context). The accuracy in this scenario correlates with the number of used attributes. In this way, the framework benchmarks event-activity mapping approaches.

5   Conclusion and Implications

Events need to be contextualized through the use of context information for a successful mapping to activity instances. However, a systematic discussion on the use of context information for event-activity mappings is missing. To fill this gap, we conducted a comprehensive literature review on existing event-activity mappings as well as on the general use of context properties in process mining methods. The literature was structured according to four context dimensions: personal and social context, task context, environmental context, and spatial-temporal context, which we identified from work on context taxonomies. As a result, we identified 14 context properties that should be recorded in event logs and that should be used by event-activity mapping methods to improve the mapping accuracy. We found that the context properties causality and history, which belong to the task context dimension, are supported most frequently. However, other properties such as, e.g., activity, ability, goal, equipment, performance, and location are not or only rarely described in the literature on event-activity mappings. Thus, it remains challenging to consider the wider context for event-activity mapping problems.

References

1. Soffer, P., et al.: From event streams to process models and back: challenges and opportunities. Information Systems (2018)
2. van der Aalst, W.M.P.: Process Mining - Data Science in Action, 2nd edn. Springer, Heidelberg (2016). https://doi.org/10.1007/978-3-662-49851-4
3. Mannhardt, F., de Leoni, M., Reijers, H.A., van der Aalst, W.M.P., Toussaint, P.J.: Guided process discovery - a pattern-based approach. Inf. Syst. **76**, 1–18 (2018)
4. Günther, C.W.: Process Mining in Flexible Environments. PhD thesis, Technische Universiteit Eindhoven (2009)
5. Folino, F., Guarascio, M., Pontieri, L.: Mining predictive process models out of low-level multidimensional logs. In: Jarke, M., Mylopoulos, J., Quix, C., Rolland, C., Manolopoulos, Y., Mouratidis, H., Horkoff, J. (eds.) CAiSE 2014. LNCS, vol. 8484, pp. 533–547. Springer, Cham (2014). https://doi.org/10.1007/978-3-319-07881-6_36
6. Ferreira, D.R., Szimanski, F., Ralha, C.G.: Improving process models by mining mappings of low-level events to high-level activities. J. Intell. Inf. Syst. **43**(2), 379–407 (2014)
7. Eyers, D.M., Gal, A., Jacobsen, H., Weidlich, M.: Integrating process-oriented and event-based systems. Dagstuhl Rep. **6**(8), 21–64 (2016)
8. van der Aa, H., Leopold, H., Reijers, H.A.: Checking process compliance on the basis of uncertain event-to-activity mappings. In: Dubois, E., Pohl, K. (eds.) CAiSE 2017. LNCS, vol. 10253, pp. 79–93. Springer, Cham (2017). https://doi.org/10.1007/978-3-319-59536-8_6
9. Dey, A.K.: Understanding and using context. Pers. Ubiquitous Comput. **5**(1), 4–7 (2001)
10. Trunko, R.: Kontextsensitive Ausnahmebehandlung in Geschftsprozessen. Verlag Dr. Hut (2011)

11. Bose, R.J.C., Van der Aalst, W.M.: Context aware trace clustering: towards improving process mining results. In: Proceedings of the 2009 SIAM International Conference on Data Mining, SIAM, pp. 401–412 (2009)
12. Rosemann, M., Recker, J.: Context-aware process design exploring the extrinsic drivers for process flexibility. In: BPMDS, CEUR Workshop Proceedings, vol. 236 (2006)
13. Zimmermann, A., Lorenz, A., Oppermann, R.: An operational definition of context. In: Kokinov, B., Richardson, D.C., Roth-Berghofer, T.R., Vieu, L. (eds.) CONTEXT 2007. LNCS (LNAI), vol. 4635, pp. 558–571. Springer, Heidelberg (2007). https://doi.org/10.1007/978-3-540-74255-5_42
14. Mounira, Z., Mahmoud, B.: Context-aware process mining framework for business process flexibility. In: iiWAS 2010, pp. 421–426. ACM (2010)
15. Folino, F., Guarascio, M., Pontieri, L.: Discovering context-aware models for predicting business process performances. In: Meersman, R., et al. (eds.) OTM 2012, Part I. LNCS, vol. 7565, pp. 287–304. Springer, Heidelberg (2012). https://doi.org/10.1007/978-3-642-33606-5_18
16. Saidani, O., Nurcan, S.: Towards context aware business process modelling. In: BPMDS 2007 (2007)
17. Kofod-Petersen, A., Cassens, J.: Using activity theory to model context awareness. In: Roth-Berghofer, T.R., Schulz, S., Leake, D.B. (eds.) MRC 2005. LNCS (LNAI), vol. 3946, pp. 1–17. Springer, Heidelberg (2006). https://doi.org/10.1007/11740674_1
18. Michael, J., Steinberger, C.: Context modeling for active assistance. In: ER Forum/Demos, CEUR Workshop Proceedings, vol. 1979, pp. 207–220 (2017)
19. Becker, T., Intoyoad, W.: Context aware process mining in logistics. Procedia CIRP 63, 557–562 (2017). Manufacturing Systems 4.0, Proceedings of the 50th CIRP Conference on Manufacturing Systems
20. Schnig, S., Cabanillas, C., Jablonski, S., Mendling, J.: A framework for efficiently mining the organisational perspective of business processes. Decis. Support Syst. 89, 87–97 (2016)
21. Mărușter, L., Weijters, A.J.M.M.T., Van Der Aalst, W.M.P., Van Den Bosch, A.: A rule-based approach for process discovery: Dealing with noise and imbalance in process logs. Data Min. Knowl. Discov., 13(1), 67–87 (2006)
22. Hornung, T., Koschmider, A., Oberweis, A.: Rule-based auto completion of business process models. In: CAiSE Forum, CEUR Workshop Proceedings, vol. 247 (2007)
23. van der Aalst, W.M.P., Reijers, H.A., Song, M.: Discovering social networks from event logs. Comput. Support. Coop. Work 14(6), 549–593 (2005)
24. Song, M., van der Aalst, W.M.P.: Towards comprehensive support for organizational mining. Decis. Support Syst. 46(1), 300–317 (2008)
25. Jin, T., Wang, J., Wen, L.: Organizational modeling from event logs. In: Sixth International Conference on Grid and Cooperative Computing, pp. 670–675 (2007)
26. Rinderle-Ma, S., Wil, M.: Life-cycle support for staff assignment rules in process-aware information systems. Technical report (2007)
27. Cheng, H.J., Kumar, A.: Process mining on noisy logs can log sanitization help to improve performance? Decis. Support Syst. 79, 138–149 (2015)
28. Deneckère, R., Hug, C., Khodabandelou, G., Salinesi, C.: Intentional process mining: Discovering and modeling the goals behind processes using supervised learning. IJISMD 5(4), 22–47 (2014)

29. Koschmider, A., Song, M., Reijers, H.A.: Advanced social features in a recommendation system for process modeling. In: Abramowicz, W. (ed.) BIS 2009. LNBIP, vol. 21, pp. 109–120. Springer, Heidelberg (2009). https://doi.org/10.1007/978-3-642-01190-0_10

30. Caron, F., Vanthienen, J., Baesens, B.: Rule-based business process mining: applications for management. In: Management Intelligent Systems, vol. 171, pp. 273–282. Springer, Heidelberg (2012). https://doi.org/10.1007/978-3-642-30864-2_26

31. Schönig, S., Gillitzer, F., Zeising, M., Jablonski, S.: Supporting rule-based process mining by user-guided discovery of resource-aware frequent patterns. In: Toumani, F., et al. (eds.) ICSOC 2014. LNCS, vol. 8954, pp. 108–119. Springer, Cham (2015). https://doi.org/10.1007/978-3-319-22885-3_10

32. Tax, N., Sidorova, N., Haakma, R., van der Aalst, W.M.P.: Event abstraction for process mining using supervised learning techniques. In: Bi, Y., Kapoor, S., Bhatia, R. (eds.) IntelliSys 2016. LNNS, vol. 15, pp. 251–269. Springer, Cham (2018). https://doi.org/10.1007/978-3-319-56994-9_18

33. Song, M., Günther, C.W., van der Aalst, W.M.P.: Trace clustering in process mining. In: Ardagna, D., Mecella, M., Yang, J. (eds.) BPM 2008. LNBIP, vol. 17, pp. 109–120. Springer, Heidelberg (2009). https://doi.org/10.1007/978-3-642-00328-8_11

34. Blank, P., Maurer, M., Siebenhofer, M., Rogge-Solti, A., Schonig, S.: Location-aware path alignment in process mining. EDOCW **2016**, 1–8 (2016)

35. Fernandez-Llatas, C., Lizondo, A., Monton, E., Benedi, J.M., Traver, V.: Process mining methodology for health process tracking using real-time indoor location systems. Sensors **15**(12), 29821–29840 (2015)

36. Koschmider, A., Reijers, H.A.: Improving the process of process modelling by the use of domain process patterns. Enterp. IS **9**(1), 29–57 (2015)

37. Folino, F., Guarascio, M., Pontieri, L.: Miningmulti-variant process models from low-level logs. In: Abramowicz, W. (ed.) BIS 2015. LNBIP, vol. 208, pp. 165–177. Springer, Cham (2015). https://doi.org/10.1007/978-3-319-19027-3_14

38. Mannhardt, F., de Leoni, M., Reijers, H.A., van der Aalst, W.M.P., Toussaint, P.J.: From low-level events to activities - a pattern-based approach. In: La Rosa, M., Loos, P., Pastor, O. (eds.) BPM 2016. LNCS, vol. 9850, pp. 125–141. Springer, Cham (2016). https://doi.org/10.1007/978-3-319-45348-4_8

39. Baier, T., Mendling, J., Weske, M.: Bridging abstraction layers in process mining. Inf. Syst. **46**, 123–139 (2014)

40. Tax, N., Alasgarov, E., Sidorova, N., Haakma, R.: On generation of time-based label refinements. arXiv preprint arXiv:1609.03333 (2016)

41. Diamantini, C., Genga, L., Potena, D.: Behavioral process mining for unstructured processes. J. Intell. Inf. Syst. **47**(1), 5–32 (2016)

42. Goedertier, S., Martens, D., Baesens, B., Haesen, R., Vanthienen, J.: A new approach for discovering business process models from event logs. Technical report, SSRN (2007)

43. Zang, C., Fan, Y.: Complex event processing in enterprise information systems based on RFID. Enterp. Inf. Syst. **1**(1), 3–23 (2007)

44. Alpers, S., Pilipchuk, R., Oberweis, A., Reussner, R.H.: Identifying needs for a holistic modelling approach to privacy aspects in enterprise software systems. ICISSP, SciTePress **18**, 74–82 (2018)

45. Fazzinga, B., Flesca, S., Furfaro, F., Masciari, E., Pontieri, L.: Efficiently interpreting traces of low level events in business process logs. Inf. Syst. **73**, 1–24 (2018)

46. Tax, N., Sidorova, N., van der Aalst, W.M.P.: Discovering more precise process models from event logs by filtering out chaotic activities. J. Intell. Inf. Syst., 1–33 (2018)
47. Lu, X., et al.: Semi-supervised log pattern detection and exploration using event concurrence and contextual information. In: Panetto, H., et al. (eds.) On the Move to Meaningful Internet Systems. OTM 2017 Conferences. OTM 2017. Lecture Notes in Computer Science, vol. 10573, pp. 154–174. Springer, Cham (2017). https://doi.org/10.1007/978-3-319-69462-7_11
48. Begicheva, K., Lomazova, I.A.: Discovering high-level process models from event logs. Model. Anal. Inf. Syst. **24**, 125–140 (2017)
49. Fazzinga, B., Flesca, S., Furfaro, F., Pontieri, L.: Online and offline classification of traces of event logs on the basis of security risks. J. Intell. Inf. Syst. **50**(1), 195–230 (2018)

A Classification Framework for IoT Scenarios

Sankalita Mandal$^{(\boxtimes)}$, Marcin Hewelt, Maarten Oestreich, and Mathias Weske

Hasso Plattner Institute, University of Potsdam, Potsdam, Germany
{sankalita.mandal,marcin.hewelt,maarten.oestreich,mathias.weske}@hpi.de

Abstract. The Internet-of-Things (IoT) is here to stay, as applications increasingly make use of IoT devices to deliver value to customers and organizations. Smart home, predictive maintenance, asset tracking are just a few examples of business scenarios that employ the IoT. As concepts from the domain of Business Process Management (BPM) are used to realize IoT scenarios, the need arises to classify which scenarios can profit from BPM concepts. In this contribution, we present a range of IoT scenarios and discuss the dimensions to classify them. Further, we suggest the BPM concepts that might be advantageous to use for realizing IoT scenarios.

Keywords: Internet of Things · Business Process Management

1 Introduction

It is estimated that in 2020 around 20 billion devices will be connected to the internet, forming the Internet-of-Things (IoT). Traditionally, networks connected computers with each others and facilitated the exchange of data among distributed nodes. With the advent of the IoT many heterogeneous devices joined these networks that were not computers in the traditional sense: GPS sensors built into cars, smart meters, fitness tracking wristbands, smart fridges, NFC readers, and many more. These devices provided capabilities that computers did not offer: sensing the world around them and acting towards the world. These connected devices facilitated business scenarios that were not possible before, like tracking of containers with GPS sensors, remotely controlling the heating at home, or coordinating thousands of devices in a smart factory. Most of these scenarios center around collecting data employing distributed sensors, exchanging, processing and visualizing this data, as well as acting on it.

In several implemented IoT scenarios concepts from the domain of Business Process Management (BPM) have been successfully used [1]. However, not all IoT scenarios benefit equally from BPM concepts. The single app-controlled Phillips Hue lamp[1] will not profit from BPM concepts, while a scenario that schedules maintenance appointments for a fleet of cars might. The diversity of

---

[1] Philips Hue. https://www2.meethue.com/en-us.

© Springer Nature Switzerland AG 2019
F. Daniel et al. (Eds.): BPM 2018 Workshops, LNBIP 342, pp. 458–469, 2019.
https://doi.org/10.1007/978-3-030-11641-5_36

existing scenarios poses a problem, when trying to decide whether to apply BPM concepts for a projected IoT application [2]. To answer that, we need to understand the IoT world better, with several features it might offer. Unfortunately, a classification of IoT scenarios with regard to supporting them with BPM concepts has not been undertaken so far.

Looking at the other direction – supporting business processes with IoT – one obvious scenario is to use the events produced by sensors to drive execution of processes, e.g. by starting a new case, adding case data, or making routing decisions [3]. The process model can be considered "IoT-enabled" and is enhanced by the events. In this case, the process model just needs to know the expected events, which "abstract" from the devices, while the devices themselves are transparent. The situation changes, when actuators are added to the mix, which need to be triggered by the case. Here, it is important to find a good abstraction layer, instead of communicating with multiple, heterogeneous devices directly. The possibilities as well as the challenges of integrating complex event processing and BPM are discussed extensively in [4].

In this contribution we describe a variety of implemented or projected IoT scenarios and analyze them for commonalities and differences in Sect. 2. From this description we derive criteria, e.g. number and type of involved devices or locus of control, that enable a classification of IoT scenarios. The classification framework with these criteria is presented in Sect. 3. Next, in Sect. 4, we look at how various BPM concepts [5] can help to realize them. In addition, we consider flexible process execution as provided by the fragment-based case management approach Chimera [6] as a mean to support the realization of IoT applications. Finally, we conclude in Sect. 5.

2   Evolution of IoT

Internet of things started gaining popularity during 1980s. During the last two decades, many more applications based on IoT are coming into the play. However, even if the term was coined later in 1999 [7], the concept of interconnected objects has been introduced much earlier. This section gives a brief history of IoT and also sketches the future trends of IoT with the help of example applications.

2.1   The Beginning of the IoT

A good way to get an overview of the possibilities of the IoT is to have a look at various scenarios and how these evolved over time. Therefore, this section covers the beginnings of connected devices and their more recent developments in different domains. While the possibility to access funds almost everywhere is taken for granted today, the first ATMs were based on static tokens, which could be exchanged for cash [8]. In 1972, IBM developed a platform [9] that allowed cash machines to work as connected devices enabling the current system of accessing customer's bank account via machines. Therefore, *ATMs* can be considered as the earliest IoT devices [10]. The concept of 'product as a service' proposes a shift to usage based payment instead of ownership of products [11], although the concept existed before sensors could transmit operational data to the manufacturer in real-time. Examples include *turbines as a service*[2], *product based car sharing*[3] and *bike sharing*[4]. The introduction of connectivity in these services enabled much more convenient versions afterwards, e.g. by tracking the location via phone.

2.2 Towards Smarter Solutions – the Status Quo of the IoT

While *phone-based mobile payment*[5] platforms are a much more recent development than the credit card, their functionality still closely resembles their ancestors. However, information about the account balance or past transactions is additionally available through the smartphone apps. The *smart car cockpit*[6] takes this a step further: instead of numerous analogue dials and gauges, there is only one display for all relevant information from various systems. As part of the sharing economy, *platform based car sharing*[7] uses the existing smartphones of its users for data collection (e.g. location) and interaction (e.g. find a car). Unlike product based car sharing, where the operator owns the vehicles, all kinds of cars can be seamlessly integrated with the platform service. Therefore, supply and demand for mobility can be matched in a novel way. A recurring theme among IoT applications is the so-called *smart home*, where previously unconnected devices are equipped with additional remote monitoring and control capabilities. Examples include the *smart fridge*[8], *smart oven*[9], or the *smart thermostat*[10], where one can check and set the temperature from anywhere. Instead of monitoring the internal operations of one device, other scenarios rather follow a tangible asset. This ranges from tracking individual parts in *intelligent automotive manufacturing*[11] over the completed product to arbitrary valuable objects in *luxury freight tracking*[12]. On the other hand, it can also be desirable to use a *smart parking meter*[13] to find and manage a spot for the car.

---

[2] Rolls Royce Power-by-the-Hour. http://www.rolls-royce.com/media/press-releases/yr-2012/121030-the-hour.aspx.

[3] Car2go. https://www.car2go.com/.

[4] Call a Bike. https://www.callabike-interaktiv.de/.

[5] Apple Pay. http://www.apple.com/apple-pay/.

[6] Bosch Smart Car Cockpit. https://www.bosch-presse.de/pressportal/de/en/bosch-unclutters-vehicle-cockpit-139008.html.

[7] Uber. https://www.uber.com/.

[8] Samsung Family Hub. http://www.samsung.com/us/explore/family-hub-refrigerator/.

[9] June Oven. https://juneoven.com/.

[10] Nest Learning Thermostat. https://store.nest.com/product/thermostat/.

[11] Daimler Smart Production. http://www-05.ibm.com/de/pmq/_assets/pdf/IBM_SPSS_Case_Study_Daimler_de.pdf.

[12] DHL Luxury Freight Tracking. http://www.dhl.fr/en/logistics/industry_sector_solutions/luxury_expertise.html.

[13] Smart Parking Limited. http://www.smartparking.com/.

In agriculture domain, some plants require a narrow range of conditions for optimal growth. However, it is not feasible for farmers to constantly check their fields in person. Hence, it makes sense to automate the *monitoring of environmental conditions for crops*[14] by leveraging sensors to collect real-time information about weather conditions. By setting a threshold value for changes in temperature and humidity, farmers receive notifications via an app so that they can react quickly to ensure the well-being of their crops. When using more than one sensor, it is possible to correlate multiple input streams to infer more accurate predictions – e.g. while a single high temperature might be an outlier, rising temperatures across the field indicate a high probability of reaching the threshold value. By recognizing this, the farmer can be notified beforehand so he can proactively protect his plants.

In healthcare, medical professionals as well as personal fitness enthusiasts leverage small wearable devices (e.g. watches, bracelets) to constantly monitor vital information (e.g. hearth rate). For example, *tele-ECG*[15] for heart patients sends the current status to the doctor regularly to prevent accidents and for immediate medical support on demand. For medical professionals as well as personal fitness enthusiasts, *real-time fitness tracking*[16] gained popularity. In a physicians practice, the use of *medical devices as a service*[17] allows doctors to avoid high upfront payments and the logistics associated with an external laboratory. Instead, they can perform the analysis of samples directly in office. Combined with constant remote monitoring of the machines by the manufacturer, this provides sufficient data to create good estimates of downtimes. Consequently, repairs can be scheduled before problems occur. The same principle applies to *smart grid*[18] where utility providers can run a hosted grid, while their supplier can monitor the usage of all participants to support maintenance and product development.

2.3   Living in a Connected World – the Future IoT

Looking at predictions and products currently in development, it can be said that in spite of having several technical and social challenges, in near future, connectivity will become truly ubiquitous [12]. The start-up *Myriota*[19] creates robust micro units to track the position and motion of assets, which are not covered by traditional means of communication. While not operating in real time, the system uses "Low Earth Orbit Satellites" to gather data from many devices at once every 90 min, thus enabling the monitoring of previously unavailable assets.

---

[14] Bosch Deepfield Connect. http://www.deepfield-robotics.com/en/Deepfield-Connect.html.

[15] Getemed PM100. http://www.pm100.de/.

[16] Fitbit Flex. https://www.fitbit.com/flex.

[17] Sysmex Medical Analysis. https://www.sysmex.com/us/en/Pages/Beyond-A-Better-Box.aspx.

[18] GE Grid IQ. http://www.gegridsolutions.com/DemandOpt/Catalog/GridIQ.htm.

[19] Myriota. http://myriota.com/.

The EU project *ecall*[20] leverages the information from many cars on the road to find new opportunities to improve security and efficiency of traffic in cities, as well as to enable the adaptive redirection of drivers when a congestion threshold is met. Taking effect in 2018, *ecall* is concerned with the combination of emergency detection in cars with an automated notification of response personal. If the car's sensors indicate a crash, the *ecall* device will establish a connection to an emergency hotline – depending on the situation of the driver, a voice call can clarify the situation. In all cases, the last available information from the car's sensors is transmitted via a cellular data connection in order to enable an appropriate response. Another big application of IoT is the extended *smart factory* concept, namely the strategic initiative *Industry 4.0* [13] where German Government promotes the digital structural change in industrial manufacturing and offers a framework to realize it.

3   IoT Classification Framework

In Sect. 2, several IoT scenarios have been introduced. Definitely, many other scenarios are already implemented or will be materialized in future. The classification framework is based on an analysis of the presented scenarios. It revealed the specific features shared by the IoT scenarios which are needed to be considered to understand the scenarios better, or one step ahead, to implement a scenario. In this section, the framework containing the features for classifying the IoT scenarios is presented.

Participants. The first aspect to be considered is to explore the participants in a particular scenario. This includes the number of components present as well as the type of components. In most of the cases, an IoT scenario includes a subset of the following type of participants:

- *Sensor:* The thing or device that detects the change in the environment and sends the information to other thing(s) or a processor.
- *Actuator:* The thing or device that receives information from a processor or other thing(s) and reacts on that to manipulate the environment.
- *Display:* The device responsible for visualizing relevant information about the scenario such as the interactions among the things, change of status in the environment, failure of a thing and so on.
- *Controller:* A central processor that sends and receives data and processes the information to control the next operations. It can be a central computer or a platform where the logic for interactions reside.
- *Complex Device:* A device which can perform certain combination of the above functionality. For example, a smartphone can act as a sensor, an actuator, a display, and a controller at the same time.
- *Web Service:* The services provided by the web applications which can be outsourced by the controller for processing data, if needed.

---

[20] EU Project ecall. https://ec.europa.eu/digital-single-market/en/ecall-time-saved-lives-saved/.

- *Human Beings:* The human resources responsible for operating the controller and the end users receiving the benefit of the internet of things.

The scope of IoT applications ranges from single devices to complex systems spanning across continents and this influences the dynamics of executing interactions among them. Table 1 shows different scaling possibilities.

Table 1. Levels of participation in an IoT scenario

Level	Description	Example
Device	A use case can be realized with a single device	Fitness tracker
Group	The scenario is enabled by the collaboration of a few (2–5) devices	Smart car cockpit
Site	Multiple devices operate within a well defined location	Crop monitoring
City	Loose combination of many things within a large yet continuous area	Bike sharing
Wide	Collaboration across multiple remote location	Industry 4.0

Control. The control logic for the interaction among the devices can be placed as per the need of the scenario. It can be distributed in the following ways:

- *Central:* A central controller takes the decision. The devices execute instructions from a remote location. For example, a sensor in *crop monitoring* field only measures the temperature and sends it to the processor.
- *On Device:* A local program directs the operations of the device. Note that this does not restrict external communication, it only means that decisions are made locally. An example can be a *smart car.*
- *Distributed:* Though the controller is in charge, the devices have minimal logic to perform certain tasks by themselves. A *smart fridge* can serve as an example; it is part of a smart home, but can itself control the temperature inside the refrigerator.

Interaction. The interaction among the participants of an IoT scenario can be of the following types. Note that the interfaces of the display, web service or UI are not discussed, since they follow the same technicalities as a non-IoT setup.

- *Things to Things:* Things (sensors/actuators) in one net talk to each other.
- *Things to Controller:* The sensors send data to the controller and the controller sends instruction to the actuators. Often a gateway is used for such communication which is an entry point for the cumulative data gathered from sensors. Generally the protocol used to transfer information from the things to the gateway is different than the protocol used to communicate with the controller.

Data. Every IoT application is based on the data from the participating IoT devices. We distinguish between different types of data that can be communicated from the IoT device to the controller: (1) identifier, (2) location, and

(3) sensor values. An identifier allows to connect the device with further data related to the device stored in the controller, e.g. the ID of an NFC card establishes contact with data of the card holder and can be used for access control. Location data allows to track devices in a spatial context, thus enabling asset tracking. Finally, data produced by sensors allows to gather information about the environment of a device. The data from devices is collected and stored, e.g. in a log file, either on the device itself or on the controller. Collected data can be further processed, as detailed in Table 2. The third aspect regarding data in IoT scenarios is their further usage.

- *Dashboard:* the potentially aggregated data is visualized in a dashboard to provide information to human participants. Those will make decisions based on the data and trigger a reaction, which might potentially involve other IoT devices. We consider scenarios that involve sending notifications to human participants as part of this category.
- *Automated decisions:* the collected and potentially aggregated data is used by the controller to make automated decisions, e.g. sending commands to actuators or starting business processes.

Table 2. Levels of data processing in an IoT scenario

Level	Description	Example
None	Data from sensors is just collected for further manual processing	Device error log
Group	Information from multiple similar devices is used together to detect spatial trends or outliers	Crop monitoring
Temporal	Data from a device is collected over time which are not feasible to be done manually	Freight tracking
Temporal group	Historical information across multiple dimensions is collected for a group of devices	Smart parking
Complex	Advanced aggregation supplements sensor input with additional data sources	Dashboard

Automation. Modern technology connected with IoT devices can often ease or replace manual labor. This work considers the degrees of automation shown in Table 3.

Table 3. Levels of automation in an IoT scenario

Level	Description	Example
Minimal	Little or no replacement of manual activity i.e. the process closely resembles the analogue version	ApplePay
Incremental	Increases efficiency in tasks that can be done by humans	Crop monitoring
Enabling	Creates new functionality or enables tasks which are not feasible to be done manually	Real-time fitness tracking

Ownership. The ownership details are needed to make the right design decisions like the data or device access control. The owner of an IoT scenario can be *homogeneous* where the end user decides the policies, such as a *smart home* application. In contrast, the ownership can belong to one or more *heterogeneous* private or public corporation. For example, a *smart factory* scenario would be controlled by the factory owner(s).

4   Application of BPM Concepts to IoT

This section discusses the relevant concepts from BPM that might be beneficial for implementing and managing specific IoT scenarios. Based on the analysis of scenarios in Sect. 2 and the classification dimensions discussed in Sect. 3, examples are provided for which certain BPM concepts will be suitable. Often, the traditional process model languages are not enough to represent the context-adaptiveness in a dynamic scenario [14]. We propose the case management approach for scenarios demanding high flexibility. Since BPM has more human-centric perspective than the automated device interactions required for IoT, the BPMN extension proposed in [15] can be applied for an efficient integration of IoT and BPM.

Business Process Management. Business process management (BPM) is an established mean for modeling, executing and improving organizational business processes. BPMN 2.0 [16] is the industry standard for modeling, implementing, and enacting business processes. A process has a specific set of goals and *activities* are executed, either automatically or by process participants, in a specific order to achieve those goals [5]. Communication with the environment is represented as *events* in a process model and a process execution can be hugely influenced by such environmental interactions [17]. Processes can consume events (*catching events*) as well as produce events (*throwing events*). These events can trigger a process (*start event*), abort certain activities (*boundary event*), and choose between many possible execution branches (*event-based gateway*). The information carried by events can be used to make *decisions* in the course of the process execution.

Often, several participants collaborate in a process. These participants can belong to one organization or can be separate entities, shown as *swim-lanes* and *pools*, respectively. Process participants can interact with each other by means of *message exchange*. In case only the interaction is needed to be visualized, *choreography diagrams* are useful.

Now, process models can be used for several purposes in an IoT setup. Being an expressive language, BPMN artifacts can be used to model the interactions among the things and with the controller. This allows designers to better understand and communicate the IoT scenario they are developing. For example, processes can show the internal behavior of an IoT setup with a *group* of devices, having a *homogeneous* owner, like in the *smart car* scenario. For scenarios like *Car Sharing*, the interactions in a *city* context can be modeled with message exchanges between different pools. On the contrary, with a *wide* range of devices owned by *heterogeneous* stakeholders, a choreography diagram will be more appropriate to show the interactions while abstracting from the internal processes.

Going beyond representation of the IoT scenario, a business process management system (BPMS) can be used to implement the application logic of the scenario. In this case the BPMS acts as the controller: It receives the environmental occurrences from the sensors using the catching event construct, stores the event payload for further processing, chooses the appropriate execution branch based on event data, and sends instructions to the actuators via the throwing event construct. For exceptional situations, the error event can be executed with the semantics of a boundary event to abort the ongoing activities and follow the exception handling path.

The authors previously suggested to decouple event sources, e.g. the sensors in an IoT scenario, from the process logic acting on event data that is implemented in the BPMS [18]. Instead, the BPMS should be connected to a complex event processing (CEP) system that analyzes the raw events received from sensors and aggregates them to generate the higher level business events required for the process. The benefit of this approach is that the BPMS does not have to deal with the management of subscriptions, which is a challenging task due to the heterogeneous technologies used by different event sources. This approach is also well suited for IoT scenarios, which include many, heterogeneous sensors that frequently send events, for which a reaction is not always required. For example in a *smart home* scenario, only if five consecutive thermostat events report the temperature to be above a certain threshold, should the air conditioner be turned on.

Case Management. Case management has been proposed to support flexible and knowledge-intensive business processes [19] that cannot be represented well using standard approaches like BPMN [16] and traditional process engines. We focus the discussion on the Chimera approach proposed in [6]. The Chimera approach captures business scenarios in a *case model* that consists of (a) a domain model, (b) a set of object lifecycles, (c) a set of process fragments, and (d) a goal state, also called termination condition. During runtime a case model is instantiated into a *case* that at any time is in a certain *case state*, which changes through knowledge workers performing activities, but also due to external events. Cases are similar to process instances in traditional workflow systems, however, contrary to those, cases can contain several concurrently running fragment instances, as well as data objects. Each fragment, just like a BPMN process model, consists of event, gateway, activity, and data nodes, connected by sequence and data flow arcs. The *domain model* defines the business objects relevant for the scenario as a set of data classes and their relations. To each data class an *object lifecycle (OLC)* is associated specifying valid behavior of its instances, i.e. data objects.

Since IoT is about integrating the physical world into digital systems, exceptions and uncertainty are a significant part of IoT scenarios [20]. Depending on the occurrence of (sensor) events, as well as user decisions for some IoT scenarios, the execution path gradually emerges over time. Thus, predefined processes might not be a good match for them. Instead, process fragments as defined in the Chimera approach can be used to represent possible execution variants, that are triggered by certain events.

When considering the high degree of uncertainty involved in some of the use cases, it is sensible to include case management in more advanced versions that go beyond plain tracking in scenarios like *freight tracking* or *crop monitoring*. If the distributed participants run their internal processes in individual process engines, a case management approach can be used to have an overview of the whole scenario.

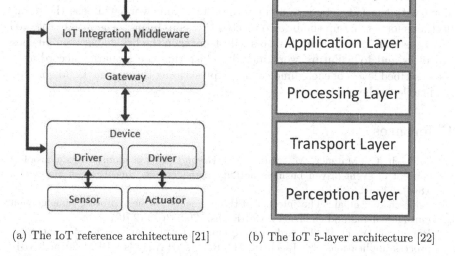

(a) The IoT reference architecture [21]   (b) The IoT 5-layer architecture [22]

Fig. 1. The IoT architecture

The IoT Architecture. There are many variations of the architectural layers, the components and the interactions among them. In [21], the authors compared different architectures for several IoT applications and came up with a reference architecture, shown in Fig. 1a. The architecture contains the drivers where sensors and actuators are embedded, the gateway, the middleware where the processing logic is executed and the application that uses the processed information. The architecture (shown in Fig. 1b) proposed in [22] gives a similar overview of the IoT layers. The perception layer includes sensors and actuators whereas the transport layer can be mapped to the gateway. The additional business layer here takes care of the ownership and is responsible for the application management. Based on the IoT application scenario and required participants; the components, the layers or the interfaces can change. However, if the BPM concepts are to be applied in an IoT scenario, the processing layer can be mapped to the CEP engine and the application layer can be mapped to the BPMS.

5   Conclusion and Future Work

The explosion of IoT devices have been significant in past two decades. This new technology tries to digitize the physical world with the help of sensors and actuators embedded in the things around us. These things talk to each other and are controlled by one or more logical unit. New business scenarios emerged due to IoT can benefit from the existing concepts from the area of business process management. However, to implement the suitable BPM concepts to IoT, first it is needed to realize the IoT scenarios better. This work provides an elaborate analysis of the IoT scenarios implemented currently and going to be implemented in near future. The classification framework covers the dimensions to be considered while analyzing or realizing an IoT setup.

To this end, BPM concepts that might be beneficial for IoT are discussed along with the IoT reference architecture. The framework will ease the design decisions for setting up the interconnected things in future. The insight into IoT scenarios with its important features will strengthen the bridge between business processes and IoT. Future work includes clustering the scenarios according to the described levels of each dimension and prescribing the specific BPM concepts applicable to the clusters.

References

1. Janiesch, C., Matzner, M., Müller, O.: Beyond process monitoring: a proof-of-concept of event-driven business activity management. Bus. Process Manage. J. **18**(4), 625–643 (2012)
2. Janiesch, C., et al.: The internet-of-things meets business process management: mutual benefits and challenges. CoRR abs/1709.03628 (2017)
3. Baumgraß, A., et al.: Towards a methodology for the engineering of event-driven process applications. In: Reichert, M., Reijers, H.A. (eds.) BPM 2015. LNBIP, vol. 256, pp. 501–514. Springer, Cham (2016). https://doi.org/10.1007/978-3-319-42887-1_40
4. Soffer, P., et al.: From event streams to process models and back: challenges and opportunities. Inf. Syst. (2017)
5. Weske, M.: Business Process Management - Concepts, Languages, Architectures, 2nd edn. Springer, Heidelberg (2012). https://doi.org/10.1007/978-3-642-28616-2
6. Hewelt, M., Weske, M.: A hybrid approach for flexible case modeling and execution. In: La Rosa, M., Loos, P., Pastor, O. (eds.) BPM 2016. LNBIP, vol. 260, pp. 38–54. Springer, Cham (2016). https://doi.org/10.1007/978-3-319-45468-9_3
7. Ashton, K.: That 'internet of things' thing. RFID J. **22**, 97–114 (2009)
8. Batiz-Lazo, B.: A brief history of the ATM, March 2015
9. Marr, B.: 17 "internet of things" facts everyone should read, October 2015
10. Chapman, L.: Happy 40th birthday to the modern day cash machine, December 2012
11. Stackpole, B.: IoT-enabled product as a service could transform manufacturing (2015)
12. Mattern, F., Floerkemeier, C.: From Active Data Management to Event-Based Systems and More, pp. 242–259. Springer, Heidelberg (2010). https://doi.org/10.1007/978-3-642-17226-7

13. BMBF: Germany: Industrie 4.0. (2017)
14. Hu, J., Aghakhani, G., Hasić, F., Serral, E.: An evaluation framework for design-time context-adaptation of process modelling languages. In: Poels, G., Gailly, F., Serral Asensio, E., Snoeck, M. (eds.) PoEM 2017. LNBIP, vol. 305, pp. 112–125. Springer, Cham (2017). https://doi.org/10.1007/978-3-319-70241-4_8
15. Meyer, S., Ruppen, A., Magerkurth, C.: Internet of things-aware process modeling: integrating IoT devices as business process resources. In: Salinesi, C., Norrie, M.C., Pastor, Ó. (eds.) CAiSE 2013. LNCS, vol. 7908, pp. 84–98. Springer, Heidelberg (2013). https://doi.org/10.1007/978-3-642-38709-8_6
16. OMG: Business process model and notation (BPMN), Version 2.0., January 2011. http://www.omg.org/spec/BPMN/2.0/
17. Etzion, O., Niblett, P.: Event Processing in Action. Manning Publications, Shelter Island (2010)
18. Mandal, S., Hewelt, M., Weske, M.: A framework for integrating real-world events and business processes in an IoT environment. In: Panetto, H., et al. (eds.) OTM 2017. LNCS, vol. 10573, pp. 194–212. Springer, Heidelberg (2017). https://doi.org/10.1007/978-3-319-69462-7_13
19. Marin, M.A., Hauder, M., Matthes, F.: Case management: an evaluation of existing approaches for knowledge-intensive processes. In: Reichert, M., Reijers, H.A. (eds.) BPM 2015. LNBIP, vol. 256, pp. 5–16. Springer, Cham (2016). https://doi.org/10.1007/978-3-319-42887-1_1
20. Marrella, A., Mecella, M.: Adaptive process management in cyber-physical domains. In: Grambow, G., Oberhauser, R., Reichert, M. (eds.) Advances in Intelligent Process-Aware Information Systems. ISRL, vol. 123, pp. 15–48. Springer, Cham (2017). https://doi.org/10.1007/978-3-319-52181-7_2
21. Guth, J., Breitenbücher, U., Falkenthal, M., Leymann, F., Reinfurt, L.: Comparison of IoT platform architectures: a field study based on a reference architecture. In: 2016 Cloudification of the Internet of Things (CIoT), pp. 1–6. IEEE, November 2016
22. Wu, M., Lu, T.J., Ling, F.Y., Sun, J., Du, H.Y.: Research on the architecture of Internet of Things. In: 2010 3rd International Conference on Advanced Computer Theory and Engineering (ICACTE), vol. 5, pp. V5-484–V5-487, August 2010

First Declarative/Decision/Hybrid Mining and Modeling for Business Processes (DeHMiMoP)

First Declarative/Decision/Hybrid Mining and Modeling for Business Processes (DeHMiMoP)

The recent years have witnessed a rising interest of the business process management community in the investigation of the decisional and normative aspects of workflows. Processes and business process models involve rules and decisions describing the premises and possible outcomes of specific situations. However, important though they are, rules and decisions are often hidden in process flows, process activities, or in the head of employees (tacit knowledge), so that they need to be discovered using state-of-art intelligent techniques. For knowledge-intensive processes, it is common that rules and decisions, as opposed to the process-flow, define the allowed behavior of a process. To cater for it, a new declarative modeling paradigm has been proposed that aims to directly capture the business rules or constraints underlying the process. The approach is recently attracting considerable interest, and several declarative notations have been developed such as declare, dynamic condition response (DCR) graphs, decision modeling and notation (DMN), case management model and notation (CMMN), guard-stage-milestone (GSM), and extended compliance rule graphs (eCRG). Lately, there has been a rapidly growing interest in hybrid approaches, which combine the strengths of different modeling paradigms. The main focus of the workshop is thus on the application and challenges of decision- and rule-based modeling in all phases of the BPM lifecycle (identification, discovery, analysis, redesign, implementation, and monitoring).

The sixth edition of the Workshop on Declarative/Decision/Hybrid Mining and Modeling for Business Processes (DeHMiMoP 2018) began with the keynote of Rik Eshuis, entitled "Modeling Decision-Intensive Processes with Declarative Business Artifacts." Starting with a discussion centered around the need of modern business processes to support knowledge workers in making decisions, he showed how declarative business artifacts can be a promising ingredient to support such decision-intensive processes. His inspirational talk illustrated some challenges to be tackled to that end, and promising solutions to overcome them.

Three full papers and two short papers were subsequently presented at the workshop. The reported research endeavors spanned over all the topics of the workshop, as they focussed on declarative process models, declarative specifications as a means to generate imperative models, decision-aware processes, and decision models.

In particular, Andaloussi et al. investigate in how far the understandability of DCR Graphs can be improved through the combined adoption of different visual artifacts. Their paper entitled "Evaluating the Understandability of Hybrid Process Model Representations Using Eye Tracking: First Insights" also shows the initial results of their experiments conducted with users. In "Generating Decision-Aware Models and Logs: Towards an Evaluation of Decision Mining" by Jouck et al., a novel simulation-based framework is proposed that allows for the comparison of decision-mining techniques. The software tool that implements their framework is presented in the paper as well. Haarmann et al. describe their approach and software prototype to check compliance rules expressed in temporal logic against process models endowed with decision models. Their solution is described in the paper entitled "Compliance Checking for Decision-Aware Process Models." The generation of imperative process models based on a declarative partial specification of their activities, input and output data dependencies, and process states is the aim of the approach proposed by Wiśniewski et al. Their paper "Towards Automated Process Modeling Based on BPMN Diagram Composition" describes their approach and discusses its application in the context of process modeling and mining. In "Measuring the Complexity of DMN Decision Models," Hasić et al. propose a set of metrics to assess the complexity of DMN models. Their paper also presents the results of an empirical evaluation conducted as an exploratory survey.

November 2018

Claudio Di Ciccio
Søren Debois
Dennis Schunselaar
Jan Vanthienen Tijs Slaats

Organization

Program Committee

Marco Montali   Free University of Bozen-Bolzano, Italy
Jorge Munoz-Gama   Pontificia Universidad Católica de Chile, Chile
Artem Polyvyanyy   Queensland University of Technology, Australia
Hajo A. Reijers   Vrije Universiteit Amsterdam, The Netherlands
Stefan Schönig   University of Bayreuth, Germany
Lucinéia H. Thom   Universidade Federal do Rio Grande do Sul, Brazil
Han van der Aa   Vrije Universiteit Amsterdam, The Netherlands
Wil M. P. van der Aalst RWTH Aachen University, Germany
Barbara Weber   Technical University of Denmark, Denmark
Matthias Weidlich   Humboldt University of Berlin, Germany
Mathias Weske   Hasso Plattner Institute, University of Potsdam,
                        Germany

Evaluating the Understandability of Hybrid Process Model Representations Using Eye Tracking: First Insights

Amine Abbad Andaloussi[1](✉), Tijs Slaats[2](✉), Andrea Burattin[1](✉),
Thomas T. Hildebrandt[2](✉), and Barbara Weber[1](✉)

[1] Software and Process Engineering, Technical University of Denmark,
2800 Kongens Lyngby, Denmark
{amab,andbur,bweb}@dtu.dk
[2] Department of Computer Science, University of Copenhagen,
2100 Copenhagen, Denmark
{slaats,hilde}@di.ku.dk

Abstract. The EcoKnow project strives to promote flexible case management systems in the public administration and empower end-users (i.e., case workers) to make sense of digitized models of the law. For this, a hybrid representation combining the declarative DCR notation with textual annotations depicting the law text and a simulation tool to simulate the execution of single process instances was proposed. This hybrid representation aims to overcome the notorious limitations of existing declarative notations in term of understandability. Using eye tracking, this paper investigates how users engage with the different artifacts of the hybrid representation.

1 Introduction

The Ecoknow project aims at integrating hybrid technologies in public administration as part of the effective digitization of knowledge work processes. In this context, a hybrid representation combining the declarative DCR (Dynamic Condition Response) notation [1] with textual annotations depicting the law and a simulation tool allowing to simulate the possible process executions was proposed (cf. Fig. 1). While the use of declarative notations (i.e., DCR graphs) enables flexibility and higher adaptability, their understandability is controversial especially with regards to novice users [3]. This hybrid representation (called "hybrid DCR representation") aims to overcome the understandability limitation by offering a multi-artifact representation to help end-users to make sense of digitized models of the law.

Work supported by the Innovation Fund Denmark project *EcoKnow* (7050-00034A); the second author additionally by the Danish Council for Independent Research project *Hybrid Business Process Management Technologies* (DFF-6111-00337).

F. Daniel et al. (Eds.): BPM 2018 Workshops, LNBIP 342, pp. 475–481, 2019.
https://doi.org/10.1007/978-3-030-11641-5_37

As part of an exploratory study investigating the understandability of the hybrid DCR representation, this paper uses eye tracking to provide some first insights about the way end-users engage with the different artifacts of this hybrid representation. The outcome of this study will provide insights into the use of the DCR platform. Section 2 presents briefly the related work. Section 3 describes the research method pursued to plan and conduct the exploratory study. Section 4 unveils some first insights into the study results, and finally Sect. 5 highlights the future work and concludes the paper.

2   Related Work

The literature about hybrid process model representations can be categorized into the following groups: *(a)* The first group comprises hybrid representations that combine two or more notations into a single artifact. For instance, in [2] and [7] the authors combine imperative and declarative notations into a single process model. This part of the literature is beyond the scope of this paper. Instead, this work focuses on a hybrid representation that combines several artifacts. *(b)* The second group considers approaches that combine a graphical model with textual annotations. For instance, Pinggera et al. in [5] proposed the Literate Process Modeling technique (LiProMo). Inspired by the dual coding theory, LiProMo aims at fostering the communication at the process modeling stage by fusing textual descriptions with a process model. Likewise, Wang et al. in [8] proposed a hybrid representation that combines a process model with linked rules expressed as textual annotations. Both approaches demonstrated higher comprehension accuracy and lower mental effort. *(c)* The third group of the literature comprises hybrid representations that combine a graphical model with a tool that allows the execution of single traces. For instance, Zugal et al. in [10] proposed a test driven approach to support modelers in maintaining declarative process models. Hereafter, the authors introduced a tool that allows to test whether the process model complies with a set pre-defined (positive and negative) test cases. The evaluation (cf. [9]) demonstrated an increased process maintainability with reduced mental effort. Similarly, this work combines a graphical model in DCR notation with textual annotations referring to the law. In addition, the scrutinized hybrid representation includes a simulation tool that allows to simulate the execution of single traces and perceive the allowed behaviour.

3   Research Method

To investigate the factors affecting the understandability of the hybrid DCR representation, we have planned and conducted an eye tracking experiment. This section highlights the key aspects considered in the design phase of the experiment and provides insights into the measurements deployed in the analysis.

Research Question. To obtain a better understanding about the way the hybrid DCR representation is used, we formulate the following research question: **"How end-users engage with the different artifacts proposed by the hybrid DCR representation?"** To answer to this question we analyze the following: *(a)* the distribution of attention between the different artifacts, and *(b)* the common reading patterns seen in different groups of end-users.

Subjects and Objects. The end-users (called "subjects" in an experimental context) who took part in this study have varying levels of expertise in using DCR. They have been recruited among case-workers from Syydjurs municipality in Denmark and students/employees from Technical University of Denmark (DTU) and IT university of Copenhagen (ITU). The DCR model used in this study originates from Section §45 of the "Consolidation Act on Social Services"[1].

Design. The experiment begins with a brief training session where all the subjects receive basic guidance about the hybrid DCR representation. Throughout the experiment eight comprehension tasks are displayed sequentially. The tasks evaluate the subjects' capacity to understand the semantics of the DCR graph as well as engage them to read the law fragments and to use the simulation. After each comprehension task, a set of questions investigating the artifacts used, the cognitive load and the subject's emotional state are prompted. By the end of the experiment, a think aloud session is held. Finally, a post-experiment questionnaire is used to collect data about the subjects' demographics, domain knowledge and experience[2]. For the sake of brevity, this paper will put emphasis only on the gaze data collected from the eye tracker, and the artifacts data. The rest of the data will be investigated in upcoming work. The experiment material used for this paper is available online at http://andaloussi.org/papers/DeHMiMoP2018/Material.pdf.

Measures. To investigate the distribution of attention between the different artifacts, we have used the following fixation-derived measures [6]: *(a) fixation count* which quantifies the number of fixations on a specific area of the stimulus [4, pp. 412–415], and *(b) total fixation duration* which sums up the duration of all the fixations on a specific area of the stimulus [4, pp. 377–386]. Indeed, as shown in Fig. 1, the stimulus can be divided into several areas called "Areas Of Interest" (AOIs), such that each AOI corresponds to a different artifact (DCR graph, law text, simulation). Finally, we have compared the distribution of attention with the subjective artifact data used by the subjects to answer the comprehension tasks.

---

[1] http://english.sm.dk/media/14900/consolidation-act-on-social-services.pdf (Eng), https://www.retsinformation.dk/Forms/R0710.aspx?id=197036 (Dan).

[2] The post-experiment data is available online at http://andaloussi.org/papers/DeHMiMoP2018/demographicsAndBackground.xlsx.

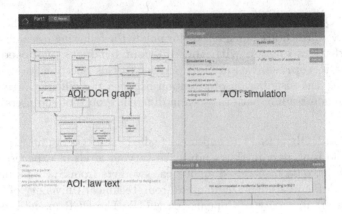

Fig. 1. A view showing the hybrid DCR representation. At the analysis phase, this view is split into 3 areas of interest, each representing a distinct artifact.

To investigate the common reading patterns, we have developed a new approach to analyze the transitions between the different AOIs. Given a timestamped log file containing the sequence of fixations and their corresponding AOIs, we have generated an event log and used the process mining tool Disco[3] to discover and analyze the underlying reading pattern. More insights about the experimental settings and the measures are demonstrated in the video available online at https://youtu.be/8OsY9PYAs3I.

4   Early Results

This section provides insights about the understandability of the hybrid DCR representation. Section 4.1 investigates the distribution of attention, and Sect. 4.2 analyzes the different reading patterns.

4.1   Distribution of Attention

The data used for the analysis and the tables described in this section are available in an online spreadsheet at http://andaloussi.org/papers/DeHMiMoP2018/Analysis.xls. By looking at the total fixation duration for all participants, we have noticed that the DCR graph was the most focused artifact (duration: 881.767 s, proportion: 0.511), followed by the law text (duration: 443.251 s, proportion: 0.257), then the simulation (duration: 398.954 s, proportion: 0.231). Comparing these values with the fixation count for all participants, we have noticed a similar distribution of attention with almost similar proportions. The same observation holds with the subjective artifact data. Overall, this comparison between the artifacts shows that the graph caught most of the subjects' attention. However,

---

[3] Available online at https://fluxicon.com/disco/.

this observation does not provide enough insights into the usability of the law text and the simulation, since the different AOIs differ in size and content.

As an alternative, we have compared the subjects based on their proportions of fixation duration and fixation count on each artifact (cf. AOI artifacts Sheet). Hereafter, we have observed the following: *(a)* subjects with highest fixation duration and fixation count proportions on the graph (P03, P07, P09 and P10), *(b)* subjects with highest fixation duration and fixation count proportions on the law text (P01 and P05), and *(c)* subjects with highest fixation duration and fixation count proportions on the simulation (P02, P06 and P08). These observations allow distinguishing between three user profiles with varying artifacts preference (DCR graph, law text, simulation), which lead to the question: "Do the three user profiles exhibit different reading patterns?"

4.2 Reading Patterns

To investigate the reading patterns for the different user profiles, we have analyzed the transitions between the different AOIs using the process mining tool Disco. For sake of brevity, this analysis was conducted on a single question (Question 3, cf. online material). Figure 2 depicts the reading patterns for the different profiles. The activities in the model represent the different AOIs of the stimulus, the digits on the activities refer to the absolute frequency of AOI visits, the arcs between the activities indicate the transitions between the AOIs, and the digits on the arcs count the number of transitions. The arcs looping around the same activity refer to intermediate transitions to areas that are not relevant to our analysis.

Figure 2a shows that the graph profile has high abs. visit frequency on the graph and few transitions from the graph to the law text and simulation. By looking at the gaze video recordings, we have noticed that the graph profile subjects were constantly checking the graph semantics to identify the answering clues. However, since the graph is not providing clear answering clues to all the questions, some subjects have used the law text and the simulation to complement their understanding. Figure 2b shows that the simulation profile has balanced abs. visit frequency on the graph and the simulation, and more transitions between the graph and simulation compared to the transitions between the graph and the law text. By looking at the gaze video recordings, we have noticed that the simulation profile subjects were constantly checking the DCR relations surrounding the activities targeted by the simulation, which explains the high frequency of transitions between the graph and simulation AOIs. Finally, Fig. 2c shows that the law text profile has two thirds of the abs. visit frequency on the graph while the remaining is on the law text. By looking at the gaze video recordings, we have noticed that the subjects from this profile have not used the simulation, thus, the fixations on the simulation AOI for this profile were ignored. Hereafter, frequent transitions were between the graph and the law text. This observations demonstrate that each user profile has a distinct reading pattern.

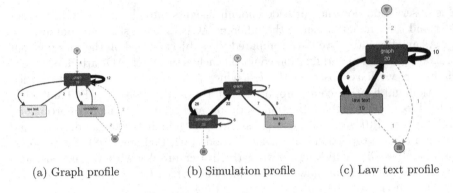

(a) Graph profile   (b) Simulation profile   (c) Law text profile

Fig. 2. Reading patterns for the different user profiles. Higher resolution at http://andaloussi.org/papers/DeHMiMoP2018/userprofiles.pdf

5   Conclusion and Future Work

This paper describes the approach used to investigate the understandability of the hybrid DCR representation. The analysis depicts the distribution of attention between the different artifacts, and shows that each user profile exhibits a different reading pattern. It has to be noted that the reported results are subject to limitations due to the small number of subjects, and the use of domain knowledge by some subjects (i.e., case workers from the municipality) to answer to some of the experiment questions.

As future work, the user interactions and the verbal data transcribed from the think aloud will be used to explain the transitions between the different artifacts, and to identify the circumstance of using each of them. Moreover, the verbal data will be analyzed to spot the typical challenges faced by the subjects, and the questionnaire data will be used to measure the comprehension accuracy. Finally, the subjects' cognitive load and emotional reactions will be examined using the pupil data and the Galvanic Skin Response (GSR) data.

References

1. De Smedt, J., De Weerdt, J., Serral, E., Vanthienen, J.: Improving understandability of declarative process models by revealing hidden dependencies. In: Nurcan, S., Soffer, P., Bajec, M., Eder, J. (eds.) CAiSE 2016. LNCS, vol. 9694, pp. 83–98. Springer, Cham (2016). https://doi.org/10.1007/978-3-319-39696-5_6
2. De Smedt, J., De Weerdt, J., Vanthienen, J., Poels, G.: Mixed-paradigm process modeling with intertwined state spaces. Bus. Inf. Syst. Eng. **58**(1), 19–29 (2016)
3. Fahland, D., et al.: Declarative versus imperative process modeling languages: the issue of understandability. In: Halpin, T., et al. (eds.) BPMDS/EMMSAD 2009. LNBIP, vol. 29, pp. 353–366. Springer, Heidelberg (2009). https://doi.org/10.1007/978-3-642-01862-6_29

4. Holmqvist, K., Nyström, M., Andersson, R., Dewhurst, R., Jarodzka, H., van de Weijer, J.: Eye Tracking: A Comprehensive Guide to Methods and Measures. OUP, Oxford (2011)
5. Pinggera, J., Porcham, T., Zugal, S., Weber, B.: LiProMo-literate process modeling. In: CEUR Workshop Proceedings, vol. 855, pp. 163–170 (2012)
6. Poole, A., Ball, L.J.: Eye tracking in human-computer interaction and usability research: current status and future. In: Encyclopedia of Human-Computer Interaction. Idea Group Inc., Pennsylvania (2005)
7. Slaats, T., Schunselaar, D.M.M., Maggi, F.M., Reijers, H.A.: The semantics of hybrid process models. In: Debruyne, C., et al. (eds.) OTM 2016. LNCS, vol. 10033, pp. 531–551. Springer, Cham (2016). https://doi.org/10.1007/978-3-319-48472-3_32
8. Wang, W., Indulska, M., Sadiq, S., Weber, B.: Effect of linked rules on business process model understanding. In: Carmona, J., Engels, G., Kumar, A. (eds.) BPM 2017. LNCS, vol. 10445, pp. 200–215. Springer, Cham (2017). https://doi.org/10.1007/978-3-319-65000-5_12
9. Zugal, S., Pinggera, J., Weber, B.: The impact of testcases on the maintainability of declarative process models. In: Halpin, T., et al. (eds.) BPMDS/EMMSAD 2011. LNBIP, vol. 81, pp. 163–177. Springer, Heidelberg (2011). https://doi.org/10.1007/978-3-642-21759-3_12
10. Zugal, S., Pinggera, J., Weber, B.: Creating declarative process models using test driven modeling suite. In: Nurcan, S. (ed.) CAiSE Forum 2011. LNBIP, vol. 107, pp. 16–32. Springer, Heidelberg (2012). https://doi.org/10.1007/978-3-642-29749-6_2

A Framework to Evaluate and Compare Decision-Mining Techniques

Toon Jouck[1]([⊠]), Massimiliano de Leoni[2], and Benoît Depaire[1]

[1] UHasselt - Hasselt University, Hasselt, Belgium
{toon.jouck,benoit.depaire}@uhasselt.be
[2] Eindhoven University of Technology, Eindhoven, The Netherlands
m.d.leoni@tue.nl

Abstract. During the last decade several decision mining techniques have been developed to discover the decision perspective of a process from an event log. The increasing number of decision mining techniques raises the importance of evaluating the quality of the discovered decision models and/or decision logic. Currently, the evaluations are limited because of the small amount of available event logs with decision information. To alleviate this limitation, this paper introduces the 'DataExtend' technique that allows evaluating and comparing decision-mining techniques with each other, using a sufficient number of event logs and process models to generate evaluation results that are statistically significant. This paper also reports on an initial evaluation using 'DataExtend' that involves two techniques to discover decisions, whose results illustrate that the approach can serve the purpose.

Keywords: Decision mining · Evaluation · Log generation

1 Introduction

Automated process discovery from event logs has mainly focused on the control-flow perspective of processes. The control-flow perspective can be considered as the process backbone; however, many other perspectives should also be considered to ensure that the model is sufficiently accurate. The decision perspective (a.k.a. the data or case perspective) focuses on how the routing of process-instance executions is affected by the characteristics of the specific process instance, such as the amount requested for a loan, and by the outcomes of previous execution steps, e.g. the verification result. The representation of this decision perspective on a process in an integrated model or as separate tables is nowadays gaining momentum due to the introduction of the Decision Model and Notation (DMN) standard [1]. Decision mining focuses on discovering the decision perspective of a process from an event log. Some techniques (e.g. [2,3]) augment the routing decisions of a control-flow model with the decision logic induced from the data attributes in the event log. Other techniques (e.g. [4,5]) have focused on discovering a decision model in the form of a Decision Requirements Diagram.

© Springer Nature Switzerland AG 2019
F. Daniel et al. (Eds.): BPM 2018 Workshops, LNBIP 342, pp. 482–493, 2019.
https://doi.org/10.1007/978-3-030-11641-5_38

The increasing number of decision mining techniques raises the importance of evaluating the quality of the discovered decision models and/or decision logic. Currently, no standard evaluation framework has been proposed in literature. The techniques presented in [2,4,5] have been evaluated informally: by showing it can rediscover some example routing decision logic [2], an example decision model [4] or by demonstrating it on a real-life data set [5]. In contrast, the techniques of [3,6] have been formally evaluated, yet they have applied their techniques on a small number of data sets. Due to their small scale, no statistical tests can be applied to determine the significance of the results.

Current decision mining evaluations have used only 5 of the publicly available repository of event logs[1]. Moreover, the repository does not contain a reference process and decision model and thus the characteristics of the real underlying process are unknown. As a consequence, the evaluations cannot generalize the results to conclude that technique A on average outperforms technique B when confronted with certain process and decision characteristics. However, this is necessary to gain insights in the strengths and weaknesses of the existing decision mining techniques.

This paper introduces a framework to evaluate and compare decision mining techniques with some guarantee that the results are statistically significant, and, hence, generally valid. The framework, which is presented in Sect. 2, extends the existing method in [7] to generate random artificial control-flow models with a decision dimension and simulates those into event logs. It allows users to control for the process and decision characteristics of the generated event logs as needed during evaluation. An initial validation was conducted on two techniques in Sect. 3. Related work is discussed in Sects. 4 and 5 summarizes the paper with conclusions and future work.

2   Evaluation Framework

Decision mining algorithms discover the decision perspective of a process based on an event log that contains both control-flow and decision perspectives. Artificially generating such event logs is challenging. It requires to generate an artificial sound process model and a decision model with decision rules (i.e. data dependencies) first and then simulate the process model and rules into a multiperspective event log. Soundness is a necessary condition for the generated models with rules as otherwise runtime errors can occur during simulation [8]:

- Deadlock: a case gets stuck in the middle of the process where it is not possible to execute any activities.
- Dead activity: an activity part of the process can never be executed for any case.

---

[1] https://data.4tu.nl/repository/collection:event_logs_real.

Simply adding decision rules to random process models generated by existing control-flow based techniques, e.g. [7], would not guarantee soundness. On the other hand, the only existing method for generating multiperspective models and logs [9] is not tailored for adding decision information to the control-flow model (see Sect. 4). Therefore, this paper introduces the 'DataExtend' approach for generating artificial event logs with both control-flow and decision perspectives. In the next part of this Sect. 2.1 we will present the idea behind the 'DataExtend' using an example. The final part of this Sect. 2.2 formally describes the steps of 'DataExtend'.

2.1   Illustration of Generating Multiperspective Logs

The make-to-order process as illustrated in Fig. 1 will be used as an example throughout the rest of the paper. The process handles the production of a customer order: it starts with issuing the customer order, then materials are prepared, the products are produced, possibly followed by an inspection, then products are packaged, and finally, the products are delivered or the order is canceled when something went wrong. It contains three XOR-splits (indicated in the figure as 'choice x') where choices between multiple activities need to be made:

- the first choice is whether to use new materials or mixed (recycled and new) materials,
- the second choice is about the inspection of the produced products: no inspection, a normal inspection or a thorough inspection,
- the third choice is whether the products will be delivered or canceled.

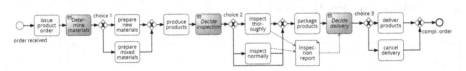

Fig. 1. Make-to-order process

The process model in Fig. 1, without the tasks in grey and the data object, presents the control-flow perspective of the process, i.e. it does not contain information on the decision perspective of the process. In this paper we assume that the decision perspective consists of a decision model and rules that explain the choices in the process. From DMN [1] we adopt the DRD as a decision model that visualizes the dependencies between decisions and the inputs (here case attributes). This paper assumes that each exclusive choice (XOR-split) in the process is preceded by a decision that is modelled in Fig. 1 using a business rule task: 'Determine materials' (choice 1), 'Decide inspection' (choice 2), and

'Decide delivery' (choice 3). Each decision in the DRD corresponds to one business rule task in the process model. For each of the decisions in the DRD we can specify the logic as rules in a decision table.

In the example, the 'Determine materials' (see Fig. 1) decision cannot depend on an earlier made decisions in the process as it is the first decision in the process. It can depend on case attributes, but it is not necessary. Suppose that in this process the decision between new and mixed materials relies upon some contextual information not embedded in the underlying information system. In that case we do not generate decision rules but rather represent the decision stochastically by assigning a probability of choosing each decision output: on average for 50% of the orders the activity 'prepare new materials' is executed and for 50% of the orders the activity 'prepare mixed materials' is performed.

The 'Decide inspection' decision (see Fig. 1) can depend on the decision about the used materials and case attributes. In this example the inspection decision depends on the outcome of the first decision and a case attribute 'premium' which is related to the type of the customer placing the order. This results in the DRD in Fig. 2 where the 'Determine matrials' decision and the customer type are inputs of the inspection decision. The policy is that products produced with mixed materials always need to be inspected thoroughly regardless of what the customer type is. Products consisting of new materials are only inspected for premium customers, otherwise the inspection is skipped to save costs. Such decision logic can be represented as rules as illustrated in Table 1.

Table 1. Decision tables for 'Decide inspection' (left) and 'Decide delivery' (right)

Rule	Materials	Premium?	Inspection
1	new material	True	inspect normally
2	new material	False	(skip inspection)
3	mixed material	True	inspect thoroughly
4	mixed material	False	inspect thoroughly

Rule	OK quality?	Delivery
1	True	deliver
2	False	cancel

Finally, the 'Decide delivery' decision could depend on the outcome of the first and second decision and some case attribute(s). Suppose that the inspection results in an inspection report (the data object in Fig. 1) based on which the quality of the products is labeled as acceptable or non-acceptable. If an inspection was skipped, acceptable quality of the products is assumed. A delivery will only be executed if the quality of the products are acceptable, otherwise the order is cancelled. These decision dependencies can be illustrated in the DRD in Fig. 2 and the decision logic as shown in Table 1.

The control-flow model together with the above decision model and rules can then be simulated into an event log. The simulator evaluates the rules tied to each decision in the process in order to decide which outcome, i.e. choice branch, to activate. An example case is shown in Table 2. Notice that we only displayed the inputs of each decision as case attributes to save space.

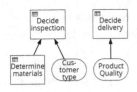

Fig. 2. Decision requirements diagram of example process

Table 2. Example case of the produce order process

Event ID	Activity	Materials	OK quality?	Premium?
1	issue			True
2	prep. mixed mater.	mixed material		True
3	produce	mixed material		True
4	inspect thoroughly	mixed material	True	True
5	package	mixed material	True	True
6	deliver	mixed material	True	True

2.2 Formal Method for Generating Multiperspective Logs

This subsection formalizes the steps of the 'DataExtend' method we propose to generate an event log containing both control-flow and decision perspectives. 'DataExtend' involves the following five steps:

1. Generate a random control-flow model from a population of models
2. Randomly build a decision model
3. Randomly generate decision logic for decisions in the decision model
4. Simulate control-flow model with decision model and logic into event log

Step 1: Generating Random Control-Flow Models. For this step we rely on the existing method in [7] to generate a random control-flow model from a process model population. This method always generates block-structured models that are guaranteed to be sound. A block-structured model can be decomposed in properly nested subprocesses such that each subprocess has a single entry and singly exit point, e.g. the model in Fig. 1. The population definition describes the behavioral characteristics of the models: the size and the control-flow patterns of the models. The user defines a population of models by setting the following parameters:

- Model Size Parameters = {mode, min, max}.
- Control-flow Characteristic Probabilities = {sequence (WCP-1), parallelism (WCP-2/3), exclusive choice (WCP-4/5), multi-choice loop (WCP-6/7), loop (WCP-21), silent activity, reoccurring activity, infrequent path}.

The model size parameters define a triangular distribution that will be used to determine the number of activities that will be present in the model. The WCP-x between brackets refers to the standard control-flow workflow patterns as defined by Russell [10]. A silent activity is used for modelling a skip (e.g. the third branch of the second choice in Fig. 1). Reoccurring activities allow the same activity to appear in different parts of the process. Infrequent paths make some outgoing branches of an exclusive choice more likely to occur than others. The control-flow characteristic probabilities influence the probability for each characteristic to be included in the resulting model. For example, if the probability of an exclusive choice is 0.2, then on average 20% of the nested subprocesses will be exclusive choices.

Step 2: Randomly Build Decision Model. This step will initiate the decision perspective that is added on top of the generated control-flow model. This paper assumes that the routing of the cases throughout the process depends on decision outcomes. This corresponds to adding the business rule tasks before each exclusive choice in the model of the example in Fig. 1, e.g., the outcome for 'Determine materials' influences whether activity 'prepare new materials' or 'prepare mixed materials' is executed for a particular case. Adding these decisions initiates the Decision Requirement Diagram that will contain a decision for each exclusive choice in the model. As such, we will not add decisions before multi-choice and loop constructs in this paper. In a next step, we randomly determine the inputs of the decisions in the DRD. Here we assume that a decision can depend on a case attribute or a previously made decision.

'DataExtend' generates a decision $d_i \in D$ for each exclusive choice in the generated control-flow model. Then, it assigns zero or more attributes as inputs, either a case attribute or a previous decision, to each decision. A decision without attributes means that it is based on some information not embedded in the information system (e.g. the 'Determine materials' decision in the above example).

Definition 1 (Assign). *Given a set D of decisions and a set V of case attributes (including previous decisions), Assign: $D \mapsto \mathbb{P}(V)$ is a function that labels each decision d_i with a set $V' \subseteq V$ of attributes which d_i is based upon.*

The attribues $\text{Assign}(d_i)$ used in decision d_i can take on values on the basis of decisions that 'precede' d_i:

Definition 2 (Precedence). *Precedence: $D \mapsto \mathbb{P}(D)$ is a function that labels each decision with a set of preceding decisions. The precedence is based on the control-flow semantics of the model.*

Consider again the example in Sect. 2.1. The 'Decide inspection' decision (d_2) is preceded by the 'Determine materials' decision (d_1): Precedence(d_2)$\mapsto \{d_1\}$. Then, in the example, the 'Decide inspection' decision is assigned the 'Determine materials' decision and 'premium' as attributes: Assign(d_2) $\mapsto \{d_1, premium\}$. The assigned attributes of each decision are visualized in the DRD as shown in Fig. 1 where 'Determine materials' and 'Customer type' are inputs of the 'Decide inspection' decision.

Step 3: Randomly Generate Decision Logic. This step will specify the decision logic for each decision in the decision model generated in the previous step. More specifically, each decision influences a choice between multiple alternative branches in the process model. The values of the assigned attributes of each decision restrict the possible branches that can be activated. These restrictions, also called decision dependencies, can be expressed as decision rules. A decision rule is defined as a mapping:

Definition 3 (Decision Rule). A decision rule is a mapping

$$V_1 \bowtie q_1, \ldots, V_w \bowtie q_w \mapsto \times_{jk}$$

where $V_i \in V$ is the set of attributes, $\bowtie$ is a relational operator $\in \{<, \leq, >, \geq, =, \neq\}$, $q_1, \ldots, q_w$ are constants, $\times_{jk}$ denotes outgoing branch k of choice $\times_j$ after decision j.

The set of all decision rules related to a routing decision can be represented as a decision table such as Table 1, where rule 1 of the left table expresses the decision rule: materials = 'new materials', premium? = 'True' $\mapsto$ 'inspect normally' (i.e. the second outgoing branch of the second decision 'Decide inspection').

'DataExtend' initially generates all possible decision rules, i.e. each possible combination of attribute values can lead to any of the outgoing branches.[2] For example, consider Table 3 that shows the initial set of decision rules for decision 'Decide inspection' in the make-to-order example process. When a case has the following attribute values: materials = 'new materials' and premium? = 'True', then the three outgoing choice branches containing activities 'inspect thoroughly', 'inspect normally', and skip inspection are all possible according to rules 1, 2 and 3.

Table 3. Example complete decision table for the 'Decide inspection' decision in the make-to-order example process.

Rule	Materials	Premium?	Inspection
1	new material	True	inspect thoroughly
2	new material	True	(skip inspection)
3	new material	True	inspect normally
4	new material	False	inspect thoroughly
5	new material	False	(skip inspection)
6	new material	False	inspect normally
7	mixed material	True	inspect thoroughly
8	mixed material	True	(skip inspection)
9	mixed material	True	inspect normally
10	mixed material	False	inspect thoroughly
11	mixed material	False	(skip inspection)
12	mixed material	False	inspect normally

Randomly removing rules from the initial set of decision rules restricts the decision outcomes, i.e. the possible outgoing branches at the choice impacted by each decision. In this way, 'DataExtend' creates decision dependencies. However, it cannot restrict the behavior too much as this could create unsound behavior in the form of deadlocks and dead parts which break the simulator in the next step. Therefore, the following soundness constraints are imposed on the rule removal step:

- each decision table has at least one rule for each possible outgoing choice branch to prevent *dead activities*
- each decision table has at least one rule for each value combination of the attributes values to prevent *deadlocks*.

Additionally, the user can set a stopping criterion for the removal of random decision rules. Without such a stopping criterion, 'DataExtend' will remove rules until no removal can happen without violating the soundness constraints. This results in fully deterministic decisions, i.e. for any combination of attribute values there is only one outgoing branch possible. However, business rules are often non-deterministic and this 'cannot be solved until the business rule is instantiated in a particular situation' [11]. This ambiguity can occur due to conflicting

---

[2] Impossible combinations happen when a decision depends on two other decisions that are mutually exclusive. Such combinations are removed from the decision table.

rules or missing contextual information. Therefore, the approach allows users to set a determinism level as stopping criterion. The determinism level is defined as the number of decision rules removed relative to the maximum amount of decision rules that could possibly be removed (without violating the soundness constraints). The maximum determinism level of 1 results in a fully deterministic decisions. The minimum value of 0 denotes the initial state, i.e. any combination of attribute values can lead to all possible decision outcomes. The user specifies the target determinism level, which is the average determinism level over all decisions *with* input attributes after the removal of rules. We explicitly leave out decisions without assigned attributes, e.g. 'Determine materials' decision in the make-to-order example, because these decisions always have a determinism level of 0, i.e. no rules can be removed, which makes it impossible to reach an average determinism level of 1.

Definition 4 (Determinism level). *Let d_i be a decision and $\#rule(d_i)$ be the number of rules in d_i. Let $\bar{d}_i$ be a decision obtained from d_i after removing a number of rules, then:*

$$DeterminismLevel(d_i, \bar{d}_i) = \frac{\#rule(d_i) - \#rule(\bar{d}_i)}{\#rule(d_i) - \#minimum(d_i)}$$

where $\#minimum(d_i)$ is the minimum number of rules to ensure soundness and is determined by taking the maximum of the number of possible decision outcomes for d_i and the number of attribute value combinations of $Assign(d_i)$.

In the make-to-order example (see Sect. 2.1) the desired determinism level is set to 1. This means that as much rules as possible have to be removed from the decision table of the 'Decide inspection' and 'Decide delivery' decisions. The initial decision table for 'Decide inspection' (see Table 3) contains 12 rules. The soundness constraints imply that at least one rule for each of the three possible decision outcomes should remain to avoid dead activities. Additionally, the soundness constraints require that the decision table should contain at least one rule for unique combination of case attribute values: {'prepare new', 'prepare mix'} × {'True', 'False'} = 4 thus $\#minimum = max(3, 4) = 4$. Removing rules 1, 2, 4, 6, 8, 9, 11 and 12 from Table 3 results in the decision Table 1 with a maximum determinism level: $\frac{12-4}{12-4} = 1$. Similarly, decision rules are removed for 'Decide quality' which ends in the decision Table 1 with determinism level 1. This makes the average determinism level equal to 1 as all routing decisions with assigned attributes are fully deterministic.

Algorithm 1 summarizes the steps 2 and 3 of 'DataExtend'.

Step 4: Simulate Control-Flow Model with Decision Perspective into Event Log. 'DataExtend' will simulate the models with decision rules into an event log. It takes a user specified number of cases to be generated, the process model, and the set of decision rules as input. Then, each attribute, except the ones that correspond to a previous decision[3], are initialized with a random value.

---

[3] These attributes are initialized with the decision outcome when it is executed.

Algorithm 1. Extend process model with decision perspective

1: **Input:**
2:      M : process model
3:      dl : target determinism level
4: **Output:**
5:      M : process model
6:      $\mathcal{R}$: set of decision rules
7: **Start ExtendModel**(M, dl)
8: **for** each exclusive choice in M **do**
9:      $Assign(d_i) \mapsto V_{random}$
10:     $R_{d_i} \leftarrow$ initial decision table d_i
11:     $\mathcal{R} \leftarrow \mathcal{R} \cup R_{d_i}$
12: **end for**
13: **while** $AverageDeterminismLevel(PT) < dl$ **do**
14:     Remove random rule from $\mathcal{R}$ without violating soundness constraints
15: **end while**
16: **return** $\mathcal{R}$

For example, in the make-to-order process the 'premium' case attribute gets value 'True'.

The simulation algorithm will execute each activity in the model according to the control-flow semantics and include this in the resulting event log. When it encounters a decision d_i (business rule task) it will execute the decision using the generated decision rules $R_{d_i} \in \mathcal{R}$. Therefore, 'DataExtend' will collect the values of each of the assigned attributes $\{V_1, \ldots, V_w\}$ to make a state of the current case. Then it will iterate over all the decision rules to collect the possible decision outcomes. A decision outcome is possible if a rule condition matches with the state. Finally, the decision outcome leads to a particular outgoing choice branch to be executed after the decision.

The simulation of a process model with the decision perspective yields an event log with both control-flow and case information as needed for decision mining evaluation.

3 Demonstration

This section presents an empirical analysis of two decision mining algorithms to validate 'DataExtend'. 'DataExtend' has been implemented in the ProM framework as part of the 'PTandLogGenerator' package.[4] It enables the evaluation of decision mining techniques that discover a decision model from an event log (e.g. [4,5]) or techniques that discover the decision logic from an event log (e.g. [3,6]). Due to a lack of space, the experiments will focus on the latter type of techniques.

---

[4] See https://svn.win.tue.nl/repos/prom/Packages/PTAndLogGenerator/.

Experiment Setup. The experiment will evaluate the *mutually-exclusive* technique [3] that discovers fully-deterministic decision rules based on case attributes in the input event log, and the *overlapping* technique [6] that allows to discover non-deterministic decision rules. The goal is to determine the effect of different determinism levels of the generated decision rules by 'DataExtend' on the quality of the discovered decision rules. We have generated a random sample of 129 process models for each miner from six model populations shown in Table 4, i.e. one population for each value combination for the determinism level {0.5, 0.75, 1} and infrequent paths {0 (False), 1 (True)} parameters. The other process characteristics are fixed for each model population. The probability of the sequence, exclusive choice and parallelism patterns is fixed at values 46%, 35% and 19%, respectively based on the analysis of a large collection of models by Kunze et al. [12]. The size of the models varies between 6 and 10 activities, with a mode of 8 activities. Furthermore, we have specified that the case attributes introduced by 'DataExtend' are of three different types: boolean, string and numerical. Each numerical attribute is discretized to a random number of intervals, between 1 and 4, following a uniform distribution to make a finite number of decision rules for each decision. Secondly, the number of case attributes that are assigned to a decision varies between 0 and 3 following a discrete uniform distribution. Each model with rules is simulated into an event log containing between 200 and 1000 cases.

Table 4. Model population parameters for the experiments, where X and Y are assigned all 6 combinations of values in {0, 5.75, 1} and {0 (False), 1 (True)} respectively.

Parameter	MP_{data}
No visible activities	(6,8,10)
Sequence ($\Pi^{\rightarrow}$)	0.46
Parallel ($\Pi^{\wedge}$)	0.19
Choice ($\Pi^{\times}$)	0.35
Infrequent paths (Π^{In})	Y
Sample size (# models)	129
Logs per model	1
Number of cases	[200,1000]
Determinism level	X
Attribute type	$\in$ {bool, string, numerical}
# intervals	$\sim uniform(1,4)$
# assigned attributes	$\sim uniform(0,3)$

For each generated control-flow model, 'DataExtend' will first generate decision rules and an event log. Each generated log is split into a training log (90% of the cases) and a test log (10% of the cases). The generated control-flow model and the training log are used as input of the decision mining techniques to discover decision rules. Then, we evaluate the discovered rules using a classification approach. We first alter the attributes of half of the cases in the test log such that they do not comply with the generated decision rules. Next, the discovered rules are used to classify the cases in the test log as fitting or non-fitting (violating the discovered rules). This enables us to quantify the quality of the discovered rules using the quality metrics recall (how much fitting cases are classified as fitting), precision (how much cases classified as fitting are actually fitting) and their harmonic average the F_1 score.

Analysis of Results. The graph in Fig. 3 illustrates the average F_1 scores for the two decision mining techniques over different determinism levels. The bars indicate the 95% confidence interval for the averages. The graph indicates a positive trend, i.e. increasing the determinism level has a positive effect on F_1 scores. The Kruskall-Wallis test [13] shows that the differences in

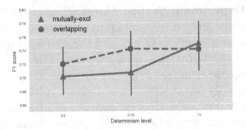

Fig. 3. F_1 scores for decision mining techniques for different levels of determinism

F_1 scores between fully-deterministic (determinism of 1) and non-deterministic decision rules (determinism of 0.5 or 0.75) are statistically significant for the *mutually-exclusive* technique. For the *overlapping* technique only the differences in F_1 score between the largest and the smallest determinism levels are statistically significant. Therefore we can conclude that the determinism level has an effect on the quality of the two decision mining techniques. This effect is smaller for the *overlapping* technique than the *mutually-exclusive* technique. This result is not surprising given the fact that the *overlapping* technique specifically focuses on discovering non-deterministic decision rules as included in the experiments.

4   Related Work

PLG2 [9] allows for extending control-flow models with data attributes, but in a more general sense. It can add case attributes to activities such that an activity can either generate a case attribute or require a case attribute. The latter is implemented by automatically generating the required case attribute before the execution of that activity. The user cannot control that this case attribute requirement happens to activities after a decision in the process.[5] Nevertheless, this is necessary for the evaluation of decision mining techniques as they focus on discovering the decision model and rules.

5   Conclusion and Future Work

This paper has introduced the 'DataExtend' framework that allows evaluating and comparing decision-mining techniques using a sufficient amount of event logs and process models to detect statistically significant quality differences. For the generation of event logs enriched with data attributes we developed a novel approach, because the only technique to generate such event logs, namely PLG2 [9] does not allow to control the generation of attributes values to influence the decision perspective. A demonstration of 'DataExtend' involved two decision

---

[5] This is for the random model generator. PLG2 allows users to add the requirements also manually, however, that would not lead to random samples and thus obstruct the generalization of evaluation results.

mining techniques and its results illustrated that the novel approach can serve the evaluation purpose. Future work needs to provide a more extensive evaluation that includes more techniques, such as [4,5]. Furthermore, we want to extend the framework so as to incorporate the loop and multi-choice patterns that involve decisions. The initial evaluation is also based on an implementation that requires a lot of tedious and manual repetition of the application of the techniques for each of the generated event logs. As future work, we aim to integrate it in a scientific-workflow tool, which would automate the currently-tedious work.

References

1. Object Management Group: Decision Model And Notation 1.1, June 2016
2. Rozinat, A., van der Aalst, W.M.P.: Decision mining in ProM. In: Dustdar, S., Fiadeiro, J.L., Sheth, A.P. (eds.) BPM 2006. LNCS, vol. 4102, pp. 420–425. Springer, Heidelberg (2006). https://doi.org/10.1007/11841760_33
3. de Leoni, M., van der Aalst, W.M.P.: Data-aware process mining: discovering decisions in processes using alignments. In: Proceedings of the 28th Annual ACM Symposium on Applied Computing, pp. 1454–1461. ACM (2013)
4. Bazhenova, E., Buelow, S., Weske, M.: Discovering decision models from event logs. In: Abramowicz, W., Alt, R., Franczyk, B. (eds.) BIS 2016. LNBIP, vol. 255, pp. 237–251. Springer, Cham (2016). https://doi.org/10.1007/978-3-319-39426-8_19
5. De Smedt, J., Hasić, F., vanden Broucke, S.K.L.M., Vanthienen, J.: Towards a holistic discovery of decisions in process-aware information systems. In: Carmona, J., Engels, G., Kumar, A. (eds.) BPM 2017. LNCS, vol. 10445, pp. 183–199. Springer, Cham (2017). https://doi.org/10.1007/978-3-319-65000-5_11
6. Mannhardt, F., de Leoni, M., Reijers, H.A., van der Aalst, W.M.P.: Decision mining revisited - discovering overlapping rules. In: Nurcan, S., Soffer, P., Bajec, M., Eder, J. (eds.) CAiSE 2016. LNCS, vol. 9694, pp. 377–392. Springer, Cham (2016). https://doi.org/10.1007/978-3-319-39696-5_23
7. Jouck, T., Depaire, B.: Generating artificial data for empirical analysis of control-flow discovery algorithms: a process tree and log generator. Bus. Inf. Syst. Eng. 18 (2018)
8. Van Der Aalst, W.M.P., Ter Hofstede, A.H.: Verification of workflow task structures: a petri-net-baset approach. Inf. Syst. 25(1), 43–69 (2000)
9. Burattin, A.: PLG2: multiperspective process randomization with online and offline simulations. In: BPM (Demos), pp. 1–6 (2016)
10. Russell, N., ter Hofstede, A.H.M., van der Aalst, W.M.P., Mulyar, N.: Workflow controlflow patterns: a revised view. Technical report 06-22 (2006)
11. Rosca, D., Wild, C.: Towards a flexible deployment of business rules. Expert Syst. Appl. 23(4), 385–394 (2002)
12. Kunze, M., Luebbe, A., Weidlich, M., Weske, M.: Towards understanding process modeling – the case of the BPM academic initiative. In: Dijkman, R., Hofstetter, J., Koehler, J. (eds.) BPMN 2011. LNBIP, vol. 95, pp. 44–58. Springer, Heidelberg (2011). https://doi.org/10.1007/978-3-642-25160-3_4
13. Siegel, S., Castellan Jr., N.J.: Nonparametric Statistics for the Behavioral Sciences, 2nd edn. Mcgraw-Hill Book Company, New York (1988)

Compliance Checking for Decision-Aware Process Models

Stephan Haarmann[✉], Kimon Batoulis, and Mathias Weske

Hasso Plattner Institute, University of Potsdam,
Prof.-Dr.-Helmert-Str. 2-3, 14482 Potsdam, Germany
{stephan.haarmann,kimon.batoulis,mathias.weske}@hpi.de
https://bpt.hpi.uni-potsdam.de/

Abstract. The business processes of an organization are often required to comply with domain-specific regulations. Such regulations can be checked based on the models of the respective processes. These models' main focus is on the operational part of the process. However, also decisions play a major role in the execution behavior of processes, and they are expressed in separate decision models. In this paper, we investigate the influence of decision models on business process compliance checking. To this end, we formalize decision-aware processes as colored Petri nets, extract the state space, and check compliance rules using temporal logic model checking. The approach improves the quality of existing compliance checking by reducing the risk of false negatives. We provide a prototype and discuss advantages and disadvantages.

Keywords: Business process management ·
Business process compliance · Decisions

1 Introduction

Business process compliance is the topic of ensuring that an organization's business processes comply with internal and external domain-specific regulations. Such regulations often refer to the sequence of certain activities that must or must not occur. Given that the organization documented its processes in respective models, it is possible to uncover potential violations of the regulations at design time already to ensure a compliant execution of the process.

Compliance checking for business process models has been given a lot of attention in the literature in recent years [1–3,18,19]. These approaches focus on process models specifying aspects such as control flow and high level data dependencies, which may be subject to internal and external domain-specific regulations. However, particular instances of these processes often depend on additional decision logic defining fine-grained data dependencies, which are not specified in the process model, but in a separate decision model.

This paper presents a semi-automated approach to design-time compliance checking of decision-aware process models. To this end, we formally capture the

© Springer Nature Switzerland AG 2019
F. Daniel et al. (Eds.): BPM 2018 Workshops, LNBIP 342, pp. 494–506, 2019.
https://doi.org/10.1007/978-3-030-11641-5_39

execution semantics of decision-aware processes (i.e., the logic of the process and the decision as well as the data dependencies between the two). Subsequently, we show that considering decisions increases the size of the processes' state space by adding more information to the data.objects. However, at the same time, the amount of traces is limited, since there may be interdependencies between decisions that rule out certain traces. Therefore, compliance checking with decision-aware processes may actually lead to more accurate results since in the decision-unaware process rules were violated by traces that could actually never occur.

The remainder of this paper is structured as follows. Section 2 presents related work. In Sect. 3, we provide definitions of the structures used and an example. The paper presents the decision-aware compliance checking approach in Sect. 4. We evaluate the approach using a prototypical implementation in Sect. 4.4. Finally, Sect. 5 concludes the paper and discusses future work.

2   Related Work

Business Process Compliance received increasing attention of BPM research from 2000 to 2007 and is still actively researched [15]. Recently, Hashmi, Governatori, and Lam summarized the developments and gave possible directions for future work in a survey paper [15]. They distinguish between design-time, run-time, and auditing approaches—based on their application during the process life cycle. This paper contains a design-time approach that checks models for potential violations using a model checking approach [4,11]. It addresses an open issue [15]: consideration of activities' effects.

Various approaches for design-time compliance checking have been developed. Awad et al. introduce *BPM-Q* (and its visual counterpart *BPMN-Q*) to formally (and visually) model queries for process models by reusing BPMN elements and annotating them with additional information [1]. Later they used it for modeling compliance rules [2], and they added data support [3]. The compliance rules are formalized using temporal logic and checked with the model checker NuSMV[1]. Awad et al. only investigate compliance rules based on control-flow relationships and data; however, data based rules can only constrain the state of the data object.

In contrast, *(Extended) Compliance Rule Graphs (eCRGs)* are capable of expressing fine grained data conditions and additional perspectives such as time and resources [19,20,26]. While most eCRG based approaches are used for run-time or auditing compliance approaches, Knuplesch et al. present an approach for checking data-aware rules on process models [18], which is strongly related to this paper. In contrast to Awad, fine grained data conditions can be evaluated. Knuplesch et al. infer respective knowledge from arc-conditions and derive *abstraction predicates* (contraints for possible values) for all data attributes.

---

[1] NuSMV's web page: http://nusmv.fbk.eu/ (retrieved 4/10/2018).

They embedded their approach in the SeaFlow compliance checker, which models rules as *Compliance Rule Graphs (CRGs)* [23]. However, our method additionally considers decision models and supports operations on data.

The approach of this work uses a *colored Petri net (CPN)* based formalism for process models. The translational semantics are based on Dijkman's et al. approach of mapping processes to Petri nets [13] and Lee's et al. method for modeling decisions as Petri nets [22]. Instead of classical Petri nets, we use CPNs [17]. CPNs have been used for analyzing data aware processes in [27]. In our paper, CPNTools[2] is used for implementation: the formalization, the state space extraction, and the compliance checking using the ASK-CTL extension [10].

Decisions and decision models receive increasing attention from BPM research. Recently Jansen et al. and Batoulis et al. defined criteria for consistent integration of process and decision models [7,16]. Further, Batoulis et al. investigates soundness notions for decision-aware processes and thereby domain independent correctness criteria [6,8]. Compliance is based on domain specific rules. Therefore, our approach complements existing correctness criteria for decision-aware processes.

3   Foundations

A decision-aware process model contains imperative and declarative parts of a business process. The imperative parts are captured by a traditional process model (e.g., a BPMN model [24]) while the declarative parts are captured by decision models (e.g., a DMN model [25]). Process models link decision models through decision tasks, which refer to a decision in a decision model. This section contains a description of decision-aware processes, an running example, and a brief description of compliance checking.

3.1   Decision-Aware Process Models

A business process consists of a set of tasks that contribute to a common business goal and are executed in a technical and organizational environment [28]. A process model describes these tasks and their temporal and causal dependencies. Further, it has one start event and one end event. The model contains gateways to express exclusiveness (XOR gateways) and concurrency (AND gateways) of tasks. Additional dependencies can be expressed by using data objects and data flow: an activity can read data objects in specific states and write data objects in specific states. Each activity has, therefore, a set of input sets and a set of output sets. At least one input set must be available to enable the activity. One output set is chosen and its elements are written by the activity.

Consider the sample process model in Fig. 1. It depicts the inquire process of a car rental company. If an order is received, then the company automatically checks if discounts apply and grants them. Afterwards, additional fees are

---

[2] CPNTools' web page: http://cpntools.org (accessed 4/10/2018).

determined and calculated. Eventually, the final price is set and the updated order is sent to the customer. Throughout the process, various data objects are used: *Offer* comprises the key information, *Special Offer* contains information about potential discounts, *Fees* contains the determined fees which are saved in separate objects when calculated (*Young Driver Fee* and *Last Minute Fee*).

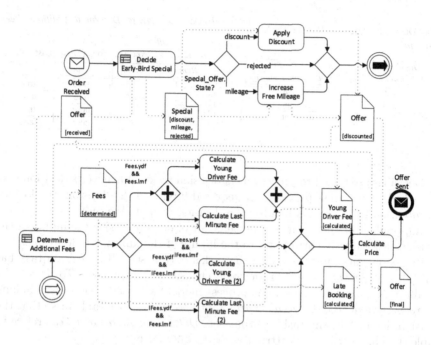

Fig. 1. Sample process model (BPMN) of a car rental company

A decision model consists of two layers: the decision requirements and the decision logic. The former comprises high-level information about the necessary inputs the data and the preceding decisions that are required to execute a certain decision. The logic level contains a specification of how decisions are made. These can be informal or formal. We assume that all decisions are formally specified by a decision table. A decision table comprises a set of rules, which consist of conditions for the inputs and expressions for producing outputs. Although rules can, in general, be overlapping, we only consider unique decision tables where only one rule matches an input. For unique tables, the order of rules is irrelevant. It has been shown, that all DMN decision tables can be transformed into unique ones [9].

The car rental scenario contains two decisions, the logic of which is given by Table 1a and b. A rule is represented horizontally. The first decision (Table 1a) has the input *Offer.Lead Time*, which is the time in weeks between booking and picking the rental up. The output is either *rejected* (no discount), *mileage* (increased free mileage), or *discount* (monetary discount). The decision in Table 1b considers the lead time and the driver's age to determine the fees: a driver younger than 25 must pay a young driver's fee. If the booking occurs less than one week in advance, then a last minute fee is due.

Table 1. Decision tables for the sample process

(a) Decision logic for *Decide Early-Bird Special*

U	Offer.Lead Time	Special Offer.State
	Number	*{rejected, discount, mileage}*
1	< 2	rejected
2	[2..4)	mileage
3	≥ 4	discount

(b) Decision logic for *Determine Additional Fees*

U	Offer		Fees	
	Lead Time	Driver Age	Young Driver	Last Minute
	Number	[18..99]	*bool*	*bool*
1	< 1	< 25	true	true
2	≥ 1	< 25	true	false
3	< 1	≥ 25	false	true
4	≥ 1	≥ 25	false	false

Decisions are linked to processes via decision tasks. Whenever a decision task is reached, the process provides the required inputs to the decision, the logic is executed, and the process handles the decision output. We make the following assumption about the input-output behavior of decision tasks: the inputs of the decision task directly correspond to the inputs of the linked decision table. The decision's outputs are reflected in the task's output sets. If a decision task has multiple output sets, the decision logic chooses an output set. To do so, the attribute *State* is set. The sample process model has two decision tasks linking the respective decisions. *Decide Early-Bird Special* links to Table 1a so that the decision determines the task's output set. *Determine Additional Fees* refers to Table 1b. The decision sets attributes of the only output *Fees*.

3.2 Compliance Checking

A decision-aware process model is a blueprint for process instances: it describes the possible behavior. The model structures the process and constrains it (e.g., limits the order of activities). The process can be implemented, for example by using a process engine, to support and control instances. However, real world process instances are subject to laws, guidelines, and regulations, which might or might not be captured in the process model. Violating these constraints can carry penalties and jurisdictional consequences. Thus, it is important to assert compliant behavior.

One step towards business process compliance is the verification of process models with respect to compliance regulations. The compliance regulations are expressed as so called compliance rules (properties that must not be violated). Table 2 contains rules for the car rental process. In general, we consider the occurrence and order of activities (rules c1, c2, c3) and might use data conditions for further restrictions (c4, c5).

Table 2. Compliance rules for the example process

c1	Every received order will eventually be sent back to the customer
c2	The price will only be calculated after additional fees have been determined
c3	If a discount applies, the order must not be subject to a last minute fee
c4	The young driver fee must be calculated if the driver is younger than 25
c5	If an early-bird special applies, calculate price reads the offer in state *discounted*

4 Decision-Aware Compliance Checking

Compliance checking can take place during different phases of the business process lifecycle. The decision-aware compliance checking approach of this paper takes place during design-time, i.e., models are checked for compliance violations. Although different methods for verifying models (e.g., theorem-proofing and simulation) exist [15], *model checking* is the most common one for compliance checking [15]. Therefore, our approach follows the general model checking paradigm depicted in Fig. 2 [21]. The decision-aware process model describes the system and defines a state space. We use colored Petri nets (CPNs) to assign formal behavioral semantics to such models. The compliance rules are properties that must not be violated, and we formalize them as *Computational Tree Logic (CTL)* formulas. The formal rules and the formal model are then consumed by a model checker, which verifies the properties.

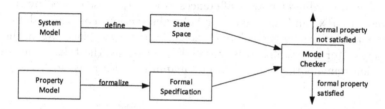

Fig. 2. Schematic description of the general model checking approach (cf. [21])

4.1 Requirements and Challenges for Formalizing Decision-Aware Process Models

A formalism for a decision-aware process comprises the behavior of the process as well as the logic of the decisions, which is why we chose CPNs for this task. CPNs can model conditions and operations on data. Since we assume decision tables to be unique—i.e., its rules are non-overlapping—the set of rules of a table extensionally define a function, mapping an input to exactly one output. Therefore, decisions are just data operations. Further, the structure and temporal and casual constraints of process models can be captured in a (colored) Petri net [13]. Consequently, CPNs are suited for formalizing decision-aware process models.

To model check a Petri net for compliance rules, one needs to investigate its state space which in case of Petri nets is called occurrence graph. This graph must be complete in order to find all possible violations of a compliance rule, i.e., *every* possible trace has to be represented. However, decisions operate on data attributes, which may have large domains (such as the integers). Representing every possible instance explicitly leads to large state spaces (i.e., $2^{32} \times |states|$, if a single integer is involved). Thus, model checking can become infeasible and a proper abstraction is required.

Some model checkers, such as NuSMV, support symbolic abstraction [12]. Instead of considering each possible instance separately, an attribute is represented by a symbol (i.e., *Offer.Lead_Time*= $\mathbb{N}$). Whenever the state space extraction reaches a condition or operation, it is applied to the abstract symbol (i.e., *Offer.Lead_Time*= $\{n \in \mathbb{N} | n < 2\}$ after executing the first rule the decision). If alternative conditions exist, the state space branches. Each branch considers one alternative [5]. But no symbolic model checker for CPNs exists. To overcome this, we incorporate the abstraction into our formalization and implemented required operations and conditions for symbolic execution [12].

4.2 Mapping Decision-Aware Process Models to CPNs for Symbolic Execution

This paper's mapping of decision-aware process models to CPNs builds upon the mapping of process models to Petri nets given by Dijkman et al. in [13]. In this section, we highlight major differences especially those caused by the data-awareness of CPNs and the use of symbolic abstraction. For one, a data object is represented by *exactly one* colored token on *exactly one* place. When the state of the object (or an attribute) is updated, the color of the token changes, but the location remains the same. A formal mapping is described in [14].

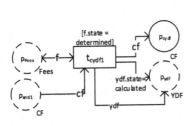

(a) Mapping for the task *calculate young driver fee (cydf)* writing the object *young driver fee (ydf)*

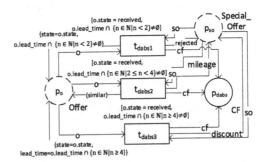

(b) Mapping for the decision task *decide early bird special* and the respective logic

Fig. 3. Mapping of a sample task and a sample decision task linking a decision table

An activity has a set of input sets and a set of output sets. Consequently, the chosen output set is reflected in the process state. For this reason, we map an activity to a set of transitions—one for each output set. The precondition given by the input set is checked by the transitions' guards. Figure 3a shows the mapping of the task *calculate young driver fee*. It has exactly one input set (*Fees* in state *determined*) and one output set (*Young Driver Fee* in state *calculated*) and is therefore mapped to one transition.

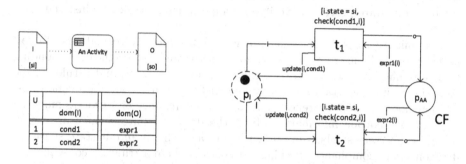

Fig. 4. Generic mapping of a decision task including the decision logic

The outcome of a decision task depends on the decision logic. At this point, we assume that the decision logic determines the output set and may also set other attributes. Figure 4 contains the generic mapping: we create a CPN transition for each row. The transitions can fire if the condition of the rule is fulfilled. Since data objects are described by symbols, it is enough if only one instance fulfills the condition. However, we need to update the input's symbolic abstraction respectively. We denote this by *update(i,cond)* where i is an input and *cond* the corresponding condition. Further, we update the symbols for the outputs according to output value of the corresponding rules (*expr(i)* where i is the input and *expr* the output expression).

Consider the decision task *Decide Early-Bird Special* and its corresponding decision table (Table 1a). The formalization is depicted in Fig. 3b: for the three rules we create three transitions. Each transition corresponds to one rule and has a respective guard. For the first rule, the symbol for *Offer.Lead_Time* must comprise at least one value that is less than two. If the transition fires, the token o for *Offer* is updated so that *Offer.Lead_Time* describes only the values less than two. Further, the output data objects are updated. For the first rule, the transition sets the state of *Special_Offer* to *rejected*.

Finally, also XOR splits are treated differently than in [13], namely similar to decision tasks: they are like decisions with no output. Thus, they must read all the data required for the branching and may update the respective symbols. In contrast to decision tasks, each transition created for the XOR split has a separate control flow place.

4.3 Compliance Checking for CPN Formalism

Since CPNs have formal semantics, it is possible to extract their state spaces, given that they are finite. In case of (colored) Petri nets, the state space corresponds to the net's occurrence graph. That is to say, the current state of the net is given by its current marking, and by firing a transition in the net a state transition in the state space is performed. Our approach checks the compliance of the decision-aware process by verifying temporal logic queries against the state space.

Compliance rules can originate in laws, guidelines, or regulations. Table 2 lists some compliance rules for the sample process, and Table 3 the corresponding CTL formulas. A rule can restrict the occurrence or order, and a rule can be conditional: its restriction must only be satisfied given certain data conditions. So, how does decision-awareness influence the compliance of a process model?

Decisions encode information about the data attributes, which are represented as abstract symbols in our CPN and accordingly in our state space. In decision-aware compliance checking, we consider these attributes; consequently, the state space grows (it is larger than the decision-independent one). However, decisions also encode instance-level dependencies. These dependencies restrict the possible traces of the process further to the existing control flow. For this reason, a decision-independent state space can have more traces than its decision-aware counterpart.

If compliance rules only constrain the occurrence and order of activities (with or without data conditions), each trace contributes to the result. If traces are removed, a rule that previously held will still hold, but a rule that did not hold must be reevaluated because all violations could be part of the removed traces. As an effect, decision-awareness can reduce the number of false negatives (compliance rules that are violated in the model, but not in reality).

For example, consider rule c3: if a discount applies, a last minute fee must not apply. The CTL formula says that in all traces, if *Apply Discount* is executed, *Calculate Last Minute Fee* must not be executed. However, if we ignore decisions it is impossible to infer whether the two XOR-splits are independent. Hence, we have to consider the trace in which $Special\_Offer.State = discount$ and $Fees.lmf=true$. In that trace, the rule is violated. If, however, we take decisions into consideration, this would imply that $Offer.Lead\_Time \geq 4$ and $Offer.Lead\_Time < 1$. This is a conflict, and such a trace is not part of the state space. Hence, the decision-aware process model is compliant to c3.

Furthermore, based on decision-aware processes, we are able to check rules based on data object attribute values (cf. rule c4). This was previously not possible, because the process model does not reference this attribute and every knowledge about its valuation is based on the decision model. Since our mapping of decision-aware processes includes decision models, we can now verify rules involving data attribute conditions. We only need to check if one value contained in the symbolic abstraction satisfies the data condition. For instance, the sample process is compliant to rule c4.

Table 3. Compliance rules for the sample process expressed as CTL formulas

c1 **AF**(end)
c2 **A**(**NOT**(Calculate Price) **U** Determine Additional Fees) ∨ **AG**(**NOT**(Calculate Price) ∧ **NOT**(Determine Additional Fees))
c3 **AG**(Apply Discount ⟹ **AG**(**NOT**(Calculate Last Minute Fee)))
c4 **AG**((Calculate Young Driver Fee ⟹ dataCondition(Offer.Driver Age < 25))
c5 **AG**((writes(Early Bird-Special,Special Offer, discount) ∨ writes(Early Bird-Special,Special Offer, mileage)) ⟹ **AF**(reads(Calculate Price,Offer,discounted)))

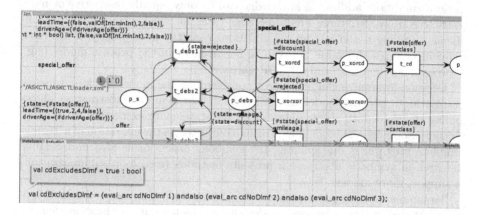

Fig. 5. Screenshot showing partly the colored Petri net and a compliance rule including the compliance checking result

4.4 Prototype

This paper's approach is a CPN based method for compliance checking of decision-aware processes. We applied this approach using CPNTools. Process models were manually translated to CPNs to extract the state space, and formal compliance rules were specified as ASK-CTL queries to be checked (cf. Fig. 5). ASK-CTL is an extension of CPNTools, that allows to evaluate CTL formulas on the state spaces of CPNs.

CPNTools does not support symbolic execution. Therefore, respective data types, comparisons, and operations need to be defined. We added the functionality for *int*, *bool*, and *real* to show the feasibility of the approach. Since large sets (such as *int*) cannot be represented explicitly, we use intervals to describe the current abstraction of a symbol. Examples (including compliance rules) are provided online[3].

5 Discussion and Conclusion

In comparison to other design-time compliance checking approaches, decision-aware compliance checking reduces the number of false alarms. Since decisions

---

[3] Example CPNs: https://owncloud.hpi.de/index.php/s/negAQyTLYPj45xH.

encode relationships that might not directly be visible in the process model, the possible traces are further restricted. If less traces exist, fewer violations of occurrence-based and order-based compliance rules can occur.

Although the knowledge about decision logic allows inferring attribute-level information on used data objects leading to more states, it is impossible that the decision-aware process model produces traces that are not part of the decision-independent model. Since we treat everything that is unknown using the open-world assumption, additional knowledge can be only equally restrictive or more restrictive. However, there are some edge-cases in which decision-aware compliance checking is too restrictive: violations can stay undiscovered if the environment changes the value between to occurrences in the process model (e.g., between a decision task and an XOR gateway).

Decision-awareness can also lead to problems in the model checking process. Translating a process model to a CPN that uses symbolic abstraction allows finding the right data abstraction during the state space extraction. In general, decision-aware process models are Turing-complete. As the execution of a Turing-complete program is undecidable, the symbolic execution is also undecidable. Research in symbolic abstraction presents methods (e.g., loop summarization) to support more models [5].

To summarize, this paper presents a decision-aware compliance checking approach. At design-time a process model and complementary decision-models are formalized as a CPN and model checking is applied to verify the model with respect to compliance rules. Thereby, symbolic abstraction is used to reduce the state space.

Tools, such as CPNTools, can model and analyze (e.g., apply model checking) to CPNs. However, it uses proprietary formats and requires expert knowledge. The manual formalization is an error-prone step. Future work should automate formalizing decision-aware process models and checking compliance, respectively.

CPNTools model checking extension provides only Boolean feedback. A rule holds or it is violated, but the cause of the violation is not exposed. Future work can overcome this by extending the model checking capabilities, using a different model checker, or integrating other approaches such as anti patterns. The latter finds all violating traces in a process model, which is a super set of the violations in a decision-aware setting [3].

Altogether, we showed that decision-aware compliance checking can improve the results compared to traditional design-time compliance checking.

References

1. Awad, A.: BPMN-Q: a language to query business processes. In: Reichert, M., Strecker, S., Turowski, K. (eds.) Enterprise Modelling and Information Systems Architectures - Concepts and Applications. Proceedings of the 2nd International Workshop on Enterprise Modelling and Information Systems Architectures (EMISA 2007), St. Goar, Germany, 8–9 October 2007. LNI, vol. P-119, pp. 115–128. GI (2007). http://subs.emis.de/LNI/Proceedings/Proceedings119/article1957.html

2. Awad, A., Decker, G., Weske, M.: Efficient compliance checking using BPMN-Q and temporal logic. In: Dumas, M., Reichert, M., Shan, M.-C. (eds.) BPM 2008. LNCS, vol. 5240, pp. 326–341. Springer, Heidelberg (2008). https://doi.org/10.1007/978-3-540-85758-7_24

3. Awad, A., Weidlich, M., Weske, M.: Specification, verification and explanation of violation for data aware compliance rules. In: Baresi, L., Chi, C.-H., Suzuki, J. (eds.) ICSOC/ServiceWave -2009. LNCS, vol. 5900, pp. 500–515. Springer, Heidelberg (2009). https://doi.org/10.1007/978-3-642-10383-4_37

4. Baier, C., Katoen, J.: Principles of Model Checking. MIT Press, Cambridge (2008)

5. Baldoni, R., Coppa, E., D'Elia, D.C., Demetrescu, C., Finocchi, I.: A survey of symbolic execution techniques. CoRR (2016). http://arxiv.org/abs/1610.00502

6. Batoulis, K., Haarmann, S., Weske, M.: Various notions of soundness for decision-aware business processes. In: Mayr, H.C., Guizzardi, G., Ma, H., Pastor, O. (eds.) ER 2017. LNCS, vol. 10650, pp. 403–418. Springer, Cham (2017). https://doi.org/10.1007/978-3-319-69904-2_31

7. Batoulis, K., Meyer, A., Bazhenova, E., Decker, G., Weske, M.: Extracting decision logic from process models. In: Zdravkovic, J., Kirikova, M., Johannesson, P. (eds.) CAiSE 2015. LNCS, vol. 9097, pp. 349–366. Springer, Cham (2015). https://doi.org/10.1007/978-3-319-19069-3_22

8. Batoulis, K., Weske, M.: Soundness of decision-aware business processes. In: Carmona, J., Engels, G., Kumar, A. (eds.) BPM 2017. LNBIP, vol. 297, pp. 106–124. Springer, Cham (2017). https://doi.org/10.1007/978-3-319-65015-9_7

9. Batoulis, K., Weske, M.: Disambiguation of DMN decision tables. In: Abramowicz, W., Paschke, A. (eds.) BIS 2018. LNBIP, vol. 320, pp. 236–249. Springer, Cham (2018). https://doi.org/10.1007/978-3-319-93931-5_17

10. Cheng, A., Christensen, S., Mortensen, K.H.: Model checking coloured petri nets-exploiting strongly connected components. DAIMI Rep. Ser. **26**(519) (1997)

11. Clarke, E.M., Grumberg, O., Long, D.E.: Model checking and abstraction. In: Sethi, R. (ed.) Conference Record of the Nineteenth Annual ACM SIGPLAN-SIGACT Symposium on Principles of Programming Languages, Albuquerque, New Mexico, USA, 19–22 January 1992, pp. 342–354. ACM Press (1992)

12. Clarke, E.M., Grumberg, O., Peled, D.A.: Model Checking. MIT Press, Cambridge (2001). http://books.google.de/books?id=Nmc4wEaLXFEC

13. Dijkman, R.M., Dumas, M., Ouyang, C.: Semantics and analysis of business process models in BPMN. Inf. Softw. Technol. **50**(12), 1281–1294 (2008)

14. Haarmann, S.: Decision-aware compliance checking. Master's thesis, University of Potsdam, March 2018. https://owncloud.hpi.de/index.php/s/GvZ3AJjdo6MJUr8

15. Hashmi, M., Governatori, G., Lam, H.P., Wynn, M.T.: Are we done with business process compliance: state-of-the-art and challenges ahead. Knowl. Inf. Syst. **57**, 79–133 (2018)

16. Janssens, L., Bazhenova, E., Smedt, J.D., Vanthienen, J., Denecker, M.: Consistent integration of decision (DMN) and process (BPMN) models. In: España, S., Ivanovic, M., Savic, M. (eds.) Proceedings of the CAiSE 2016 Forum, at the 28th International Conference on Advanced Information Systems Engineering (CAiSE 2016), Ljubljana, Slovenia, 13–17 June 2016. CEUR Workshop Proceedings, vol. 1612, pp. 121–128. CEUR-WS.org (2016). http://ceur-ws.org/Vol-1612/paper16.pdf

17. Jensen, K., Kristensen, L.M.: Coloured Petri Nets: Modelling and Validation of Concurrent Systems. Springer, Heidelberg (2009). https://doi.org/10.1007/b95112

18. Knuplesch, D., Ly, L.T., Rinderle-Ma, S., Pfeifer, H., Dadam, P.: On enabling data-aware compliance checking of business process models. In: Parsons, J., Saeki, M., Shoval, P., Woo, C., Wand, Y. (eds.) ER 2010. LNCS, vol. 6412, pp. 332–346. Springer, Heidelberg (2010). https://doi.org/10.1007/978-3-642-16373-9_24

19. Knuplesch, D., Reichert, M.: A visual language for modeling multiple perspectives of business process compliance rules. Softw. Syst. Model. 16(3), 715–736 (2017)

20. Knuplesch, D., Reichert, M., Ly, L.T., Kumar, A., Rinderle-Ma, S.: On the formal semantics of the extended compliance rule graph. Technical report, Ulm University (2013). http://dbis.eprints.uni-ulm.de/1147/1/TR_UIB_2013_05.pdf

21. Kunze, M., Weske, M.: Behavioural Models - From Modelling Finite Automata to Analysing Business Processes. Springer, Switzerland (2016). https://doi.org/10.1007/978-3-319-44960-9

22. Lee, S., O'Keefe, R.M.: Developing a strategy for expert system verification and validation. IEEE Trans. Syst. Man Cybern. 24(4), 643–655 (1994)

23. Ly, L.T.: SeaFlows - a compliance checking framework for supporting the process lifecycle. Ph.D. thesis, University of Ulm (2013). http://vts.uni-ulm.de/docs/2013/8664/vts_8664_12857.pdf

24. OMG: Business process model and notation (BPMN), v2.0, January 2011. http://www.omg.org/spec/BPMN/2.0

25. OMG: Decision model and notation (DMN), v1.1, May 2016. http://www.omg.org/spec/DMN/1.1

26. Reichert, M., Weber, B.: Enabling Flexibility in Process-Aware Information Systems: Challenges, Methods, Technologies. Springer, Heidelberg (2012). https://doi.org/10.1007/978-3-642-30409-5

27. Wang, Z., Wang, J., Wen, L., Luo, G.: Formally modeling and analyzing data-centric workflow using WFCP-net and ASK-CTL. In: Zhang, R., Cordeiro, J., Li, X., Zhang, Z., Zhang, J. (eds.) ICEIS 2011 - Proceedings of the 13th International Conference on Enterprise Information Systems, Beijing, China, 8–11 June 2011, vol. 3, pp. 139–144. SciTePress (2011)

28. Weske, M.: Business Process Management: Concepts, Languages, Architectures, 2nd edn. Springer, Heidelberg (2012). https://doi.org/10.1007/978-3-642-28616-2

Towards Automated Process Modeling Based on BPMN Diagram Composition

Piotr Wiśniewski(✉) , Krzysztof Kluza , and Antoni Ligęza

AGH University of Science and Technology,
al. A. Mickiewicza 30, 30-059 Krakow, Poland
{wpiotr,kluza,ligeza}@agh.edu.pl

Abstract. Modeling a business process is a complex task which involves different participants who should be familiar with the chosen modeling notation. In this paper, we propose an idea of generating business process models based on a declarative specification. Given an unordered list of process activities along with their input and output data entities, our method generates a synthetic, complete log of a process. The generated task sequences can then serve as an input to a selected process mining method or be processed by an algorithm constructing a BPMN model directly based on the log and additional information included in the declarative process specification.

Keywords: Business processes · BPMN · Automated planning · Constraint programming · Process mining · Business process composition

1 Introduction

One of the challenges within the area of business process management is the constant improvement and optimization of business processes. Manual redesign of a workflow is a time-consuming activity which requires close cooperation of a business analyst or a process engineer with a domain expert aware of goods production or service delivery.

As a solution to this problem, the use of a process composition technique is proposed. Composition, as one of the twenty Business Process Management use cases [1], may be regarded as a set of methods which, based on the identified tasks or subprocesses will allow the operator to generate a correct business process model. Our approach uses declarative activity specifications that include initial conditions and execution effects of the process activities, as well as rules for task repetition. Such a specification also focuses on the goal of the modeled process represented by the produced output data.

The proposed approach is based on Business Process Model and Notation (BPMN), which is one of the most widely recognized languages for business workflow modeling. In addition, the composition method presented in this paper is modeled as a Constraint Satisfaction Problem (CSP), which ensures the correct

© Springer Nature Switzerland AG 2019
F. Daniel et al. (Eds.): BPM 2018 Workshops, LNBIP 342, pp. 507–513, 2019.
https://doi.org/10.1007/978-3-030-11641-5_40

order of tasks and the compliance with business process modeling guidelines. The set of solutions generated by a CSP solver can be then translated into a business process model using the existing process mining tools or by executing a dedicated algorithm for graph-based model construction.

This position paper is organized as follows: Sect. 2 presents the state of the art solutions in the area of business process planning as well as the application of constraint programming to process modeling. Section 3 describes our research methodology and techniques being applied in the proposed approach. The composition algorithm presented in Sect. 4 is followed by concluding remarks and plans for future work included in Sect. 5.

2   Related Works

The composition of business process models is present in the literature in combination with different approaches. Process models can be composed using reusable process parts called Relevant Process Fragments [2] which are stored in a component repository. Another approach is the service composition problem, which aims to combine existing functionalities into a new sequence flow. Meyer and Weske [3] proposed the approach that uses a heuristic search to produce a list of possible event logs. Another approach provides a conceptual framework for task composition [4] including structure suggestion and validation regarding task parallelism and preconditions.

The use of automated planning within the area of business process management is dedicated to design and rebuild phases of the modeling procedure. Providing a higher level of automation for a process can be regarded as a way to overcome issues such as constant changes of requirements and unpredictable environmental factors [5]. One of the main challenges of applying common AI planning techniques for service composition is the necessity to consider complex workflow structures such as conditions and loops [6]. Process planning may focus on the goal of the analyzed workflow by finding a set of models whose task postconditions are compliant with the desired final state or use partial-order planning in order to resolve potential concurrency conflicts [7]. Business process planning can also be used to optimize the workflow by removing unnecessary redundancies in branches which follow an exclusive gateway [8].

Business processes are represented by their declarative models which specify relations between tasks and their execution conditions instead of describing the workflow explicitly [9]. According to the research conducted by Mrasek et al. [10], automated generation of a process model from such data significantly increases time efficiency compared to the manual model design. If a process modeling task is defined as a constraint satisfaction problem, this type of process specification can be used to create optimal execution plans [11] as well as to modify actual workflow traces where artifacts appear or relevant events are missing [12].

Our idea also refers to process mining which includes algorithms for generating BPMN models based on event logs. Although mining tools were created to process imperfect data from IT systems, there exist several performance measures which can help to identify the optimal technique for complete logs [13].

3   Approach Overview

Our research aims to determine how constraint programming may improve the process model generation and to discuss using the process mining for discovering BPMN models based on artificially generated logs.

The proposed method can simplify the process of knowledge acquisition from domain experts who may not be familiar with the appropriate business process modeling notation. Therefore, a user-friendly form of the input process description has been taken into consideration. One of the possible solutions for this problem is based on a spreadsheet specification of the process. However, in the case of process planning, there are no requirements for any ordering of tasks in the specification of the process. The specification includes:

- an unordered list of activities,
- task input and output data entities,
- maximum number of executions for each task,
- initial state of the process,
- a set of final states: one goal state and a number of error states.

Since business process models usually involve multiple participants, parts of the process specification can be created independently by each of the contributors. The reason of basing on a tabular specification is caused by the fact that during the phase of collecting process data the participants may not be aware of the interdependencies between activities performed by different actors of the process. Another idea behind supporting a tabular form of input data is the popularity of spreadsheet editors being accessible by users. This task is performed manually by filling a dedicated form or worksheet.

In the next step, all the files are gathered from the process participants and processed by an automated tool which generates a formal specification, as required by the constraint programming solver. Figure 1 shows a general illustration of the proposed semi-automated business process composition approach.

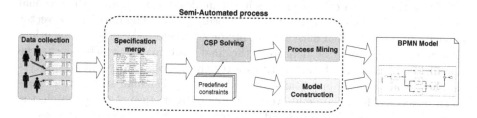

Fig. 1. Method overview (it partly uses the method presented in [14]).

According to our preliminary solution [14], a process model is composed by generating all admissible execution sequences of tasks, based on the input data and the set of predefined constraints. In the current phase of the research, we have included the possibility to represent complex flow structures such as loops and multiple final states [15]. The generated workflow traces are then used as an input to one of the process mining algorithms that generates a workflow net which is then translated into a process model. Another possible solution is based on constructing a process model by merging the generated workflow traces into an activity graph [16] and transforming it directly into a BPMN model.

4   Generating Process Traces

A business process can be described as a set of activities that produce a specific service or product. Therefore, in the initial phase of process composition, it is necessary to identify the tasks being executed within the workflow. Each task is assigned a set of triggering conditions and generates an effect of its execution. This input and output information can be represented as data entities, defined as variables of the primitive or complex data type.

The proposed approach can be used when a set of tasks and a set of data entities were identified. Then, it is necessary to define the dependencies between these sets. Table 1 presents an illustrative example of a task list which can be used for process composition. Data entities included in parentheses are considered as optional for the corresponding tasks.

Table 1. An example list of tasks and data entities. The maximum number of executions for each task is given in parentheses.

Task Id	Task name	Task inputs	Task outputs
01	Create Offer (1)	RequestForOffer	InitialOffer
02	Include Remarks (1)	OfferReviewed	RemarksIncluded
03	Review Offer (2)	InitialOffer, (RemarksIncluded)	OfferReviewed
04	Send Offer (1)	OfferReviewed, (RemarksIncluded)	OfferSent

If m is the number of data entities in the modeled process, an m-dimensional vector has to be defined along with the process model. The initial state vector explicitly indicates which data entities are present before the process execution. The other structure to be defined is the final state matrix which reflects the possible combinations of data entities after the process end event. A process should contain exactly one goal state and several other final states which represent error and terminate end events. At the current stage of the research, the following constraints were proposed to ensure the correctness of the process:

1. The number of executions for each task should be lower than or equal to the value included in the specification.
2. The maximum length of the workflow is equal to the number of defined tasks multiplied by their number of executions.

3. Data entities required for the first executed task satisfy the initial state.
4. If the goal state is achieved, then the process ends.
5. The output data entities of the last task satisfy one of the final states.
6. A task can be executed when the current state satisfies the input conditions.
7. The presence of data entities can be changed only by a task execution.

The synthetic workflow log is generated using the Gecode solver with the assumption that each solution of the CSP is a different trace. Given the simple specification presented in Table 1 and a final state represented by data entity OfferSent, a synthetic log of two distinct traces was generated:

$$W_S = \{\{01, 03, 04\}, \{01, 03, 02, 03, 04\}\}. \tag{1}$$

In the next step, the list of generated solutions was converted to an XES file and processed in ProM environment using ILP miner. The result of process mining algorithm is shown in Fig. 2.

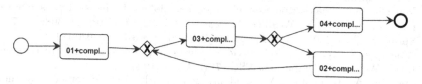

Fig. 2. The mined BPMN model representing the generated log.

5   Conclusion and Future Works

The purpose of this paper is to give an overview of the composition approach and to trigger a discussion on the concept of automated process modeling based on the existing approaches such as declarative languages, workflow trace generation, process mining and process constructing algorithms.

In this paper, we briefly discussed the concept of business process composition based on a partially structured specification. As a contrast to many existing process planning techniques, our method does not require the ordering relations of tasks being declared explicitly. Thus, in this approach, no knowledge of specific notation and modeling guidelines is needed to design business process models.

Since we based our method on different phases, it is more flexible and it may be controlled in different stages. For example, inconsistencies in the specification may be discovered before the final process model is generated, e.g. if the constraints are unsatisfiable. Every part of the method can be analyzed separately and replaced with another algorithm.

In the further development, we plan to evaluate other approaches such as Answer Set Programming and perform a comparative analysis of implementable techniques. In addition, we would like to analyze and evaluate the suitability of different process mining techniques for processing the generated event logs. These results are going to be compared with the dedicated algorithm for constructing BPMN models in terms of model fitness and the ability to discover complex workflow structures.

References

1. van der Aalst, W.M.P.: Business process management: a comprehensive survey. ISRN Softw. Eng. **2013**, 37 (2013)
2. Skouradaki, M., Andrikopoulos, V., Leymann, F.: Representative BPMN 2.0 process models generation from recurring structures. In: Proceedings of the 23rd IEEE International Conference on Web Services, pp. 468–475. IEEE, June 2016
3. Meyer, H., Weske, M.: Automated service composition using heuristic search. In: Dustdar, S., Fiadeiro, J.L., Sheth, A.P. (eds.) BPM 2006. LNCS, vol. 4102, pp. 81–96. Springer, Heidelberg (2006). https://doi.org/10.1007/11841760_7
4. Weber, I.M.: Semantic Methods for Execution-level Business Process Modeling: Modeling Support Through Process Verification and Service Composition. LNBIP, vol. 40. Springer, Heidelberg (2009). https://doi.org/10.1007/978-3-642-05085-5
5. Marrella, A.: What automated planning can do for business process management. In: Teniente, E., Weidlich, M. (eds.) BPM 2017. LNBIP, vol. 308, pp. 7–19. Springer, Cham (2018). https://doi.org/10.1007/978-3-319-74030-0_1
6. AlSedrani, A., Touir, A.: Web service composition processes: a comparative study. Int. J. Web Serv. Comput. (IJWSC) **7**(1), 1–21 (2016)
7. Marrella, A., Lespérance, Y.: A planning approach to the automated synthesis of template-based process models. Serv. Oriented Comput. Appl. **11**(4), 367–392 (2017)
8. Heinrich, B., Schön, D.: Automated planning of process models: the construction of simple merges. In: European Conference on Information Systems (ECIS), Research Papers (2016)
9. Barba, I., Del Valle, C., Weber, B., Jimenez, A.: Automatic generation of optimized business process models from constraint-based specifications. Int. J. Coop. Inf. Syst. **22**(02), 1350009 (2013)
10. Mrasek, R., Mülle, J., Böhm, K.: Automatic generation of optimized process models from declarative specifications. In: Zdravkovic, J., Kirikova, M., Johannesson, P. (eds.) CAiSE 2015. LNCS, vol. 9097, pp. 382–397. Springer, Cham (2015). https://doi.org/10.1007/978-3-319-19069-3_24
11. Barba Rodriguez, I.: Constraint-based planning and scheduling techniques for the optimized management of business processes. Ph.D. thesis, Universidad de Sevilla (2011)
12. De Giacomo, G., Maggi, F., Marella, A., Sardina, S.: Computing trace alignment against declarative process models through planning. In: International Conference on Automated Planning and Scheduling (ICAPS 2016), pp. 367–375 (2016)
13. Buijs, J.C.A.M., van Dongen, B.F., van der Aalst, W.M.P.: On the role of fitness, precision, generalization and simplicity in process discovery. In: Meersman, R., et al. (eds.) OTM 2012. LNCS, vol. 7565, pp. 305–322. Springer, Heidelberg (2012). https://doi.org/10.1007/978-3-642-33606-5_19
14. Wiśniewski, P., Kluza, K., Ślażyński, M., Ligęza, A.: Constraint-based composition of business process models. In: Teniente, E., Weidlich, M. (eds.) BPM 2017. LNBIP, vol. 308, pp. 133–141. Springer, Cham (2018). https://doi.org/10.1007/978-3-319-74030-0_9

15. Wiśniewski, P., Kluza, K., Ligęza, A.: An approach to participatory business process modeling: BPMN model generation using constraint programming and graph composition. Appl. Sci. **8**(9), 1428 (2018)
16. Wiśniewski, P., Ligęza, A.: Constraint-based identification of complex gateway structures in business process models. In: Rutkowski, L., Scherer, R., Korytkowski, M., Pedrycz, W., Tadeusiewicz, R., Zurada, J.M. (eds.) ICAISC 2018. LNCS (LNAI), vol. 10842, pp. 788–798. Springer, Cham (2018). https://doi.org/10.1007/978-3-319-91262-2_69

Measuring the Complexity of DMN Decision Models

Faruk Hasić[(✉)], Alexander De Craemer, Thijs Hegge, Gideon Magala,
and Jan Vanthienen

Department of Decision Sciences and Information Management,
KU Leuven, Leuven, Belgium
faruk.hasic@kuleuven.be

Abstract. Complexity impairs the maintainability and understandability of conceptual models. Complexity metrics have been used in software engineering and business process management (BPM) to capture the degree of complexity of conceptual models. A vast array of metrics has been proposed for processes in BPM. The recent introduction of the Decision Model and Notation (DMN) standard provides opportunities to shift towards the Separation of Concerns paradigm when it comes to modelling processes and decisions. However, unlike for processes, no studies exist that address the representational complexity of DMN decision models. In this paper, we provide a first set of ten complexity metrics for the decision requirements level of the DMN standard by gathering insights from the process modelling and software engineering fields. Additionally, we offer a discussion on the evolution of those metrics and we provide directions for future research on DMN compexity.

Keywords: Decision modelling · Decision Model and Notation ·
DMN · Complexity · Complexity metrics

1 Introduction

Decision modelling has seen a surge in scientific literature, as illustrated by the vast body of recent work on DMN [1–7]. DMN consists of two levels. Firstly, the decision requirement level in the form of a Decision Requirement Diagram (DRD) is used to portray the requirements of decisions and the dependencies between the different constructs in the decision model. Secondly, the decision logic level is used to specify the underlying decision logic, usually in the form of decision tables. The standard also provides an expression language FEEL (Friendly Enough Expression Language), as well as boxed expressions and decision tables for the notation of the decision logic. In DMN rectangles are used to depict decisions, corner-cut rectangles for business knowledge models, and ovals to represent data input. The arrows represent information requirements (from data or decisions). DMN aims at providing a clear and simple representation

© Springer Nature Switzerland AG 2019
F. Daniel et al. (Eds.): BPM 2018 Workshops, LNBIP 342, pp. 514–526, 2019.
https://doi.org/10.1007/978-3-030-11641-5_41

of decisions in a declarative form and offers no decision resolution mechanism of its own. Rather, the invoking context, e.g. a business process, is responsible for ensuring a correct invocation and enactment of the decision, as well as ensuring data processing and the storage and propagation of data and decision outcomes throughout the process. This makes DMN particularly interesting for a *Service-Oriented Architecture*, as DMN is independent of the applications and the invoking context. That way, DMN is able to capitalise on the benefits that are inherent to service-orientation in terms of maintainability, scalability, understandability, and flexibility both for modelling and mining decisions.

Complexity metrics have been adopted in the BPM field for process model complexity and applied on for instance the Business Process Model and Notation (BPMN) standard [8]. Despite the adoption of the DMN standard in the BPM field, a discussion on DMN model complexity is still lacking in literature. This paper aims at addressing that research gap and at proposing a set of metrics for the decision requirements level of the DMN standard.

This paper is structured as follows. In Sect. 2, relevant works on complexity are provided, as well as a running example that will be used throughout the paper. Section 3 provides a first set of ten DRD metrics for DMN models, while Sect. 4 outlines a discussion on the evolution of the proposed metrics. Section 5 provides an initial empirical evaluation of the metrics. In Sect. 6 a research agenda for DMN model complexity is contributed. Finally, Sect. 7 provides the conclusions.

2   Related Work and Running Example

In this section we provide an overview of related work for DMN, complexity metrics in the BPM field, and complexity assessments to the DMN standard in particular. Additionally, we provide a running example which will be used to illustrate the proposed complexity metrics in the subsequent sections.

2.1   Related Work

Recent BPM literature moves towards accommodating decision management into the paradigms of The Separation of Concerns (SoC) [3,9,10] and Service-Oriented Architecture (SOA) [6], by externalising decisions and encapsulating them into separate decision models, hence implementing decisions as externalised services. Literature proposes several conceptual decision service platforms and frameworks [6,11,12] and industry has adopted this trend, as several decision service systems have appeared, e.g. SAP Decision Service Management [13]. This externalisation of decisions from processes provides a plethora of advantages regarding maintainability and flexibility of both process and decision models [3,6,10,14–17].

A plethora of works on software complexity metrics exists [18–20]. Additionally, software metrics have been transformed and applied on processes and workflow nets in a vast array of studies [21–27]. Most of these studies focus on the BPMN standard. A systematic literature review of process metrics is provided in [28], where the authors identify and discuss 65 process metrics found in BPM literature.

Unlike for processes and BPMN, few works on complexity metrics for DMN models exist. In [29], the meta model complexity of the DMN modelling method is assessed according to the theory specified by [30]. Additionally, an explorative study of the notational aspects of DMN was conducted in [31]. In this study the authors focus on the cognitive analysis of the DMN notation in the light of theories on effective visual design. Hence, DMN complexity was assessed on a meta model level, i.e. the theoretical complexity of the modelling method as a whole, and on the cognitive visual level. However, no works on the complexity of DMN decision models are present in literature. In the following sections, we propose a set of complexity metrics for the DRD level of DMN.

2.2 Running Example

Figure 1 provides a running example of a DRD model that will be used in the coming section to illustrate the complexity metrics. The DRD represents an event selection decision based on the preferred location and the food and drinks that are offered, while taking into consideration the season, the number of guests, whether children are allowed, the sleeping facilities, and the budget. The value of every proposed metric will be calculated for this DRD.

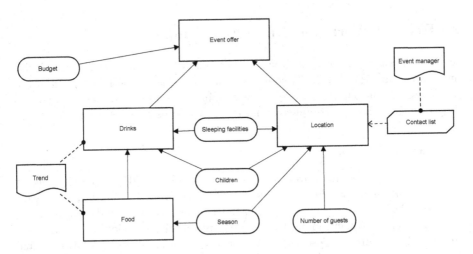

Fig. 1. DRD running example.

3 DRD Metrics

In this section we provide a set of ten DRD metrics that are capable of representing graph complexity in analogy with business process or software engineering literature. For every metric, a brief explanation is provided. Additionally, we calculate the value of every proposed metric for the running example provided in Fig. 1. An overview of the metrics and the metric values for the running example is given in Table 1. Later on, we will discuss the evolution of the metrics and validate them through an exploratory survey.

3.1   Number of Decisions (NOD)

As proposed by [23], BPMN complexity can be measured by counting the number of activities. They called this metric *number of activities* (NOA), which is a summation of all activity elements in the model. A similar metric can be worked out for DMN, counting the number of decision nodes in the DRD model instead of counting the activities, thus arriving to the metric of *number of decisions* (NOD). Applied to the running example of Fig. 1, the NOD is 4. As the models grow larger, they tend to have more decision elements. Thus, this metric will go up if a decision element is added to the model, indicating that the model has become more complex according to this metric. Given that DMN is a standard for modelling decisions, we assume that the number of decisions that are modelled within one DRD model will be indicative of the complexity of the model. However, note that the granularity of the DRD will play a crucial role as well, as one decision node can possibly be decomposed into a number of decision nodes each containing a portion of the underlying decision logic. Therefore, it will be of paramount importance to develop complexity metrics for the logic layer of DMN as well to capture these changes in granularity of the DRD model.

3.2   Number of Elements (NOE)

The *number of elements* (NOE) is the sum of all building blocks of the DRD. Hence, NOE takes into account all elements of the DRD rather than only the decision nodes, as is the case in the NOD metric. More specifically:

$$NOE = \#decisions + \#inputs + \#knowledgesources$$
$$+ \#businessknowledgemodels + \#informationrequirements$$
$$+ \#knowledgerequirements + \#authorityrequirements$$

Applied to the running example model in Fig. 1, the NOE is 27. The larger the DRD model, the higher NOE will be. This is self-explanatory, as DRD models are solely made up out of these elements.

3.3   Number of Basic Elements (NOBE)

The most basic elements of a DRD model are decisions nodes, input nodes, and information requirements. They form the spine of a DRD model. Therefore, the *number of basic elements* (NOBE) probably is a good metric. NOBE can be calculated as follows:

$$NOBE = \#decisions + \#inputs + \#informationrequirements$$

Applied to the running example in Fig. 1, the NOBE is 20. Clearly, the NOBE metric will be higher as basic elements are added to the DRD model.

3.4   Total Number of Data Objects (TNDO)

As discussed in [28], the *total number of data objects* (TNDO) represents all data objects in the BPMN diagram. For a DRD, data objects are represented by all input data objects present in the model. The running example in Fig. 1 has 5 input data elements, hence the TNDO is 5.

Similar to the explanation of NOD and NODIR, bigger models tend to have more input data elements. By adding a data input element to a model, the TNDO metric will grow.

3.5   Sequentiality (SEQ)

As pointed out by [28], the sequentiality (SEQ) of BPMN is equal to one minus the percentage of nodes with no more than one incoming and outgoing arrow. In other terms: the percentage of nodes that have more than one successor or predecessor. This metric can be used in the same way for the nodes of a DRD. Sequentiality is expressed as a number between one and zero. If the DRD looks more like a sequence rather than a parallell network, the value of the sequentiality metric will be low. This corresponds with a less complex model, and vice versa. Applying this to the running example in Fig. 1, we get the formula

$$SEQ = 1 - 4/12 = 0.6667.$$

Models with a lot of single-path sequences will have a low complexity value. Also note that sequentiality in DMN will be greatly impacted by leaf elements since DRD models usually have multiple input data elements, thus increasing the SEQ metric.

3.6   Longest Path (LP)

Unlike BPMN, a DRD model does not allow loops. Therefore, the *longest path* (LP) can be measured unambiguously. Calculating the longest path of a DRD graph, which in essence is a directed acyclic graph (DAG), is done by topologically sorting the graph [32]. In the running example of Fig. 1, the LP of 4 is found. This is the length of the path going from *Season* to *Event offer* through *Food* and *Drinks*. It is not possible to find a longer path in the model. Typically, the longest path and the sequentiality metrics of a DRD will oppose in value. When a model is more sequential, it will have a low complexity according to the *sequentiality* metric. However, the model will typically have longer paths and thus higher complexity according to the *longest path* metric. This proves the importance of using multiple complexity metrics to assess DRD model complexity from different perspectives.

As DRD models are getting bigger and more arcs, i.e. requirements in the DRD graph, are introduced, the value of LP will indicate a high complexity.

3.7   Average Vertex Degree (AVD)

The *average vertex degree* (AVD) [33] is calculated as the average of all incoming and outgoing connections across all nodes of the DRD. This can be applied directly to DRD graphs. The bigger the AVD, the more complex the model. Applying this to the running example in Fig. 1, we get the following result:

$$AVD = (1 + 3 + 1 + 5 + 2 + 6 + 2 + 2 + 2 + 3 + 2 + 1)/12 = 2.5$$

The average vertex degree heavily relies on the number of connections between DRD model elements. In other terms, the more the decision requirements diagram resembles a strongly connected network, the more complex it is. Additionally, it might be interesting to look at the modular behaviour of the DRD model trough fan-in and fan-out metrics that are heavily dependent on the average vertex degree.

3.8   Coefficient of Network Complexity (CNC)

The *coefficient of network complexity* (CNC) was proposed to measure the degree of complexity of a critical pass network [34]. It was adapted by [23] to measure the degree of complexity in processes by dividing the number of arcs by the number of the activities, splits and joins in the BPMN diagrams. It is possible to have identical values of the coefficient of network complexity for different models but with different comprehensibility due to a different set of used node types. This metric can be adopted for the DMN standard by focusing on the nodes and arcs in the decision requirements diagram. Applying this to the running example in Fig. 1 gives the following result:

$$CNC = 15/12 = 1.25$$

Clearly, if an arc is added to the DRD graph, the CNC value will increase because of the increasing effect on the numerator.

3.9   Knot Count (KC)

In decision models, some components, more specifically requirement associations, may be forced to cross each other. This is captured in the *knot count* (KC) metric. Each occurrence of a crossing is expressed as a knot and each knot occurrence in a DRD means an increase in the complexity of understanding the model. Unlike the metrics used by [18] which focused on counting the knots created by the crossing of only arrows, counting all requirement relation crossings regardless of their types is suggested for DMN adoption. The higher the knot count value, the higher the complexity assumed. The running model in Fig. 1 does not have knot occurrences, and hence has a knot count value of 0.

As more arcs and nodes are introduced in the DRD model, the more difficult it becomes to avoid crossing arcs, i.e. knots. This will thus likely result in a higher knot count.

3.10   Cyclomatic Complexity (CC)

In [23] the adaptation of McCabes *cyclomatic complexity* (CC) metric [35] for process is proposed. According to [19], this is one of the most widely used complexity metrics. The cyclomatic complexity formula for non-strongly connected graphs, such as a DRD graph, is the number of edges (E) minus the number of nodes (N) plus two times the number of connected components. Since a decision requirements diagram is one connected component, the formula can be reduced to the following calculation for the running example in Fig. 1:

$$CC = E - N + 2 = 15 - 12 + 2 = 5$$

Thus, larger models, especially those that contain many arcs, are likely to have a higher CC value.

Table 1. DRD metrics as calculated for the running example in Fig. 1.

Metric	Value
Number of decisions (NOD)	4
Number of elements (NOE)	27
Number of basic elements (NOBE)	20
Total number of Data Objects (TNDO)	5
Sequentiality (SEQ)	0.6667
Longest Path (LP)	4
Average Vertrex Degree (AVD)	2.5
Coefficient of Network Complexity (CNC)	1.25
Knot Count (KC)	0
Cyclomatic Complexity (CC)	5

4   Expected Evolution of the Metrics

In this section we concisely discuss the evolution of the metric values when a certain element is added to the DRD model. We limit our discussion to adding arc requirements and decision nodes to the DRD model respectively. When a decision requirements diagram gets larger in terms of number of elements, most metrics will indicate that the representational complexity of the decision model has increased. Model size is what most of the proposed metrics rely on. NOD, NOE, NOBE, and TNDO are simple count metrics that grow larger as relevant elements are added to the model. The remaining metrics, i.e. SEQ, LP, AVD, CNC, KC, and CC, are also indirectly dependent on the number of DRD elements. Here too, adding an element to the DRD model is likely to result in an increase in complexity metric values. Note that all metrics have a lower value when indicating simpler models and a higher value when indicating more complex DRD models.

4.1   Requirement Arcs

The following metrics all rely on the number of arcs in the model, i.e. information requirements, authority requirements, and knowledge requirements. When a requirement is added to a DRD model:

- NOE increases since it is the summation of all elements.
- NOBE increases if that arc is an information requirement.
- LP is not affected or can increase, depending on whether the added arc results in a longer or another longest path.
- AVD will definitely increase because the new connection will always positively impact exactly two model elements, which in turn increases the overall average vertex degree.
- CNC will increase given the increasing effect on the numerator in the formula.
- KC can never decrease as the new arc can either cross existing arcs or not.
- CC will increase given the increasing effect on the first term in the formula.

4.2   Decision Nodes

The following metrics all rely on the number of decision nodes in the model. When a decision node is added to a DRD model:

- NOD increases by definition.
- NOE increases since it is the summation of all elements.
- NOBE increases since a decision node is a basic element.
- LP is either not affected or it is likely to increase, depending on whether the added decision node results in a longer or another longest path.
- CC stays unchanged or decreases. While the formula suggests that CC would increase, this is not the case in reality. When a decision node is added, at least one edge is added as well to connect the node to the rest of the graph.

5   Empirical Evaluation

An exploratory survey was held during the master's course of *Knowledge Management and Business Intelligence* at KU Leuven. Students were presented with 11 DRD models ranging from simple to complex in an arbitrary order. The students were asked to indicate on a visual analogue scale how complex they perceived each of the DRD models to be. In total 22 students with previous knowledge about DMN took part in the survey.

To detect how well the proposed metrics describe the perceived complexity as indicated by the survey, the metric values were compared to the survey results. This was done by calculating the correlation and the sum of squared differences (SSD) of the metric values for all the DRD models and the survey averages. In order to calculate a valid sum of squared differences, the metric values were first scaled to a range from zero to ten, i.e. reflecting the complexity range of the visual analogue scale of the survey.

Initial results are presented in Table 2. For all the 11 tested DRD models, the metric value of all 10 proposed metrics are calculated and included in the table. Higher (lower) values of the metrics represent a higher (lower) degree of complexity. The final two rows of the table give the average degree of complexity as indicated by the students in the survey on the visual analogue scale (on a scale of 10) and the standard deviation of the complexity as indicated in the survey. The final two columns in Table 2 depict the correlation and the sum of squared differences (SSD) of the metric values and the survey results respectively.

By examining the sum of squared differences we can conclude that basic metrics such as Number of Elements (NOE), Number of Decisions (NOD), and Total Number of Data Objects (TNDO) measure the perceived complexity quite well, indicated by the low values in SSD. Additionally, the Cyclomatic Complexity (CC) also showcases a low SSD, indicating that the popular CC metric might be a good measure for DRD model complexity as well.

6   Future Work

In future work, we will expand the set of DRD metrics to include other metrics from the software engineering and BPM fields. Additionally note that DMN contains two levels: the DRD and the decision logic level, usually specified in the form of decision tables. Thus, we plan to constitute a set of complexity metrics for the decision tables by capitalising on complexity metrics of database tables. Furthermore, we will look into combining DRD and decision table metrics into aggregated and holisitc complexity metrics, thus denoting the complexity of the entire DMN decision model.

Next to this theoretical metric discourse on DMN complexity, we will look into additional empirical validation for the proposed metrics through additional surveys. Finally, inquiries into the complexity of integrated process and decision models will be conducted, by combining and integrating complexity metrics of DMN decision models and BPMN process models.

Table 2. Survey results and analysis

Metric	Model 1	Model 2	Model 3	Model 4	Model 5	Model 6	Model 7	Model 8	Model 9	Model 10	Model 11	Cor	SSD
NOD	2.00	2.00	5.00	4.00	7.00	4.00	8.00	7.00	5.00	5.00	11.00	-0.09	106.65
NOE	9.00	16.00	17.00	27.00	29.00	27.00	46.00	42.00	28.00	37.00	33.00	0.20	129.88
NOBE	9.00	8.00	13.00	20.00	29.00	20.00	43.00	21.00	20.00	29.00	40.00	0.01	122.56
TNDO	3.00	2.00	2.00	5.00	4.00	5.00	14.00	2.00	3.00	9.00	4.00	0.05	92.49
SEQ	0.40	0.50	0.22	0.67	0.82	0.67	0.43	0.58	0.58	0.33	0.55	0.14	133.42
LP	3.00	3.00	3.00	4.00	6.00	4.00	5.00	6.00	5.00	5.00	7.00	0.07	146.87
AVD	1.60	1.89	1.78	2.50	3.45	2.50	2.09	2.42	2.67	2.11	3.10	-0.03	157.74
CNC	0.80	1.00	0.89	1.25	1.73	1.25	1.04	1.16	1.33	1.06	1.52	-0.06	156.03
KC	0.00	0.00	0.00	0.00	8.00	3.00	1.00	4.00	0.00	0.00	28.00	0.10	169.32
CC	1.00	2.00	1.00	4.00	13.00	5.00	3.00	7.00	6.00	3.00	8.00	-0.12	113.63
Survey AVG	3.5364	3.7182	1.6182	4.0409	1.2286	7.1909	2.8909	4.8909	3.4136	4.9318	4.5636		
STDEV	2.2311	1.9549	1.1636	1.7928	1.7373	2.0672	1.7949	1.5405	1.5388	2.3204	1.9182		

7 Conclusion

This paper provides a first discussion on complexity metrics for individual DMN decision models. Ten complexity metrics for decision requirement graphs were proposed and illustrated on a running example. Furthermore, metric evolution was discussed and an agenda for future inquiry into DMN decision model complexity was suggested. The emphasis was put on expanding the body of metrics to the decision logic level of DMN and later on combining the metrics of both the logic level and the requirements level into aggregate metrics for the DMN model as a whole. Finally, a survey was set up to empirically evaluate the proposed complexity metrics and initial results revealed that the simple metrics were the most suitable for capturing DRD complexity.

References

1. OMG: Decision Model and Notation 1.1 (2016)
2. Horita, F.E., de Albuquerque, J.P., Marchezini, V., Mendiondo, E.M.: Bridging the gap between decision-making and emerging big data sources: an application of a model-based framework to disaster management in Brazil. Decis. Support Syst. **97**, 12–22 (2017)
3. Hasić, F., De Smedt, J., Vanthienen, J.: Augmenting processes with decision intelligence: principles for integrated modelling. Decis. Support Syst. **107**, 1–12 (2018)
4. Estrada-Torres, B., del-Río-Ortega, A., Resinas, M., Ruiz-Cortés, A.: On the relationships between decision management and performance measurement. In: Krogstie, J., Reijers, H.A. (eds.) CAiSE 2018. LNCS, vol. 10816, pp. 311–326. Springer, Cham (2018). https://doi.org/10.1007/978-3-319-91563-0_19
5. De Smedt, J., Hasić, F., vanden Broucke, S.K.L.M., Vanthienen, J.: Towards a holistic discovery of decisions in process-aware information systems. In: Carmona, J., Engels, G., Kumar, A. (eds.) BPM 2017. LNCS, vol. 10445, pp. 183–199. Springer, Cham (2017). https://doi.org/10.1007/978-3-319-65000-5_11
6. Hasić, F., De Smedt, J., Vanthienen, J.: A service-oriented architecture design of decision-aware information systems: decision as a service. In: Panetto, H., et al. (eds.) On the Move to Meaningful Internet Systems. OTM 2017. LNCS, vol. 10573, pp. 353–361. Springer, Cham (2017). https://doi.org/10.1007/978-3-319-69462-7_23
7. Calvanese, D., Dumas, M., Laurson, Ü., Maggi, F.M., Montali, M., Teinemaa, I.: Semantics, analysis and simplification of DMN decision tables. Inf. Syst. **78**, 112–125 (2018)
8. OMG: Business Process Model and Notation (BPMN) 2.0 (2011)
9. Campos, J., Richetti, P., Baião, F.A., Santoro, F.M.: Discovering business rules in knowledge-intensive processes through decision mining: an experimental study. In: Teniente, E., Weidlich, M. (eds.) BPM 2017. LNBIP, vol. 308, pp. 556–567. Springer, Cham (2018). https://doi.org/10.1007/978-3-319-74030-0_44
10. Hasić, F., Devadder, L., Dochez, M., Hanot, J., De Smedt, J., Vanthienen, J.: Challenges in refactoring processes to include decision modelling. In: Teniente, E., Weidlich, M. (eds.) BPM 2017. LNBIP, vol. 308, pp. 529–541. Springer, Cham (2018). https://doi.org/10.1007/978-3-319-74030-0_42

11. Zarghami, A., Sapkota, B., Eslami, M.Z., van Sinderen, M.: Decision as a service: separating decision-making from application process logic. In: EDOC, pp. 103–112. IEEE Computer Society (2012)
12. Mircea, M., Ghilic-Micu, B., Stoica, M.: An agile architecture framework that leverages the strengths of business intelligence, decision management and service orientation. In: Business Intelligence-Solution for Business Development (2011)
13. SAP: SAP Decision Service Management. http://www.sap.com/pc/tech/business-process-management/software/decision-service-management/index.html
14. Biard, T., Le Mauff, A., Bigand, M., Bourey, J.-P.: Separation of decision modeling from business process modeling using new "decision model and notation" (DMN) for automating operational decision-making. In: Camarinha-Matos, L.M., Bénaben, F., Picard, W. (eds.) PRO-VE 2015. IAICT, vol. 463, pp. 489–496. Springer, Cham (2015). https://doi.org/10.1007/978-3-319-24141-8_45
15. van der Aa, H., Leopold, H., Batoulis, K., Weske, M., Reijers, H.A.: Integrated process and decision modeling for data-driven processes. In: Reichert, M., Reijers, H.A. (eds.) BPM 2015. LNBIP, vol. 256, pp. 405–417. Springer, Cham (2016). https://doi.org/10.1007/978-3-319-42887-1_33
16. Hasić, F., De Smedt, J., Vanthienen, J.: An illustration of five principles for integrated process and decision modelling (5PDM). Technical report, KU Leuven (2017)
17. Figl, K., Mendling, J., Tokdemir, G., Vanthienen, J.: What we know and what we do not know about DMN. Enterp. Model. Inf. Syst. Archit. **13**(2), 1–16 (2018)
18. Woodward, M.R., Hennell, M.A., Hedley, D.: A measure of control flow complexity in program text. IEEE Trans. Softw. Eng. **5**(1), 45–50 (1979)
19. Sharma, A., Kumar, R., Grover, P.: Empirical evaluation and critical review of complexity metrics for software components. In: Proceedings of the 6th WSEAS International Conference on SE, Parallel and Distributed Systems, pp. 24–29 (2007)
20. Queiroz, R., Passos, L., Valente, M.T., Hunsen, C., Apel, S., Czarnecki, K.: The shape of feature code: an analysis of twenty C-preprocessor-based systems. Softw. Syst. Model. **16**(1), 77–96 (2017)
21. Kluza, K., Nalepa, G.J., Lisiecki, J.: Square complexity metrics for business process models. In: Mach-Król, M., Pełech-Pilichowski, T. (eds.) Advances in Business ICT. AISC, vol. 257, pp. 89–107. Springer, Cham (2014). https://doi.org/10.1007/978-3-319-03677-9_6
22. Gruhn, V., Laue, R.: Adopting the cognitive complexity measure for business process models. In: 2006 5th IEEE International Conference on Cognitive Informatics. ICCI 2006, vol. 1, pp. 236–241. IEEE (2006)
23. Cardoso, J., Mendling, J., Neumann, G., Reijers, H.A.: A discourse on complexity of process models. In: Eder, J., Dustdar, S. (eds.) BPM 2006. LNCS, vol. 4103, pp. 117–128. Springer, Heidelberg (2006). https://doi.org/10.1007/11837862_13
24. Kluza, K.: Measuring complexity of business process models integrated with rules. In: Rutkowski, L., Korytkowski, M., Scherer, R., Tadeusiewicz, R., Zadeh, L.A., Zurada, J.M. (eds.) ICAISC 2015. LNCS (LNAI), vol. 9120, pp. 649–659. Springer, Cham (2015). https://doi.org/10.1007/978-3-319-19369-4_57
25. Sánchez-González, L., García, F., Mendling, J., Ruiz, F., Piattini, M.: Prediction of business process model quality based on structural metrics. In: Parsons, J., Saeki, M., Shoval, P., Woo, C., Wand, Y. (eds.) ER 2010. LNCS, vol. 6412, pp. 458–463. Springer, Heidelberg (2010). https://doi.org/10.1007/978-3-642-16373-9_35
26. Lassen, K.B., van der Aalst, W.M.: Complexity metrics for workflow nets. Inf. Softw. Technol. **51**(3), 610–626 (2009)

27. Gruhn, V., Laue, R.: Complexity metrics for business process models. In: 9th International Conference on Business Information Systems (BIS 2006), vol. 85, pp. 1–12. Citeseer (2006)

28. Polančič, G., Cegnar, B.: Complexity metrics for process models-a systematic literature review. Comput. Stan. Interfaces **51**, 104–117 (2017)

29. Hasić, F., De Smedt, J., Vanthienen, J.: Towards assessing the theoretical complexity of the decision model and notation (DMN). In: Enterprise, Business-Process and Information Systems Modeling, vol. 1859, pp. 64–71. Springer, Heidelberg (2017)

30. Rossi, M., Brinkkemper, S.: Complexity metrics for systems development methods and techniques. Inf. Syst. **21**(2), 209–227 (1996)

31. Dangarska, Z., Figl, K., Mendling, J.: An explorative analysis of the notational characteristics of the decision model and notation (DMN). In: 2016 IEEE 20th International Enterprise Distributed Object Computing Workshop (EDOCW), pp. 1–9. IEEE (2016)

32. Kahn, A.B.: Topological sorting of large networks. Commun. ACM **5**(11), 558–562 (1962)

33. Cardoso, J.: Business process quality metrics: log-based complexity of workflow patterns. In: Meersman, R., Tari, Z. (eds.) OTM 2007. LNCS, vol. 4803, pp. 427–434. Springer, Heidelberg (2007). https://doi.org/10.1007/978-3-540-76848-7_30

34. Kaimann, R.A.: Coefficient of network complexity. Manag. Sci. **21**(2), 172–177 (1974)

35. McCabe, T.J.: A complexity measure. IEEE Trans. Softw. Eng. **2**(4), 308–320 (1976)

Joint Requirements Engineering and Business Process Management Workshop/Education Forum (REBPM/EdForum)

Joint Requirements Engineering and Business Process Management Workshop/Education Forum (REBPM/EdForum)

This joint workshop brings together practitioners and researchers interested in requirements engineering and education in BPM. Successful business process management (BPM) involves the ability to create value through effectively orchestrating, communicating, and transforming business processes across the organization. These efforts require a combination of a plethora of knowledge, skills, and abilities that many organizations find lacking in their current workforce. As BPM continues to evolve as a discipline, there is a need to study the current state of BPM's body of knowledge and how it addresses the BPM capabilities needed in practice. A deeper appreciation on how process-centric concepts are expressed, understood, and consumed by diverse learner groups is needed to design well-grounded and applicable BPM-related curricula.

The first part of the workshop concentrates on the knowledge behind interrelations between RE and BPM, with a focus on agile and flexible BPM as well addressing the need of providing users with the required flexibility in defining processes, which has also risen in importance. Two papers were selected. The first paper, "From Requirements to Data Analytics Process: An Ontology-Based Approach" by Madhushi Bandara et al. from UNSW, investigates the use of process patterns and ontologies for expressing frequently used sets of analytics requirements. The second paper, "Process Weakness Patterns for the Identification of Digitalization Potentials in Business Processes" by Florian Rittmeier et al. from Paderborn University, is concerned with the systematic identification of digitalization potentials in business processes via process weakness patterns in BPMN diagrams. The two papers had some similarities by putting more emphasis on the use of knowledge management techniques and the need to capture domain knowledge during analysis. In terms of proposed solutions, both papers advocate the use some form of patterns to capture the essence of the solution.

This second part of the workshop aims to share innovative BPM education approaches and to promote a dialog between the academy and practice. This year, we invited two papers: The first paper, "An Assignment on Information System Modeling: On Teaching Data and Process Integration," was presented by Jan Martijn Van Der Werf from Utrecht University. The paper presented the design specification and experience of running of an assignment specifically aimed at teaching the inter-related nature of data and process and how the two concepts should be integrated in an information system. The second paper, "Motivational and Occupational Self-Efficacy Outcomes of Students in a BPM Course: The Role of Industry Tools vs Digital Games," was presented by Jason Cohen from the University of the Witwatersrand. The paper presented the design and experimental results of student learning activities aimed at assessing the effectiveness of digital games vs. industry tools in promoting occupational self-efficacy and motivation.

Overall, the presentations provoked many comments and feedback, which created an active discussion session for all participants.

November 2018

Banu Aysolmaz
Rüdiger Weibach
Onur Demirörs
Fethi Rabhi
Wasana Bandara
Hye-young Paik
Cesare Pautasso

Organization

Program Committee

Boualem Benatallah	University of New South Wales, Australia
Maria Bielikova	Slovak University of Technology in Bratislava, Slovakia
Fabio Casati	University of Trento, Italy
Ahmet Coskuncay	Middle East Technical University, Turkey
Maya Daneva	University of Twente, The Netherlands
Florian Daniel	Politecnico di Milano, Italy
Peter Dolog	Aalborg University, Denmark
Mahdi Fahmideh	University of Technology Sydney, Australia
Hakim Hacid	Zayed University, UAE
Nico Herzberg	SAP, Dresden, Germany
Jennifer Horkoff	Chalmers University of Technology, Gothenburg, Sweden
Ralf Klamma	RWTH Aachen University, Germany
Matthias Kunze	Zalando SE, Berlin, Germany
Marcello La Rosa	The University of Melbourne, Australia
Ralf Laue	Zwickau University of Applied Sciences, Germany
Michael Leyer	University of Rostock, Germany
Sinisa Neskovic	University of Belgrade and Dorius d.o.o, Serbia
Olivera Marjanovic	University of Technology, Sydney, Australia
Paul Mathiesen	Queensland Health, Australia
Juergen Moormann	Frankfurt School of Finance and Management, Germany
Hannes Schlieter	TU Dresden, Germany
Rainer Schmidt	Munich University of Applied Sciences, Germany
Heiko Schuldt	University of Basel, Switzerland
John Shepherd	University of New South Wales, Sydney, Australia
Inge van de Weerd	Vrije Universiteit Amsterdam, The Netherlands

Process Weakness Patterns
for the Identification of Digitalization
Potentials in Business Processes

Florian Rittmeier[1]([✉]), Gregor Engels[1], and Alexander Teetz[2]

[1] SI-Lab, Paderborn University, Fürstenallee 11, 33102 Paderborn, Germany
{florianr,engels}@uni-paderborn.de
[2] Department of Computer Science, Paderborn University, Fürstenallee 11,
33102 Paderborn, Germany
alexander.teetz@uni-paderborn.de

Abstract. An important element of digital transformation is the digitalization of processes within enterprises. A major challenge is the systematic identification of digitalization potentials in business processes. Existing approaches require process analysts who identify these potentials by using the time-consuming method of pattern catalogs or by relying on their professional experiences. In this paper, we classify potentials of digitalization and derive corresponding patterns for a future pattern-based analysis procedure. This shall enable the automated identification of digitalization potentials in BPMN diagrams. Those patterns were derived from our work with five companies from different sectors. In comparison to existing approaches, our proposed method could support a more efficient and effective identification of digitalization potentials by process analysts.

Keywords: Digitalization potentials · Process weakness patterns ·
BPI · Digital transformation · Information flow modeling ·
Requirements engineering

1 Motivation

Digitalization is on everyone's mind as it changes many areas of life. Digitalization also changes the general conditions for companies, for example when competitors make existing products and services more attractive for customers by exploiting digitalization potentials. Those digitalization potentials for processes arise, for example, if existing processes can be improved through the use of new assistance systems or digital interfaces. Correspondingly, companies have to adapt by identifying and exploiting digitalization potentials in their own company and market in order to remain competitive.

When discussing digitalization potentials, it is important to emphasize what is meant by digitalization. The differences between digitization, digitalization and digital transformation are explained using an example from Fischer et al. (2017). If the business process of an industrial picking scenario is performed

© Springer Nature Switzerland AG 2019
F. Daniel et al. (Eds.): BPM 2018 Workshops, LNBIP 342, pp. 531–542, 2019.
https://doi.org/10.1007/978-3-030-11641-5_42

using a "Paper-based Clipboard" and this is replaced by a "Digital checklist on a tablet" this change represents the change by doing digitization, i.e. replacing paper by bits and bytes. The "Digital checklist on a tablet ordered dynamically based on a big data analysis" is an example of the change in the process through the use of digital technology, which we call digitalization. Thus, digitization is a more basic and technical transformation, which forms the foundation for more complex transformations in which digitalization enables news types of processes. While we will mostly talk about digitalization, this usually also includes digitization. The digital transformation is an even broader term and includes the transformation of business process, competencies, activities and models.

It is a challenge for companies to systematically identify the digitalization potentials of their processes. Figure 1 illustrates how process analysts identify digitalization potentials today. As a common methodical basis, practitioners and scholars recommend modeling a business process to document the current business processes using a language for describing business processes (Process discovery). Process analysts then identify process weaknesses by analyzing the business processes (Process analysis) and provide digitalization recommendations, which then are discussed with the process stakeholders and lead to improved business processes (Process redesign). Thus digitalization potentials are exploited.

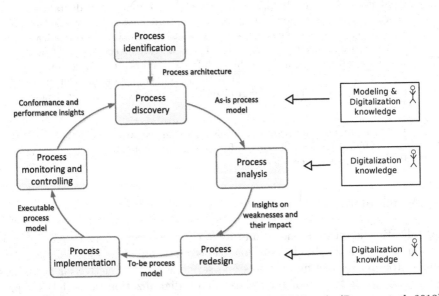

Fig. 1. Work of process analysts today related to BPM Lifecycle (Dumas et al. 2013)

Accordingly, process analysts and their knowledge of modeling and digitization play an important role today. Such process analysts are in demand and their use entails considerable costs for small and medium-sized enterprises (SMEs). Therefore, the question arises as to how the number of process analysts who deliver high-quality results can be increased and support from them can become more cost-effective for SMEs.

Other approaches address these challenges by describing process weaknesses, such as digitalization potentials, using so-called process weakness patterns. These allow process analysts to discuss the characteristics of weaknesses and also to identify weaknesses based on these patterns. This is how the quality of the results of the process analysts is decoupled from the knowledge they have gathered in practice. Unfortunately, the systematic application of these patterns in existing approaches is very time-consuming. Others are more efficient, but focus on digitalization potentials for the public sector instead on those for SMEs.

We propose using an assistance system that supports process analysts by automatically identifying digitalization potentials in business process models using process weakness patterns for digitalization. In addition, we provide guiding questions that support process analysts in capturing relevant digitalization aspects and describe a relevant language extension for BPMN 2.0 in order to be able to model these digitalization aspects.

The patterns, guiding questions and insights of our approach are based on our work in the project "Business 4.0 – New business models and value chains with ICT"[1]. The aim of the project is to support small and medium-sized companies in developing digitalization strategies. Within the scope of this project, workshops were conducted with five companies in order to identify digitalization potentials in processes relevant for SMEs. The companies came from different industries and acted as research subjects.

The rest of the paper is structured as follows: We first discuss the related work (Sect. 2). Afterwards, our solution approach is explained (Sect. 3) and examples for digitalization potentials are given (Sect. 3.1). In addition, the guiding questions are presented (Sect. 4) followed by the introduction of the information carrier type (Sect. 5). Based on this language extension, the process weakness patterns for digitalization are described (Sect. 6). The article is concluded with a summary and outlook (Sect. 7).

2 Related Work

The result of the analyses of process analysts is highly dependent on the experience and interpretation of the process analyst (Phalp and Shepperd 2000). Less experienced process analysts produce less effective results. Vergidis et. al. (2008) emphasize that for analyses of business process models, which should not be primarily based on experience of the process analyst, support from the business process modeling language is necessary. This enables implicit knowledge to be documented explicitly.

Language support would make it possible to identify digitalization potentials using process weakness patterns. Such a pattern is a formalized description of a process weakness. As a rule, such a description refers to a part of a process

---

[1] http://owl-morgen.de/projekte/business-40/.

model that can be described on the basis of concrete structural properties. Based on such a pattern, comparable constellations in other process models can be identified on the basis of this pattern.

Existing approaches such as Falk (2017) already work with patterns, but the process analyst must check manually whether a pattern is applicable. For many patterns this check is very time-consuming (Falk 2017). Our approach focuses on the use of an assistance system which is intended to identify applicable patterns and thereby not only make the work of the process analyst more effective, but at the same time make it as efficient as possible.

Other approaches like Höhenberger and Delfmann (2015) use automated matching of process weakness patterns to analyze existing process models from the public sector. Most of the SMEs we had contact with do not have existing process models. Therefore, modeling the business process usually is the starting point. This allows to use guiding questions when modeling to take relevant information required for the later analysis into account and also to model digitalization aspects in more detail if the process modeling language allows for that. Therefore, our approach is tailor-made for digitalization and, to the best of our knowledge, more holistic.

3   Solution Approach

The resulting question is how digitalization potentials in business process models can be identified on the basis of language elements. We employ the business process modeling language BPMN 2.0 (Object Management Group 2011), since it is widely used in practice (Dumas et al. 2013) and comes with a precise definition of syntax and execution semantics. Therefore, the following research questions should be noted:

1. How can process analysts be supported in modeling all aspects of a given process that are relevant to the detection of process weaknesses in a digitalization context?
2. Can digitalization potentials be identified by using language elements of BPMN 2.0?
3. Which process weakness patterns describe digitalization potentials and which recommendations can be given on the basis of these potentials?

From the previous remarks it follows that it is necessary to identify digitalization potentials and patterns for these. Furthermore, it is necessary to describe the patterns in machine-readable form so that an algorithm can then check whether an application of a corresponding pattern in an (extended) BPMN diagram exists. With regard to such an algorithm, approaches such as Förster et al. (2007) can be used. Based on this, individual patterns can then also be assigned to recommendations that exploit the digitization potential. In a first step, these recommendations can be formulated in textual descriptions. In the future, these recommendations should be made directly applicable through suitable model transformations.

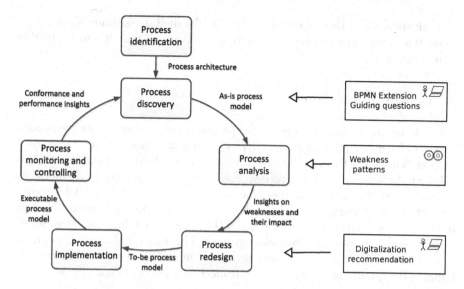

Fig. 2. Solution approach related to BPM Lifecycle (Dumas et al. 2013)

Putting our solution approach in the context of the BPM Lifecycle described by Dumas et al. (2013), we address the phases *Process discovery*, *Process analysis* and *Process redesign*. Guiding questions support the process analyst in capturing as many relevant aspects as possible in the course of the *Process discovery*. This supports the *Process analysis* using process weakness patterns for digitalization, as these can only be found automatically by the assistance system if the quality of the model reaches an appropriate level. As those patterns have digitalization recommendations associated they also provide relevant input for the *Process redesign*. Figure 2 illustrates the relationship between our solution approach and the BPM Lifecycle.

3.1 Digitalization Potentials

In the Business 4.0 project, we identified digitalization potentials in workshops on the digitalization of processes. These digitalization potentials relate to situations in which

- only nondigital information carriers are used in a process step,
- a nondigital information carrier is linked to the copy of the information in an IT system and this link is not simple[2] or not efficient[3],
- information between process steps is not transferred through digital information carriers,

---

[2] It is not considered simple if the digital twin has to be searched for, for example because no primary key exists or it cannot be used for selection.
[3] It is not efficient to type in a primary key, e.g. a customer or order number. Scanning the primary key with a reader would be considered efficient here.

- work steps could be supported by the use of a digital assistance system,
- unstructured data can be structured in such a way that it can be further processed.

4   Guiding Questions

The guiding questions shall support the process analyst in modelling the process to cover all aspects relevant for identifying process weaknesses in a digitalization context. Therefore, the structure used in Turban and Schmitz-Lenders (2017) is followed. Although they derive software requirements from the process model on the basis of guiding questions, the structuring on the basis of the model elements seems to be a promising procedure, since it easily allows a systematic check-up by the process analyst. We have identified the following guiding questions, which we have grouped according to model elements. This first set of guiding questions should be asked by a process analyst during modeling per model element to improve the quality of the model with respect to aspects of digitalization.

Process

- Are all decisions explicitly modeled?

Task

- Are all data inputs for the tasks covered? On which data does the task work?
- Did you specify whether a human user (manual task), a human user using an application (user task), or a service (service task) is performing the task?
- If a human user is using an application, when performing the task, did you model the application as data store, if the application has data store characteristics, or as artefact, if it does not have data store characteristics?

Decision (data driven)

- Is there a task in front of the decision node which prepares the decision?
- Does the task preparing the decision makes this based on data? Is this data modelled?

Data (input and output)

- Did you capture the type of information carrier?

Data input

- Is the source of the data modeled? Is it another task or a data store?

Data store

- Did you capture the type of information carrier?

When analyzing how a process analyst can use these guiding questions to model a business process, we faced the problem, that there is no existing good way in modeling the type of information carrier. The sole existing option is to model this aspect by writing it into labels of the relevant model elements. But this leads to ambiguity. We therefore choose to extend BPMN 2.0 to model this aspect as it is described in the next section.

5   Modeling the Information Carrier Type

In BPMN 2.0 you can only describe textually what kind of information carrier provides the information. The extension mechanism of BPMN 2.0 has to be used to add an attribute that formally describes the type of information carrier. Alternatively, you can insert new types of data objects that represent the values of the attribute. It would be good to visualize the additional information content as this supports process stakeholders during the discussion with the process analyst. Depending on the tool support, you can define a separate display for either one or the other.

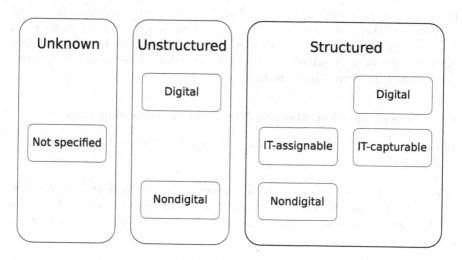

Fig. 3. Taxonomy of information carrier type

Our approach formulates the following possible values for the attribute with which we describe the information carrier formally. These values build a taxonomy illustrated in Fig. 3:

Not specified. This is the default value. Software systems/tools should encourage the modeler to set one of the other values.

Nondigital Unstructured. This classifies the data as unstructured data such as notes and original sounds that require interpretation.

Nondigital Structured. Although this data has a nondigital information carrier, the data is structured in such a way that it can be easily mapped in an IT system for further processing, e.g. a form.

IT-assignable. Information carriers that have the properties of *Nondigital Structured* and where the same data exists in an IT system and can be found there with little effort using a key, e.g. using an order number.

IT-capturable. For such information carriers, the data in the corresponding IT system can not only be found with little effort, but the key on the nondigital information carrier can be machine-recorded, e.g. with a barcode.

Digital Unstructured. It is a digital information carrier but the data has no or no useful meta-model, e.g. a PDF file.

Digital Structured. It is a digital information carrier and the data has a useful meta-model, e.g. a Excel file.

The assistance system can support the process analyst in choosing the right type by providing hints and examples.

6   Process Weakness Patterns for Digitalization

In order to identify the patterns, we have clarified which model elements and properties of these were used by process analysts to identify the digitalization potentials in the five SMEs analysed in the project Business 4.0. The first four patterns have been identified as follows. The headings Intent, Motivation, etc. are standard structures for patterns.

6.1   Pattern 1: Information on Nondigital Information Carrier

Intent. Enforce the processing of information using *Digital Structured* information carriers.

Motivation. This pattern identifies situations in which information is required for a task and that information is available on an information carrier that is not the best choice in terms of digitalization and processing of information by IT systems.

Applicability. This pattern can be applied if information from order forms, goods without barcode and scanned documents is used. The use of PDF files whose contents must be typed for further processing is also an application case.

Structure. This pattern is illustrated in Fig. 4a. A task *T1* has an assigned data input *DI1*. *DI1* is provided by an information carrier that is not *Digital Structured*. There is no information in the model on how *DI1* is generated, i.e. *DI1* is not assigned to any other model element as data output.

Recommendation(s). Taking into account the type of information carrier of the found *DI1*, there are different potentials at occurrences of this pattern, depending on the type of information carrier encountered in each case.

If the data is *Nondigital Unstructured* it would be recommended to first examine the data to structure or formalize it. This is to be understood as a preparatory step for a later digitalization. If it is already *Nondigital Structured* it would be recommended to select a digital data store in which this data will be stored and managed digitally in the future. The corresponding data store would have to be added to the diagram if this potential were to be realized. If the data is already *IT-assignable*, it is recommended to make the key capturable. Examples of different common solutions to make the key capturable, such as the use of barcodes, QR codes or RFID tags, can be given here. If the data is already *IT-capturable*, the recommendation would be to check whether the data cannot be obtained via a digital interface, because a digital data store must already exist by definition. In the case the data is *Digital Unstructured*, it should be checked whether it can also be provided or being automatically transformed in a structured format.

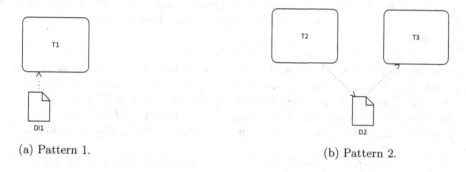

(a) Pattern 1.   (b) Pattern 2.

Fig. 4. Patterns of digitalization potentials

6.2   Pattern 2: Information Transmission via Nondigital Information Carriers

Intent. Enforce the transfer of information using digital structured information carriers.

Motivation. The second pattern describes the transfer of information from one task to another using a nondigital information carrier.

Applicability. This is the case, for example, when a clerk transfers information from one application to another by typing.

Structure. This pattern is illustrated in Fig. 4b. Data *D2* is data output of task *T2* and data input for task *T3*. *D2* is provided by an information carrier that is not *Digital Structured*.

Recommendation(s). The recommendations are similar to those of pattern 1, but are formulated in relation to the transfer from *T2* to *T3*. They are therefore somewhat more specific in terms of their wording than in application of pattern 1 in these cases.

6.3   Pattern 3: Nondigital Information Transmission Between Digital Data Stores

Intent. Enforce the use of digital interfaces between digital data stores.

Motivation. The third pattern describes the transmission between two tasks, each supported by a digital data store. Transmission takes place using a nondigital information carrier or digital unstructured information carrier. Those transmissions should be done using a digital structured information carrier as this simplifies data processing.

Applicability. This is the case, for example, when information is typed from one application to another or information is printed from one application so that a colleague can enter this information from the printout into another business application.

Structure. This pattern is illustrated in Fig. 5a. A task *T4* has a data store *DS1* as data input. *T4* has data *D3* as data output, which is also data input for *T5*. Task *T5* has a data store *DS2* as data output. *T4* and *T5* have no other data inputs or data outputs. The information carrier type of *D3* is not *Digital Structured*. The information carrier type of *DS1* and *DS2* is *Digital Structured*.

Recommendation(s). The recommendation is to switch the information carrier for the transfer from *T4* to *T5* to a *Digital Structured* information carrier, which usually is done by establishing a digital interface between *DS1* and *DS2*.

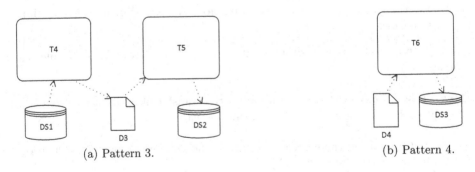

(a) Pattern 3.   (b) Pattern 4.

Fig. 5. More patterns of digitalization potentials

6.4   Pattern 4: Storage of Digital Information in Nondigital Data Store

Intent. Enforce the use of digital data stores.

Motivation. This pattern describes storing digital information in a nondigital information store.

Applicability. This may indicate, for example, that information is printed out to be archived in a file folder.

Structure. This pattern is illustrated in Fig. 5b. There is a task $T6$, which has data $D4$ as data input. Furthermore, $T6$ has data store $DS3$ as data output. $D4$ is *Digital Structured* and $DS3$ is *Nondigital Unstructured* or *Nondigital Structured*.

Recommendation(s). Substitute $DS3$ by a data store, which is *Digital Structured*.

7   Conclusion and Outlook

A number of digitalization potentials were classified and it was shown that BPMN 2.0 has to be extended if digitalization potentials are to be automatically identified via process weakness patterns in BPMN diagrams. The information that has to be expressed in BPMN has been described for this purpose. Appropriate patterns and associated recommendations were also explained as examples.

Further guiding questions and patterns are to be developed in the future, particularly as patterns have not been identified for all the digitalization potentials presented. For example, when it comes to identifying tasks that need to be supported by assistance systems. It can be assumed that some of these patterns require additional extensions of BPMN. It is also necessary to investigate how patterns can be made even more precise in order to identify recommendations that are even more specific to the respective process context.

Also, the procedure for identifying the pattern matches shall be explained in detail. The efficiency and effectiveness of the approach will also be evaluated.

Acknowledgements. This article was written in the context of the project Business 4.0 (http://owl-morgen.de/projekte/business-40/), which is part of the integrated action concept "OWL 4.0 - Industry, Labour, Society". The project is supported by the European Regional Development Fund (ERDF).

References

Dumas, M., et al.: Fundamentals of Business Process Management. Springer, Berlin (2013). https://doi.org/10.1007/978-3-642-33143-5

Falk, T.: Evaluation of a pattern-based approach for business process improvement. In: Leimeister, J.M., Brenner, W. (eds.) Proceedings der 13. Internationalen Tagung Wirtschaftsinformatik (WI 2017), pp. 241–255 (2017)

Fischer, H., Engler, M., Sauer, S.: A human-centered perspective on software quality: acceptance criteria for work 4.0. In: Marcus, A., Wang, W. (eds.) DUXU 2017. LNCS, vol. 10288, pp. 570–583. Springer, Cham (2017). https://doi.org/10.1007/978-3-319-58634-2_42

Förster, A., et al.: Verification of business process quality constraints based on visual process patterns. In: First Joint IEEE/IFIP Symposium on Theoretical Aspects of Software Engineering (TASE 2007). IEEE 2007, pp. 197–208. https://doi.org/10.1109/TASE.2007.56

Höhenberger, S., Delfmann, P.: Supporting business process improvement through business process weakness pattern collections. In: Thomas, O., Teuteberg, F. (eds.) Smart Enterprise Engineering: 12. Internationale Tagung Wirtschaftsinformatik, WI 2015, Osnabrück, Germany, 4–6 March 2015, pp. 378–392 (2015)

Object Management Group. Business process model and notation (BPMN), Version 2.0. (2011)

Phalp, K., Shepperd, M.: Quantitative analysis of static models of processes. J. Syst. Softw. 52(2–3), 105–112 (2000). https://doi.org/10.1016/S0164-1212(99)00136-3

Turban, B.M., Schmitz-Lenders, J.: A Pattern-Based Question Checklist for Deriving Requirements from BPMN Models. In: Teniente, E., Weidlich, M. (eds.) BPM 2017. LNBIP, vol. 308, pp. 630–641. Springer, Cham (2018). https://doi.org/10.1007/978-3-319-74030-0_50

Vergidis, K., Tiwari, A., Majeed, B.: Business process analysis and optimization: beyond reengineering. IEEE Trans. Syst. Man Cybern. Part C (Appl. Rev.) 38(1), 69–82 (2008). https://doi.org/10.1109/TSMCC.2007.905812

From Requirements to Data Analytics Process: An Ontology-Based Approach

Madhushi Bandara[1(✉)], Ali Behnaz[1], Fethi A. Rabhi[1], and Onur Demirors[1,2]

[1] University of New South Wales, Sydney, Australia
[2] Department of Computer Engineering, Izmir Institute of Technology, Izmir, Turkey
{k.bandara,ali.behnaz,f.rabhi,o.demirors}@unsw.edu.au

Abstract. Comprehensively describing data analytics requirements is becoming an integral part of developing enterprise information systems. It is a challenging task for analysts to completely elicit all requirements shared by the organization's decision makers. With a multitude of data available from e-commerce sites, social media and data warehouses selecting the correct set of data and suitable techniques for an analysis itself is difficult and time-consuming. The reason is that analysts have to comprehend multiple dimensions such as existing analytics techniques, background knowledge in the domain of interest and the quality of available data. In this paper, we propose to use semantic models to represent different spheres of knowledge related to data analytics space and use them to assist in analytics requirements definition. By following this approach users can create a sound analytics requirements specification, linked with concepts from the operation domain, available data, analytics techniques and their implementations. Such requirements specifications can be used to drive the creation and management of analytics solutions, well aligned with organizational objectives. We demonstrate the capabilities of the proposed method by applying on a data analytics project for house price prediction.

Keywords: Analytics process · Requirements · Ontology

1 Introduction

Analytic projects are complex with large investments being made on data preparation, tools and knowledge workers. Data analytics processes differ from traditional repeatable processes, requiring frequent intervention from knowledge-workers and flexibility to adapt the process when new insights emerge. To engineer correct analytics solutions that match the respective business objectives is challenging [11]. From the high-level requirements declared by management to the final analytics model adopted there is a complex process involving different decision making such as selecting suitable tools, algorithms, data sets and how to generate results and report them accurately. If the outcome is not accurate enough, nor considered the appropriate context, nor incorporated the correct datasets, nor satisfied the stakeholders' requirements, their time, resources and money spent on the study are wasted [10].

© Springer Nature Switzerland AG 2019
F. Daniel et al. (Eds.): BPM 2018 Workshops, LNBIP 342, pp. 543–552, 2019.
https://doi.org/10.1007/978-3-030-11641-5_43

As a solution to these issues, we propose an approach to express analytics requirements accurately via semantic models in order to support decision making and drive the process composition. In semantic modelling, ontologies are used to capture the domain knowledge, to evaluate constraints over domain data, to prove the consistency of domain data and to guide domain model engineering [2]. As ontologies provide a representation of knowledge and the relationship between concepts, they are malleable models good at tracking various kinds of software development artifacts ranging from requirements to implementation code [9]. Though there is research devoted to utilizing semantic web technologies in requirements engineering, most of it concentrates on specific requirements artifacts such as goals and use-cases and does not support reasoning over relations between all concepts [12].

To study this further, we conducted a systematic mapping study [3] to identify how existing literature leverages semantic web technologies to realize different stages of the data analytics process. The findings reveal a gap between defining analytics goals and requirements and linking them to the actual process realizations. The goals or requirements defined in existing literature are either at a very high level, not linked to operational level or they are expressed at the query level, limiting their capabilities for declarative analysis.

In this paper, we discuss the potential and limitations of using semantic models to capture data analytics requirements, relating them to different domain knowledge spheres, and how such an approach can lead to a requirements driven process composition in data analytics. Those requirements can be used as a communication tool, an artifact that enables traceability between requirements and the analytics solution implementation and a knowledge-base to guide semi-automated analytics process composition. Section 2 of the paper discusses the background and related work. Section 3 presents our proposed solution- a system that uses ontologies to drive the requirements capturing of the analytics process. Section 4 presents the system capabilities via an application to a predictive analytics process and paper concludes in Sect. 5.

2   Background and Related Work

In this section some background on the space of requirements related to data analytics is discussed in details, followed by the related work on how different existing systems capture or manage requirements to support data analytics.

2.1   Analytic Process Requirements Space

The data analytics requirements an organization should define can be identified at strategic, operational or tactical levels [14]. Taylor [13], in his work on framing requirements for predictive analytics projects, presents multiple dimensions that need to be captured in order to define requirements of an analytics project.

1. Performance Measures - metrics or business objectives
2. Decision requirements - What decisions are aided by the analysis, which in return can affect the performance measures, metrics or business objectives
3. Business context - Details of the processes and systems impacted by analysis, organizational units and roles involved, measures and metrics that may impact the scope of analysis
4. Data requirements - Input data and information useful for the analysis
5. Knowledge requirements - External regulations, internal policies, organizational practices and existing analytics insights

Among those performance measures, decision requirements and business context fall under strategic level requirements while data and knowledge requirements fall under the operational requirements. Wiegers provides a set of guidelines to define strategic requirements around an analytics project [14]. Brijs, in his work [7], presents a list of dimensions to express data requirements with respect to the data source, data storage, extraction, management and governance.

When linking the strategic level requirements to the operational level, another dimension we need to consider in detail is the type of analysis necessary to support certain decisions. There is no complete classification of analytics needs or problems and it may vary by context.

In addition, there are non-functional requirements that need to be specified for an analytics process at the operational level such as the model training time, expected accuracy and memory footprint which may impact the strategic level requirements as well.

2.2   Role of Requirement Models in Engineering Analytics Processes

Semantic web technologies have been used to engineer different stages of the data analytics process [11] as well as to compose the analytics process. In the systematic mapping study mentioned earlier we found that data analytics related literature use different ontologies to capture four concept classes: domain related, analytics specific, service related and intent concepts. Domain concepts described application specific information such as health care and sensor data. Analytics concepts were focusing on representing data pre-processing, mining and statistical algorithms. Service related concepts captured the implementation details such as data importing or computing services and work-flow composition. Intent concepts were the category that captures data analyst's requirements or goals.

We identified 10 studies that used intent concepts. 8 of them are used to express requirements at the execution level such as the Analytica Queries (AnQ) model in [8] that facilitates expressing user queries that need to be performed on data. Two studies were focusing on representing the high-level goals of the analysts: the Scientist's Intent Ontology [10] and the Goal Oriented Model [6]. They facilitate the modelling of strategic level user goals such as the desired outcomes of analytics tasks yet fail to link that to the operational level requirements or tasks related to an analytics process.

Moreover, we observed that existing techniques only focus on facilitating the selection of data providers, web services, and computational software modules. Hence to a large extent, analytics requirements are still a part of the mental model of the developer or the analyst who performs these tasks. In practice, several iterations of data cleansing, reformatting, the model selecting and process composition may be required in order to optimally serve the analytics problem. This may result in less effective data analytics solutions whose performance is likely to degrade with time.

As the data analytics community is extending wider into different industries and organizations and with analytics contexts and requirements changing rapidly, it is necessary to explore techniques that consider all dimensions such as business requirements, context and constraints. Hence a potential research area is a study of how high-level user goals and context can be represented and incorporated into data analytics solution engineering through data integration, process construction, and result interpretation. Approaches that link the user intentions and context into analytics processes have the potential of changing static analytics models deployed today into dynamic and adaptable analytics models that change the behavior, responding to changes in user goals or the operational context.

As discussed throughout Sect. 2, there are multiple levels of requirements associated with an analytics process. Capturing them accurately via models, linking them to existing analytics knowledge can improve the analytics solution design, enabling consistency and constraint checking as well as the requirement-driven design of data analytics processes.

3   Proposed Solution

We propose a system for managing data analytics requirements supported by a semantic knowledge-base. Figure 1 illustrates the main components of the proposed system organized into three layers: user interface (UI), business logic and data. Each component is described in more detail below.

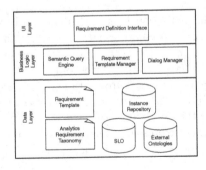

Fig. 1. Proposed system design

– **Data Layer**

At the core of the Data Layer is the Statistical Learning Ontology (SLO) [5] we designed to capture knowledge related to the data analytics space such as variables, prediction models and their relationships. The main components of the SLO are shown in Fig. 2.

Another component of the Data Layer is the Analytics Requirement Taxonomy (ART) which extends the different dimensions of analytics requirements discussed in Sect. 2.2 to suit the organization perspectives. Figure 3 represents the ART we use in the application discussed in Sect. 4. Requirements are classified as strategic and operational. Strategic requirements may be defined by the management of the organization and passed into the analysts, who will define operational requirements in line with them. As illustrated, users can extend the taxonomy to represent a set of requirements related to their domain of analytics.

The next component is the Instance Repository which stores actual data instances of the ontologies related to the analytics process. It contains details about the actual variables, existing prediction models and related publications, links between variables and models, available datasets and accessible data sources. This repository will be kept up to date, so the requirements definitions are generated based on the latest information.

A requirements template provides the structure for a set of requirements definitions. It is defined by selecting relevant requirements from the ART and linking them with the concepts of the ontology repository. Figure 4 provides an example requirements template we defined for predictive analytics with pre-trained models. Each requirement in the template is linked to the associated concept in SLO and captures the dependencies and constraints imposed by one requirement on the others. In Sect. 4, Fig. 5 illustrates an instance of this requirements template. What concepts from the instance repository are mapped to define each requirement in the template is shown by associated numbers.

– **Business Logic Layer**

The semantic query engine is responsible for fetching the information from the instance repository to fulfill each requirement defined in the template. It will capture the decisions made by the user at each stage of requirements definition and use them to enrich queries in the following steps. For example, if the user selects a prediction model, independent variables will be selected to match that model.

The requirements template manager provides the ability to update or create new requirement templates.

The dialog manager coordinates the UI layer and communicates with template manager and the semantic query engine to drive requirements definition process. When a user wants to define particular requirements defined in the requirements template, the dialog manager fetches different options available in the instance repository through the semantic query engine.

– **UI Layer**

User interface layer provides an interface for requirements definition. It uses a dialog-based approach to capture strategic and operational requirements of an analytics process. The dialogs are driven by dialog manager, supported by the requirements template manager and the instance repository. This interface guides users to create an instance of the provided requirements template that matches his analytics needs.

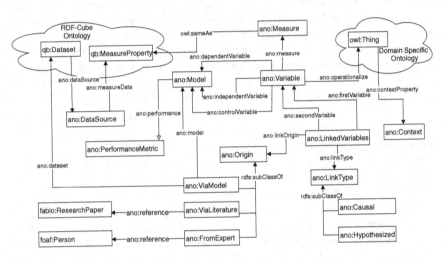

Fig. 2. Main components of the statistical learning ontology [5]

4   Application of the Proposed System

To illustrate how the proposed system can be used to define analytics process requirements specifications, we applied it to a process of developing a house price prediction model.

4.1   Application Description

This application aims to develop a framework which builds on existing predictive house price models and advance them through a range of approaches: studying and applying other econometric models, testing a range of additional economic variables, re-focusing predictive model on a 20-year horizon for Australian real estate prices in metropolitan areas. One key aspect of this study is voluminous heterogeneous datasets used in the study as house price prediction problem is addressed by the different experts with diverse perspectives. In addition, a plethora of analytics techniques (statistical learning techniques in this case) is used by users. As a result, knowledge acquisition plays a pivotal role in defining analytics requirements when implementing the project. The analytics

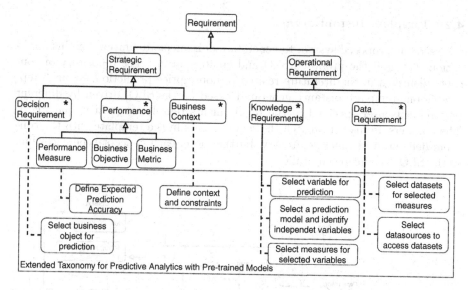

Fig. 3. Analytics requirement taxonomy - extension of the dimensions (represented by *) proposed by Taylor [13]

team comprised a group of academics from the two fields: computer science and economics. They had to go through literature and survey different prediction models, variables etc. used for house price prediction in different time spans and various countries. They use spreadsheets to accumulate findings from the literature. They focused on 30 previous studies and it was difficult for them to link those studies together, and to identify what studies are using similar models or variables. There was no naming convention, so the same variable was named differently or different names were used to refer to the same variable in different studies. Navigating through such spreadsheets and understanding was difficult and time-consuming. They needed a better approach to accumulate all that knowledge in 30 studies and pick the insights that are useful for study-at-hand.

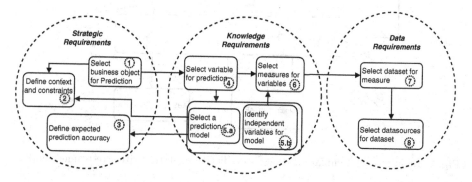

Fig. 4. Requirement template for predictive analytics with Pre-trained models

4.2   Template Instantiation

We used the proposed system to generate requirements definition for this application. We used the proposed SLO and created an instance repository of concepts identified in the literature related to house price prediction. Figure 5 represents one template instance generated based on the SLO, driven by different requirements user specified in the requirements template shown in the Fig. 4. The numbers are used to match the requirements in the template with the concepts defined in instance repository. Furthermore, each concept type according to the SLO is indicated in italic.

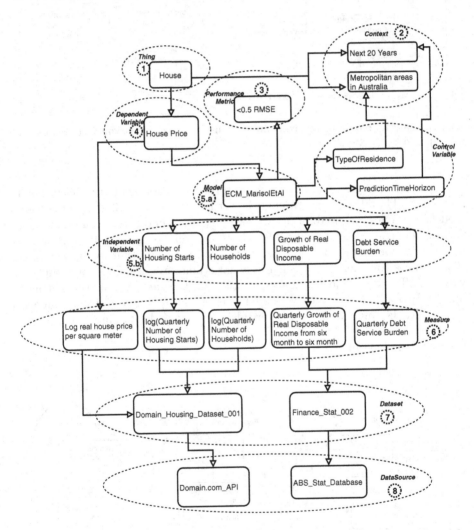

Fig. 5. Requirement template instance - predictive analytics with Pre-trained models

4.3   Prototype Implementation

We are developing a complete prototype of the proposed system. As the first step, the ontologies and instance repository are created on a semantic triple store. The ART and requirements templates are also defined in OWL XML syntax. A semantic query engine is being implemented as a REST API where results are generated when the requirements and parameters are passed.

A web-based front end will be designed to drive the interaction for requirements definition based on the requirements templates. Requirement template manager and dialog manager will be developed on the server side of the web-based application.

5   Conclusion and Future Work

In this paper, we propose a system for defining requirements related to data analytics process via ontologies and a requirements taxonomy. The main advantage of leveraging ontologies is that it provides traceability between different strategic and operational requirements as well as related domain or contexts. Once an organization develops a rich knowledge repository following the proposed system design, it becomes easy and efficient to define new analytics requirements. We express the capabilities of the system by applying it to developing an analytics solution for house price prediction.

By using the proposed method in the house price prediction application we observed that it provided a sound approach to define and link the essential properties of the tacit knowledge of the domain experts from which requirements can be easily generated. It also enabled us to link the known ontologies in the domain with the requirements which enabled us to utilize well established knowledge in the filed. In addition, as requirements generation is automated we can update the requirements specifications as the domain knowledge changes.

Next step of our research is to design an effective interface that enables users to communicate with requirements templates, ontologies and instance repository. We are experimenting with a web-based application that drives dialog-based communication with the user to finalize the requirements definition [4]. Further, we are looking at tools that capture and catalog business models through ontologies and how they can be extended to support our system design.

Potential extension of this work is mapping the requirements taxonomy with the traditional requirements definitions expressed in natural languages, enabling requirements definition for users in more intuitive fashion. We are also planning to integrate the system into our business process model based requirements generation model [1]. As the outcome of this system is a well-defined requirements definition connected to low-level artifacts of the analytics process such as data sets, incorporating it with a semantic service model such as SA-REST this system can be enhanced to support requirements driven service orchestration to realize execution level analytics processes.

Acknowledgments. We are grateful to Capsifi, especially Dr. Terry Roach, for sponsoring the research which led to this paper.

References

1. Aysolmaz, B., Leopold, H., Reijers, H.A., Demirörs, O.: A semi-automated approach for generating natural language requirements documents based on business process models. Inf. Softw. Technol. **93**, 14–29 (2018)
2. Baader, F.: The Description Logic Handbook: Theory, Implementation and Applications. Cambridge University Press, Cambridge (2003)
3. Bandara, M., Rabhi, F.A.: Semantic modeling for engineering data analytic solutions (2018, Under Review)
4. Bandara, M., Rabhi, F.A., Meymandpour, R.: Semantic model based approach for knowledge intensive processes. In: Stamelos, I., O'Connor, R.V., Rout, T., Dorling, A. (eds.) SPICE 2018. CCIS, vol. 918, pp. 215–229. Springer, Cham (2018). https://doi.org/10.1007/978-3-030-00623-5_15
5. Behnaz, A., Bandara, M., Rabhi, F.A., Maurice, P.: A statistical learning ontology for managing analytics knowledge. In: Proceedings of Workshop on Enterprise Applications, Markets and Services in the Finance Industry (2018)
6. Bellatreche, L., Khouri, S., Berkani, N.: Semantic data warehouse design: from ETL to deployment à la Carte. In: Meng, W., Feng, L., Bressan, S., Winiwarter, W., Song, W. (eds.) DASFAA 2013. LNCS, vol. 7826, pp. 64–83. Springer, Heidelberg (2013). https://doi.org/10.1007/978-3-642-37450-0_5
7. Brijs, B.: Business Analysis for Business Intelligence. Auerbach Publications, Boca Raton (2016)
8. Colazzo, D., Goasdoué, F., Manolescu, I., Roatiş, A.: RDF analytics: lenses over semantic graphs. In: Proceedings of the 23rd International Conference on World Wide Web, pp. 467–478. ACM (2014)
9. Pan, J.Z., Staab, S., Aßmann, U., Ebert, J., Zhao, Y.: Ontology-Driven Software Development. Springer, Berlin (2012). https://doi.org/10.1007/978-3-642-31226-7
10. Pignotti, E., Edwards, P., Gotts, N., Polhill, G.: Enhancing workflow with a semantic description of scientific intent. Web Semant.: Sci. Serv. Agents World Wide Web **9**(2), 222–244 (2011)
11. Rabhi, F., Bandara, M., Namvar, A., Demirors, O.: Big data analytics has little to do with analytics. In: Beheshti, A., Hashmi, M., Dong, H., Zhang, W.E. (eds.) ASSRI 2015/2017. LNBIP, vol. 234, pp. 3–17. Springer, Cham (2018). https://doi.org/10.1007/978-3-319-76587-7_1
12. Siegemund, K., Thomas, E.J., Zhao, Y., Pan, J., Assmann, U.: Towards ontology-driven requirements engineering. In: Workshop Semantic Web Enabled Software Engineering at 10th International Semantic Web Conference (ISWC), Bonn (2011)
13. Taylor, J.: Framing requirements for predictive analytic projects with decision modeling (2015)
14. Wiegers, K., Beatty, J.: Business analytic projects. Software Requirements (2013)

An Assignment on Information System Modeling
On Teaching Data and Process Integration

Jan Martijn E. M. van der Werf[1(✉)] and Artem Polyvyanyy[2]

[1] Department of Information and Computing Science, Utrecht University,
P.O. Box 80.089, 3508 TB Utrecht, The Netherlands
`j.m.e.m.vanderwerf@uu.nl`
[2] The University of Melbourne, Parkville, VIC 3010, Australia
`artem.polyvyanyy@unimelb.edu.au`

Abstract. An *information system* is an integrated system of components that cooperatively aim to collect, store, manipulate, process, and disseminate data, information, and knowledge, often offered as digital products. A *model* of an existing or envisioned information system is its simplified representation developed to serve a purpose for a target audience. A model may represent various aspects of the system, including the structure of information, data constraints, processes that govern information, and organizational rules. Traditionally, the teaching of information system modeling is carried out in a fragmented way, i.e., modeling of different aspects of information systems is taught separately, often across different subjects. The authors' teaching experience in this area suggests the shortcomings of such fragmented approach, evidenced by the lack of students' ability to exploit the synergy between data and process constraints in the produced models of information systems.

This paper proposes an assignment for undergraduate students which requests to model an information system of an envisioned private teaching institute. The assignment comprises a plethora of requirements grounded in the interplay of data and process constraints, and is accompanied by a tool that supports their explicit representation.

Keywords: Data and process modeling ·
Information system modeling ·
Computer science and information systems education

1 Introduction

In the information age we live, information systems provide core mechanisms for supporting operational business processes of organizations. Hence, leading Computer Science and Information Systems curricula comprise courses that teach students the art and rigor of designing information systems. Traditionally, modeling of each aspect of an information system, e.g., data and process constraints,

© Springer Nature Switzerland AG 2019
F. Daniel et al. (Eds.): BPM 2018 Workshops, LNBIP 342, pp. 553–566, 2019.
https://doi.org/10.1007/978-3-030-11641-5_44

is taught separately, often across different subjects. The authors have independently taught the foundations of information systems modeling to undergraduate students at Utrecht University, The Netherlands, and Queensland University of Technology, Australia (for five and seven consecutive semesters, respectively). In this paper, the authors report on identified drawbacks of such a fragmented approach to teaching information system modeling, and argue for the need in educating students on data and process integration.

As an example, consider a task of designing a learning management system that keeps track of course offering, and corresponding lecturers and student enrollments. A decision to start by developing a high-quality data model for the proposed scenario may result in a design which requires that every course offering is assigned at least one lecturer. This design may contradict the corresponding business processes that require to assign a lecturer to a course offering only once it reaches the minimum number of student enrollments. Conversely, a decision to introduce a process constraint may limit the number of solutions to the design of the data model in a way that excludes the required solution. Note that even if all the data and process requirements of the desired solution are laid out prior to embarking into modeling, they may lead to a contradiction that does not manifest neither in a data model nor in a process model that satisfies the respective requirements. Thus, an effective approach to modeling an information system should allow a designer to experience the interplay between data and process constraints. Building from this understanding, the paper at hand contributes:

1. An assignment to model an information system of an envisioned private teaching institute;
2. A systematic analysis of challenges experienced by students when solving the assignment in a traditional way, i.e., by tackling modeling of information constraints and business processes of the system separately;
3. A proposal to address the identified challenges by using a new tool capable of representing an interplay between the data and process constraints in an integrated model of an information system.

The remainder of this paper is organized as follows. The next section examines how data and process modeling skills are recognized in the curricula of undergraduate degrees in Information Systems. Section 3 proposes an assignment that aims to teach data and process modeling skills in an integrated way. Section 4 shares our experience, while Sect. 5 proposes a tool support for designing data and process constraints in an integrated way. The paper closes with conclusions.

2   Teaching Data and Process Modeling in IS Curricula

In 2010, the Association for Information Systems (AIS) and the Association for Computing Machinery (ACM) have released IS 2010, the latest in a series of proposed *model curricula* for undergraduate degrees in Information Systems [15].

IS 2010 provides guidance regarding the core content of a curriculum in Information Systems and suggests possible electives and career tracks.

IS 2010 comprises seven core and several elective courses, among which Data and Information Management (IS 2010.2) and Systems Analysis and Design (IS 2010.6) are recognized to play a central role. Next, we examine these two courses with respect to the proposed learning outcomes and topics that contribute to data and process modeling skills, taking a close look at the skills that are grounded in the interplay of data and process constraints in the designs of information systems.

2.1 Data and Information Management

According to IS 2010, the Data and Information Management (IS 2010.2) course provides students with an introduction to the core concepts in data and information management. Concretely, this course teaches students methods and techniques for identifying organizational information requirements, constructing conceptual models of these requirements, converting the conceptual data models into logical models, e.g., relational data models, verifying the correctness of the models, and implementing the models, e.g., using a Relational Database Management System (DBMS) [11,14].

Among the 21 suggested learning objectives of this course, we identify three core objectives[1] that specifically target the data modeling skills of a student:

- Use at least one conceptual data modeling technique (such as entity-relationship modeling) to capture the information requirements for an enterprise domain;
- Design high-quality relational databases;
- Understand the concept of database transaction and apply it appropriately to an application context.

The topics of the course that contribute to these skills are conceptual, logical, and physical data models, for example entity-relationship model, relational data model, and data types, respectively. The curriculum suggests that the focus should be on conceptual and logical data modeling skills, while "students should understand the basic nature of the DBA tasks and be able to make intelligent decisions regarding DBMS choice and the acquisition of DBA resources."

Two learning objectives of the IS 2010.2 course may be interpreted as such that suggest an interplay between the data and process modeling skills:

- Apply information requirements specification processes in the broader systems analysis and design context;
- Link to each other the results of data/information modeling and process modeling.

---

[1] Note that several other proposed learning objectives can be seen as refinements of the core ones, e.g., the objective of "Design a relational database so that it is at least in 3NF" can be seen as a refinement of "Design high-quality relational databases".

None of the proposed course topics explicitly contributes to the integration of data and process modeling skills of a student. One may argue that such skills are implicit in the topic of "Using a database management system from an application development environment". Still, this topics advocates for a compartmented approach to data and process modeling. At the same time the curriculum acknowledges that "information requirements specification processes must be firmly linked to the organizational systems analysis and design processes".

2.2   Systems Analysis and Design

The curriculum suggests that the Systems Analysis and Design (IS 2010.6) course should contribute to 13 learning objectives, among which only two implicitly target process modeling skills, namely:

– Use at least one specific methodology for analyzing a business situation (a problem or opportunity), modeling it using a formal technique, and specifying requirements for a system that enables a productive change in a way the business is conducted.
– Within the context of the methodologies they learn, write clear and concise business requirements documents and convert them into technical specifications.

We identify that the topics of the course that can contribute to these objectives are Business Process Management and analysis of business requirements. The curriculum contains an elective course entitled Business Process Management [1,2,8], which refines the learning objectives that address process modeling skills. The main focus of this elective course is on understanding and designing of business processes, which manifests in four learning outcomes (out of 11):

– Model business processes;
– Understand different approaches to business process modeling and improvement;
– Use basic business process modeling tools;
– Simulate simple business processes and use simulation results in business process analysis.

Two proposed learning objectives of the IS 2010.6 course address the integration of data and process modeling skills, namely:

– Use contemporary CASE tools for the use in process and data modeling.
– Design high-level logical system characteristics (user interface design, design of data and information requirements).

However, again, similar to IS 2010.2, none of the proposed topics of IS 2010.6, or those of the elective Business Process Management course, explicitly contributes to the integration of data and process modeling skills of a student.

3   Assignment: Supporting the Private Teaching Institute

An effective assignment to modeling an information system should allow students to experience the interplay between data and processes. The assignment should have a sufficiently challenging and realistic case description, while being manageable in size.

3.1   Learning Objectives

As a first step, we crafted the learning objectives, following the IS 2010 guidelines, and the Bloom Taxonomy [4]. As the assignment focuses on learning to apply techniques, we assume that once the assignment starts, students already have an initial understanding of data modeling e.g. with ERM [6], and process modeling, e.g., with Petri nets [13] and BPMN [8]. In other words, we assume students to start at level 2 (comprehension) of the Bloom Taxonomy. The learning objectives of the assignment cover the next levels, being application, analysis, synthesis and evaluation. After the assignment, the students should be able to:

- Model and analyze process and information requirements using formal techniques;
- Critically assess models and make well-informed design decisions to solve real world problems related to information systems;
- Write clear and concise requirements and convert these into technical specifications using formal techniques;
- Manage the complexity of contemporary and future information systems and the domains in which these systems are used; and
- Use contemporary off-the-shelf components to integrate models into an information system.

Experience from a previous assignment [10], where students had to design and build an information system for an online shop, showed that students had difficulties in understanding the underlying problems of the domain. Therefore, the context of this assignment should be geared to the students' perception of their environment. For this purpose, we designed a case around a fictive educational institute, the Private Teaching Institute (PTI). Several requirements have been left implicit, or are even underspecified to allow students to reflect and perform a proper context analysis. In this way, students can use their own experience to better understand the situation.

3.2   The Case: The Private Teaching Institute

The Private Teaching Institute (PTI) offers education tracks. Each education track consists of several mandatory courses, and some optional courses. PTI consists of a small team per track, the track management, and a small student administration for all tracks together. To deliver the courses, PTI has a pool of lecturers who are qualified to deliver several courses. Everybody is entitled to enroll for a track. As soon as somebody registered themselves, and they are accepted by the track management, they become a student of that track. Students enrolled have to create an educational plan, consisting of the courses they want to follow. This plan has to be approved by the appropriate track management, and filed by the administration.

As soon as the plan is approved, students may register for courses. Once there are sufficient registrations for a course, the management creates a tender and sends it out to the lecturers who are qualified to give that course. After the response offers by the lecturers, the management selects the best offer and appoints the corresponding lecturer for that course. Every course at PTI consists of several lectures, either in a classical class room setting or on-line, practical assignments, and one or more exams, depending on the wishes of the appointed lecturer. Once the student meets all criteria set by the lecturer, i.e., passing a sufficient number of assignments and exams, the student receives a certificate of passing. In all cases, the result is filed by the administration.

Once a student passed all the courses agreed upon in the educational plan, the student is eligible to receive a diploma for that track. The track management verifies the course certificates and the plan, after which the management can award the diploma. Students can choose for a formal ceremony, or to receive their diploma by post.

PTI wants a process-aware information system that supports them in their primary processes, to ease the administrative burden.

3.3   Phases and Deliverables

The information system should be designed and implemented, while ensuring that all deliverables remain consistent. The assignment identifies two phases: the specification phase, and the implementation phase. Instead of following the traditional waterfall approach, the phases run concurrently, and the deliverables of the two phases should be synchronized regularly. Having small cycles assist in keeping the problem at hand manageable, and also allows the teaching staff to provide the students with early feedback.

During the first phase, the students have to analyze the assignment, and identify the involved stakeholders and their interactions with the to-be-designed information system. For this analysis, students may apply different techniques. Some students prefer to create use cases [5], other students perform a PACT analysis [3]. A PACT analysis studies the People involved, their Activities, the Context in which these activities are performed, and the main Technologies used to support these.

Once the context of the assignment has been analyzed to gain a better understanding of the environment, the students have to derive the information requirements and build a specification. Part of the specification is a data model in ERM notation. Many choices have been left implicit in the case description, such as the number of courses a track consists of, whether courses are mandatory for the complete institute, or only for tracks, etc. Students have to discover these choices, and make and document their design decisions. To model the flow of information, the different processes in the case have to be identified, analyzed and modeled using Petri nets. The resulting models should be analyzed for correctness using formal approaches, such as weak termination (i.e., absence of deadlocks and livelocks) and boundedness. Additionally, the different models created should be consistent, and validated with the context analysis, i.e., the use cases and scenarios created initially should be supported by the models.

The context description, information model and process models together with their analyses are captured in the *Specification Document* that the students have to deliver. The resulting document should be concise, clear and contain all important requirements of the case.

Once an initial version of the specification document, containing one or two processes, is being created, the implementation phase starts. The goal of the implementation phase is to use packaged solutions, rather than implement a system from scratch. The assignment relies on the Business Process Management Suite (BPMS) ProcessMaker[2], which has both an open source edition, as well as a commercial cloud service. For the implementation of the information system, each process designed in the specification document should be converted into a BPMN model, together with the forms and triggers for each activities. As the complete information system comprises several processes, the data model has to be implemented, and the forms and activities of the different processes should manipulate the data model. This phase results in two deliverables: the *Implementation Guide*, and the *implementation* itself.

Table 1. Grading schema for the assignment

Specification document	Points	Implementation guide	Points
Context analysis	15	Quality BPMN models	25
Data model	10	Model descriptions	5
Quality process models	30	Gateway logic	5
Documentation of models	15	Forms per activity	10
Verification and validation	25	Reflection	10
Layout	5	Layout	5
		Implemented functionality	25
		Demonstration	15
Total	100	**Total**	100

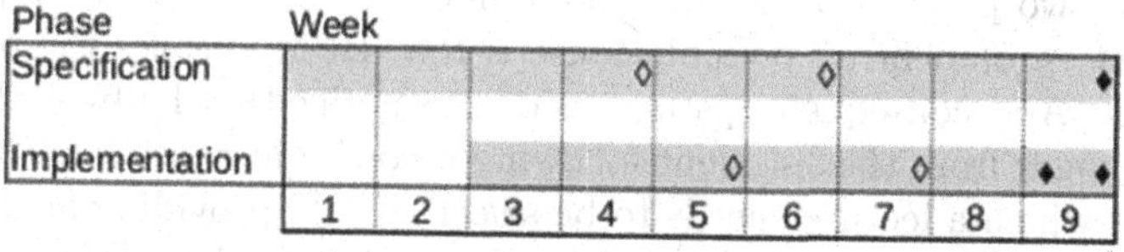

Phase	Week								
Specification				◇		◇			◆
Implementation					◇		◇	◆	◆
	1	2	3	4	5	6	7	8	9

Fig. 1. Gantt chart of the assignment. The open diamonds are feedback moments, the filled diamonds are official deadlines, including a demonstration.

---

[2] http://www.processmaker.com/.

As in real life, processes may be altered, updated or completely revised during the implementation. Therefore, during the different phases, the specification document and implementation guide need to be updated together, ensuring that the revised models remain correct, and the documentation consistent.

For grading, the schema shown in Table 1 is used. The schema addresses the different learning objectives. For feedback and grading a rubric based on this schema is used[3]. Part of the implementation phase is a demonstration of the system to the teaching staff, simulating the role of a stakeholder at PTI.

4   First Experiences with the Assignment

Last year, the assignment has been executed for the first time during the Information Systems course at Utrecht University, with about 170 first year Information Science Bachelor students. Although the group is quite large, we decided to have the students to create pairs, instead of larger groups. In this way, students are able to cooperate, and discuss design options, at the same time preventing free riders.

The course is taught in the final block of the year, and runs over a period of 10 weeks. As a 7,5 EC credit course[4], students are expected to work 20 h per week on the subject, including lectures on process modeling and analysis. In total, each student is expected to dedicate in total 100 h to the assignment. Each phase had two intermediary deadlines for feedback, and a final deadline at the end of the period (see Fig. 1). The demonstrations were in the same week as the final deadline.

Process Identification. During the first feedback moment, we noticed that many students found it challenging to discover the different processes in the assignment. Many groups had problems in dividing the case description into smaller, manageable components. Several authors acknowledge the difficulty of discovering the processes in an organisation (cf. [8]), and point e.g. at categories of Processes according to Porter, to assist in this activity. However, as these categories are tailored towards businesses, students found it difficult to apply them on a different context.

Some students delivered a single large model that covered all facets of the institute. For example, the student's enrollment and the tender process for lecturers were combined in a single process. They failed to recognize that by combining these two processes, the complete tender process had to be repeated for each student enrollment. A possible cause is that BPMN leaves the notion of a case implicit. As a consequence, students do not notice that halfway the process the case changes from the "student following a course instance", to "the course instance for which a lecturer needs to be selected". By providing feedback after the first round on how to read the case description, and by posing questions

---

[3] The rubric can be found at http://www.architecturemining.org/publications/ WerfP18a.pdf.

[4] https://ec.europa.eu/education/resources/european-credit-transfer-accumulation-system_en.

like "what is the subject of this process?" explicitly in the feedback, students understood the notion of cases and processes much better.

Other groups divided the assignment in many small processes, such as "do assignment", which comprised two activities: the student creating an assignment, and a lecturer grading the assignment. Although in essence this is not wrong, the finer the granularity of the processes identified, the more challenging it is to understand the interplay of the different processes. For example, is a student allowed to receive a grade if one of the assignment processes is still running? Having a too fine-grained solution simplifies modeling and analyzing the separate models, but complicates the overall design of the information system.

In the end, most student groups delivered an information system that implemented two to four business processes. These processes capture different aspects of the information system, from enrolling in an educational track, following a course instance, the lecturer tendering process, and obtaining the diploma. Some students combined the enrollment and obtaining the diploma, i.e., the process a student follows in an educational track. Others combined the students following a course instance process with the lecturer tendering process, by taking the course instance as a case, rather than a student following a course instance.

Process Modeling. Although having Petri nets as the primary modeling notation helps students in making the state, and thus the case, explicit, it turned out to be difficult for students to give proper meaning to tokens and places. Tokens resembling a single object, such as a lecturer or a student were often found at a first round. However, combining different notions, like "a token in this place resembles a student that is following a course" turns out to be more difficult than initially anticipated. After the first round of feedback, students were taught the concept of place invariants. This increased the students' understanding of the idea of tokens and places resembling combinations of elements, rather than just being single elements representing the state of the net.

As in a previous course on information modeling, students learned to design forms to populate their data model, several groups created "screen-based" processes. Each activity represented a screen a user would see in the system, and the process flow depicted the possible orders in which the screens would be displayed. Discussing their solution after the first feedback round, revealed that these student groups had similar problems in understanding the notion of a case.

Another challenge many students faced is the level of abstraction in activities. For example, several groups produced process models with small activities like "fill in address", "fill in telephone number", and "select education track", rather than having a larger activity "enroll for education track", leaving the details of what data is needed for an enrollment to a later stage in the process. These small activities appeared either in a large parallel construct, or were modeled consecutively, in a fixed order.

In the final deliverable, all student groups delivered process models with each containing ten to twenty activities. Each activity had a clear form and roles assigned. The interplay between the different processes was expressed both in Petri nets, and implemented using triggers on the activities, and by connecting the data model to the different activities in the process models.

Process Analysis. During the lectures of the course, many different analysis techniques, such as reachability and invariant calculus are discussed. Relating these abstract properties, like liveness, boundedness and place invariants to properties turns out to be a good exercise in understanding why these properties help in improving their solutions.

The students had to analyze their solution in different dimensions. The first dimension is intra-process versus inter-process. Within a single process, all properties are relatively easy to verify, especially if their solution contains many small processes. The challenge is in analyzing the interplay between different processes. For example, dependencies may exist, like in the example of the small assignment process: who is allowed to start this process, and when? Similarly, to model a check whether a course instance has sufficient students enrolled, can be challenging if each student enrolls in a separate process instance.

A second dimension is verification within the models versus validation with the context. Verification of the models, i.e., checking whether the models satisfy properties like liveness, boundedness and weak termination, was performed by all students. Validation, i.e., checking whether the models are appropriate for the problem at hand turns out to be more difficult. Most students delivered initially reports containing many, large user stories, but no analysis whether their solution can actually replay the scenarios they described earlier in the same document.

Implementation. Another challenge remains in transforming the formal process models designed with Petri nets into BPMN models that are executable by Business Process Management Suites (BPMSs) like ProcessMaker. On the one hand, the formal semantics of Petri nets allow the students to simulate and analyze their processes, and test their dependencies by composing all models into a large Petri net. On the other hand, a BPMS requires the model to be divided into small processes, in which the state is left implicit. In addition, several constructs are needed in Petri nets to keep models analyzable, e.g. the amount of lecturers available to teach a course. In BPMN specialized constructs exist, such as parallel repetition via multi-instance activities, that are designed to solve such situations, as an example shows in Fig. 2. This requires the students to be creative in their solutions on how to move from a formal specification into a technical implementation, while showing that their ideas remain consistent with the specification.

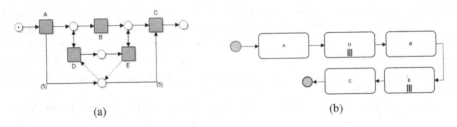

(a)   (b)

Fig. 2. Situation modeled in Petri nets (a) for which the multi-instance activity in BPMN (b) gives a more natural solution.

Balancing Data and Processes. An important observation we made during the assignment is how subtle the connection between processes and data is. Although these subjects are being taught in different courses, these go hand in hand in an integrated information system.

To give an example, most students create a data model in which a course instance always has a lecturer (a one-to-many relation), has one exam and one assignment. However, in the process of running a course instance, the track management first decides that a course instance, for which students already could subscribe, will start, and only then decide to start a tender for which lecturers can apply. Hence, although the course instance already exists, no lecturer is assigned to it. Consequently, the data model is violated, as the one-to-many relationship is not valid, whereas adding a lecturer while creating a course instance violates the process model. This results in a deadlock caused by the integration of the two models. Although the example seems trivial, it turns out that many such integration issues occur in the assignment.

The interplay between processes and data is very difficult to analyze and discover at design time, and is mostly found only while testing the information system, which is already difficult and challenging in itself. This debugging and "bug hunting", as some students named it, is a very time-consuming and frustrating process, as it is scattered over the different forms, triggers and database handling in all processes.

Overall Perception. All student groups delivered an integrated information system that supported most functionality. The specification document and implementation guide typically were consistent. Reduction rules [13] combined with reachability graphs were the most used analysis tool to verify the models, and several groups used place invariants to show that their resources, such as lecturers, courses and students remained constant in the system.

Afterwards, the course was evaluated by the students (n = 41) using closed questions on a 1–5 likert scale. Students pointed out that the lectures were well usable for the assignment (85% scored ≥4), and that they learned "a great deal" (83% scored ≥4). Although labor intensive, the students valued the early feedback rounds and stated that the feedback helped improving their results (73%

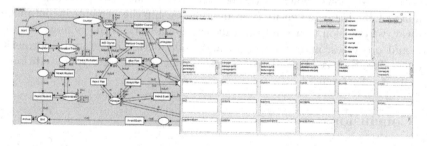

Fig. 3. ISModeler. The tool combines CPN Tools with a theorem prover for the data model.

scored ≥ 4). In the open feedback questions, students posed that the used system has its problems and peculiarities. This made it often difficult to understand what went wrong, and how this could be mitigated. However, the students valued the freedom the assignment provides, ensuring that everybody has a different solution, enabling them to discuss alternatives among each other.

5   Next Steps

Based on the results of the first run of the assignment, we found that integrating data and processes is experienced as challenging by the students. For many practitioners, experience plays an important role in knowing how to adapt processes and data, and when. In some cases it is better to alter the data model, in other cases the process model. This requires experience, and practice.

In our view, integrating processes and data is given too little attention in current curricula. The assignment shows that students find it very difficult to analyze the specification on deadlocks caused by the integration of data and process models. To our knowledge, hardly any analysis technique taught in textbooks is grounded in both data and processes. At the same time, we see that courses on Data and Information Management (IS2010.2) focus on information requirements and data modeling. Processes are acknowledged, but play a very small role in the IS 2010 guideline. Similarly, process modeling courses, like the elective on BPM, focus on processes, but tend to ignore that these processes manipulate (structured) data.

A course on information system modeling should not only focus on these two aspects, but also show the synergy between the two modeling paradigms. We therefore developed the tool *ISModeler* that makes this synergy explicit [16]. It combines a process model in the form of a Petri net in which tokens carry identifiers [10,12], a data model, and a transition specification that defines how each transition manipulates the data model through transactions. The tool builds upon CPN tools [17], and a theorem prover to validate the transactions on populations of the data model. In *ISModeler*, a transition is enabled if it is both enabled in the Petri net, and the transaction yields a valid population. Figure 3 shows a screenshot of the system. In the top part of the window, the enabled transitions are shown, whereas the bottom part depicts the population of the data model, by listing per entity type and relationship the elements it contains. In this way, we envision that students will better understand the synergy between data and processes, and thus design and build better integrated information systems. The tool is planned to be put into action in next year's edition to evaluate its effectiveness.

6   Conclusions

In this paper, we propose an assignment that allows students to experience the design and implementation of an information system using a BPMS. The proposed assignment combines data and process modeling, forcing students to design and analyze their solution using formal techniques, and translate their solution into an information system.

Running the assignment for the first time shows that the assignment helps students to experience design issues that arise while studying the case description. Students discovered that abstract properties used in verification can be linked to actual properties in the case description, and assist them in improving their solution.

However, the run also shows that students find it difficult to understand the synergy between data and processes. Although in scientific literature several approaches exist that allow to model this (cf. [7,9,12]), experiences with the assignment show that these have not yet been embedded sufficiently in our education curricula.

References

1. van der Aalst, W.M.P., Stahl, C.: Modeling Business Processes - A Petri Net-Oriented Approach. MIT Press, Cambridge (2011)
2. van der Aalst, W.M.P., van Hee, K.M.: Workflow Management: Models, Methods and Systems. Academic Service, Schoonhoven (1997)
3. Benyon, D.: Designing Interactive Systems: A Comprehensive Guide to HCI, UX and Interaction Design, 3rd edn. Pearson, Edinburgh (2014)
4. Bloom, B.S., Engelhart, M.D., Furst, E.J., Hill, W.H., Krathwohl, D.R.: Taxonomy of Educational Objectives: The Classification of Educational Goals. David McKay Company, New York (1956). Handbook I: Cognitive domain
5. Booch, G., Rumbaugh, J., Jacobson, I.: The Unified Modeling Language User Guide. Addison-Wesley, Upper Saddle River (2005)
6. Chen, P.P.: The entity-relationship model: towards a unified view of data. ACM Trans. Database Syst. **1**, 9–36 (1976)
7. Deutsch, A., Hull, R., Patrizi, F., Vianu, V.: Automatic verification of data-centric business processes. In: ICDT 2009, pp. 252–267. ACM (2009)
8. Dumas, M., La Rosa, M., Mendling, J., Reijers, H.: Fundamentals of Business Process Management. Springer, Heidelberg (2018). https://doi.org/10.1007/978-3-662-56509-4
9. van Hee, K.M.: Information System Engineering - A Formal Approach. Cambridge University Press, New York (1994)
10. van Hee, K.M., Keiren, J., Post, R., Sidorova, N., van der Werf, J.M.: Designing case handling systems. In: Jensen, K., van der Aalst, W.M.P., Billington, J. (eds.) Transactions on Petri Nets and Other Models of Concurrency I. LNCS, vol. 5100, pp. 119–133. Springer, Heidelberg (2008). https://doi.org/10.1007/978-3-540-89287-8_8
11. Kroenke, D.M., Auer, D.J.: Database Concepts, 7th edn. Pearson, London (2015). Global Edition
12. Montali, M., Rivkin, A.: DB-Nets: on the marriage of colored petri nets and relational databases. In: Koutny, M., Kleijn, J., Penczek, W. (eds.) Transactions on Petri Nets and Other Models of Concurrency XII. LNCS, vol. 10470, pp. 91–118. Springer, Heidelberg (2017). https://doi.org/10.1007/978-3-662-55862-1_5
13. Murata, T.: Petri nets: properties, analysis and applications. Proc. IEEE **77**(4), 541–580 (1989)

14. Silberschatz, A., Korth, H., Sudarshan, S.: Database System Concepts, 5th edn. McGraw Hill, New York (2006)
15. Topi, H., Kaiser, K.M., Sipior, J.C., Valacich, J.S., Nunamaker Jr., J.F., de Vreede, G.J., Wright, R.: Curriculum guidelines for undergraduate degree programs in Information Systems. Technical report, ACM (2010)
16. van der Werf, J.M.E.M., Polyvyanyy, A., Overbeek, S.J., Brouwers, R.A.C.M.: On a synergy between data and processes. Technical Report UU-CS-2018-004, Utrecht University (2018)
17. Westergaard, M., Kristensen, L.M.: The Access/CPN framework: a tool for interacting with the CPN tools simulator. In: Franceschinis, G., Wolf, K. (eds.) PETRI NETS 2009. LNCS, vol. 5606, pp. 313–322. Springer, Heidelberg (2009). https://doi.org/10.1007/978-3-642-02424-5_19

Motivational and Occupational Self-efficacy Outcomes of Students in a BPM Course: The Role of Industry Tools vs Digital Games

Jason Cohen[✉][iD] and Thomas Grace

University of the Witwatersrand, Johannesburg, South Africa
jason.cohen@wits.ac.za

Abstract. While past studies have considered the educational benefits of industry tools and simulated games within BPM courses, the relative efficacy of these interventions for student outcomes has not yet been established. In this study we sought to determine whether added exposure to industry tools would be more, or less, effective at influencing students' perceived competence, intrinsic motivation and occupational self-efficacy for BPM than exposure to a digital BPM game. An experimental study was carried out on 38 students and revealed that students exposed to additional industry tools reported increased levels of occupational self-efficacy, while those exposed to the digital game reported lower levels of perceived competence. Results have useful implications for BPM educators.

Keywords: Digital-game based learning · Occupational self-efficacy · Motivation

1 Introduction

In response to a growing demand from industry for individuals with business process capability, universities continue to invest in the design and delivery of Business Process Management (BPM) curricula [43]. Much effort has been devoted by the BPM academic community to innovating course curricula, textbooks and teaching materials. However, BPM remains a particularly challenging subject as students learn skills across the lifecycle of process identification, discovery, analysis, redesign, execution, monitoring and control [18].

Unsurprisingly, BPM educators have devoted much attention to the question of how such curricula can best be delivered. Efforts include presenting content and syllabi for BPM courses [4, 5, 43] and mapping course content to the BPM capabilities required in practice [16]. Importantly, past work has demonstrated that BPM education requires an emphasis on both theory and practical application [33, 42]. Authors discuss the use of case studies and tools [37], and the importance of ensuring students develop practical skills and ability to apply BPM knowledge to complex problems and modeling scenarios [39].

One stream of research has brought attention to the importance of using professional BPM tools in practical components of course delivery, e.g. ARIS [4]; MS Visio, ARIS and

© Springer Nature Switzerland AG 2019
F. Daniel et al. (Eds.): BPM 2018 Workshops, LNBIP 342, pp. 567–579, 2019.
https://doi.org/10.1007/978-3-030-11641-5_45

YAWL [43]; BPMS platforms such as Adonis, Tibco and Bizagi [5]; ARIS and Web-Sphere [16] and, in some cases, ERP software [22]. Industry software tools have clearly come to play an important role in the teaching of BPM concepts. Hands-on exposure to industry systems develops skills, helps student understand how software actually works and improves their business process knowledge [41]. They are clearly important to building skills, but little empirical research exists to identify best practice [5].

Another stream of research has focused on how educational games can be used to provide students an environment in which to practice BPM. IBM's Innov8[1] has been argued as particularly useful given its professional orientation (e.g. [7, 27, 30, 50]). Students benefit from experiencing hands-on, role-play within the game environment [51]. Games might also stimulate interaction, teamwork and promote more social and communicative learning [51]. Games are argued to be useful because of their potential to illustrate the relevance of what students have been learning as they navigate the simulation to apply their knowledge in realistic scenarios.

Taken together, these two streams of research provide evidence that the incorporation of hands-on training in the use of industry software as well as the incorporation of digital games can be used to help students learn BPM. However, past work does not explore the relative efficacy of these tools for student outcomes. This presents a dilemma for BPM educators who must decide with the limited resources available, which types of intervention will yield the best returns for student outcomes.

In this study we thus sought to determine whether exposing students to additional workplace tools to enhance their knowledge of BPM practice would be more, or less, effective at influencing outcomes than exposing students to a digital BPM game. We focus on three outcomes, namely students' perceived competence in performing BPM tasks, intrinsic motivation toward performing BPM and their occupational self-efficacy i.e. belief in their ability to succeed in a BPM career. We considered these outcomes appropriate because BPM curricula should not only develop student skills but also their confidence and motivation to pursue careers in BPM.

We have previously explored the effect of the inclusion of digital game-based learning on students' motivation and perceived competence in an earlier offering of our BPM course [21]. This study extends that work to examine within a new cohort of students the relative efficacy of games versus industry tools in the teaching and learning of BPM.

2 Study Design

2.1 Context of the Study

Our study context was a core (required) undergraduate class in BPM for information systems majors. The students were enrolled in a seven-week BPM course in their third year of study. The course consisted on 28 lecture hours, 7 workshop hours, and 14 hands-on computer laboratory hours. The course uses project-based learning where students work on an assigned project case in small teams. The project reinforces

---

[1] IBM's "Innov8 2.0" (http://www-01.ibm.com/software/solutions/soa/innov8/index.html).

concepts by requiring students to apply knowledge to assigned process modelling, analysis, redesign and automation tasks in the case study. In practical laboratory sessions, students use Bizagi to learn to model processes in BPMN and then to enact processes through Bizagi's BPMS engine (Bizagi Suite).

2.2   Conceptual Background and Research Variables

Digital game-based learning (DGBL) can be defined as the use of computer games (or digital games) within an educational context to supplement learning and support a particular learning outcome [2, 49, 54]. Digital games have also been termed "serious games" [20], "simulation games" [29] and "digital learning games" [40].

Digital learning games can be effective in addressing knowledge and skill and motivational outcomes. They can be used to illustrate the relevance of what students have been learning, provide them an opportunity to apply their knowledge in realistic scenarios that mimic real world problems, practice skills in a "virtual" real life situation, experience hands-on role-play and immersion in the virtual world, stimulate interaction among students and promote more social and communicative learning. Moreover, they provide an environment where students can express their autonomy, make their own choices, try out new skills and feel a sense of mastery or competence [6, 12, 19, 28, 34, 38, 46, 48]. Digital games should allow users to experience an optimally challenging environment but to be effective must contain positive feedback that is free from judgmental evaluation [14]. However, the efficacy of DGBL is being questioned [10]. Several recent studies, reviews and meta-analyses of DGBL have shown no impact on motivation and learning achievement or it has been shown to perform worse than a traditional classroom setting [3, 17, 20, 24, 53, 54]. The impact of DGBL on outcomes thus remains an open question.

Industry software tools have come to play an important role in the teaching of BPM concepts. Emerging curricula are emphasizing practical application and use of modelling tools in course delivery (e.g. [33, 42]). The use of industry software platforms has been considered among the requirements for 'state of the art' in BPM education [5]. It is difficult to argue that a course in BPM is relevant if it does not ensure students develop practical skills in use of professional industry tools [37]. BPM educators are often making choices about which BPM tools to include, such as ARIS; MS Visio; YAWL; Adonis; Bizagi; WebSphere [4, 5, 16, 43].

The incorporation of tools helps students develop practical skills and ability to apply BPM knowledge to complex problems and modeling scenarios [39], improve their knowledge by understanding how software actually works [41], experience the modelling grammars and notation as implemented within marketing leading process tools [43], and obtain up-to-date skills highly valued by industry [42]. Undoubtedly the incorporation of professional tools into the curriculum can help students to assimilate concepts by connecting theory and practice. However, the question remains as to whether 'more is better' when it comes to the use of these tools in the classroom. Unfortunately, little empirical research exists to identify best practice [5]. More specifically, BPM educators are left wondering which specific tools are useful to include and how many tools should one incorporate and for which types of outcomes.

Consequently, we contribute by examining the relative effects of digital games versus exposure to additional industry tools. The study was grounded in social cognitive career theory [31] along with Deci and Ryan's [15] self-determination theory. Together, they suggest an important set of constructs that represent higher-order outcomes that educators should look to achieve in a course on BPM.

Deci and Ryan [13–15] introduced their self-determination theory (SDT) and its underpinning of cognitive evaluation theory (CET) in an effort to explain and understand human motivation in any context. According to the theory humans actively seek to improve themselves by pursuing and engaging with challenges that allow them to realize their potential and capacity [15]. Moreover, the social environment that the individual is in will either support or diminish this process of self-realisation [15]. The theory views intrinsic motivation as the highest level of human motivation.

Intrinsic Motivation can be defined as an individual's internal motivation towards performing a task. When someone is intrinsically motivated they are willing to devote extra effort to an activity because of the interest and enjoyment derived from its performance [15]. Students with higher levels of intrinsic motivation are more engaged in the learning process and put more effort into their academic activities to achieve higher levels of academic performance [23, 44]. According to SDT, intrinsic motivation is the most powerful form of motivation that an individual can feel towards an activity and should therefore affect behaviour in a far more powerful way than external rewards. Consequently, we considered that a student's intrinsic motivation toward BPM as reflected in their expression of enjoyment and interest in BPM is an important outcome of interest when considering the efficacy of curriculum and classroom interventions.

According to SDT and CET, for an individual to be motivated, there are basic psychological needs that must be supported by the social environment. These needs can either be reinforced by the social environment to result in high levels of motivation or can be diminished by it. In an educational context, competence is considered the most influential of these psychological needs for developing intrinsic motivation. Prior work supports perceived competence as a predictor of intrinsic motivation [13, 15].

Perceived Competence can be defined as a person's perception of how skilled they are at performing an activity or task. Previous studies have found that perceived competence felt towards a subject domain has a positive relationship with learning achievement in that particular subject domain [8, 25, 35]. When perceived competence is low, academic performance diminishes [36]. Therefore, BPM coursework should be designed to so that students perceive themselves as growing in competence and skill along the BPM lifecycle, and across the theoretical and practical components of BPM.

Social cognitive career theory (SCCT) purports that there are sociocognitive determinants of career and academic interests, where such interests subsequently promote career-related activity involvement and skill acquisition [31]. The core construct within SCCT is occupational self-efficacy, which is considered to mediate between occupationally relevant abilities and occupational interest. Drawing on the work of Bandura, SCCT defines self-efficacy as an individual's judgment of their capabilities to successfully execute a course of action and is considered a central mechanism of personal agency determining one's choices and effort expenditure [31]. According to SCCT, self-efficacy is prominent in the formation of career interest such that people form enduring interests in activities in which they view themselves as efficacious [31]. Moreover, individuals are likely to perceive greater rewards and anticipate greater satisfaction from pursuing those activities in which they consider themselves more able to succeed.

Occupational Specific Self-efficacy is considered to have effects on career choice [45] and is defined as an individual's belief in their ability to succeed in a given career. Lent et al. [32] consider self-efficacy to influence choice of major directly and indirectly through effects on interest and outcome expectations. In the IT education context, past work has shown that feeling "not suited for IT type work" is among the reasons for students not wanting to pursue IT studies or an IT career [52]. Occupational self-efficacy has been shown to correlate with IT career intention and choice to major in IS [11] and is important to forming positive attitudes toward IT jobs [26]. This link between occupational self-efficacy and career interest and choice is likely to also hold in the more specific case of BPM. It is unlikely that students will pursue BPM as a career option or even learn more about the opportunities available to them for careers in BPM if they have low levels of BPM occupational self-efficacy.

2.3   Experimental Design

To achieve our research objective, our study followed a between group randomized experimental design [47]. This design is characterized by the comparison of multiple groups that are randomly assigned participants. Random assignment ensures groups are on average equal and that any effects observed in the study can be attributed to the interventions [47]. Data collection followed a pre-test post-test design, which allows for the comparison of groups before and after interventions have been implemented [1]. The comparison of scores from a pretest ensures that there are no differences between groups prior to the interventions, and the comparison of pre-test and post-test scores enables the calculation of any differences in each group after the interventions have taken place [1].

As our study was exploratory, we did not establish *a priori* hypotheses as to which between DGBL and industry tools would be more effective than the other. However, our general expectation, based on the literature presented above, was for both DGBL and the additional exposure to industry tools to have positive impacts on perceived competence, intrinsic motivation and occupational self-efficacy. This is because digital games have demonstrated potential to impact learning and motivational outcomes in an educational context [12], while exposure to industry tools may help signal career readiness and build relevant skills [33, 37, 42].

Although all students had the same exposure to lectures, project work and Bizagi laboratory sessions, the experimental design allowed one randomly selected group (Group B) to extend their knowledge of industry BPM tools by additional exposure to modelling and simulation tasks using the Signavio Editor and the online BIMP simulation tool. These students had knowledge of BPMN and modeling experience from Bizagi, but now learned to extend that to a new editor (Signavio). They could either use their own Signavio bpmn file for simulation or use an instructor provided solution. The second randomly selected group (Group I) played the IBM Innov8 game as an opportunity to apply their understanding of BPM in a virtual business environment. The call-centre scenario within the Innov8 game was used in the session.

Students were randomly allocated to one of the two groups at the beginning of the experimental session, at the end of 5-weeks into the 7-week course. A baseline questionnaire was administered to each group to establish their existing levels of motivation (5 items), competence (6 items) and occupational self-efficacy (3 items). All items were measured on 7-point Likert scales. The motivation and competence items were adapted from Deci and Ryan's intrinsic motivation inventory, and occupational self-efficacy adapted from Lent [32]. Negatively phrased items were reverse-coded.

3   Results

We carried out pre (baseline) and post (after exposure) comparisons of both groups. Results are summarized in Table 1. Baseline tests show that the randomization resulted in two equivalent groups prior to our intervention with no significant differences across the 14 questionnaire items.

Table 1. Pre- and post-test comparisons.

	Before/After (Group I, n = 20)	Before/After (Group B, n = 15)	Group I/B (Pre) (I = 21, B = 17)	Group I/B (Post) (I = 21, B = 17)
1. I think I am pretty good at BPM	4.677***	−0.323	−0.045	−2.674 *
2. I am pretty skilled at BPM	3.577**	0.899	−0.554	−1.725
3. I am satisfied with my performance at BPM	4.924***	−1.451	0.841	−3.723**
4. I think I do pretty well at BPM, compared to others	3.269**	−0.823	1.582	−0.444
5. I think I am good at BPM	3.335**	−0.323	0.494	−2.432*
6. After working at BPM for a while, I felt pretty competent.	2.99**	−1.099	0.347	−2.353*
7. I think BPM is quite enjoyable	3.199**	1.835#	−0.29	−0.277
8. I think BPM is very interesting	2.557*	1.103	−0.494	−0.373
9. I think BPM is fun	1.831#	0.764	0.135	0.125

(continued)

Table 1. (*continued*)

	Before/After (Group I, n = 20)	Before/After (Group B, n = 15)	Group I/B (Pre) (I = 21, B = 17)	Group I/B (Post) (I = 21, B = 17)
10. While doing BPM I often think about how much I enjoy it	0.547	0.676	−1.028	−0.411
11. I think BPM is boring (-)	−0.309	−0.774	0.349	−0.535
12. I believe that I poses the necessary skills to pursue a career in BPM	5.08***	−0.979	0.863	−2.226*
13. I would prefer a career in BPM over other careers in IS	1.560	−1.247	0.457	−0.548
14. I would prefer a career in BPM over a career in any other field	−0.438	−2.168*	0.915	0.198

***p < 0.001 **p < 0.01 *p < 0.05 #p < 0.10

Within Group I, significant decreases were observed from pre-test to post-test in perceived competence, aspects of intrinsic motivation such as interest and enjoyment, and in occupational self-efficacy indicators (see Fig. 1).

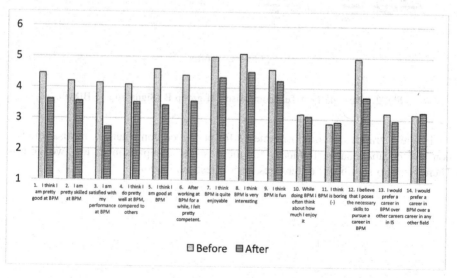

Fig. 1. Pre- and Post-Test comparison of group I (Innov8).

On the other hand, Fig. 2. illustrates no significant differences for Group B, except for a slight marginal decrease in enjoyment (p < 0.10), but a significant increase in career preference for BPM. An examination of post-test differences shows that Group I is lower than Group B on perceptions of competence. Their lower levels and drop in perceived competence could be due to the immediate feedback provided by Innov8 when students failed to adequately redesign the process models to achieve improvements in the KPAs as envisaged by the game. Games appear to embed feedback mechanisms that allowed students to recognize limitations and work to improve them, which is absent in industry toolsets. Frustration with success in the game also spilled-over into motivational and career outcomes. By contrast, Group B students believed post-test that they possessed more skills needed for a BPM career and reported increased feelings of competence. This suggests that exposure to industry tools signals workplace readiness to students far more than the game experience.

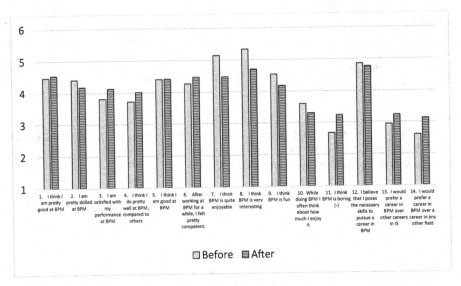

Fig. 2. Pre- and Post-Test comparison of group B (Signavio and BIMP).

Figure 3 graphs the composite means for perceived competence, intrinsic motivation and occupational self-efficacy for the post-test comparison of the two groups. There were no significant differences in post-test motivation.

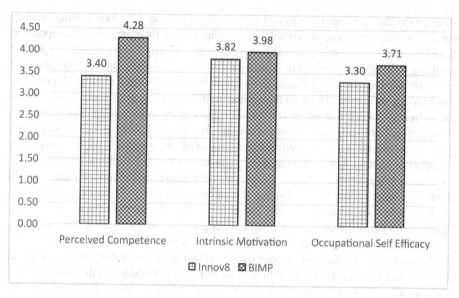

Fig. 3. Post-Test comparison of groups.

4 Discussion and Conclusion

Given the demand for additional BPM skills in the market, educators must focus on strengthening students' beliefs in their ability to succeed in BPM careers. BPM course evaluations should therefore collect data on outcomes such as occupational self-efficacy along with traditional motivation and competence-related factors.

In our study, the group selected to learn an additional set of industry-relevant tools scored higher on post-test perceptions of competence, motivation and occupational self-efficacy than those in the DGBL group. These results suggest that students appreciated the opportunity to gain hands-on experience with industry tools and that such exposure is important, particularly for increased occupational self-efficacy outcomes. However, industry tools may not increase outcomes beyond a specific point. Our course had introduced students to Bizagi and the subsequent exposure to additional tools did not significantly increase perceived competence from pre-test baseline of this group. There were small non-significant increases in most perceived competence items possibly because students could see how their skills transferred across tools. Motivations were not significantly affected, but were slightly lower possibly due to the experimental tasks adding little additional enjoyment.

In the DGBL group, we observed an unexpected general decrease in perceived competence along with indicators of motivation and self-efficacy among the DGBL group. One explanation for this finding could be a "Dunning-Kruger Effect" that occurs when individuals are unaware of their own incompetence, or are in a stage of unconscious incompetence, where they have a tendency to over-estimate their level of competence [9]. This means that perceived competence could be inflated before students have been truly 'tested' by the game. A digital game such as Innov8 can provide students the chance to apply skills and to move them from unconscious incompetence to conscious incompetence, observable in a reduction of their perceived competence. Based on this, we do not necessarily conclude that a drop in perceived competence resulting from digital game play is a negative outcome to be avoided, but rather we speculate whether the direct feedback mechanism embedded in the game allowed students an opportunity to reflect on their actual skills. These students were exposed to an initial 'reality' check that escaped their colleagues in the other group. Thus, although games should not substitute in the curriculum for exposing students to industry tools, games can usefully provide a feedback mechanism absent from BPM tools. However, it is important to note that digital games are considered to work best when they provide feedback that is free from negative judgements [46]. However, the Innov8 game resulted in many students being 'fired'. Ending the game as a 'loser' will very likely having a negative effect on motivation and cause a student to question their self-efficacy. Moreover, because perceived competence is a predictor of motivation with SDT, the drop in perceived competence would subsequently result in reduced motivation.

An alternative explanation for our finding is that there was a confounding learning problem stemming from lack of sufficient background knowledge in the business case of the game [43]. Innov8 game was based on a call-centre context and students may have faced additional learning in attempting to understand and relate to the specific processes and key performance targets.

Taken together, we recommend the incorporation of both industry-relevant tools and games into the practical components of BPM courses as they have differential effects on competence, motivation and occupational self-efficacy. Industry tools increase occupational self-efficacy and games present a 'reality' check against perceived competence. Games must however be implemented to ensure students receive positive and constructive feedback. Where BPM educators do not have control over game design, they may wish to make use of a post-game briefing to deal with any unintended consequences of negative judgement experienced during game play. Enhancing student background knowledge of the context of the game prior to game play can also help to reduce any confounding learning problems.

Much work also remains to determine best-practices in the use of industry tools and future work should determine which tools produce the best outcomes and in what sequences. Mastery of one tool may be sufficient for development of interest and motivation toward BPM. However, if educators combine multiple professional tools along with digital games, they may wish to consider how to position game play within the course so that students have an added opportunity to build confidence in marketplace readiness and mastery over real-world BPM tools following the 'reality' check of game play.

A major limitation of our work is that it was restricted to a single session with a pre- and post-test design. Future work should consider comparing outcomes in longitudinal studies with ongoing exposure to different interventions.

References

1. All, A., Castellar, E.P.N., Van Looy, J.: Assessing the effectiveness of digital game-based learning: best practices. Comput. Educ. **92**, 90–103 (2016)
2. Ariffin, M.M., Sulaiman, S.: Evaluating game-based learning (GBL) effectiveness for higher education (HE). In: 2nd International Conference on Advanced Computer Science Applications and Technologies (ACSAT 2013), pp. 485–489. IEEE (2013)
3. Backlund, P., Hendrix, M.: Educational games-are they worth the effort? A literature survey of the effectiveness of serious games. In: 5th International Conference on Games and Virtual Worlds for Serious Applications (VS-GAMES 2013). IEEE, Bournemouth (2013)
4. Bandara, W., Rosemann, M., Davies, I., Tan, C.: A structured approach to determining appropriate content for emerging information systems subjects: an example for BPM curricula design. In: 18th Australasian Conference on Information Systems, pp. 1132–1141 (2007)
5. Bandara, W., Chand, D.R., Chircu, A.M., Hintringer, A., Karagiannis, D.: Business process management education in academia: status, challenges, and recommendations. Commun. Assoc. Inf. Syst. **27**(1), Article 41 (2010)
6. Boeker, M., Andel, P., Vach, W., Frankenschmidt, A.: Game-based e-learning is more effective than a conventional instructional method: a randomized controlled trial with third-year medical students. PLoS One **8**(12), e82328 (2013)
7. Boughzala, I., Tantan, O., Lang, D.: Feedback on the integration of a serious game in the business process management learning. In: 21st Americas Conference on Information Systems. AIS (2015)
8. Chan, S.H., Song, Q., Hays, L.E., Trongmateeru, P.: The roles of intrinsic and extrinsic motivation and perceived competence in enhancing system use and performance. In: Proceedings of PACIS 2014, p. 382. AIS (2014)
9. Chapman, A.: Conscious Competence Learning Model (2007). http://www.businessballs.com/consciouscompetencelearningmode.htm. Accessed Nov 2015
10. Chiu, Y.H., Kao, C.W., Reynolds, B.L.: The relative effectiveness of digital game-based learning types in English as a foreign language setting: a meta-analysis. Br. J. Edu. Technol. **43**(4), E104–E107 (2012)
11. Cohen, J.F., Parsotam, P.: Intentions to pursue a career in information systems and technology: an empirical study of South African students. In: Reynolds, N., Turcsányi-Szabó, M. (eds.) KCKS 2010. IAICT, vol. 324, pp. 56–66. Springer, Heidelberg (2010). https://doi.org/10.1007/978-3-642-15378-5_6
12. Connolly, T.M., Boyle, E.A., Macarthur, E., Hainey, T., Boyle, J.M.: A systematic literature review of empirical evidence on computer games and serious games. Comput. Educ. **59**(2), 661–686 (2012)
13. Deci, E.L., Ryan, R.M.: Intrinsic Motivation and Self-Determination in Human Behavior. Springer, New York (1985). https://doi.org/10.1007/978-1-4899-2271-7
14. Deci, E.L., Ryan, R.M.: The "what" and "why" of goal pursuits: human needs and the self-determination of behavior. Psychol. Inq. **11**(4), 227–268 (2000)
15. Deci, E.L., Ryan, R.M.: Overview of self-determination theory: an organismic dialectical perspective. In: Handbook of Self-Determination Research, pp. 3–33. University of Rochester Press (2002)
16. Delavari, H., Bandara, W., Marjanovic, O., Mathiesen, P.: Business process management (BPM) education in Australia: a critical review based on content analysis. In: Proceedings of ACIS 2010, Paper 80. AIS (2010)

17. Divjak, B., Tomić, D.: The impact of game-based learning on the achievement of learning goals and motivation for learning mathematics-literature review. J. Inf. Organ. Sci. **35**(1), 15–30 (2011)

18. Dumas, M., La Rosa, M., Mendling, J., Reijers, H.: Fundamentals of Business Process Management. Springer, Berlin Heidelberg (2013). https://doi.org/10.1007/978-3-662-56509-4

19. Felicia, P.: Motivation in games: a literature review. Int. J. Comput. Sci. Sport **11**(1), 4–14 (2012)

20. Girard, C., Ecalle, J., Magnan, A.: Serious games as new educational tools: how effective are they? A meta-analysis of recent studies. J. Comput. Assist. Learn. **29**(3), 207–219 (2013)

21. Grace, T., Cohen, J.F.: Business process management and digital game based learning. In: Proceedings of the 22nd Americas Conference on Information Systems (AMCIS 2016), San Diego, CA, USA (2016). ISBN 978-0-9966831-2-8. http://aisel.aisnet.org/amcis2016/ISEdu/

22. Hawking, P., McCarthy, B., Stein, A.: Integrating ERP's second wave into higher education curriculum. In: PACIS 2005 Proceedings, Paper 83 (2005)

23. Hess, T., Gunter, G.: Serious game-based and nongame-based online courses: learning experiences and outcomes. Br. J. Edu. Technol. **44**(3), 372–385 (2013)

24. Hou, H.T., Li, M.C.: Evaluating multiple aspects of a digital educational problem-solving-based adventure game. Comput. Hum. Behav. **30**, 29–38 (2014)

25. Jansen, B.R., Louwerse, J., Straatemeier, M., Van Der Ven, S.H., Klinkenberg, S., Van Der Maas, H.L.: The influence of experiencing success in math on math anxiety, perceived math competence, and math performance. Learn. Individ. Differ. **24**, 190–197 (2013)

26. Johnson, R.D., Stone, D.L., Phillips, T.N.: Relations among ethnicity, gender, beliefs, attitudes, and intention to pursue a career in information technology. J. Appl. Soc. Psychol. **38**(4), 999–1022 (2008)

27. Joubert, P., Roodt, S.: The relationship between prior game experience and digital game-based learning: an INNOV8 case-study. In: SIGED: IAIM Conference Proceedings, p. 25 (2010)

28. Kang, J.: Attributes and motivation in game-based learning: a review of the literature. In: World Conference on Educational Media and Technology (EdMedia 2013), Victoria, Canada (2013)

29. Kikot, T.N., Costa, G.J.M.D., Fernandes, S.C.P.B.: Business simulators and lecturers perception! The case of University of Algarve. In: Conferência ETHICOMP 2014, Paris, France (2014)

30. Lawler, J.P., Joseph, A.: Educating information systems students on business process management (BPM) through digital gaming metaphors of virtual reality. Inf. Syst. Educ. J. **8**(6), 1–22 (2010)

31. Lent, R.W., Brown, S.D., Hackett, G.: Toward a unifying social cognitive theory of career and academic interest, choice, and performance. J. Vocat. Behav. **45**(1), 79–122 (1994)

32. Lent, R.W., Lopez, A.M., Lopez, F.G., Sheu, H.: Social cognitive career theory and the prediction of interests and choice in the computing disciplines. J. Vocat. Behav. **73**(1), 52–62 (2008)

33. Levina, O.: Teaching business process management for heterogeneous audience. In: Proceedings of the 19th Americas Conference on Information Systems, pp. 1–9. AIS (2013)

34. Liu, T.Y.: Using educational games and simulation software in a computer science course: learning achievements and student flow experiences. Interact. Learn. Environ. **24**(4), 724–744 (2016)

35. Liu, E.S., Carmen, J.Y., Yeung, D.Y.: Effects of approach to learning and self-perceived overall competence on academic performance of university students. Learn. Individ. Differ. **39**, 199–204 (2015)

36. Miserandino, M.: Children who do well in school: Individual differences in perceived competence and autonomy in above-average children. J. Educ. Psychol. **88**(2), 203 (1996)
37. Pal, R., Sen, S.: Relevance of business process management (BPM) course in business school curriculum & course outline. In: Proceedings of the Seventeenth Americas Conference on Information Systems, Paper 199. AIS (2011)
38. Papastergiou, M.: Digital game-based learning in high school computer science education: impact on educational effectiveness and student motivation. Comput. Educ. **52**(1), 1–12 (2009)
39. Pasha, M.: Developing business process management capabilities of undergraduate IT students. Int. J. Comput. Appl. **53**(17), 45–50 (2012)
40. Prensky, M.: Digital game-based learning. Comput. Entertain. (CIE) **1**(1), 21 (2003)
41. Pridmore, J., Deng, J., Prince, B., Turner, D.: Enhancing student learning on ERP and business process knowledge with hands-on ERP exercises. In: SAIS 2014 Proceedings, Article 31 (2014)
42. Ravesteyn, P., Versendaal, J.: Design and implementation of business process management curriculum: a case in dutch higher education. In: Reynolds, N., Turcsányi-Szabó, M. (eds.) KCKS 2010. IAICT, vol. 324, pp. 310–321. Springer, Heidelberg (2010). https://doi.org/10.1007/978-3-642-15378-5_30
43. Recker, J., Rosemann, M.: Teaching business process modelling: experiences and recommendations. Commun. Assoc. Inf. Syst. **25**(32), 379–394 (2009)
44. Reyes, M.R., Brackett, M.A., Rivers, S.E., White, M., Salovey, P.: Classroom emotional climate, student engagement, and academic achievement. J. Educ. Psychol. **104**(3), 700 (2012)
45. Rottinghaus, P.J., Lindley, L.D., Green, M.A., Borgen, F.H.: Educational aspirations: the contribution of personality, self-efficacy, and interest. J. Vocat. Behav. **61**(1), 1–19 (2002)
46. Ryan, R.M., Rigby, C.S., Przybylski, A.: The motivational pull of video games: a self-determination theory approach. Motiv. Emot. **30**(4), 344–360 (2006)
47. Shadish, W.R., Cook, T.D., Campbell, D.T.: Experimental and Quasi-Experimental Designs for Generalized Causal Inference. Wadsworth Cengage learning, Boston (2002)
48. Soflano, M., Connolly, T.M., Hainey, T.: Learning style analysis in adaptive GBL application to teach SQL. Comput. Educ. **86**, 105–119 (2015)
49. Vogel, J.J., Vogel, D.S., Cannon-Bowers, J., Bowers, C.A., Muse, K., Wright, M.: Computer gaming and interactive simulations for learning: a meta-analysis. J. Educ. Comput. Res. **34**(3), 229–243 (2006)
50. Vuksic, V.B., Bach, M.P.: Simulation games in business process management education. Int. J. Ind. Syst. Eng. **6**(9), 2424–2429 (2012)
51. Vuksic, V.B., Bach, M.P., Hernaus, T.: Educating students in business process management with simulation games. Int. J. Ind. Syst. Eng. **8**(5), 1245–1250 (2014)
52. Walstrom, K.A., Jones, K.T., Crampton, W.J.: Why are students not majoring in information systems? J. Inf. Syst. Educ. **19**(1), 43–52 (2008)
53. Wang, T.L., Tseng, Y.F.: Learning effect for students with game-based learning on meta-analysis. In: 6th International Conference on Computer Science and Education (ICCSE 2011). IEEE (2011)
54. Wouters, P., Van Nimwegen, C., Van Oostendorp, H., Van Der Spek, E.D.: A meta-analysis of the cognitive and motivational effects of serious games. J. Educ. Psychol. **105**(2), 249 (2013)

Author Index

Printed in the United States
By Bookmasters

Printed in the United States
By Bookmasters